An Introduction to
Statistical Signal Processing
with Applications

An Introduction to Statistical Signal Processing With Applications

M. D. SRINATH

Southern Methodist University
Dallas, Texas

P. K. RAJASEKARAN

Teledyne Brown Engineering
Huntsville, Alabama

A WILEY-INTERSCIENCE PUBLICATION

JOHN WILEY & SONS
New York • Chichester • Brisbane • Toronto

Library of Congress Cataloging in Publication Data:

Srinath, M D 1935–
An introduction to statistical signal processing with applications.

"A Wiley-Interscience publication."
Bibliography: p.
Includes Index.
1. Signal processing. 2. Signal theory (Telecommunication) 3. Estimation theory. I. Rajasekaran, P. K., 1942– joint author. II. Title.

TK5102.5.S68 621.38′0432 78-15417
ISBN 0-471-04404-0

Printed in the United States of America

10 9 8 7 6 5 4 3 2

PREFACE

Signal processing may broadly be considered to involve the recovery of information from physical observations. In the presence of random disturbances, we must resort to statistical techniques for the recovery of the information. Applications of statistical signal processing arise in a wide variety of areas such as communications, radio location of objects, seismic signal processing, and computer-assisted medical diagnosis.

This book is intended to serve as a first-year graduate level text in statistical signal processing and aims at covering certain basic techniques in the processing of stochastic signals and illustrating their use in a few specific applications. The motivation for writing this book arose from the instructional experience of the authors at Southern Methodist University (M.D.S.), Dallas, Texas, the Indian Institute of Technology (P.K.R.), Madras, India, and the Indian Institute of Science (M.D.S.), Bangalore, India, where they have been involved in teaching graduate level courses in statistical communication theory, radar systems, system identification, pattern recognition, stochastic control, and biomedical signal processing. An important realization of this experience was that an essentially common background in certain techniques for statistical signal processing is required for these apparently different courses. In particular, the topics of detection and estimation theory constitute a common foundation in many of these courses. Most available texts provide a treatment of these two topics in varying detail, but usually cover only one of the application areas in depth while making only a passing reference to the others. It was felt that there was a need for a text that presented these two basic topics in a clear and concise fashion and illustrated their use in several areas. The particular areas chosen for coverage in this text are communications, radar, pattern recognition, and system identification. These four areas are by no means exhaustive but were selected because of their generality and widespread use. Many of the techniques in the text can be either applied directly or extended to other application areas such as speech or image processing. It is expected that first-year graduate students as well as practicing engineers in different disciplines will use the text to acquire a background suitable for advanced study and a better understanding of the fields in which they are involved.

The background required to use this book is a course on systems theory and a course on stochastic processes. The mathematical rigor takes a middle path so as not to obscure intuitive reasoning. The first chapter presents an introduction and overview of the text. Chapter 2 is intended to serve as a quick review of certain results in systems theory and stochastic processes that are useful in later developments. The basic background in detection and estimation theory that is required for the latter applications chapters is provided in Chapters 3 through 6. The applications chapters, Chapters 7 through 10, can be covered selectively and in any order, depending on the interests of the class. For a class with strong interests in communication, a good suggestion is to review Chapter 2 quickly, follow up with Chapters 3 through 6 on theory, and emphasize Chapter 7 on communication systems and Chapter 8 on radar systems. On the other hand, for a class with strong interests in pattern recognition or control, Chapter 9 may be emphasized instead. Many exercises are included at the end of each chapter, serving the dual purpose of obtaining a better understanding of the text material, and presenting new applications or results not covered in the text.

We wish to express our gratitude to our graduate students and the reviewers of the manuscript for their comments which were extremely helpful in bringing this book to publication. Much of the contribution of one of the authors (P.K.R.) was accomplished during his stay at the Instituto Nacional de Astrofisica, Optica y Electronica (INAOE), Puebla, Mexico. He is grateful to Dr. G. Haro, Director-General and other authorities of INAOE for providing him the necessary facilities.

M. D. Srinath
P. K. Rajasekaran

Dallas, Texas
Huntsville, Alabama
March 1978

CONTENTS

Chapter 1. Introduction **1**

1.1 Organization, 3

Chapter 2. Signals and Systems **6**

2.1 Introduction, 6
2.2 System Theory, 6
2.3 Stochastic Processes, 25
2.4 Gauss–Markov Models, 36
2.5 Complex Envelopes of Band-limited Signals, 45
2.6 Summary, 53

Chapter 3. Detection Theory **56**

3.1 Introduction, 56
3.2 Classical Detection Theory and Hypothesis Testing, 57
3.3 Decision Criteria, 60
3.4 Multiple Measurements, 73
3.5 Multiple Hypotheses Testing, 79
3.6 Composite Hypothesis Testing, 83
3.7 Sequential Detection: Wald's Test, 88
3.8 Summary, 94

Chapter 4. Detection of Signals in Noise **104**

4.1 Introduction, 104
4.2 Detection of Known Signals in White Noise: The Correlation Receiver, 105
4.3 Detection of Known Signals in Colored Noise, 120
4.4 Detection of Known Signals in Noise: Maximum SNR Criterion, 128
4.5 Solution of Integral Equations, 136
4.6 Detection of Signals with Unknown Parameters, 141
4.7 Summary, 145

Chapter 5. Estimation Theory **156**

5.1 Introduction, 156
5.2 Estimation of Parameters, 156
5.3 Random Parameters: Bayes Estimates, 158
5.4 Maximum Likelihood Estimates, 166
5.5 Properties of Estimators, 168
5.6 Linear Mean Square Estimation, 182
5.7 Reproducing Densities, 190
5.8 Summary, 196

Chapter 6. Estimation of Waveforms **202**

6.1 Introduction, 202
6.2 Linear MMSE Estimation of Waveforms: Preliminaries, 203
6.3 Estimation of Stationary Processes: The Wiener Filter, 208
6.4 Estimation of Nonstationary Processes: The Kalman Filter, 224
6.5 Stationary Processes with Semi-infinite Observation Intervals: Relation between the Kalman and Wiener Filters, 249
6.6 Nonlinear Estimation, 254
6.7 Summary, 265

Chapter 7. Applications to Communication Systems **277**

7.1 Introduction, 277
7.2 Digital Communication, 278
7.3 Analog Communication Systems, 316
7.4 Summary, 336

Chapter 8. Applications to Radar Systems **348**

8.1 Introduction, 348
8.2 Radar Target Models, 349
8.3 Target Detection, 356
8.4 Parameter Estimation in Radar Systems, 373
8.5 Dynamic Target Tracking, 382
8.6 Summary, 390

Chapter 9. Miscellaneous Applications **398**

9.1 Introduction, 398
9.2 Application to Pattern Classification, 398
9.3 Application to System Identification, 418

Appendix A: Bilateral Transforms **463**

Appendix B: Calculus of Extrema **466**

Appendix C: Vectors and Matrices **470**

Supplementary Bibliography **478**

Author Index **491**

Subject Index **495**

An Introduction to Statistical Signal Processing with Applications

CHAPTER 1

INTRODUCTION

Statistical signal processing has application in a wide variety of human activity. Such applications range from processing of seismic signals to computer-assisted medical diagnosis and treatment, tracking of objects in space, and traffic control.

Signal processing generally involves the recovery of information from physical observations. The processing required is relatively simple if the observation contains the information explicitly, and any interference present is exactly described. Often, the physical characteristics and limitations of the devices used for observation, and/or the media through which the information is observed or communicated, make this impossible. In fact, the interference is usually random in nature, and can only be described in terms of its average properties or statistics. The processing of such an observation in order to recover information can be termed *statistical signal processing*. This book is devoted to this topic and some applications.

To illustrate the type of problems that arise in statistical signal processing, let us consider the radio location of a flying object (target) in space. We can do this by beaming a packet of electromagnetic energy in the direction of the target and by observing the reflected electromagnetic wave. We have two problems to consider at this point. The primary problem is to decide whether there is an object present at all (the detection problem). If we decide that the object is present, we may desire to determine certain parameters associated with the object, say its range or velocity (the estimation problem). If there is no interference and if the reflected wave is not distorted through the transmission medium, the solution is straightforward. We monitor the reflected signal and observe the time delay τ between the transmitted and reflected waves, by noting the time at which a peak occurs. If the target is not present, there will be no peak. If a target is present, we can estimate its range as $R = \frac{\tau c}{2}$, where c is the velocity of propagation of the electromagnetic wave.

In the presence of interference (noise), the solution is no longer simple. The interference could be due to distortions through the transmission media, or thermal noise in the measurement device. The effect of the interference is to mask the peak that we are monitoring. We may get a spurious peak when no target is present or we may not be able to discern a peak when a target is present. In either case, the presence of noise can cause erroneous decisions. Our problem is to monitor the signal for a certain length of time, and decide about the presence of the object. This is the detection problem and falls into the general topic of statistical decision making.

If we decide that a target is present and seek to determine its range by observing the delay, we may still have difficulties because of the interference causing the peak to appear at the wrong time. We then have the problem of recovering the information (target range) from the "noisy" observation and is the estimation problem referred to earlier.

These two problems of detection and estimation arise in many other areas besides the radio location problem and are basic to all statistical signal processing techniques. Similar problems are encountered in areas such as communication, pattern recognition, and system identification. In analog communication systems, the transmitted message usually undergoes distortion during transmission. This distortion is usually characterized at the receiver as noise in the observations. In many cases, the message, for example, an audio or visual signal, may be modeled as a stochastic signal. The problem of recovering the message at the receiver can then be formulated as one of estimating a random signal in the presence of random noise.

In a digital communication system, the message is encoded into a sequence of binary digits (more generally into a sequence of several symbols). Typically these digits (bits), represented by a 1 or 0 are transmitted by sending suitably chosen pulses which again are distorted during transmission. The effect of this distortion is that, at the receiver it is no longer possible to determine which waveform was transmitted. We can again model the distortion during transmission as random noise in the receiver. The problem is one of deciding on the basis of noisy observations, whether the transmitted waveform corresponds to 1 or 0.

Recent work in efficient speech transmission involves characterizing the speech waveform by means of certain parameters that describe its spectrum. These parameters are then transmitted to the receiver which synthesizes the speech waveform from a knowledge of these paramters. The problem of extracting the parameters is an identification problem and is essentially one of estimating the parameters in a suitably chosen model for the speech waveform. These parameters undergo distortion during transmission so that at the receiver we have the problem of recovering (estimating) the parameters from noisy observations.

Pattern recognition systems also provide examples of systems involving signal processing. Consider that we want to design an automatic machine to distinguish between two handwritten characters, say "a" and "b". The characteristics of these letters will vary with the person writing them. Typically, a single specimen can be characterized as a sample from a population with a known statistical description. The machine is to be designed to distinguish between the two classes or categories—the letters "a" and "b"—when a sample is presented to it. Often the statistics of the population will be unknown and will have to be determined from samples from each of the two classes. Another pattern classification example arises in electroencephalograph(EEG) analysis. The EEG is a recording of electrical brain signals and may be used in a variety of clinical purposes. A typical application is in the determination of the sleep state of a patient. The EEG signals are corrupted by noise introduced by the measuring device so that the characteristic features of the EEG will have to be estimated in order to determine the sleep state. Other applications involve the study of evoked potentials or response to stimuli and can be used for determining sensory perception. The evoked response is superimposed on the normal EEG, which in this case may be treated as interference. The useful information from the evoked response is obtained in terms of certain characteristic lines in the power spectrum which can be used for diagnosis and study of any abnormalties.

The preceding has served to point out that the problems of detection and estimation arise either seperately or jointly in a wide variety of applications. While there is a seeming dichotomy between the two, in fact, there is an underlying similarity in the structure of the two problems, which facilitates the solution of many signal-processing problems. The aim of this book is to develop the fundamentals of detection and estimation theory as the basis of statistical signal processing and illustrate the application of the concepts and techniques in various areas. The application areas chosen are communications, radar systems, pattern recognition, and system identification. These areas are by no means exhaustive, but have been chosen because they are general enough to be of interest to a fairly wide audience.

1.1 ORGANIZATION

The book is addressed to the first-year graduate student in the disciplines of electrical and systems engineering. It also will serve the purpose of a graduate engineer in industry who is exposed to a wide variety of signal-processing problems but has not had an opportunity to obtain a cohesive background. Typically the student of this test may be expected to have had a prior course in systems and a course in probability and stochastic processes. The exposure to systems should preferably have included state variable and transform techniques for both continuous-time and discrete-time systems.

The book essentially consists of two parts. The first part consisting of Chapters 2 through 6 provides the basic framework of detection and estimation theory, while the second part which consists of Chapters 7 through 9 covers applications to the four areas mentioned earlier. Specifically our coverage is as follows. In Chapter 2, we provide a quick review of basic concepts from systems theory and stochastic processes which is useful in the developments in the later chapters. In particular we discuss Gauss–Markov models for signal generation and the complex representation of band-limited signals.

Chapter 3 is concerned with the first class of statistical signal-processing problems, namely, decision problems. In this chapter we discuss classical decision theory using statistical hypotheses testing methods. Various decision criteria are introduced and decision rules obtained. The performance of the decision rules are also investigated. The sequential decision test of Wald is also included for the sake of completeness. These concepts are extended in Chapter 4 to the detection of waveforms observed in noise. The results are primarily derived in terms of a binary hypotheses testing problem in which we are required to detect which one of two known waveforms is present in the observations. The correlation receiver and the matched filter are obtained for optimum processing when the interference is additive white noise. Receiver structures for signals observed in colored noise are derived. The chapter concludes with a brief section on the detection of signals with unknown parameters.

Chapter 5 deals with the problem of estimating parameters from observations that are corrupted by noise. The Bayes' estimators for parameters are obtained. Bounds on the performance of these estimators are presented. The concept of a reproducing density is introduced and applied to the estimation of parameters of probability density functions.

Chapter 6 considers the minimum-mean-square estimation of signals observed in noise. For stationary processes, the classical Wiener filter is derived for both continuous-time and discrete-time processes. The Kalman filter which provides an attractive recursive solution to the estimation problem in nonstationary processes is discussed. Estimation techniques for certain nonlinear signal models are also considered.

After developing the powerful tools of detection and estimation theory in these four chapters, in Chapter 7 we consider applications to communication systems. We first consider digital communication systems and develop optimum receiver structures for a variety of digital transmission schemes. We consider the performance of these receivers in various situations, including fading. Techniques of synchronization are presented and the problem of intersymbol interference discussed. The second part of this chapter considers demodulator structures for demodulation of analog modulated signals. Two different techniques, one using the maximum a posteriori estimation technique and the other using the minimum-mean-square estimation technique are used in deriving optimum demodulator structures.

In Chapter 8, we study applications to radar systems. The models for radar targets are derived using the complex representation of signals developed in Chapter 2. In addition to studying the detection of fluctuating and nonfluctuating targets, the problem of estimating such radar parameters as delay and Doppler frequency is investigated, and the concept of signal design to improve the accuracy in estimating these parameters is presented. The chapter concludes with an application of the Kalman filter for dynamic target tracking and maneuver detection.

Chapter 9 considers applications to pattern classification and system identification. The problem of pattern classification is posed as a hypothesis testing problem by associating the various pattern classes with hypotheses. The problem of learning the parameters of the associated density functions under both supervised and unsupervised conditions is discussed. The important problem of extracting the features of a pattern is also presented.

The second part of this chapter discusses the identification of systems based on input and output measurements. We first consider the identification of the parameters of an autoregressive (AR) or autoregressive-moving-average (ARMA) model of a stochastic process. Identification in signal models using state augmentation techniques is discussed. The maximum likelihood technique for identification in linear time-invariant models is also presented.

Three appendices are included. Appendix A provides a brief review of the bilateral Laplace and z-transforms which are used widely in the text. Appendix B summarizes pertinent features of certain optimization techniques which are used in Chapter 4 while Appendix C discusses vector and matrix operations which are useful. In particular, a useful matrix lemma is presented in this appendix.

While Chapters 3 through 6 are essential for an understanding of the remainder of the text, the text is structured such that the applications chapters can be covered selectively and in any order depending on the interest and background of the intended audience. The exercises at the end of each chapter serve a dual purpose—to better understand the material covered and to present new applications or results not covered in the text. While each chapter contains several references, a supplementary bibliography of about 200 references which contain many pertinent contributions, is provided at the end of the text.

CHAPTER 2

SIGNALS AND SYSTEMS

2.1 INTRODUCTION

In the previous chapter we have presented several examples in which statistical signal processing plays an important role. The role of the signal processor is to extract information from the observed signals and present it in a useful form. The proper design of the processor requires an appropriate description or characterization of the signals as well as of the systems generating these signals. For example, in radar tracking, we would like to know the location and velocity of the object being tracked. The signal of interest in this case is the radar return which is usually corrupted by various noises such as thermal noise in the processor. A description of the signal can be given in terms of the equations of motion of the object being tracked while the noise is usually modeled as a sample function of a stochastic process.

In this chapter we briefly review models for the generation of random signals and discuss some representations of these signals. We will find these results to be of use in our discussions in later chapters.

2.2 SYSTEM THEORY [1,2]

By a *system* we mean a model of a physical object or a collection of physical objects. A model is essentially an idealization of a physical object which retains the salient features of the system, but is easier to study. In this section we discuss some concepts and techniques which are useful in studying the behavior of systems.

A given physical system may give rise to many different models and the choice of a model depends on the use to which it is to be put. As an example, in studies of the trajectory of a satellite, we may use a particle as the model. In maneuvering, however, a rigid body model would be more appropriate.

Once a model has been determined, the next step is to obtain a mathematical representation of it. This requires the selection of pertinent variables, reference directions and coordinate axes and the application of appropriate physical laws. The equations that describe a system may assume many forms; they may be algebraic equations, integral or differential equations, and so on. The choice of a particular representation again depends on the application. Once the mathematical description of a system is obtained, the next step is to evaluate the behavior of the system and to modify the system if the behavior is not acceptable. The evaluation may be carried out by observing the general properties of the system, or by determining specific responses to typical inputs.

In obtaining a mathematical model we define certain quantities (variables) as the inputs to the system and certain other quantities as the outputs. For example, in an electric motor, we may consider the armature voltage or the field current as inputs and the shaft speed as the output. In general the inputs and output variables will be functions of time. If the output at any instant t, is dependent only on the value of the inputs at that instant, the system is a *zero-memory* system. Most systems, however, have memory. That is the output at t_1 depends not only on the inputs applied at t_1 but also on the inputs applied prior to and/or after t_1. Such systems are called dynamical systems. For dynamical systems, if an input $u(t)$ is applied for $t<t_1$, the output $y(t)$ for $t>t_1$ cannot in general, be uniquely determined, unless $u(t)$ is known for $t<t_1$. If two inputs $u_1(t)$ and $u_2(t)$ which are identical for $t>t_1$ but distinct for $t<t_1$ are applied to the system, the output $y(t)$ for $t>t_1$ will be different for the two inputs. For an input to give rise to a unique output, the system must be *initially relaxed* or *at rest*. That is if an input is applied for $t \geqslant t_1$, there must be no stored energy in the system prior to t_1. For systems initially relaxed at $t=-\infty$, we can write the relation between the input and the output as

$$y(t)=\mathbf{T}\{u(t)\} \tag{2.2.1}$$

where $\mathbf{T}$ is an operator or function describing this relationship.

We consider only systems in which the input–output relation is a differential equation (differential systems) or a difference equation (discrete systems). The system is linear or nonlinear depending on whether these equations are linear or nonlinear.

All physical systems are *causal* or *nonanticipative*. That is, if the system is initially relaxed, there can be no output prior to an input being applied. More precisely, if $[u(t), y(t)]$ and $[u_1(t), y_1(t)]$ correspond to two input–output pairs, the system is causal if $u(t) = u_1(t)$ for $t < t_1$ implies that $y(t) = y_1(t)$ for $t < t_1$. In some of our applications, however, we will have occasion to use noncausal models. An example is the ideal lowpass filter.

Linear Systems

The system described by Eq. (2.2.1) is said to be linear if, for any two arbitrary inputs $u_1(t)$ and $u_2(t)$ and any two scalars α_1 and α_2

$$\mathbf{T}\{\alpha_1 u_1(t) + \alpha_2 u_2(t)\} = \alpha_1 \mathbf{T}\{u_1(t)\} + \alpha_2 \mathbf{T}\{u_2(t)\} \tag{2.2.2}$$

It easily follows from Eq. (2.2.2) that for any finite n

$$\mathbf{T}\left\{\sum_{i=1}^{n} \alpha_i u_i(t)\right\} = \sum_{i=1}^{n} \alpha_i \mathbf{T}\{u_i(t)\} \tag{2.2.3}$$

We will assume that Eq. (2.2.3) holds also for infinite sums and integrals. In regard to input–output relationships, the terms *linear system* and *linear operator* are equivalent. Equation (2.2.2) is also referred to as the principle of superposition. We note that for the principle to hold, the system must be initially relaxed. That is, all initial conditions in the system are zero.

A *linear differential system* is characterized in terms of its input–output relation by a linear differential equation.

A system is *time invariant* or *stationary* if its parameters do not vary with time. For example the differential equation characterizing the system will be invariant with respect to a shift in the time-axis. That is, if the response to an input $u(t)$ is $y(t)$, then the response to the input $u(t)$ delayed by τ equals $y(t-\tau)$:

$$\mathbf{T}\{u(t-\tau)\} = y(t-\tau) \tag{2.2.4}$$

Impulse Response and the Convolution Integral

A linear system can be characterized in terms of its response to certain elementary inputs. One such input is the impulse function. Let us denote by $h(t, \tau)$ the response of the system at time t to a unit impulse applied at time τ. We note that $h(t, \tau)$ denotes a family of functions indexed by the parameter τ.

The output $y(t)$ to an arbitrary input $u(t)$ is then obtained from the convolution integral as

$$y(t) = \int_{-\infty}^{\infty} h(t, \tau)u(\tau)\, d\tau \tag{2.2.5}$$

We say that $y(t)$ is obtained as the convolution of the two functions $h(t, \tau)$ and $u(t)$ and write

$$y(t) = h(t, \tau) * u(t) \tag{2.2.6}$$

where $*$ refers to the convolution operation of Eq. (2.2.5).

For a causal system, the response is zero prior to an input being applied. Thus for a causal system we can write

$$h(t, \tau) = 0 \qquad t < \tau \tag{2.2.7}$$

It follows that the output of a linear causal system is given by

$$y(t) = \int_{-\infty}^{t} h(t, \tau)u(\tau)\, d\tau \tag{2.2.8}$$

If the input is zero prior to a finite time t_0, Eq. (2.2.8) reduces to

$$y(t) = \int_{t_0}^{t} h(t, \tau)u(\tau)\, d\tau \tag{2.2.9}$$

For time-invariant systems the convolution integral, Eq. (2.2.5), becomes

$$y(t) = \int_{-\infty}^{\infty} h(t-\tau)u(\tau)\, d\tau = \int_{-\infty}^{\infty} h(\tau)u(t-\tau)\, d\tau \tag{2.2.10}$$

If the system is causal, then $h(t) = 0$ for $t < 0$ and the output is given by

$$y(t) = \int_{-\infty}^{t} h(t-\tau)u(\tau)\, d\tau = \int_{0}^{\infty} h(\tau)u(t-\tau)\, d\tau \tag{2.2.11}$$

We will refer to any function $f(t)$ of t which is zero for negative values of t as a *causal function*. Similarly if $f(t) = 0$ for $t > 0$, we will refer to $f(t)$ as an *anticausal* function. If

$$f(t) = \begin{cases} f_a(t) & t < 0 \\ f_c(t) & t \geqslant 0 \end{cases} \tag{2.2.12}$$

we will refer to $f_c(t)$ as the *causal part* of $f(t)$ and $f_a(t)$ as the *anticausal* part of $f(t)$.

Let us now consider a linear multivariable system with p inputs $u_1(t)$, $u_2(t), \ldots, u_p(t)$ and q outputs $y_1(t) \ldots y_q(t)$. Let us denote by $h_{ij}(t, \tau)$, the response at time t at the ith output terminal due to a unit impulse applied at time τ at the jth input terminal with all other inputs being set equal to zero. It then follows from Eqs. (2.2.3) and (2.2.5) that the response $y_i(t)$ at the ith output terminal is given by

$$y_i(t) = \sum_{j=1}^{p} \int_{-\infty}^{\infty} h_{ij}(t, \tau) u_j(\tau)\, d\tau \qquad \text{for } i = 1 \ldots q \tag{2.2.13}$$

Expressions for the outputs of causal or time-invariant systems follow similar to those for the single-input single-output system.

For linear time-invariant systems, we can define a transfer function (a transfer matrix for multi-input multi-output systems) as the transform of the impulse response of the system. Two of the commonly used descriptions are in terms of the Fourier and Laplace transforms [3]. Laplace transformation of both sides of Eq. (2.2.10) yields

$$Y(s) = H(s) U(s) \tag{2.2.14a}$$

where $Y(s)$ and $U(s)$ represent the transforms of $y(t)$ and $u(t)$, respectively, and $H(s)$ represents the system transfer function. We note that since the signals are defined for all $t \in (-\infty, \infty)$, the appropriate transform to use is the bilateral Laplace transform (see Appendix A). Similarly in terms of the Fourier transform, we can write

$$Y(\omega) = H(\omega) U(\omega) \tag{2.2.14b}$$

State-Space Characterization of Systems

In determining the response of a linear system by the convolution operation, the initial conditions in the system must be zero. That is, the convolution integral yields only the forced response of the system. We now consider an alternative description of a system which will enable us to take into account the initial conditions of the system. We recall that in a dynamic system, the output at any instant t depends not only on the input at t but is also dependent on the prior history of the system. Any set of variables that summarizes the past history of the system affecting its future behavior is said to be the *state variables* of the system. More precisely, we define the *state* of a system at t_0 as the amount of information at t_0, that, together with the input, determines uniquely the output for all $t \geqslant t_0$. Thus the choice of state variables in a system is not unique.

Given at time t_0 a set of n state variables $x_1(t_0), x_2(t_0), \ldots, x_n(t_0)$, we can define a state vector $\mathbf{x}(t_0)$ as

$$\mathbf{x}(t_0) = [x_1(t_0)x_2(t_0) \ldots x_n(t_0)]^T$$

The number of state variables used in a description of a system is the *order* of the system. If the number of state variables is the minimum possible, the state vector has *minimum order*. For example, for a system described by an Nth order differential equation, we need N independent initial conditions to determine the output of the system. Thus, the minimal order of the state vector is N.

We confine our discussion in the rest of this chapter to systems in which the number of state variables is finite. Such systems are called finite-dimensional systems. This excludes certain types of systems from our consideration, for example distributed parameter systems or systems involving time delays.

From the previous discussion, it follows that the output $y(t)$ at any instant t can be computed from the values of the state variables together with the values of the input $u(s)$ for $s \geqslant t$. In functional form, we can therefore write the *output equation* for the system as

$$y(t) = h(x_1(t) \ldots x_n(t); u(s), s \geqslant t; t) \tag{2.2.15}$$

The explicit dependence on the parameter t is introduced to account for the fact that in nonstationary systems, the functional relationship between $y(t)$ and the system states and input may vary from instant to instant. For causal systems $y(t)$ will not depend on values of the input $u(s)$ for $s > t$ and Eq. (2.2.15) becomes

$$y(t) = h(x_1(t) \ldots x_n(t), u(t), t) \tag{2.2.16}$$

Since the history of the system changes as time progresses, the values of the system states change and must be updated. For differential systems the equations describing the evolution of the states can be written as

$$\dot{x}_i(t) = f_i(x_1(t) \ldots x_n(t), u(t), t) \qquad i = 1 \ldots n \tag{2.2.17}$$

The state equations form a set of simultaneous first-order differential equations.

The extension to multiple inputs and outputs is quite straightforward. Given a system with p inputs $u_1(t) \ldots u_p(t)$ and q outputs $y_1(t) \ldots y_q(t)$, we define as the state variables of the system any set of variables $x_1(t) \ldots x_n(t)$ which, together with a knowledge of the values of the inputs, is sufficient to uniquely determine the values of the outputs. We write the output and state equations as

$$y_i(t) = h_i(x_1(t) \ldots x_n(t), u_1(t) \ldots u_p(t), t) \quad i = 1 \ldots q \tag{2.2.18a}$$

$$\dot{x}_j(t) = f_j(x_1(t) \ldots x_n(t), u_1(t) \ldots u_p(t), t) \quad j = 1 \ldots n \tag{2.2.18b}$$

By defining the vectors

$$\mathbf{x}(t) = [x_1(t) \ldots x_n(t)]^T$$

$$\mathbf{y}(t) = [y_1(t) \ldots y_q(t)]^T$$

$$\mathbf{u}(t) = [u_1(t) \ldots u_p(t)]^T$$

we can write these equations compactly as

$$\mathbf{y}(t) = \mathbf{h}(\mathbf{x}(t), \mathbf{u}(t), t) \tag{2.2.19a}$$

$$\dot{\mathbf{x}}(t) = \mathbf{f}(\mathbf{x}(t), \mathbf{u}(t), t) \tag{2.2.19b}$$

where the vectors $\mathbf{h}(\cdot,\cdot,\cdot)$ and $\mathbf{f}(\cdot,\cdot,\cdot)$ have the obvious definitions. We will assume that the initial condition for solving Eq. (2.2.19) is specified at initial time t_0, $\mathbf{x}(t_0) = \mathbf{x}_0$. Once Eq. (2.2.19b) has been solved for $\mathbf{x}(t)$, the determination of $\mathbf{y}(t)$ from Eq. (2.2.19a) is fairly straightforward. Unfortunately, the analytical solution of Eq. (2.2.19b) is in general a formidable problem. In fact, such solutions can be obtained only for particular cases. If we can, however, establish that this equation has a unique solution, we can attempt to find this solution approximately by numerical integration of the equation. However, for nonlinear differential equations, the question of the existence and uniqueness of the solution is a fundamental problem. For example, consider the nonlinear equation.

$$\dot{x}(t) = [x(t)]^{1/3} \qquad x(0) = 0$$

It can easily be verified by direct substitution that the following two functions are both solutions of this equation:

$$x(t) = 0 \qquad t \geqslant 0$$

$$x(t) = \left(\frac{2}{3}t\right)^{3/2} \qquad t \geqslant 0$$

We can, however, obtain a *sufficient* condition for the existence and uniqueness of the solution of Eq. (2.2.19b). In order to introduce this condition, let us suppress the explicit dependence on $\mathbf{u}(t)$ in Eq. (2.2.19b) and write the equation as

$$\dot{\mathbf{x}}(t) = \mathbf{f}(\mathbf{x}, t) \qquad \mathbf{x}(t_0) = \mathbf{x}_0 \tag{2.2.20}$$

Then if $\mathbf{f}(\mathbf{x}, t)$ is continuous in $\mathbf{x}$, it can be shown [4] that there exists a solution $\mathbf{x}(t)$ to Eq. (2.2.20) which is continuous and differentiable. Furthermore, if for any $\mathbf{x}$ and $\mathbf{x}^1$, $\mathbf{f}(\cdot, t)$ is such that

$$\|\mathbf{f}(\mathbf{x}, t) - \mathbf{f}(\mathbf{x}^1, t)\| < C\|\mathbf{x} - \mathbf{x}^1\| \tag{2.2.21}$$

for some constant C, the solution is unique. Here $\|\cdot\|$ indicates the norm of a vector (see Appendix C). The condition in Eq. (2.2.21) is known as the *Lipschitz* condition.

If $f(x, t)$ is continuously differentiable in x, the Lipschitz condition reduces to

$$\left\|\frac{\partial \mathbf{f}}{\partial \mathbf{x}}\right\| < C \qquad \text{for all } t, \mathbf{x} \tag{2.2.22}$$

In many nonlinear equations, the solution may be defined over only a finite interval of time. For example, consider the equation

$$\dot{x}(t) = x^2(t), \; x(0) = 0 \tag{2.2.23}$$

Its solution is

$$x(t) = \frac{1}{1 - t} \tag{2.2.24}$$

Clearly this solution cannot be carried beyond $t = 1$. Such solutions are said to have a finite escape time. Obviously such an equation will not correspond to a physical system. A condition which ensures that the solution is defined for all t is that

$$\left\|\frac{\partial \mathbf{f}}{\partial \mathbf{x}}\right\| \leqslant C(t) < \infty \qquad \text{for all } t, \mathbf{x} \tag{2.2.25}$$

for some continuous function $C(t)$.

State Equations for Linear Systems

For linear systems the functional dependence of the output on the system states and inputs is linear. We can therefore write

$$\begin{aligned} y_i(t) &= c_{i1}(t)x_1(t) + c_{i2}(t)x_2(t) + \cdots + c_{in}(t)x_n(t) \\ &\quad + d_{i1}(t)u_1(t) + \cdots + d_{ip}(t)u_p(t) \\ &= \sum_{j=1}^{n} c_{ij}(t)x_j(t) + \sum_{j=1}^{p} d_{ij}(t)u_j(t) \qquad \text{for } i = 1 \ldots q \end{aligned} \tag{2.2.26}$$

Similarly the state equations can be written as

$$\begin{aligned}\dot{x}_i(t) &= a_{i1}(t)x_1(t) + \cdots + a_{in}(t)x_n(t) \\ &\quad + b_{i1}(t)u_1(t) + \cdots + b_{ip}(t)u_p(t) \\ &= \sum_{j=1}^{n} a_{ij}(t)x_j(t) + \sum_{j=1}^{p} b_{ij}(t)u_j(t) \qquad i = 1 \ldots n\end{aligned} \tag{2.2.27}$$

and form a set of first-order *linear* differential equations. By defining the matrices $\mathbf{A}(t)$, $\mathbf{B}(t)$, $\mathbf{C}(t)$, and $\mathbf{D}(t)$ as

$$\begin{aligned}\mathbf{A}(t) &= \begin{bmatrix} a_{11}(t) \ldots & a_{1n}(t) \\ \vdots & \vdots \\ a_{n1}(t) \ldots & a_{nn}(t) \end{bmatrix}, \quad \mathbf{B}(t) = \begin{bmatrix} b_{11}(t) \ldots & b_{1p}(t) \\ \vdots & \vdots \\ b_{n1}(t) \ldots & b_{np}(t) \end{bmatrix} \\ \mathbf{C}(t) &= \begin{bmatrix} c_{11}(t) \ldots & c_{1n}(t) \\ \vdots & \vdots \\ c_{q1}(t) \ldots & c_{qn}(t) \end{bmatrix}, \quad \mathbf{D}(t) = \begin{bmatrix} d_{11}(t) \ldots & d_{1p}(t) \\ \vdots & \vdots \\ d_{q1}(t) \ldots & d_{qp}(t) \end{bmatrix}\end{aligned} \tag{2.2.28}$$

we can write these equations in vector-matrix notation as

$$\mathbf{y}(t) = \mathbf{C}(t)\mathbf{x}(t) + \mathbf{D}(t)\mathbf{u}(t) \tag{2.2.29a}$$

$$\dot{\mathbf{x}}(t) = \mathbf{A}(t)\mathbf{x}(t) + \mathbf{B}(t)\mathbf{u}(t) \tag{2.2.29b}$$

We note that for a linear system since

$$\mathbf{f}(\mathbf{x}, \mathbf{u}, \mathbf{t}) = \mathbf{A}(t)\mathbf{x}(t) + \mathbf{B}(t)\mathbf{u}(t) \tag{2.2.30}$$

it follows that

$$\left\| \frac{\partial \mathbf{f}}{\partial \mathbf{x}} \right\| = \|\mathbf{A}(t)\| \tag{2.2.31}$$

and hence a unique solution always exists for all t if $\|\mathbf{A}(t)\|$ is bounded.

For time-invariant systems, the matrices **A**, **B**, **C**, and **D** are matrices of constants and the preceding equations can be written as

$$\mathbf{y}(t) = \mathbf{C}\mathbf{x}(t) + \mathbf{D}\mathbf{u}(t) \tag{2.2.32a}$$

$$\dot{\mathbf{x}}(t) = \mathbf{A}\mathbf{x}(t) + \mathbf{B}\mathbf{u}(t) \tag{2.2.32b}$$

We can obtain alternative descriptions of the system of Eq. (2.2.29) by choosing a different set of state variables. Let us consider a second minimal-order representation of the system of Eq. (2.2.29) in terms of the state variable $\bar{\mathbf{x}}(t)$ as

$$\bar{\mathbf{x}}(t) = \bar{\mathbf{A}}(t)\bar{\mathbf{x}}(t) + \bar{\mathbf{B}}(t)\mathbf{u}(t) \tag{2.2.33a}$$

$$\mathbf{y}(t) = \bar{\mathbf{C}}(t)\bar{\mathbf{x}}(t) + \bar{\mathbf{D}}(t)\mathbf{u}(t) \tag{2.2.33b}$$

Then there exists a nonsingular matrix $\mathbf{P}(t)$ such that

$$\bar{\mathbf{x}}(t) = \mathbf{P}(t)\mathbf{x}(t)$$

and the dynamical system of Eq. (2.2.29) is said to be *equivalent* to the system of Eq. (2.2.33). It can easily be verified that

$$\bar{\mathbf{A}}(t) = \mathbf{P}(t)\mathbf{A}(t) + \dot{\mathbf{P}}(t)\mathbf{P}^{-1}(t) \tag{2.2.34a}$$

$$\bar{\mathbf{B}}(t) = \mathbf{P}(t)\mathbf{B}(t) \tag{2.2.34b}$$

$$\bar{\mathbf{C}}(t) = \mathbf{C}(t)\mathbf{P}^{-1}(t) \tag{2.2.34c}$$

$$\bar{\mathbf{D}}(t) = \mathbf{D}(t) \tag{2.2.34d}$$

Example 2.2.1 Consider the system whose input–output relation is given by the differential equation

$$\ddot{y}(t) + (a+b)\dot{y}(t) + aby(t) = \beta\dot{u}(t) + \alpha u(t) \text{ with } u(t) = 0 \text{ for } t < 0$$

Let the system initial conditions be

$$y(0) = 1,\ \dot{y}(0) = 0$$

It is clear that the minimum number of state variables needed to describe the system is two. Let us choose our states such that

$$y(t) = \alpha x_1(t) + \beta x_2(t) \text{ with } x_2(t) = \dot{x}_1(t)$$

It easily follows that in vector-matrix notation, the equations of the system are

$$\dot{\mathbf{x}}(t) = \begin{bmatrix} \dot{x}_1(t) \\ \dot{x}_2(t) \end{bmatrix} = \begin{bmatrix} 0 & 1 \\ -ab & -(a+b) \end{bmatrix} \begin{bmatrix} x_1(t) \\ x_2(t) \end{bmatrix} + \begin{bmatrix} 0 \\ 1 \end{bmatrix} u(t)$$

$$y(t) = \begin{bmatrix} \alpha & \beta \end{bmatrix} \begin{bmatrix} x_1(t) \\ x_2(t) \end{bmatrix}$$

The initial conditions easily follow from $y(0) = 1$ and $\dot{y}(0) = 0$ as

$$x_1(0) = \frac{\alpha - \beta(a+b)}{\alpha^2 + \beta^2 ab - \alpha\beta(a+b)}$$

$$x_2(0) = \frac{\beta ab}{\alpha^2 + \beta^2 ab - \alpha\beta(a+b)}$$

The state variables are in the phase-variable form. An alternative description can be obtained by choosing as our first state variable

$$\bar{x}_1(t) = y(t)$$

In this case, if we choose $\bar{x}_2(t)$ to be the derivative of $\bar{x}_1(t)$, the equation for $\dot{\bar{x}}_2(t)$ will involve terms in $\dot{u}(t)$ and hence the equations will not be in the form of Eq. (2.2.33). We will therefore define $\bar{x}_2(t)$ as

$$\bar{x}_2(t) = \dot{\bar{x}}_1(t) - \beta u(t)$$

The system equations then become

$$\dot{\bar{\mathbf{x}}}(t) = \begin{bmatrix} \dot{\bar{x}}_1(t) \\ \dot{\bar{x}}_2(t) \end{bmatrix} = \begin{bmatrix} 0 & 1 \\ -ab & -(a+b) \end{bmatrix} \begin{bmatrix} \bar{x}_1(t) \\ \bar{x}_2(t) \end{bmatrix} + \begin{bmatrix} \beta \\ \alpha - \beta(a+b) \end{bmatrix} u(t)$$

$$y(t) = \begin{bmatrix} 1 & 0 \end{bmatrix} \begin{bmatrix} \bar{x}_1(t) \\ \bar{x}_2(t) \end{bmatrix}$$

It can easily be verified that the matrix $\mathbf{P}$ relating the vectors $\bar{\mathbf{x}}(t)$ and $\mathbf{x}(t)$ is

$$\mathbf{P} = \begin{bmatrix} \alpha & \beta \\ -ab\beta & \alpha - (a+b) \end{bmatrix}$$

That is

$$\bar{x}_1(t) = \alpha x_1(t) + \beta x_2(t)$$

$$\bar{x}_2(t) = -ab\beta x_1(t) + [\alpha - (a+b)\beta] x_2(t)$$

Solution of the State Equations: The State Transition Matrix

We now consider the solution of the equations

$$\dot{\mathbf{x}}(t) = \mathbf{A}(t)\mathbf{x}(t) + \mathbf{B}(t)\mathbf{u}(t) \qquad t \geqslant t_0;\ \mathbf{x}(t_0) = \mathbf{x}_0 \tag{2.2.35}$$

We again note that these constitute a set of n simultaneous first-order linear differential equations. The set of homogeneous equations

$$\dot{\mathbf{x}}(t) = \mathbf{A}(t)\mathbf{x}(t) \tag{2.2.36}$$

has n linearly independent solutions $\mathbf{x}^1(t)$, $\mathbf{x}^2(t)$, . . . , $\mathbf{x}^n(t)$, such that any other solution of Eq. (2.2.36) can be written in terms of these solutions. Let us denote by $\mathbf{W}(t)$ the matrix formed with these solutions as columns:

$$\mathbf{W}(t) = \left[\mathbf{x}^1(t)\mathbf{x}^2(t) \ldots \mathbf{x}^n(t)\right] \tag{2.2.37}$$

Then $\mathbf{W}(t)$ is called the fundamental matrix associated with Eq. (2.2.36) [4]. It follows that since the columns of $\mathbf{W}(t)$ are linearly independent solutions of Eq. (2.2.36), $\mathbf{W}(t)$ is nonsingular and satisfies the matrix differential equation

$$\frac{d\mathbf{W}(t)}{dt} = \mathbf{A}(t)\mathbf{W}(t) \tag{2.2.38}$$

Let us now define the matrix $\psi(t, \tau)$ for fixed t and τ as

$$\psi(t, \tau) = \mathbf{W}(t)\mathbf{W}^{-1}(\tau) \tag{2.2.39}$$

Then $\psi(t, \tau)$ is called the state transition matrix associated with the system of Eq. (2.2.36). It will be shown subsequently that the state transition matrix plays an important role in determining the solution of Eq. (2.2.35). The following properties of $\psi(t, \tau)$ can easily be established:

1. For each τ, $\psi(t, \tau)$ satisfies the equation

$$\frac{\partial\psi(t, \tau)}{\partial t} = \mathbf{A}(t)\psi(t, \tau) \tag{2.2.40}$$

   This follows by postmultiplying both sides of Eq. (2.2.38) by $\mathbf{W}^{-1}(\tau)$ and using Eq. (2.2.39).

2. Substitution of $\tau = t$ in Eq. (2.2.39) yields

$$\psi(t, t) = \mathbf{I} \qquad \text{for all } t \tag{2.2.41}$$

3. Inverting Eq. (2.2.39), we obtain

$$\psi^{-1}(t, \tau) = \mathbf{W}(\tau)\mathbf{W}^{-1}(t) = \psi(\tau, t) \tag{2.2.42}$$

4. For any t, ξ, and τ, the following relation holds:

$$\begin{aligned}\psi(t, \xi)\psi(\xi, \tau) &= \mathbf{W}(t)\mathbf{W}^{-1}(\xi)\mathbf{W}(\xi)\mathbf{W}^{-1}(\tau)\\ &= \mathbf{W}(t)\mathbf{W}^{-1}(\tau) = \psi(t, \tau)\end{aligned} \tag{2.2.43}$$

5. For any t and τ, we have

$$\frac{\partial\psi(t, \tau)}{\partial\tau} = -\psi(t, \tau)\mathbf{A}(\tau) \tag{2.2.44}$$

The last equation follows from the fact that

$$\psi(t, \tau)\psi(\tau, t) = \mathbf{I}$$

Differentiating both sides with respect to τ yields

$$\frac{\partial\psi(t, \tau)}{\partial\tau}\psi(\tau, t) + \psi(t, \tau)\frac{\partial\psi(\tau, t)}{\partial\tau} = \mathbf{0}$$

so that

$$\begin{aligned}\frac{\partial\psi(t, \tau)}{\partial\tau} &= -\psi(t, \tau)\frac{\partial\psi(\tau, t)}{\partial\tau}\psi^{-1}(\tau, t)\\ &= -\psi(t, \tau)\mathbf{A}(\tau)\psi(\tau, t)\psi^{-1}(\tau, t)\end{aligned}$$

by using Eq. (2.2.40)

We note that $\psi(t, \tau)$ is uniquely determined by $\mathbf{A}(t)$. In general, the determination of $\psi(t, \tau)$ is extremely difficult and frequently recourse must be had to numerical integration techniques to determine $\psi(t, \tau)$ over a particular range of interest.

We now show that the solution to Eq. (2.2.35) can be written in terms of the state transition matrix as

$$\mathbf{x}(t) = \psi(t, t_0)\mathbf{x}(t_0) + \int_{t_0}^{t} \psi(t, \tau)\mathbf{B}(\tau)\mathbf{u}(\tau)\, d\tau \tag{2.2.45}$$

We establish this by showing that $\mathbf{x}(t)$ in Eq. (2.2.45) satisfies the differential equation (2.2.35) and the associated boundary condition. $\mathbf{x}(t)$ in Eq.

(2.2.45) certainly satisfies the initial condition since

$$\mathbf{x}(t_0) = \boldsymbol{\psi}(t_0, t_0)\mathbf{x}_0 + \int_{t_0}^{t_0} \boldsymbol{\psi}(t_0, \tau)\mathbf{B}(\tau)\mathbf{u}(\tau)\, d\tau = \mathbf{x}_0$$

If we differentiate both sides of Eq. (2.2.45) with respect to t we obtain

$$\dot{\mathbf{x}}(t) = \frac{\partial \boldsymbol{\psi}(t, t_0)}{\partial t}\mathbf{x}_0 + \boldsymbol{\psi}(t, \tau)\mathbf{B}(\tau)\mathbf{u}(\tau)\Big|_{\tau=t} + \int_{t_0}^{t} \frac{\partial \boldsymbol{\psi}(t, \tau)}{\partial t}\mathbf{B}(\tau)\mathbf{u}(\tau)\, d\tau$$

Use of Eq. (2.2.40) then yields

$$\dot{\mathbf{x}}(t) = \mathbf{A}(t)\boldsymbol{\psi}(t, t_0)\mathbf{x}_0 + \mathbf{B}(t)\mathbf{u}(t) + \mathbf{A}(t)\int_{t_0}^{t} \boldsymbol{\psi}(t, \tau)\mathbf{B}(\tau)\mathbf{u}(\tau)\, d\tau$$

which can be rewritten using Eq. (2.2.45) as

$$\dot{\mathbf{x}}(t) = \mathbf{A}(t)\mathbf{x}(t) + \mathbf{B}(t)\mathbf{u}(t)$$

so that $\mathbf{x}(t)$ satisfies the differential equation. Thus $\mathbf{x}(t)$ is the (unique) solution to Eq. (2.2.35). Note that in writing the solution as in Eq. (2.2.45), we have implicitly assumed that the system is causal. The first term on the right side represents the initial condition response and the second term, the response due to the input. The expression for the output $\mathbf{y}(t)$ follows as

$$\begin{aligned}\mathbf{y}(t) &= \mathbf{C}(t)\mathbf{x}(t) + \mathbf{D}(t)\mathbf{u}(t)\\ &= \mathbf{C}(t)\left[\boldsymbol{\psi}(t, t_0)\mathbf{x}(t_0) + \int_{t_0}^{t} \boldsymbol{\psi}(t, \tau)\mathbf{B}(\tau)\mathbf{u}(\tau)\, d\tau\right] + \mathbf{D}(t)\mathbf{u}(t)\end{aligned} \tag{2.2.46}$$

The impulse response of this system can easily be determined by setting $\mathbf{x}(t_0) = 0$ and $\mathbf{u}(t) = \boldsymbol{\delta}(t - t_0)$. We obtain for the *impulse response matrix*

$$\mathbf{H}(t, t_0) = \mathbf{C}(t)\boldsymbol{\psi}(t, t_0)\mathbf{B}(t_0) + \mathbf{D}(t)\delta(t - t_0) \tag{2.2.47}$$

For the time-invariant system of Eq. (2.2.32), the solution can be written as

$$\mathbf{x}(t) = \boldsymbol{\psi}(t - t_0)\mathbf{x}(t_0) + \int_{t_0}^{t} \boldsymbol{\psi}(t - \tau)\mathbf{B}\mathbf{u}(\tau)\, d\tau \tag{2.2.48}$$

The transition matrix in this case is easily computed as

$$\psi(t - t_0) = e^{\mathbf{A}(t - t_0)} = \sum_{i=0}^{\infty} \frac{\mathbf{A}^i (t - t_0)^i}{i!} \tag{2.2.49}$$

We therefore can write the solution, Eq. (2.2.45), as

$$\mathbf{x}(t) = e^{\mathbf{A}(t - t_0)}\mathbf{x}(t_0) + \int_{t_0}^{t} e^{\mathbf{A}(t - \tau)}\mathbf{B}\mathbf{u}(\tau)\,d\tau \tag{2.2.50}$$

The output $\mathbf{y}(t)$ is given by

$$\mathbf{y}(t) = \mathbf{C}\left[e^{\mathbf{A}(t - t_0)}\mathbf{x}(t_0) + \int_{t_0}^{t} e^{\mathbf{A}(t - \tau)}\mathbf{B}\mathbf{u}(\tau)\,d\tau \right] + \mathbf{D}\mathbf{u}(t) \tag{2.2.51}$$

For time-invariant systems, since the initial time can be arbitrarily chosen, we can set $t_0 = 0$ in Eqs. (2.2.48) through (2.2.51). We can then obtain an alternative expression for $e^{\mathbf{A}t}$ by setting $\mathbf{x}(0) = 0$ in Eq. (2.2.32b) and taking the Laplace transform to obtain

$$s\mathbf{X}(s) = \mathbf{A}\mathbf{X}(s) + \mathbf{B}\mathbf{U}(s) \tag{2.2.52}$$

so that

$$\mathbf{X}(s) = [s\mathbf{I} - \mathbf{A}]^{-1}\mathbf{B}\mathbf{U}(s) \tag{2.2.53}$$

However, transformation of both sides of Eq. (2.2.50) with $\mathbf{x}(0) = \mathbf{0}$ yields

$$\mathbf{X}(s) = \mathcal{L}[e^{\mathbf{A}t}]\mathbf{B}\mathbf{U}(s) \tag{2.2.54}$$

Comparison of Eqs. (2.2.53) and (2.2.54) yields

$$\mathcal{L}[e^{\mathbf{A}t}] = [s\mathbf{I} - \mathbf{A}]^{-1}\mathbf{B} \tag{2.2.55a}$$

or equivalently

$$e^{\mathbf{A}t} = \mathcal{L}^{-1}[s\mathbf{I} - \mathbf{A}]^{-1}\mathbf{B} \tag{2.2.55b}$$

We can obtain an expression for the impulse response matrix of the system by setting $\mathbf{x}(t_0) = 0$ and $\mathbf{u}(t) = \boldsymbol{\delta}(t - t_0)$ in Eq. (2.2.48). This yields

$$\mathbf{H}(t, t_0) = \mathbf{H}(t - t_0) = \mathbf{C}e^{A(t - t_0)}\mathbf{B} + \mathbf{D}\delta(t - t_0) \tag{2.2.56}$$

or as more commonly written

$$\mathbf{H}(t) = \mathbf{C}e^{\mathbf{A}t}\mathbf{B} + \mathbf{D}\delta(t) \tag{2.2.57}$$

The corresponding transfer matrix obtained by transformation of Eq. (2.2.57) is

$$\mathbf{H}(s) = \mathbf{C}(s\mathbf{I} - \mathbf{A})^{-1}\mathbf{B} + \mathbf{D} \tag{2.2.58}$$

where we have used Eq. (2.2.55a).

By writing

$$(s\mathbf{I} - \mathbf{A})^{-1} = \frac{\operatorname{Adj}(s\mathbf{I} - \mathbf{A})}{\det(s\mathbf{I} - \mathbf{A})} \tag{2.2.59}$$

we have

$$\mathbf{H}(s) = \frac{\mathbf{C}\operatorname{Adj}(s\mathbf{I} - \mathbf{A})\mathbf{B}}{\det(s\mathbf{I} - \mathbf{A})} + \mathbf{D} \tag{2.2.60}$$

It therefore follows that the poles (singularities) of the transfer function matrix $H(s)$ are the eigenvalues of the matrix $\mathbf{A}$.

Discrete-Time Systems [2]

The discussions preceding for continuous-time systems can easily be carried over for discrete-time systems. Given a linear discrete-time system whose input–output relationship is given by

$$\mathbf{T}\{u(k)\} = y(k) \tag{2.2.61}$$

let us denote by $h(k, j)$ the response of the system observed at time k to an impulse input applied at j. In terms of the impulse response $h(k, j)$, the output $y(k)$ to an arbitrary input $u(j)$ defined for j in $(-\infty, \infty)$ is given by the *convolution summation*

$$y(k) = \sum_{j=-\infty}^{\infty} h(k, j)u(j) \tag{2.2.62}$$

For causal systems, since

$$h(k, j) = 0 \qquad \text{for } k < j$$

Equation (2.2.62) becomes

$$y(k) = \sum_{j=-\infty}^{k} h(k,j)u(j) \tag{2.2.63}$$

For time-invariant systems, if $h(k)$ denotes the response at time k due to the impulse input applied at time zero, the output due to an arbitrary input $u(\cdot)$ can be expressed as

$$y(k) = \sum_{j=-\infty}^{\infty} h(k-j)u(j) \tag{2.2.64}$$

For a causal system this reduces to

$$y(k) = \sum_{j=-\infty}^{k} h(k-j)u(j) = \sum_{j=0}^{\infty} h(j)u(k-j) \tag{2.2.65}$$

We can define a transfer function for discrete-time systems by use of the z-transform [2]. We thus write

$$H(z) = z[h(k)]$$

so that z-transformation of both sides of Eq. (2.2.58) yields in terms of the appropriate transforms

$$\mathbf{Y}(z) = \mathbf{H}(z)\mathbf{U}(z)$$

Again the impulse response or transfer function characterizations cannot take into account initial conditions in the system. In order to do so, we consider the state variable formulation. The states again summarize the past history of the system so that the output $y(k)$ can be written in functional form as

$$y(k) = f(x_1(k), x_2(k), \ldots, x_n(k), u(k), k) \tag{2.2.66}$$

where

$$\mathbf{x}(k) = [x_1(k) \ldots x_n(k)]^T$$

represents the state vector. The state equations can now be written as

$$x_i(k+1) = f_i(x_1(k) \ldots x_n(k), u(k), k) \qquad i = 1 \ldots n \tag{2.2.67}$$

and form a set of n simultaneous first-order difference equations. For multiple-input multiple-output systems these equations can be written in vector notation as

$$\mathbf{x}(k+1) = \mathbf{f}(\mathbf{x}(k), \mathbf{u}(k), k) \tag{2.2.68a}$$
$$\mathbf{y}(k) = \mathbf{h}(\mathbf{x}(k), \mathbf{u}(k), k) \tag{2.2.68b}$$

where $\mathbf{u}(k)$ represents a p-vector of inputs and $\mathbf{y}(k)$ represents the q-vector of outputs.

In contrast to the situation with nonlinear differential equations, for the discrete-time system of Eqs. (2.2.68) the solution $\mathbf{x}(k)$ exists and is unique for all integers $k \geqslant k_0$.

We note that numerical solutions for difference equations can easily be obtained by repeated use of the Eq. (2.2.68). In fact, all techniques for numerical solution of differential equations involve some type of discretization with respect to time.

For linear systems, the state equations become

$$\mathbf{x}(k+1) = \mathbf{A}(k)\mathbf{x}(k) + \mathbf{B}(k)\mathbf{u}(k) \tag{2.2.69a}$$
$$\mathbf{y}(k) = \mathbf{C}(k)\mathbf{x}(k) + \mathbf{D}(k)\mathbf{u}(k) \tag{2.2.69b}$$

and for time-invariant systems

$$\mathbf{x}(k+1) = \mathbf{A}\mathbf{x}(k) + \mathbf{B}\mathbf{u}(k) \tag{2.2.70a}$$
$$\mathbf{y}(k) = \mathbf{C}\mathbf{x}(k) + \mathbf{D}\mathbf{u}(k) \tag{2.2.70b}$$

in which the matrices are constant.

Solution of the State Equations for Linear Discrete-Time Systems

The solution of the state equations for the linear discrete-time system of Eq. (2.2.69) is easily obtained by using the equations repeatedly. Thus given $x(k_0)$ for any k_0, the solution $\mathbf{x}(k)$ for $k \geqslant k_0$ is obtained as follows:

$$\begin{aligned}
\mathbf{x}(k) &= \mathbf{A}(k-1)\mathbf{x}(k-1) + \mathbf{B}(k-1)\mathbf{u}(k-1) \\
&= \mathbf{A}(k-1)[\mathbf{A}(k-2)\mathbf{x}(k-2) + \mathbf{B}(k-2)\mathbf{u}(k-2)] \\
&\quad + \mathbf{B}(k-1)\mathbf{u}(k-1) \\
&\quad \vdots \\
&= \mathbf{A}(k-1)\mathbf{A}(k-2)\ldots\mathbf{A}(k_0)\mathbf{x}(k_0) \\
&\quad + \sum_{j=k_0}^{k-2} \mathbf{A}(k-1)\mathbf{A}(k-2)\ldots\mathbf{A}(j+1)\mathbf{B}(j)\mathbf{u}(j) \\
&\quad + \mathbf{B}(k-1)\mathbf{u}(k-1)
\end{aligned} \tag{2.2.71}$$

If we now define for $k > j$

$$\psi(k,j) = \prod_{i=j}^{k-1} \mathbf{A}(i) = \mathbf{A}(k-1)\ldots\mathbf{A}(j+1)\mathbf{A}(j) \tag{2.2.72}$$

and

$$\psi(k,k) = \mathbf{I} \qquad \text{(identity matrix)} \tag{2.2.73}$$

we can write

$$\mathbf{x}(k) = \psi(k,k_0)\mathbf{x}(k_0) + \sum_{j=k_0}^{k-1} \psi(k,j+1)\mathbf{B}(j)\mathbf{u}(j) \tag{2.2.74}$$

Equation (2.2.74) is the discrete analog of Eq. (2.2.45). The first term on the right side of Eq. (2.2.74) is the contribution from the initial state $\mathbf{x}(k_0)$ and the second term is the contribution from the input over the interval $[k_0, k-1]$. The output of the system is obtained as

$$\mathbf{y}(k) = \mathbf{C}(k)\left[\psi(k,k_0)\mathbf{x}(k_0) + \sum_{j=k_0}^{k-1} \psi(k,j+1)\mathbf{B}(j)\mathbf{u}(j)\right] + \mathbf{D}(k)\mathbf{u}(k) \tag{2.2.75}$$

By setting $\mathbf{x}(k_0) = 0$ and $\mathbf{u}(k) = \delta_{kl}$, we easily obtain the impulse response of the system as

$$\mathbf{H}(k,l) = \mathbf{C}(k)\psi(k,l+1)\mathbf{B}(l) + \mathbf{D}(k)\delta_{kl} \tag{2.2.76}$$

For time-invariant systems, the state transition matrix becomes

$$\psi(k,j) = \prod_{i=j}^{k-1} \mathbf{A}(i) = \mathbf{A}^{k-j} \tag{2.2.77}$$

so that the expression for the state vector is

$$\mathbf{x}(k) = \mathbf{A}^{k-k_0}\mathbf{x}(k_0) + \sum_{j=k_0}^{k-1} \mathbf{A}^{k-j-1}\mathbf{B}\mathbf{u}(j) \tag{2.2.78}$$

The output is given by

$$\mathbf{y}(k) = \mathbf{C}\left[\mathbf{A}^{k-k_0}\mathbf{x}(k_0) + \sum_{j=k_0}^{k-1} \mathbf{A}^{k-j-1}\mathbf{B}\mathbf{u}(j)\right] + \mathbf{D}\mathbf{u}(k) \tag{2.2.79}$$

By setting $\mathbf{x}(k_0) = \mathbf{0}$ and $\mathbf{u}(k) = \boldsymbol{\delta}_{kl}$, we obtain the impulse response as

$$\mathbf{H}(k, l) = \mathbf{H}(k - l) = \mathbf{C}\mathbf{A}^{k-l-1}\mathbf{B} + \mathbf{D}\delta_{kl} \tag{2.2.80}$$

We can obtain a frequency domain description of the linear time-invariant system of Eq. (2.2.70). Taking the z-transform of both sides yields with $k_0 = 0$

$$z[\mathbf{X}(z) - \mathbf{x}(0)] = \mathbf{A}\mathbf{X}(z) + \mathbf{B}\mathbf{U}(z) \tag{2.2.81}$$

so that

$$\mathbf{X}(z) = z[z\mathbf{I} - \mathbf{A}]^{-1}\mathbf{x}(0) + [z\mathbf{I} - \mathbf{A}]^{-1}\mathbf{B}\mathbf{U}(z) \tag{2.2.82}$$

The output is

$$\mathbf{Y}(z) = \mathbf{C}[z(z\mathbf{I} - \mathbf{A})^{-1}\mathbf{x}(0) + (z\mathbf{I} - \mathbf{A})^{-1}\mathbf{B}\mathbf{U}(z)] + \mathbf{D}\mathbf{U}(z) \tag{2.2.83}$$

and the transfer function matrix is

$$\mathbf{H}(z) = \mathbf{C}(z\mathbf{I} - \mathbf{A})^{-1}\mathbf{B} + \mathbf{D} \tag{2.2.84}$$

Again the poles of the transfer function are the eigenvalues of matrix **A**.

2.3 STOCHASTIC PROCESSES [5–10]

In this section, we review some properties of stochastic processes that will be of use to us in our subsequent discussions. We discuss independent increment processes and introduce the concept of a white noise process, which plays a useful role in many engineering applications. We conclude the section with a discussion of the representation of stochastic process in a set of complete orthonormal functions. Our treatment of these topics is necessarily brief and the reader is referred to [5–10] for detailed discussions of these and other results.

We recall that a stochastic process $x(t, \xi)$ defined over an interval T is a mapping from a probability space Ω into the real line. such that for a fixed $t \in T$, $x(t, \xi)$ is a random variable. More precisely, $x(t, \xi)$ is a measurable function of the σ-field of events ξ on Ω.† Thus $x(t, \xi)$ represents a collection

†A class of sets $A = \{A_i\}$ is said to be a σ-field if

(i) $A_i \cup A_j \in A$ (iii) $\bar{A}_i \in A$, where $\bar{A}_i$ denotes the complement of A_i

(ii) $A_i \cap A_j \in A$ (iv) $\bigcup_{i=1}^{\infty} A_i \in A$

or ensemble of functions. Any particular function of this family is a sample function or realization of the process. For convenience we will write $x(t)$ rather than $x(t, \xi)$ to denote the process.

We can obtain a characterization of the process in terms of the joint distribution of the process evaluated at a finite number of instants $t_1, t_2, \ldots, t_n$, that is, in terms of the probability

$$P\{x(t_1) \leqslant x_1, x(t_2) \leqslant x_2 \ldots x(t_n) \leqslant x_n\} \tag{2.3.1}$$

for some real numbers $x_1, x_2, \ldots, x_n$. A complete description of the process can only be obtained if this probability is known for all n and all instants $\{t_i\}$. If we can define our process $x(t_i)$ for all t_i, given the distribution for a countable set $\{t_i\}$, the process is said to be *separable*.

An important class of processes for which Eq. (2.3.1) does provide a complete description is the Markov process.

We say that a stochastic process $x(t)$ is Markov of order one if for any set of time instants $t_1 \leqslant t_2 \leqslant \cdots \leqslant t_m$, we have

$$\begin{aligned} P[x(t_m) \leqslant x_m | x(t_1) = x_1 \ldots x(t_{m-1}) = x_{m-1}] \\ = P[x(t_m) \leqslant x_m | x(t_{m-1}) = x_{m-1}] \end{aligned} \tag{2.3.2}$$

In terms of density functions this can be written as

$$p(x_m | x_1, x_2, \ldots, x_{m-1}) = p(x_m | x_{m-1}) \tag{2.3.3}$$

If we consider t_{m-1} to be present time, we see that Eq. (2.3.3) implies that the probability density in the future, that is, at time t_m, depends only on the value of the process at present time and is completely independent of its values in the past. The right side of Eq. (2.3.3) is known as the transition probability density function of the Markov process $x(t)$. For a Markov process $x(t)$, the mth order joint density function is easily obtained if the transition probability density functions are known as also the initial density function. We can apply Bayes' rule successively to obtain

$$p(x_m, x_{m-1}, \ldots, x_1) = p(x_m | x_{m-1}) p(x_{m-1} | x_{m-2}) \ldots p(x_1) \tag{2.3.4}$$

We can define higher-order Markov processes as follows. For any set of points $t_1 \leqslant t_2 \cdots \leqslant t_m$, we say that a process is Markov-k, $k < m$ if

$$\begin{aligned} P[x(t_m) \leqslant x_m | x(t_{m-1}) = x_{m-1} \ldots x(t_1) = x_1] \\ = P[x(t_m) \leqslant x_m | x(t_{m-1}) = x_{m-1} \ldots x(t_{m-k}) = x_{m-k}] \end{aligned} \tag{2.3.5}$$

In most of our discussions we will restrict ourselves to Markov processes of order one.

Finally a process is said to be a *martingale* if

$$\mathrm{E}\{x(t_m)|x(t_{m-1}), x(t_{m-2}), \ldots, x(t_1)\} = x(t_{m-1}) \tag{2.3.6}$$

Independent Increment Processes

We have indicated earlier that the white noise process plays an important role in engineering applications. White noise can be considered to result from the formal differentiation of a class of processes known as independent increment processes. In this section, we discuss two processes of this class which have been used to model a wide variety of problems. The Wiener process is a continuous process with Gaussian statistics. The Poisson process is a discrete process following a Poisson distribution.

We say that a random process $x(t)$ is an independent increment process if for any instants $t_1 < t_2 \leqslant t_3 < t_4$, the increments $x(t_2) - x(t_1)$ and $x(t_4) - x(t_3)$ are independent of each other. We note that the increments are taken over nonoverlapping intervals.

We say that a process is Gaussian if the random variables $x(t_1)$, $x(t_2), \ldots, x(t_n)$ are jointly normal for any n and $t_1, t_2, \ldots, t_n$. That is if we define the random vector $\mathbf{x}$ as

$$\mathbf{x} = [x(t_1)x(t_2) \ldots x(t_n)]^T$$

we have

$$p(\mathbf{x}) = \frac{1}{(2\pi)^{n/2}|\mathbf{V}|^{1/2}} \exp\left\{-\frac{1}{2}(\mathbf{x}-\mathbf{m})^T \mathbf{V}^{-1}(\mathbf{x}-\mathbf{m})\right\} \tag{2.3.7}$$

where $\mathbf{m}$ is the mean of $\mathbf{x}$ and $\mathbf{V}$ is the covariance matrix defined by

$$\mathbf{V} = \mathrm{E}\{(\mathbf{x}-\mathbf{m})(\mathbf{x}-\mathbf{m})^T\} \tag{2.3.8}$$

With these definitions, we are now ready to introduce the concept of a Wiener process. We say that a process $x(t, \xi)$ defined over $t \in [0, \infty)$ is a Wiener process if

(i) $x(0, \xi) = 0$ for almost all ξ

(ii) $x(t)$ is an independent increment process

(iii) For $t_2 \geqslant t_1$

$$P[x(t_2) - x(t_1) < \lambda] = \frac{1}{\sqrt{2\pi(t_2 - t_1)}} \int_{-\infty}^{\lambda} \exp\left(\frac{-u^2}{2(t_2 - t_1)}\right) du \tag{2.3.9}$$

Hence the increments of a Wiener process have a Gaussian distribution. It can be shown that the Wiener process is continuous almost everywhere [12].

We note that in defining the Wiener process, we can eliminate condition (iii) above by requiring that $x(t, \xi)$ be continuous in t for almost all ξ. Since in this case, it can be shown that $x(t)$ is a Gaussian random variable for any t, it follows that the two definitions are equivalent [12].

We can now determine the moments of the Wiener process. It follows from Eq. (2.3.9) that $x(t)$ is zero mean. To determine its correlation function

$$\phi_x(t_1, t_2) = \mathrm{E}\{x(t_1)x(t_2)\}$$

we note that for $t_1 = t_2 = t$

$$\mathrm{E}\{x^2(t)\} = t \tag{2.3.10}$$

For $t_1 > t_2$, we can write

$$x(t_1) = x(t_2) + x(t_1) - x(t_2)$$

so that

$$\begin{aligned} \mathrm{E}\{x(t_1)x(t_2)\} &= \mathrm{E}\{[x(t_2) + x(t_1) - x(t_2)]x(t_2)\} \\ &= \mathrm{E}\{x^2(t_2)\} + \mathrm{E}\{[x(t_1) - x(t_2)][x(t_2) - x(0)]\} \\ &= t_2 \end{aligned} \tag{2.3.11}$$

where the last step follows from Eq. (2.3.10) and the fact that $x(t)$ is an independent increment process with $x(0) = 0$.

Similarly for $t_2 > t_1$, we have

$$\mathrm{E}\{x(t_1)x(t_2)\} = t_1 \tag{2.3.12}$$

so that

$$\phi_x(t_1, t_2) = \min(t_1, t_2) \tag{2.3.13}$$

Even though the Wiener process is continuous, it is not differentiable. However, suppose we formally differentiate $x(t)$ to get the process $w(t) = \dot{x}(t)$. It follows from Eq. (2.3.13) that

$$\mathrm{E}\{w(t_1)w(t_2)\} = \delta(t_1 - t_2) \tag{2.3.14}$$

Obviously $w(t)$ cannot correspond to a physical process since the energy in the signal is infinite. The process $w(t)$ is known as a white noise process. The name arises from the fact that if we define the spectrum of $w(t)$ as the Fourier transform of the correlation function of Eq. (2.3.14), we see that the spectrum is constant or "white" over the entire frequency range.

We now introduce the concept of a Poisson process which is a step process with an at most countable number of steps. We say that a process $x(t, \xi)$ defined over $t \in [0, \infty)$ is a Poisson process if

(i) for almost all ξ, $x(t, \xi)$ is a nondecreasing step function which jumps by unity at each discontinuity, with $x(0) = 0$

(ii) $x(t)$ is an independent increment process

(iii) the probability of the number of jumps between t_1 and t_2 follows a Poisson distribution:

$$P\{x(t_2) - x(t_1) = k\} = \exp[-\lambda|t_2 - t_1|]\frac{(|t_2 - t_1|)^k \lambda^k}{k!} \tag{2.3.15}$$

If we allow the jump at each discontinuity t_i to be a random variable a_i, we have the generalized Poisson process $y(t)$. Clearly $y(t)$ will also be an independent increment process.

If the random variables $\{a_i\}$ are independent, identically distributed, zero mean, Gaussian random variables with variance σ_a^2, it can be shown that the moments of $y(t)$ are given by

$$\mathrm{E}\{y(t)\} = 0$$

and

$$\mathrm{E}\{y(t_1)y(t_2)\} = \lambda\sigma_a^2 \min(t_1, t_2) \tag{2.3.16}$$

which is the same as one would obtain if $y(t)$ were a Wiener process. Again, if we differentiate $y(t)$ to obtain the process $w'(t)$, we have

$$\mathrm{E}\{w'(t_1)w'(t_2)\} = \lambda\sigma_a^2\delta(t_1 - t_2) \tag{2.3.17}$$

so that $w'(t)$ is also a white noise process. We now present an example to illustrate the use of white noise processes in signal modeling.

Example 2.3.1 Let us consider a linear time-invariant system with impulse response $h(t)$ and corresponding transfer function $H(s)$. If we apply a stationary input $x(t)$ to the system, the output $y(t)$ is given by

$$y(t) = \int_{-\infty}^{\infty} h(t-\lambda)x(\lambda)\,d\lambda = \int_{-\infty}^{\infty} x(t-\lambda)h(\lambda)\,d\lambda \tag{2.3.18}$$

In terms of the system transfer function $H(s)$, we can write

$$Y(s) = H(s)X(s) \tag{2.3.19}$$

It easily follows that the mean and autocorrelation of $y(t)$ can be obtained in terms of those of $x(t)$ as

$$m_y = m_x H(0) \tag{2.3.20a}$$

and

$$\phi_y(\tau) = \int_{-\infty}^{\infty}\int_{-\infty}^{\infty} \phi_x(\tau - u - \lambda)h(\lambda)h(-u)\,d\lambda\,du \tag{2.3.20b}$$

In terms of the respective power spectra, we can write Eq. (2.3.20b) as

$$\Phi_y(s) = H(s)H(-s)\Phi_x(s) \tag{2.3.21}$$

Equation (2.3.21) gives us a means of generating a stationary process with prescribed correlation function or power spectrum by using a white noise source and a linear system. We will assume that the required spectrum $\Phi_y(s)$ is a rational function is s (a ratio of two polynomials in s). From Eq. (2.3.21), if $x(t)$ is white noise with unity power spectrum,

$$\Phi_x(s) = 1 \tag{2.3.22}$$

so that

$$\Phi_y(s) = H(s)H(-s) \tag{2.3.23}$$

If we factor $\Phi_y(s)$ into poles and zeros, we can write

$$\Phi_y(s) = \Phi_y^+(s)\Phi_y^-(s) \tag{2.3.24}$$

where $\Phi_y^+(s)$ contains all terms in the left half-plane and $\Phi_y^-(s)$ contains terms in the right half-plane. Since $\Phi_y(s)$ is a power spectrum and consequently $\Phi_y(\omega) \geqslant 0$, it can be shown that

$$\Phi_y^+(s) = \Phi_y^-(-s) \tag{2.3.25}$$

and Eq. (2.3.24) can be written as

$$\Phi_y(s) = \Phi_y^+(s)\Phi_y^+(-s) \tag{2.3.26}$$

Identifying the right sides of Eqs. (2.3.23) and (2.3.26) gives

$$H(s) = \Phi_y^+(s) \tag{2.3.27}$$

The procedure involved in Eq. (2.3.24) is called *spectral factorization*.

Conversely, if we pass the signal $y(t)$ through a filter $G(s)$ which has a transfer function $1/\Phi_y^+(s)$, the output will be white noise. The filter $G(s)$ is called a whitening filter. We note that both $H(s)$ and its inverse $G(s)$ in this example are causal systems.

The concept of a white noise can be extended by defining a nonstationary white process as one whose correlation function is

$$\mathrm{E}\{w(t_1)w(t_2)\} = V(t_1)\delta(t_1 - t_2) \tag{2.3.28}$$

where $V(t_1)$ is a nonnegative function.

We point out that in our subsequent discussions of systems with white noise inputs, our manipulations of various equations are in a purely formal manner because of the difficulties associated with defining integrals or derivatives of white noise processes. One method of overcoming this problem is to use the concept of generalized random processes [13] which is based on the theory of distributions. Rigorous results can also be obtained by working with the Wiener process instead of with the white noise process. However, since the Wiener process is not of bounded variation, we have to use special definitions of the stochastic integral to get consistent results. The reader is referred to [7–12] for a discussion of these approaches.

Analogous to our definition of a continuous white process, we can define a discrete white process as one for which the random variables $x(k)$ and $x(j)$ at two discrete time instants k and j are uncorrelated for any k and j. That is

$$\phi_x(k, j) = \mathrm{E}\{x(k)x(j)\} = \sigma^2(k)\,\delta_{kj} \tag{2.3.29}$$

If σ^2 is independent of k, Eq. (2.3.29) becomes

$$\phi_x(k - j) = \sigma^2\delta_{kj} \tag{2.3.30}$$

In this case, if the mean is constant, we have a wide-sense stationary process. The power spectrum of this process, obtained by taking the z-transform of Eq. (2.3.30) is

$$\Phi_x(z) = \sigma^2 \tag{2.3.31}$$

and is constant for all z.

We conclude this section with a discussion of the series representations of stochastic processes.

Representation of Stochastic Processes

It will often be convenient to represent a random process $x(t)$ in terms of elementary time functions. Perhaps the most frequently encountered representation of this type is the Fourier series expansion. We will start with a discussion of the Fourier series representation of a periodic process.

Fourier Series

We say that a stationary process $x(t)$ is periodic in the mean square sense if its autocorrelation function $\phi_x(\tau)$ is periodic with period T; that is, if

$$\phi_x(\tau + T) = \phi_x(\tau) \qquad \text{for all } \tau \tag{2.3.32}$$

In this case, we can expand $\phi_x(\tau)$ in a Fourier series as

$$\phi_x(\tau) = \sum_{n=-\infty}^{\infty} \alpha_n \exp(jn\omega_0\tau) \tag{2.3.33}$$

where

$$\omega_0 = \left(\frac{2\pi}{T}\right) \tag{2.3.34}$$

and

$$\alpha_n = \frac{1}{T}\int_0^T \phi_x(\tau) \exp(-jn\omega_0\tau)\, d\tau \tag{2.3.35}$$

It can then be shown that the process $x(t)$ can also be expanded in a Fourier series as

$$x(t) = \sum_{n=-\infty}^{\infty} a_n \exp(jn\omega_0 t) \tag{2.3.36}$$

where the coefficients a_n are given by

$$a_n = \frac{1}{T}\int_0^T x(u)\exp(-jn\omega_0 u)\,du \tag{2.3.37}$$

The expansion in Eq. (2.3.36) converges in the mean square sense. The coefficients a_n are complex random variables such that

$$\mathrm{E}\{a_n\} = \begin{cases} \mathrm{E}\{x(t)\} & n = 0 \\ 0 & n \neq 0 \end{cases} \tag{2.3.38}$$

and

$$\mathrm{E}\{a_n a_k^*\} = \begin{cases} \alpha_n & k = n \\ 0 & k \neq n \end{cases} \tag{2.3.39}$$

where a_k^* denotes the complex conjugate of a_k.

Unfortunately, most stochastic processes of interest in our subsequent discussions will not be periodic and strictly speaking, the Fourier-series representation cannot be used. We can avoid this problem by assuming that the period T of such processes tends to ∞. In this case, the coefficients are approximately uncorrelated. From Eq. (2.3.37) we can write

$$\mathrm{E}\{a_n a_m^*\} = \frac{1}{T}\int_0^T\int_0^T \mathrm{E}\{x(t')x(t)\}\exp[-j\omega_0(mt - nt')]\,dt\,dt' \tag{2.3.40}$$

However, if $x(t)$ is white, $\mathrm{E}\{x(t)x(t')\} = K\delta(t - t')$ for some $K > 0$. Then Eq. (2.3.40) becomes

$$\mathrm{E}\{a_n a_m^*\} = \frac{1}{T^2}\int_0^T K\exp[-j\omega_0(m - n)t]\,dt = 0 \qquad \text{for} \quad m \neq n \tag{2.3.41}$$

That is, the Fourier coefficients a_n and a_m of the white noise process $x(t)$ are uncorrelated.

Karhunen–Loéve Expansion [14]

We can attempt to get a series expansion even for nonperiodic processes such that the coefficients of the expansion are uncorrelated. In order to do so, let $g_1(t), g_2(t), \ldots, g_n(t), \ldots$ denote a set of functions orthonormal over an interval (t_0, t_f) such that

$$\int_{t_0}^{t_f} g_n(t) g_m(t)\, dt = \delta_{nm} \tag{2.3.42}$$

We wish to represent $x(t)$ in the form†

$$x(t) = \underset{N\to\infty}{\text{l.i.m.}} \sum_{n=1}^{N} \beta_n g_n(t) \qquad t_0 \leqslant t \leqslant t_f \tag{2.3.43}$$

where the coefficients β_n are random variables given by

$$\beta_n = \int_{t_0}^{t_f} x(t) g_n(t)\, dt \tag{2.3.44}$$

We would like to choose the basis functions $\{g_i(t)\}$ such that $\{\beta_n\}$ are a set of uncorrelated variables. If we assume that $E\{x(t)\} = 0$, then $E\{\beta_n\} = 0$. In this case, if the coefficients β_n are orthogonal, they are also uncorrelated. We shall now establish that if the functions $g_n(t)$ are chosen as the solutions to the integral equation

$$\int_{t_0}^{t_f} \phi_x(t_1, t_2) g_n(t_2)\, dt_2 = \lambda_n g_n(t_1) \tag{2.3.45}$$

then

$$E\{\beta_n \beta_m\} = \lambda_m \delta_{nm} \tag{2.3.46}$$

The resulting expansion is known as the *Karhunen–Loéve expansion*. From Eq. (2.3.44), it follows that

$$\begin{aligned} E\{\beta_n\beta_m\} &= E\left\{\int_{t_0}^{t_f}\int_{t_0}^{t_f} x(t_1)x(t_2)g_n(t_1)g_m(t_2)\, dt_1\, dt_2\right\} \\ &= \int_{t_0}^{t_f}\left[\int_{t_0}^{t_f}\phi_x(t_1, t_2)g_m(t_2)\, dt_2\right] g_m(t_1)\, dt_1 \\ &= \int_{t_0}^{t_f}\lambda_m g_m(t_1)g_n(t_1)\, dt_1 \qquad \text{by Eq. (2.3.45)} \end{aligned}$$

so that Eq. (2.3.46) follows by virtue of Eq. (2.3.42).

†Limit in the mean (l.i.m.) convergence implies that

$$\lim_{N\to\infty} E\left\{\left[x(t) - \sum_{n=1}^{N}\beta_n g_n(t)\right]^2\right\} = 0 \qquad t_0 \leqslant t \leqslant t_f$$

Thus the basis functions satisfy the integral equation of Eq. (2.3.45). The kernel of the equation is the covariance function $\phi_x(t_1, t_2)$ and the values $\{\lambda_n\}$ for which the equation is satisfied are the eigenvaleus of the equation. The basis functions $\{g_n(t)\}$ are the corresponding eigenfunctions. We can therefore use results from the theory of linear integral equations [15,16] to state some properties of the expansion.

It thus follows that there is at least one real number $\lambda \neq 0$ and a function $g(t)$ satisfying Eq. (2.3.45). Also, since the kernel $\phi_x(t_1, t_2)$ is nonnegative definite, it can be uniformly expanded over $[t_0, t_f]$ in the series

$$\phi_x(t_1, t_2) = \sum_{n=1}^{\infty} \lambda_n g_n(t_1) g_n(t_2) \tag{2.3.47}$$

If $\phi_x(t_1, t_2)$ is positive definite, the eigenfunctions form a complete orthonormal set. If $\phi_x(t_1, t_2)$ is not positive definite, the eigenfunctions will not form a complete set. In this case, we can augment the basis functions with enough additional functions to form a complete set. Finally, we have

$$\mathrm{E}\left\{\int_{t_0}^{t_f} x^2(t)\, dt\right\} = \int_{t_0}^{t_f} \phi_x(t, t)\, dt = \sum_{n=1}^{\infty} \lambda_n \tag{2.3.48}$$

It only remains to show that with this choice of basis functions, the expansion in Eq. (2.3.43) converges. We note that

$$\begin{aligned}
\mathrm{E}&\left\{\left[x(t) - \sum_{n=1}^{N} \beta_n g_n(t)\right]^2\right\} \\
&= \phi_x(t, t) - 2\sum_{n=1}^{N} \mathrm{E}\{x(t)\beta_n\}\, g_n(t) \\
&\quad + \sum_{n=1}^{N}\sum_{m=1}^{N} \mathrm{E}\{\beta_n \beta_m\}\, g_n(t) g_m(t) \\
&= \phi_x(t, t) - \sum_{n=1}^{N} \lambda_n g_n^2(t)
\end{aligned}$$

where the last step follows by substituting for β_n and β_m from Eq. (2.3.44) and using Eqs. (2.3.45) and (2.3.46) in the resulting expression. We now let N go to infinity and make use of Eq. (2.3.47) to establish convergence.

The solution of the integral equation is in general a tedious task. Some solution techniques are discussed in Chapter 4. However, the solution is straightforward if $x(t)$ is a white noise process with

$$\phi_x(t_1, t_2) = V(t_1)\delta(t_1 - t_2) \tag{2.3.49}$$

for some function $V(t_1) > 0$.

Substitution in Eq. (2.3.45) then yields

$$\int_{t_0}^{t_f} V(t_1)\delta(t_1 - t_2)g(t_2)\,dt_2 = V(t_1)g(t_1) = \lambda g(t_1) \qquad \textbf{(2.3.50)}$$

Thus we see the functions $g_i(t)$ can be chosen to be any arbitrary orthonormal set. It follows from Eqs. (2.3.47) and (2.3.50) that we have

$$\phi_x(t_1, t_2) = V(t_1) \sum_{n=1}^{\infty} g_n(t_1)g_n(t_2) \qquad \textbf{(2.3.51)}$$

Comparison of Eqs. (2.3.49) with Eq. (2.3.51) shows that we can formally expand the impulse function in a series as

$$\delta(t_1 - t_2) = \sum_{n=1}^{\infty} g(t_1)g(t_2) \qquad \textbf{(2.3.52)}$$

2.4 GAUSS–MARKOV MODELS

A primary problem in many applications is to obtain a suitable model for generating stochastic signals. A complete statistical description of these signals is usually not easy to obtain. In most cases, therefore, we have to content ourselves with determining the first two moments of the process. Of course, if the process is Gaussian, these two moments completely determine the process. If the process is not Gaussian, but is Markov, the joint densities can be determined in terms of the first-order density and the transition probabilities.

If we model a stochastic signal as the output of a dynamical system driven by a suitable input, the complexity in determining the output probabilities depends on the system itself and the way that it is characterized. Some representations are better suited for manipulations than others. A useful characterization is in terms of the state-space descriptions in which we model the stochastic process as the output of a linear dynamical system driven by white noise. By imposing appropriate conditions on the model, we will establish that the output is a Markov process. If the input process is Gaussian, the output process will also be Gaussian. Thus we will refer to such models as Gauss–Markov models.

We will start our discussion of these models with the discrete time model

$$\mathbf{x}(k+1) = \mathbf{A}(k)\mathbf{x}(k) + \mathbf{B}(k)\mathbf{w}(k) \qquad \textbf{(2.4.1)}$$

where $\mathbf{x}(k)$ is an n-vector representing the state of the model $\mathbf{A}(k)$ and $\mathbf{B}(k)$ are matrices of appropriate dimension and $\mathbf{w}(k)$ is the system input which is assumed to be white. Specifically, let $\mathbf{w}(k)$, $k = 0, 1, \ldots$ be a p-dimensional Gaussian white sequence with mean

$$\mathrm{E}\{\mathbf{w}(k)\} = \mathbf{m}_{\mathbf{w}}(k) \tag{2.4.2}$$

and covariance

$$\mathrm{E}\left\{[\mathbf{w}(k) - \mathbf{m}_{\mathbf{w}}(k)][\mathbf{w}(j) - \mathbf{m}_{\mathbf{w}}(j)]^T\right\} = \mathbf{V}_w(j)\delta_{kj} \tag{2.4.3}$$

The system output is assumed to be

$$\mathbf{y}(k) = \mathbf{C}(k)\mathbf{x}(k) \tag{2.4.4}$$

We will further assume that the initial vector associated with the model of Eq. (2.4.1), $\mathbf{x}(0)$ is a Gaussian random vector with known mean

$$\mathrm{E}\{\mathbf{x}(0)\} = \mathbf{m}_{\mathbf{x}_0} \tag{2.4.5}$$

and covariance matrix

$$\mathrm{E}\left\{[\mathbf{x}(0) - \mathbf{m}_{\mathbf{x}_0}][\mathbf{x}(0) - \mathbf{m}_{\mathbf{x}_0}]^T\right\} = \mathbf{V}_{\mathbf{x}_0} \tag{2.4.6}$$

and independent of $\mathbf{w}(k)$ for all $k \geqslant 0$, such that

$$\mathrm{E}\left\{[\mathbf{x}(0) - \mathbf{m}_{\mathbf{x}_0}][\mathbf{w}(k) - \mathbf{m}_{\mathbf{w}}(k)]^T\right\} = \mathbf{0} \tag{2.4.7}$$

We now show that the process $\mathbf{x}(k)$ is Gauss–Markov. From Eq. (2.2.74), the solution to Eq. (2.4.1) is given by

$$\mathbf{x}(k) = \psi(k, j)\mathbf{x}(j) + \sum_{i=j+1}^{k} \psi(k, i)\mathbf{B}(i-1)\mathbf{w}(i-1) \tag{2.4.8a}$$

where

$$\psi(k, j) = \mathbf{A}(k-1)\mathbf{A}(k-2)\ldots\mathbf{A}(j) \tag{2.4.8b}$$

From Eq. (2.4.8) it is clear that the conditional density function of $\mathbf{x}(k)$ given $\mathbf{x}(j), \mathbf{x}(j-1), \ldots, \mathbf{x}(1), \mathbf{x}(0)$ depends only $\mathbf{x}(j)$ for any k and j. Therefore $\mathbf{x}(k)$ is Markov. To show that $\mathbf{x}(k)$ is Gaussian, by setting $j = 0$ in Eq. (2.4.8), we see that $\mathbf{x}(k)$ is a linear combination of the Gaussian variables $\mathbf{x}(0), \mathbf{w}(0), \mathbf{w}(1), \ldots, \mathbf{w}(k-1)$ and is hence also Gaussian.

We now present two ways of describing the process. In the first description we note that by using the chain rule, we can write the joint probability density function of the variables $\{\mathbf{x}(k+1), \mathbf{x}(k), \ldots, \mathbf{x}(0)\}$ as

$$p[\mathbf{x}(k+1), \mathbf{x}(k), \ldots, \mathbf{x}(0)] = p[\mathbf{x}(k+1)|\mathbf{x}(k), \mathbf{x}(k-1), \ldots, \mathbf{x}(0)] \cdot p[\mathbf{x}(k)|\mathbf{x}(k-1) \ldots \mathbf{x}(0)] \ldots p[\mathbf{x}(0)] \tag{2.4.9}$$

Since the process is Markov it follows that the joint density function can be written as

$$p[\mathbf{x}(k+1), \mathbf{x}(k), \ldots, \mathbf{x}(0)] = p[\mathbf{x}(k+1)|\mathbf{x}(k)] \cdot p[\mathbf{x}(k)|\mathbf{x}(k-1)] \ldots p[\mathbf{x}(0)] \tag{2.4.10}$$

It is thus enough to specify $p(\mathbf{x}_0)$ and the transitional densities $p[\mathbf{x}(k+1)|\mathbf{x}(k)]$. From Eqs. (2.4.5) and (2.4.6) we have

$$p(\mathbf{x}_0) = \frac{1}{(2\pi)^{n/2}|\mathbf{V}_{\mathbf{x}_0}|^{1/2}} \exp\left[-\frac{1}{2}(\mathbf{x}_0 - \mathbf{m}_{\mathbf{x}_0})^T \mathbf{V}_{\mathbf{x}_0}^{-1}(\mathbf{x}_0 - \mathbf{m}_{\mathbf{x}_0})\right] \tag{2.4.11}$$

In order to determine the transitional probability density function of $p[\mathbf{x}(k+1)|\mathbf{x}(k)]$, we note that since $\mathbf{x}(k+1)$ given $\mathbf{x}(k)$ is a Gaussian random variable, it is sufficient to determine the conditional mean and conditional covariance of $\mathbf{x}(k+1)$ given $\mathbf{x}(k)$. From Eq. (2.4.1) we can easily obtain the conditional mean as

$$\mathrm{E}[\mathbf{x}(k+1)|\mathbf{x}(k)] = \mathbf{A}(k)\mathbf{x}(k) + \mathbf{B}(k)\mathbf{m}_{\mathbf{w}}(k) \tag{2.4.12a}$$

The conditional covariance is given by

$$\begin{aligned}\mathbf{V}_{\mathbf{x}}(k+1|k) &= \mathrm{E}\{[\mathbf{x}(k+1) - \mathrm{E}[\mathbf{x}(k+1)|\mathbf{x}(k)]] \\ &\quad \cdot [\mathbf{x}(k+1) - \mathrm{E}[\mathbf{x}(k+1)|\mathbf{x}(k)]]^T\} \\ &= \mathbf{B}(k)\mathbf{V}_{\mathbf{w}}(k)\mathbf{B}^T(k)\end{aligned} \tag{2.4.12b}$$

In the second description, we obtain equations describing the time propagation of the mean and variance of the process $\mathbf{x}(k)$. These then will provide a complete description of the process as it is Gaussian. Let

$$\mathrm{E}\{\mathbf{x}(k)\} = \mathbf{m}_{\mathbf{x}}(k) \tag{2.4.13}$$

Taking expectations on both sides of Eq. (2.4.1) yields

$$\mathbf{m_x}(k+1) = \mathbf{A}(k)\mathbf{m_x}(k) + \mathbf{B}(k)\mathbf{m_w}(k) \tag{2.4.14}$$

with initial condition $\mathbf{m_x}(0) = \mathbf{m}_{\mathbf{x}_0}$.

Since $\mathbf{m}_{\mathbf{x}_0}$ and $\mathbf{m_w}(k)$ are known, $\mathbf{m_x}(k)$ can be computed for any k by solving Eq. (2.4.14) recursively. In order to determine an equation for the covariance $\mathbf{V_x}(k)$, let us define the normalized sequences

$$\bar{\mathbf{x}}(k) = \mathbf{x}(k) - \mathbf{m_x}(k) \tag{2.4.15}$$

and

$$\bar{\mathbf{w}}(k) = \mathbf{w}(k) - \mathbf{m_w}(k) \tag{2.4.16}$$

It can easily be seen that $\bar{\mathbf{x}}(k)$ and $\bar{\mathbf{w}}(k)$ are zero mean with covariances

$$\mathbf{V}_{\bar{\mathbf{x}}}(k) = \mathbf{V_x}(k) \quad \text{and} \quad \mathbf{V}_{\bar{\mathbf{w}}}(k) = \mathbf{V_w}(k) \tag{2.4.17}$$

Subtracting Eq. (2.4.14) from (2.4.1), we get

$$\bar{\mathbf{x}}(k+1) = \mathbf{A}(k)\bar{\mathbf{x}}(k) + \mathbf{B}(k)\bar{\mathbf{w}}(k) \tag{2.4.18}$$

Also from Eq. (2.4.17), we have

$$\begin{aligned}
\mathbf{V}_{\bar{\mathbf{x}}}(k+1) &= \mathrm{E}\{\bar{\mathbf{x}}(k+1)\bar{\mathbf{x}}^T(k+1)\} \\
&= \mathrm{E}\{[\mathbf{A}(k)\bar{\mathbf{x}}(k) + \mathbf{B}(k)\bar{\mathbf{w}}(k)][\mathbf{A}(k)\bar{\mathbf{x}}(k) + \mathbf{B}(k)\bar{\mathbf{w}}(k)]^T\} \\
&= \mathbf{A}(k)\mathbf{V}_{\bar{\mathbf{x}}}(k)\mathbf{A}^T(k) + \mathbf{B}(k)\mathbf{V}_{\bar{\mathbf{w}}}(k)\mathbf{B}^T(k) \\
&\quad + \mathbf{B}(k)\mathrm{E}\{\bar{\mathbf{w}}(k)\bar{\mathbf{x}}^T(k)\}\mathbf{A}^T(k) \\
&\quad + \mathbf{A}(k)\mathrm{E}\{\bar{\mathbf{x}}(k)\bar{\mathbf{w}}^T(k)\}\mathbf{B}^T(k)
\end{aligned} \tag{2.4.19}$$

To evaluate the last two terms in Eq. (2.4.19), we note from Eq. (2.4.18) that the solution $\bar{\mathbf{x}}(k)$ is given by

$$\bar{\mathbf{x}}(k) = \psi(k,0)\bar{\mathbf{x}}(0) + \sum_{i=1}^{k} \psi(k,i)\mathbf{B}(i-1)\bar{\mathbf{w}}(i-1) \tag{2.4.20}$$

where

$$\bar{\mathbf{x}}(0) = \mathbf{x}(0) - \mathbf{m_x}(0) \tag{2.4.21}$$

Postmultiplying both sides of Eq. (2.4.20) by $\bar{\mathbf{w}}^T(k)$ and taking expectations yields

$$E\{\bar{\mathbf{x}}(k)\bar{\mathbf{w}}^T(k)\} = \psi(k, 0)E\{\bar{\mathbf{x}}(0)\bar{\mathbf{w}}^T(k)\} + \sum_{i=1}^{k} \psi(k, i)\mathbf{B}(i-1)E\{\bar{\mathbf{w}}(i-1)\bar{\mathbf{w}}^T(k)\} \quad \textbf{(2.4.22)}$$

From Eqs. (2.4.7) and (2.4.3) and since $\bar{w}(k)$ is zero mean, it follows that,

$$E\{\bar{\mathbf{x}}(k)\bar{\mathbf{w}}^T(k)\} = 0 \quad \textbf{(2.4.23)}$$

Substitution in Eq. (2.4.19) then yields

$$\mathbf{V}_{\bar{\mathbf{x}}}(k+1) = \mathbf{A}(k)\mathbf{V}_{\bar{\mathbf{x}}}(k)\mathbf{A}^T(k) + \mathbf{B}(k)\mathbf{V}_{\bar{\mathbf{w}}}(k)\mathbf{B}^T(k) \quad \textbf{(2.4.24)}$$

with initial condition $\mathbf{V}_{\bar{\mathbf{x}}}(0) = \mathbf{V}_{\mathbf{x}}(0) = \mathbf{V}_{\mathbf{x}_0}$.

We now derive relations for

$$V_{\bar{\mathbf{x}}}(k, j) = E[\bar{\mathbf{x}}(k)\bar{\mathbf{x}}^T(j)] \quad \textbf{(2.4.25)}$$

For $k > j$, we can write from Eq. (2.4.18)

$$\bar{\mathbf{x}}(k) = \psi(k, j)\bar{\mathbf{x}}(j) + \sum_{i=j+1}^{k} \psi(k, i)\mathbf{B}(i-1)\bar{\mathbf{w}}(i-1) \quad \textbf{(2.4.26)}$$

Postmultiplying both sides of Eq. (2.4.26) by $\bar{\mathbf{x}}^T(j)$ and taking expectations we obtain

$$\mathbf{V}_{\bar{\mathbf{x}}}(k, j) = \psi(k, j)\mathbf{V}_{\bar{\mathbf{x}}}(j) + \sum_{i=j+1}^{k} \psi(k, i)\mathbf{B}(i-1)E\{\bar{\mathbf{w}}(i-1)\bar{\mathbf{x}}^T(j)\} \quad \textbf{(2.4.27)}$$

We note that $\bar{\mathbf{w}}(i)$ is white. Since the system is causal, it follows that for $i \geqslant j$, $\bar{\mathbf{w}}(i)$ is independent of $\bar{\mathbf{x}}(j)$. Since it is also zero mean, it follows that the second term on the right side of Eq. (2.4.27) is zero. We therefore have

$$\mathbf{V}_{\bar{\mathbf{x}}}(k, j) = \psi(k, j)\mathbf{V}_{\bar{\mathbf{x}}}(j) \quad \text{for} \quad k > j \quad \textbf{(2.4.28)}$$

For $k < j$, we can derive a similar result by writing

$$\begin{aligned} \mathbf{V}_{\bar{\mathbf{x}}}(k, j) &= [\mathbf{V}_{\bar{\mathbf{x}}}(j, k)]^T \\ &= [\psi(j, k)\mathbf{V}_{\bar{\mathbf{x}}}(k)]^T \text{ from Eq. (2.4.28)} \\ &= \mathbf{V}_{\bar{\mathbf{x}}}(k)\psi^T(j, k) \end{aligned} \quad \textbf{(2.4.29)}$$

Finally since $\mathbf{V}_{\tilde{\mathbf{x}}}(k) = \mathbf{V}_{\mathbf{x}}(k)$ we can write from Eqs. (2.4.24), (2.4.28), and (2.4.29)

$$\mathbf{V}_{\mathbf{x}}(k+1) = \mathbf{A}(k)\mathbf{V}_{\mathbf{x}}(k)\mathbf{A}^T(k) + \mathbf{B}(k)\mathbf{V}_{\mathbf{w}}(k)\mathbf{B}^T(k)$$
$$\text{with } \mathbf{V}_{\mathbf{x}}(0) = \mathbf{V}_{\mathbf{x}_0} \tag{2.4.30}$$

and

$$\mathbf{V}_{\mathbf{x}}(k,j) = \begin{cases} \psi(k,j)\mathbf{V}_{\mathbf{x}}(j) & \text{for } k > j \\ \mathbf{V}_{\mathbf{x}}(k)\psi^T(j,k) & \text{for } k < j \end{cases} \tag{2.4.31}$$

The covariance of the output is given by

$$\mathbf{V}_{\mathbf{y}}(k) = \mathbf{C}^T(k)\mathbf{V}_{\mathbf{x}}(k)\mathbf{C}(k) \tag{2.4.32}$$

We now briefly consider the extension of these results to continuous time models. For the most part, as indicated earlier, we will be performing the operations involved in a purely formal manner. We consider the linear system

$$\dot{\mathbf{x}}(t) = \mathbf{A}(t)\mathbf{x}(t) + \mathbf{B}(t)\mathbf{w}(t) \tag{2.4.33}$$

where $\mathbf{w}(t)$ is a white Gaussian process with mean $\mathbf{m}_{\mathbf{w}}(t)$ and covariance

$$\mathrm{E}\{[\mathbf{w}(t) - \mathbf{m}_{\mathbf{w}}(t)][\mathbf{w}(t') - \mathbf{m}_{\mathbf{w}}(t')]^T\} = \mathbf{V}_{\mathbf{w}}(t)\delta(t-t') \tag{2.4.34}$$

We also assume that $\mathbf{x}(t_0)$ is a Gaussian random variable with mean $\mathbf{m}_{\mathbf{x}_0}$ and covariance $\mathbf{V}_{\mathbf{x}_0}$, and is independent of $\mathbf{w}(t)$ for all $t \geqslant t_0$. The system output given by

$$\mathbf{y}(t) = \mathbf{C}(t)\mathbf{x}(t) \tag{2.4.35}$$

The solution to Eq. (2.4.33) is given by

$$\mathbf{x}(t_m) = \psi(t_m, t_{m-1})\mathbf{x}(t_{m-1}) + \int_{t_{m-1}}^{t_m} \psi(t_m, \tau)\mathbf{B}(\tau)\mathbf{w}(\tau)\, d\tau \tag{2.4.36}$$

for any set of points t_m, t_{m-1}. Here $\psi(t, \tau)$ is the state transition matrix corresponding to $\mathbf{A}(t)$. Thus given the values of $\mathbf{x}(t)$ for t equal to $t_0, t_1, \ldots, t_{m-1}, t_m$ where

$$t_0 < t_1 < \cdots < t_{m-1} < t_m$$

it follows from Eq. (2.4.36) that the density function of $\mathbf{x}(t_m)$ is dependent only on $\mathbf{x}(t_{m-1})$. Therefore $\mathbf{x}(t)$ is Markov. It is a Gaussian process since

$\mathbf{x}(t_m)$ is obtained as a linear combination of a set of Gaussian variables. To determine expressions for the mean and variance of $\mathbf{x}(t)$, we again form the normalized zero mean processes

$$\bar{\mathbf{x}}(t) = \mathbf{x}(t) - \mathbf{m}_{\mathbf{x}}(t)$$

and

$$\bar{\mathbf{w}}(t) = \mathbf{w}(t) - \mathbf{m}_{\mathbf{w}}(t) \tag{2.4.37}$$

It can easily be verified that the variances are given by

$$\mathbf{V}_{\bar{\mathbf{w}}}(t) = \mathbf{V}_{\mathbf{w}}(t) \quad \text{and} \quad \mathbf{V}_{\bar{\mathbf{x}}}(t) = \mathbf{V}_{\mathbf{x}}(t) \tag{2.4.38}$$

To directly determine the mean or variance of $\bar{\mathbf{x}}(t)$, we would need to solve for the transition matrix $\psi(t, \tau)$. Instead, we set up a differential equation to describe the evolution with time of these moments. Taking expectations on both sides of Eq. (2.4.33) yields

$$\dot{\mathbf{m}}_{\mathbf{x}}(t) = \mathbf{A}(t)\mathbf{m}_{\mathbf{x}}(t) + \mathbf{B}(t)\mathbf{m}_{\mathbf{w}}(t) \tag{2.4.39}$$

with

$$\mathbf{m}_{\mathbf{x}}(t_0) = \mathbf{m}_{\mathbf{x}_0}$$

Subtracting Eq. (2.4.39) from Eq. (2.4.33), we obtain the differential equation for $\bar{\mathbf{x}}(t)$ as

$$\dot{\bar{\mathbf{x}}}(t) = \mathbf{A}(t)\bar{\mathbf{x}}(t) + \mathbf{B}(t)\bar{\mathbf{w}}(t) \tag{2.4.40}$$

with

$$\bar{\mathbf{x}}(t_0) = \mathbf{x}(t_0) - \mathbf{m}_{\mathbf{x}_0}$$

The solution to Eq. (2.4.40) is given by

$$\bar{\mathbf{x}}(t) = \psi(t, t_0)\bar{\mathbf{x}}(t_0) + \int_{t_0}^{t} \psi(t, \tau)\mathbf{B}(\tau)\bar{\mathbf{w}}(\tau)\, d\tau \tag{2.4.41}$$

Since

$$\mathbf{V}_{\bar{\mathbf{x}}}(t) = \mathrm{E}\{\bar{\mathbf{x}}(t)\bar{\mathbf{x}}^T(t)\} \tag{2.4.42}$$

we can write for its derivative

$$\begin{aligned} \frac{d}{dt}\mathbf{V}_{\bar{\mathbf{x}}}(t) &= \frac{d}{dt}\mathrm{E}\{\bar{\mathbf{x}}(t)\bar{\mathbf{x}}^T(t)\} \\ &= \mathrm{E}\{\dot{\bar{\mathbf{x}}}(t)\bar{\mathbf{x}}^T(t)\} + \mathrm{E}\{\bar{\mathbf{x}}(t)\dot{\bar{\mathbf{x}}}^T(t)\} \end{aligned} \tag{2.4.43}$$

Substitution of Eq. (2.4.40) in Eq. (2.4.43) yields

$$\begin{aligned}\frac{d}{dt}\mathbf{V}_{\bar{\mathbf{x}}}(t) &= \mathrm{E}\{\mathbf{A}(t)\bar{\mathbf{x}}(t)+\mathbf{B}(t)\bar{\mathbf{w}}(t)\}\bar{\mathbf{x}}^T(t)\\ &\quad+\mathrm{E}\{\bar{\mathbf{x}}(t)[\mathbf{A}(t)\bar{\mathbf{x}}(t)+\mathbf{B}(t)\bar{\mathbf{w}}(t)]^T\}\\ &=\mathbf{A}(t)\mathbf{V}_{\bar{\mathbf{x}}}(t)+\mathbf{V}_{\bar{\mathbf{x}}}(t)\mathbf{A}^T(t)+\mathbf{B}(t)\mathrm{E}\{\bar{\mathbf{w}}(t)\bar{\mathbf{x}}^T(t)\}\\ &\quad+\mathrm{E}\{\bar{\mathbf{x}}(t)\bar{\mathbf{w}}^T(t)\}\mathbf{B}^T(t)\end{aligned} \tag{2.4.44}$$

To evaluate the third and fourth terms in Eq. (2.4.44) we postmultiply Eq. (2.4.41) by $\bar{\mathbf{w}}^T(t)$ and take expectations to get

$$\begin{aligned}\mathrm{E}\{\mathbf{x}(t)\bar{\mathbf{w}}^T(t)\} &= \psi(t,t_0)\mathrm{E}\{\bar{\mathbf{x}}(t_0)\bar{\mathbf{w}}^T(t)\}\\ &\quad+\int_{t_0}^{t}\psi(t,\tau)\mathbf{B}(\tau)\mathrm{E}\{\bar{\mathbf{w}}(\tau)\bar{\mathbf{w}}^T(\tau)\}\,d\tau\end{aligned} \tag{2.4.45}$$

The first term on the right-hand side of Eq. (2.4.45) is zero by virtue of the assumption that $\mathbf{w}(t)$ is independent of $\mathbf{x}(t_0)$ for all $t \geqslant t_0$. To evaluate the second term, we make use of Eq. (2.4.34) to write the second term as

$$\int_{t_0}^{t}\psi(t,\tau)\mathbf{B}(\tau)\mathbf{V}_{\mathbf{w}}(\tau)\delta(t-\tau)\,d\tau=\frac{1}{2}\mathbf{B}(t)\mathbf{V}_{\mathbf{w}}(t) \tag{2.4.46}$$

where the last step follows by assuming that the δ-function is symmetric.† Substitution of Eq. (2.4.46) in Eq. (2.4.45) yields

$$\mathrm{E}\{\bar{\mathbf{x}}(t)\bar{\mathbf{w}}^T(t)\}=\frac{1}{2}\mathbf{B}(t)\mathbf{V}_{\mathbf{w}}(t) \tag{2.4.47a}$$

It similarly follows that

$$\mathrm{E}\{\bar{\mathbf{w}}(t)\bar{\mathbf{x}}^T(t)\}=\frac{1}{2}\mathbf{V}_{\mathbf{w}}(t)\mathbf{B}^T(t) \tag{2.4.47b}$$

Substitution of Eqs. (2.4.47) in Eq. (2.4.44) yields for the variance $V_{\bar{\mathbf{x}}}(t)$

$$\frac{d}{dt}\mathbf{V}_{\bar{\mathbf{x}}}(t)=\mathbf{A}(t)\mathbf{V}_{\bar{\mathbf{x}}}(t)+\mathbf{V}_{\bar{\mathbf{x}}}(t)\mathbf{A}^T(t)+\mathbf{B}(t)\mathbf{V}_{\mathbf{w}}(t)\mathbf{B}^T(t) \tag{2.4.48}$$

†The δ-function is defined by $\int_a^b g(\tau)\delta(t-\tau)\,d\tau=\begin{cases}0 & t<a \text{ or } t>b\\ \alpha g(a) & t=a\\ (1-\alpha)g(b) & t=b\\ g(t) & a<t<b\end{cases}$

where $0 \leqslant \alpha \leqslant 1$. We obtain the symmetric δ-function if $\alpha = 1/2$.

Since $\mathbf{V}_\mathbf{x}(t) = \mathbf{V}_{\bar{\mathbf{x}}}(t)$, we also have

$$\dot{\mathbf{V}}_\mathbf{x}(t) = \mathbf{A}(t)\mathbf{V}_\mathbf{x}(t) + \mathbf{V}_\mathbf{x}(t)\mathbf{A}^T(t) + \mathbf{B}(t)\mathbf{V}_\mathbf{w}(t)\mathbf{B}^T(t) \tag{2.4.49}$$

with $\mathbf{V}_\mathbf{x}(t_0) = \mathbf{V}_{\mathbf{x}_0}$.

The mean and the variance of the output $\mathbf{y}(t)$ are easily determined as

$$\mathbf{m}_\mathbf{y}(t) = \mathrm{E}\{\mathbf{y}(t)\} = \mathbf{C}(t)\mathbf{m}_\mathbf{x}(t) \tag{2.4.50}$$

and

$$\mathbf{V}_\mathbf{y}(t) = \mathbf{C}(t)\mathbf{V}_\mathbf{x}(t)\mathbf{C}^T(t) \tag{2.4.51}$$

Example 2.4.1 Consider the simple model

$$\dot{x}(t) = -ax(t) + w(t)$$

with $\quad x(0) = 0$

The prior mean and variance are therefore given by

$$m_x(0) = 0 \quad \text{and} \quad V_{x_0} = 0$$

Let $w(t)$ be Gaussian, white with mean m_w and covariance parameter

$$V_w(t) = \lambda$$

From (2.4.39), we have for the mean

$$\dot{m}_x(t) = -am_x(t) + m_w, \qquad m_{x_0} = 0$$

The solution to this equation is

$$m_x(t) = \frac{m_w}{a}(1 - \exp(-at))$$

The variance equation as obtained from Eq. (2.4.49) is

$$\dot{V}_x(t) = -2aV_x(t) + \lambda, \qquad V_{x_0} = 0$$

with solution

$$V_x(t) = \frac{\lambda}{2a}(1 - \exp(-2at))$$

We note that the output $x(t)$ is nonstationary even though the input is stationary and the system is time invariant. However, the output is asymptotically stationary, since the mean tends to m_w/a and the variance to $\lambda/2a$ as t tends to infinity.

The result of Example 2.4.1 holds in general. That is, if the system is time invariant and $\mathbf{w}(t)$ is stationary and $t_0 = -\infty$, then $\mathbf{x}(t)$ will also be stationary. For time-invariant systems, the matrices **A**, **B**, and **C** in Eqs. (2.4.33) and (2.4.35) (Eqs. (2.4.1) and (2.4.4) for the discrete-time case) will be constant. The covariance $\mathbf{V_x}(t)(\mathbf{V_x}(k))$ can be obtained as the steady state solution to the differential equation (2.4.49) (the difference equation (2.4.30) for the discrete case). This solution can be found by setting the left-hand side of the covariance equation to zero, so that the equation becomes an algebraic one. The resulting equations are

$$\mathbf{AV_x} + \mathbf{V_x A}^T + \mathbf{BV_w B}^T = 0 \tag{2.4.52}$$

for the continuous-time case, and

$$\mathbf{AV_x A}^T + \mathbf{BV_w B}^T = 0 \tag{2.4.53}$$

for the discrete-time case.

We also point out that the equations for the propagation of the mean and variance of $\mathbf{x}(\cdot)$ and $\mathbf{y}(\cdot)$ hold even if the input process is not Gaussian but is only white. Of course, these two moments are no longer adequate to provide a complete statistical description of the system.

2.5 COMPLEX ENVELOPES OF BAND-LIMITED SIGNALS [17,18]

In many problems of interest, the signals involved are bandpass or narrow-band signals. A signal is said to be bandpass if

$$\Phi_x(\omega) = 0 \text{ except for } \omega_c - \omega_0 < |\omega| < \omega_c + \omega_0$$

where ω_c denotes the center frequency of the band. In narrow-band signals, ω_c is very large compared to ω_0 and the energy in the signal is essentially concentrated near the center frequency.

Such signals arise for example in modulation systems in which the carrier frequency is very large compared to the bandwidth of the message. In developing the mathematical techniques for processing such signals, it is often convenient to deal with only the information-bearing portion of the signal (message) without reference to the center or carrier frequency. Such baseband (modulation) processes are conveniently described in terms of complex quantities. In this section we develop the concept of the complex envelope, the modulation process associated with a narrow-band signal. We introduce the Hilbert transform as an intermediate step in obtaining the complex envelope. We show how bandpass filtering of real narrow-band signals can be interpreted in terms of associated complex envelopes. We will also introduce the concept of a complex white noise process.

We begin with a discussion of band-limited deterministic signals. Let $f(t)$ be a real signal with Fourier transform $F(\omega)$. Let us denote by $f_+(t)$ that part of $f(t)$ giving rise to the part of the spectrum corresponding to positive frequencies, that is,

$$f_+(t) = \frac{1}{2\pi}\int_0^{\infty} F(\omega)\exp[j\omega t]\,d\omega \tag{2.5.1}$$

Similarly, we let

$$f_-(t) = \frac{1}{2\pi}\int_{-\infty}^{0} F(\omega)\exp[j\omega t]\,d\omega \tag{2.5.2}$$

denote that part of $f(t)$ giving rise to the negative frequency spectrum. It is then obvious that

$$f(t) = f_+(t) + f_-(t) \tag{2.5.3}$$

where in general $f_+(t)$ and $f_-(t)$ will be complex signals. It can easily be shown from Eqs. (2.5.1) and (2.5.2) that

$$f_+^*(t) = f_-(t) \tag{2.5.4}$$

By combining Eqs. (2.5.3) and (2.5.4), it follows that $f(t)$ can be obtained from the complex signal $f_+(t)$ as

$$f(t) = 2[\operatorname{Re} f_+(t)]$$

We now introduce the complex signal $\tilde{f}(t)$ with Fourier transform $\tilde{F}(\omega)$ given by

$$\tilde{F}(\omega) = \begin{cases} \sqrt{2}\,F(\omega+\omega_c) & -\omega_c < \omega \\ 0 & \omega < -\omega_c \end{cases} \tag{2.5.5}$$

such that

$$f_+(t) = \frac{1}{\sqrt{2}}\tilde{f}(t)\exp(j\omega_c t) \tag{2.5.6}$$

It then follows that

$$f(t) = \sqrt{2}\ \text{Re}\left[\tilde{f}(t)\exp(j\omega_c t)\right] \tag{2.5.7}$$

The complex signal $\tilde{f}(t)$ is called the complex envelope of the real signal $f(t)$. The terminology arises from analog modulation. If $f(t)$ is a bandpass process with its energy concentrated near $\pm\omega_c$, we can think of $\tilde{f}(t)$ as the information bearing signal. Equivalently, the signal $\sqrt{2}\,f_+(t)$ can be thought of as the modulated signal, for which $\tilde{f}(t)$ is the envelope, a complex signal. All the information contained in $f(t)$ is contained in the signal $\tilde{f}(t)$, which is therefore appropriately called the complex envelope of $f(t)$.

We point out that the choice of the "center frequency" ω_c is somewhat arbitrary with respect to the signal $f(t)$. In many situations, a natural choice for ω_c exists. For example, if $f(t)$ is a band-limited process, ω_c can be chosen as the center frequency of the bandpass spectrum. In other cases, any convenient frequency can be chosen as the center frequency. For example with single sideband amplitude modulated (SSBAM) signals, the carrier frequency ω_c is usually chosen as the center frequency.

We can extend our previous discussions to a stationary band-limited stochastic process $x(t)$ with power spectrum $\Phi_x(\omega)$. To do so, we first need to obtain a complex process $x_+(t)$ from $x(t)$ such that its power spectrum $\Phi_{x_+}(\omega)$ corresponds to the positive frequency portion of $\Phi_x(\omega)$. Then we can again interpret $x_+(t)$ as a complex modulated process with an envelope $\tilde{x}(t)$. Or equivalently, we can express $x(t)$ in terms of its complex envelope $\tilde{x}(t)$ as

$$x(t) = \sqrt{2}\ \text{Re}\left[\tilde{x}(t)\exp(j\omega_c t)\right]$$

where now the power spectrum of $\tilde{x}(t)$ is given by

$$\Phi_{\tilde{x}}(\omega) = \begin{cases} 2\Phi_x(\omega+\omega_c) & \omega > -\omega_c \\ 0 & \omega < -\omega_c \end{cases} \tag{2.5.8}$$

which is analogous to Eq. (2.5.5). We obtain a factor of 2 instead of $\sqrt{2}$ because in this case we are dealing with a power spectrum and not the Fourier spectrum of the signal itself.

In our discussions so far, we have not considered how $\tilde{x}(t)$ or $\tilde{f}(t)$ can be obtained from the band-limited process $x(t)$ or $f(t)$. We can do this by first obtaining $x_+(t)$ or $f_+(t)$ as the case may be. An intuitive manner in which this can be done is to obtain a complex signal, $\frac{1}{2}[x(t) + jy(t)]$, which has a spectrum corresponding to $x_+(t)$ [or $f_+(t)$]. In order to illustrate this procedure, let us consider the deterministic signal case, and then check to see whether the procedure works for stochastic signals as well. Because we require that $\frac{1}{2}[f(t)+jy(t)]$ have the same spectrum as $f_+(t)$, the spectrum of $y(t)$ must satisfy the condition that

$$Y(\omega)=\begin{cases} -jF(\omega) & \omega\geqslant 0 \\ jF(\omega) & \omega<0 \end{cases} \tag{2.5.9}$$

That is, we can generate $y(t)$ by passing $f(t)$ through a linear system with transfer function

$$H(j\omega)=\begin{cases} -j & \omega\geqslant 0 \\ j & \omega<0 \end{cases} \tag{2.5.10}$$

and impulse response

$$h(t)=\frac{1}{\pi t}$$

That is,

$$y(t)=\int_{-\infty}^{\infty}\frac{f(\lambda)}{\pi(t-\lambda)}\,d\lambda \tag{2.5.11}$$

Since the Hilbert transform of a signal $z(t)$ is defined as

$$\check{z}(t)=\int_{-\infty}^{\infty}\frac{z(\lambda)}{\pi(t-\lambda)}\,d\lambda \tag{2.5.12}$$

with corresponding inverse given by

$$z(t)=-\int_{-\infty}^{\infty}\frac{\check{z}(\lambda)}{\pi(t-\lambda)}\,d\lambda \tag{2.5.13}$$

it follows that $y(t)$ is the Hilbert transform of the signal $f(t)$,

$$y(t)=\check{f}(t)$$

so that

$$f_+(t)=\tfrac{1}{2}\big[f(t)+j\check{f}(t)\big] \tag{2.5.14}$$

The signal $f_+(t)$ is called the pre-envelope of $f(t)$, or the analytic signal associated with $f(t)$. We can then define the complex envelope, $\tilde{f}(t)$, as in Eq. (2.5.6) with $f_+(t)$ given by Eq. (2.5.14). We have thus succeeded in obtaining the complex envelope of a deterministic signal. Let us now direct our attention to the stochastic case.

To check whether the same procedure will work for stochastic processes, let us define the pre-envelope

$$x_{+}(t)=\tfrac{1}{2}[x(t)+j\check{x}(t)] \tag{2.5.15}$$

similar to Eq. (2.5.14). The power spectrum of $x_{+}(t)$ is given by

$$\Phi_{x_+}(\omega)=\tfrac{1}{4}[\Phi_x(\omega)+\Phi_{\check{x}}(\omega)-j\Phi_{x\check{x}}(\omega)+j\Phi_{\check{x}x}(\omega)] \tag{2.5.16}$$

From Eq. (2.5.10) we can easily show that

$$\Phi_{\check{x}}(\omega)=|H(j\omega)|^2\Phi_x(\omega)=\Phi_x(\omega) \tag{2.5.17}$$

To obtain $\Phi_{x\check{x}}(\omega)$ and $\Phi_{\check{x}x}(\omega)$, we note that

$$\begin{aligned}\Phi_{x\check{x}}(\tau)=\mathrm{E}[x(t+\tau)\check{x}(t)]&=\int_{-\infty}^{\infty}\frac{\mathrm{E}[x(t+\tau)x(\lambda)]}{\pi(t-\lambda)}\,d\lambda\\&=\int_{-\infty}^{\infty}\frac{\Phi_x(t+\tau-\lambda)}{\pi(t-\lambda)}\,d\lambda\\&=-\check{\phi}_x(\tau)\end{aligned} \tag{2.5.18}$$

Similarly we can show that

$$\phi_{\check{x}x}(\tau)=\check{\phi}_x(\tau) \tag{2.5.19}$$

From Eqs. (2.5.18) and (2.5.19) we have

$$\Phi_{x\check{x}}(\omega)=\begin{cases}j\Phi_x(\omega) & \omega\geqslant 0\\ -j\Phi_x(\omega) & \omega<0\end{cases} \tag{2.5.20}$$

and

$$\Phi_{\check{x}x}(\omega)=\begin{cases}-j\Phi_x(\omega) & \omega\geqslant 0\\ j\Phi_x(\omega) & \omega<0\end{cases} \tag{2.5.21}$$

It follows that

$$\Phi_{x_+}(\omega)=\begin{cases}\Phi_x(\omega) & \omega\geqslant 0\\ 0 & \omega<0\end{cases} \tag{2.5.22}$$

The preceding expression shows that we have been able to use the same procedure successfully to obtain $x_+(t)$, the pre-envelope of the random signal $x(t)$. Therefore, we can now define the complex envelope $\tilde{x}(t)$ of a band-limited stochastic process $x(t)$ as

$$\tilde{x}(t) = \sqrt{2}\, x_+(t) \exp(-j\omega_c t) \tag{2.5.23}$$

with

$$x(t) = \sqrt{2}\, \operatorname{Re}\left[\tilde{x}(t) \exp(j\omega_c t)\right] \tag{2.5.24}$$

The power spectrum of the complex envelope is then given by

$$\Phi_{\tilde{x}}(\omega) = \begin{cases} 2\Phi_x(\omega + \omega_c) & \omega > -\omega_c \\ 0 & \omega < -\omega_c \end{cases} \tag{2.5.25}$$

The complex envelope $\tilde{x}(t)$ of a stationary band-limited stochastic process $x(t)$ plays an important role in radar and communication literature. We also use such a formulation for radar systems in a later chapter. It is therefore worthwhile to obtain some useful and interesting results on band-limited processes.

The first result concerns the real part of the complex envelope $\tilde{x}(t)$. Let us denote this by $a(t)$ and the imaginary part by $b(t)$. We can then use Eq. (2.5.15) and (2.5.23) to write

$$a(t) = \frac{1}{\sqrt{2}}\left[x(t)\cos\omega_c t + \check{x}(t)\sin\omega_c t\right]$$

$$b(t) = \frac{-1}{\sqrt{2}}\left[x(t)\sin\omega_c t - \check{x}(t)\cos\omega_c t\right] \tag{2.5.26}$$

We can easily show that $a(t)$ and $b(t)$ are low pass processes with power spectra

$$\Phi_a(\omega) = \Phi_b(\omega) = \tfrac{1}{2}\left[\Phi_{x_+}(\omega_c + \omega) + \Phi_{x_+}(\omega_c - \omega)\right] \tag{2.5.27}$$

Further we can prove that the random variables $a(t)$ and $b(t)$ are uncorrelated, even though the processes $a(t)$ and $b(t)$ are not. In fact, the cross power spectrum is given by

$$\Phi_{ab}(\omega) = j\left[\Phi_{x_+}(\omega_c - \omega) + \Phi_{x_+}(\omega_c + \omega)\right] \tag{2.5.28}$$

The spectra $\Phi_a(\omega)$ (or $\Phi_b(\omega)$) and $\Phi_{ab}(\omega)$ are known as the in-phase and quadrature components of $\Phi_x(\omega)$ with respect to ω_c.

We can write the band-limited process $x(t)$ in terms of the real and imaginary parts of the complex envelope as

$$x(t) = \sqrt{2}\,[a(t)\cos\omega_c t - b(t)\sin\omega_c t] \tag{2.5.29}$$

This representation is known as Rice's representation of a band-limited process [19]. It is particularly meaningful when the process $x(t)$ is zero mean and Gaussian. Then $a(t)$ and $b(t)$ are low pass, zero mean Gaussian processes (why?), whose statistics are identical and given by Eqs. (2.5.27) and (2.5.28). An alternate representation of the band-limited process is

$$x(t) = \sqrt{2}\,\epsilon(t)\cos[\omega_c t + \theta(t)] \tag{2.5.30}$$

which is called the envelope and phase form. The obvious identifications are

$$\epsilon(t) = [a^2(t) + b^2(t)]^{1/2} = |\tilde{x}(t)| \tag{2.5.31}$$

and

$$\theta(t) = \arctan\frac{-b(t)}{a(t)} = \angle\tilde{x}(t) \tag{2.5.32}$$

It is left as an exercise to the reader to prove that if $x(t)$ is Gaussian with zero mean, $\epsilon(t)$ is Rayleigh distributed, and $\theta(t)$ is uniformly distributed in $[0, 2\pi]$.

We point out that in order to describe the second-order moments of the complex process $\tilde{x}(t)$, we must be able to characterize it in terms of the single correlation function

$$\phi_{\tilde{x}}(\tau) = \mathrm{E}\{\tilde{x}(t+\tau)\tilde{x}^*(t)\}$$

This quantity as well as its transform, the power spectrum $\Phi_{\tilde{x}}(\omega)$ figured in our discussions so far. But how about the quantity $\mathrm{E}\{\tilde{x}(t+\tau)\tilde{x}(t)\}$? If we go carefully through the derivations in Eqs. (2.5.16) to (2.5.25), we will realize that this quantity is zero. Hence, if we associate a complex function $\tilde{x}(t)$ with the process $x(t)$ as discussed in the preceding, then $\tilde{x}(t)$ can be characterized in terms of a single correlation function, $\phi_{\tilde{x}}(\tau)$.

We are now ready to develop the concept of a complex white noise process. Consider a zero mean band-limited stationary process $v(t)$, with power spectrum of bandwidth W, centered around the frequency ω_c. Let the complex envelope of this process be $\tilde{v}(t)$ so that

$$v(t) = \sqrt{2}\ \mathrm{Re}[\tilde{v}(t)\exp(j\omega_c t)] \tag{2.5.33}$$

It follows that the autocorrelation function of $\tilde{v}(t)$ is given by

$$\phi_{\tilde{v}}(\tau) = N_0\left\{\frac{\sin(2\pi W\tau)}{\pi\tau}\right\} \tag{2.5.34}$$

If W, the bandwidth of the process, is larger than the other bandwidths in the system we are concerned with, we can approximate the $(\sin x)/x$ term by an impulse. That is, if we let $W \to \infty$, then

$$\phi_{\tilde{v}}(\tau) = N_0\delta(\tau) \tag{2.5.35}$$

which is also the covariance function because the mean is zero. We refer to the process $\tilde{v}(t)$ as the complex white noise process. The band-limited process $v(t)$ is often referred to as a band-limited, narrow-band, or band-pass white noise process.

An important special case is when $v(t)$ is a zero mean Gaussian process. Then $\tilde{v}(t)$ is a complex Gaussian process with the first-order density function

$$p(\tilde{v}) = \frac{1}{\pi N_0}\exp\left[-\frac{1}{2N_0}|\tilde{v}|^2\right] \tag{2.5.36}$$

Note that this is the joint density function of the real random variables, Re $[\tilde{v}(t)]$ and Im $[\tilde{v}(t)]$.†

Having discussed band-limited processes, it is only appropriate to ask whether complex representations of bandpass filtering operations exist and are useful. We now proceed to turn our attention to this question. It is reasonable to expect in view of our preceding derivations that an equivalent filtering operation in terms of the complex envelopes can be developed.

Let $h(t)$ be the impulse response of a band-limited (or bandpass) filter. Let the band-limited process $x(t)$, be its input, and $y(t)$, its output. It follows that $y(t)$ can be written as

$$y(t) = \int_{-\infty}^{\infty} h(t-\tau)x(\tau)\,d\tau \tag{2.5.37}$$

†If the original spectrum is symmetric about ω_c, Re $[\tilde{v}(t)]$ and Im $[\tilde{v}(t)]$ will be independent for any t. This independence does not hold if the spectrum is nonsymmetric about the carrier ω_c.

If we let $\tilde{h}(t)$ and $\tilde{x}(t)$ denote the complex envelopes of $h(t)$ and $x(t)$ respectively,[†] it follows that Eq. (2.5.37) can be written as

$$y(t)=\int_{-\infty}^{\infty}\frac{\left[2\tilde{h}(t-\tau)\exp\left(j\omega_c(t-\tau)\right)+2\tilde{h}^*(t-\tau)\exp\left(-j\omega_c(t-\tau)\right)\right]}{2}\cdot\frac{\left[\sqrt{2}\,\tilde{x}(\tau)\exp\left(j\omega_c\tau\right)+\sqrt{2}\,\tilde{x}^*(\tau)\exp\left(-j\omega_c\tau\right)\right]}{2}\,d\tau \tag{2.5.38}$$

In Eq. (2.5.38) we can neglect terms involving $\exp(\pm j2\omega_c\tau)$ because $\tilde{x}(t)$ and $\tilde{h}(t)$ are low pass and they integrate approximately to zero. Then we can write

$$\begin{aligned} y(t) &= \int_{-\infty}^{\infty}\frac{1}{\sqrt{2}}\left[\tilde{h}(t-\tau)\tilde{x}(t-\tau)\exp\left(j\omega_c t\right)\right. \\ &\qquad \left.+\tilde{h}^*(t-\tau)\tilde{x}^*(t-\tau)\exp\left(-j\omega_c t\right)\right]d\tau \\ &= \sqrt{2}\int_{-\infty}^{\infty}\operatorname{Re}\left[\tilde{h}(t-\tau)\tilde{x}(t-\tau)\exp\left(j\omega_c t\right)\right]d\tau \end{aligned} \tag{2.5.39}$$

It follows therefore that the complex envelope of $y(t)$ is given by

$$\tilde{y}(t)=\int_{-\infty}^{\infty}\tilde{h}(t-\tau)\tilde{x}(\tau)\,d\tau \tag{2.5.40}$$

This shows that for all practical purposes, the bandpass filtering operation of Eq. (2.5.37) can be replaced by the low pass operation of Eq. (2.5.40). Thus the processing of bandpass signals by bandpass filters can be analyzed in terms of the low pass filtering of their complex envelopes.

2.6 SUMMARY

In this chapter we have briefly reviewed some results from systems theory and stochastic processes that will be of use to us in our discussions in subsequent chapters. Section 2.2. was primarily concerned with the state space representation of dynamical systems. For linear systems, the solution

[†]We have used a slightly modified definition for the complex envelope of $h(t)$ in order to get the results in a familiar form. Thus we use $h(t)=\sqrt{2}\ \operatorname{Re}\left[\sqrt{2}\,\tilde{h}(t)\exp\left(j\omega_c t\right)\right]$

to the state equations was obtained in terms of the state transition matrix of the system.

In Section 2.3 we considered two stochastic processes of some importance, namely the Wiener and generalized Poisson processes. Formal differentiation of these processes yields the white noise process, which though not physically realizable, plays a useful role in the modeling of stochastic processes in many engineering applications. We also discussed the series expansion of stochastic processes in a set of orthonormal functions. A natural choice in many applications is the Karhunen–Loéve expansion in which the basis functions are the eigenfunctions of an integral equation whose kernel is the covariance function of the process. The advantage in this is that the statistics of the stochastic process can now be described in terms of the statistics of a set of uncorrelated random variables, namely the coefficients of expansion.

An important class of signal models namely the Gauss–Markov models were discussed in Section 2.4. For both continuous and discrete models, equations were derived which describe the propagation of the mean and variance of the signal process.

We concluded with a consideration in Section 2.5 of the representation of narrow-band stochastic processes in terms of their complex envelopes. Such processes arise naturally in modulation systems in which the message signal is used to modulate a carrier. The use of the complex envelope permits operations on these signals to be described in terms of the baseband signals.

REFERENCES

1. C. T. Chen, *Introduction to Linear System Theory*, Holt, Rinehart, and Winston, New York, 1970.
2. H. Freeman, *Discrete-Time Systems: An Introduction to the Theory*, John Wiley and Sons, New York, 1965.
3. A. Papoulis, *The Fourier Integral and Its Applications*, McGraw-Hill Book Co., New York, 1962.
4. E. A. Coddington and N. Levinson, *Theory of Ordinary Differential Equations*, McGraw-Hill Book Co., New York, 1955.
5. A. Papoulis, *Probability, Random Variables, and Stochastic Processes*, McGraw-Hill Book Co., New York, 1965.
6. W. B. Davenport, Jr. and W. L. Root, *Introduction to Random Signals and Noise*, McGraw-Hill Book Co., New York, 1958.
7. J. L. Doob, *Stochastic Processes*, John Wiley and Sons, Inc., New York, 1953.
8. E. Wong, *Stochastic Processes in Information and Dynamical Systems*, McGraw-Hill Book Co., New York, 1971.

9. K. Ito, *Lectures on Stochastic Processes*, Tata Institute of Fundamental Research, Bombay, India, 1961.
10. I. I. Gikhman and A. V. Skorokhod, *Introduction to the Theory of Random Processes*, Saunders, Philadelphia, 1969.
11. T. Hida, *Stationary Stochastic Processes*, Princeton University Press, Princeton, N. J., 1970.
12. T. P. McGarty, *Stochastic Systems and State Estimation*, Wiley–Interscience, New York, 1974.
13. A. M. Yaglom, *An Introduction to Stationary Random Functions*, Prentice-Hall, Englewood Cliffs, N. J., 1962.
14. M. Loéve, *Probability Theory*, D. Van Nostrand Co., Princeton, N. J., 1955.
15. R. Courant and D. Hilbert, *Methods of Mathematical Physics*, Interscience Publishers, New York, 1953.
16. F. G. Tricomi, *Integral Equations*, Interscience Publishers, New York, 1957.
17. A. D. Whalen, *Detection of Signals in Noise*, Academic Press, New York, 1971.
18. H. L. Van Trees, *Detection, Estimation, and Modulation Theory, Part III*, John Wiley and Sons, New York, 1971.
19. S. O. Rice, Mathematical Analysis of Random Noise, *Bell Syst. Tech. J.*, **23**, 282–332 (1944).

CHAPTER 3

DETECTION THEORY

3.1 INTRODUCTION

There are many situations in which we are required to choose among several possibilities. For example, in a radar detection problem, the radar return is observed and a decision has to be made as to whether a target was present or not. In a digital communication system, one of several possible waveforms is selected for transmission to the receiver. The waveforms may undergo some distortion in the transmission medium. Receiver noise contributes further to the distortion. The consequence is that at the receiver, it is no longer clear which of the signals was transmitted. The receiver has to make a decision as to which signal is actually present based on noisy observations. In pattern recognition problems a given pattern belongs to one of a fixed number of classes. For example, a given alphanumeric character belongs to the set of alphabets and numerals from 0 to 9. The pattern classifier must decide as to which class (alphanumeric character) a given test pattern belongs. This decision has to be made by the classifier, which is an automatic machine, based on the noisy measurements and any statistical information regarding the various pattern classes.

In each case the solution involves making a decision among several choices. In this chapter we will introduce the basic tools necessary for solving the decision problem where we assume that the observables are a set of random variables. Such problems have been discussed by statisticians for many years and fall under the general category of statistical inference. We extend these results to decision problems involving waveforms in the next chapter.

3.2 CLASSICAL DETECTION THEORY AND HYPOTHESIS TESTING [1–6]

In a statistical hypothesis-testing problem, a decision has to be made as to which of several hypothesis is the correct one. We can think of a hypothesis as a statement of a possible decision. The mechanism that generates the statements is called the source. In the simplest possible case, the output of the source is one of two possible hypotheses. For example, in a radar detection problem, the hypotheses will be the statements "Target not present" and "Target present." We label the two possible choices as H_0 and H_1. We shall refer to H_0 as the "null" hypothesis. This is a *binary detection* problem. More generally, the output of the source can be one of a set of M hypotheses, $H_0, H_1, \ldots, H_{M-1}$. We refer to this as an *M-ary detection* problem.

Corresponding to each hypothesis, we have an *observation*, a random variable which is generated according to some probabilistic law. The hypothesis-testing problem is one of deciding which hypothesis is the correct one, based on a single measurement z, of this random variable. The range of values which z takes constitutes the *observation space*. The decision problem in the case of two hypotheses thus essentially consists in partitioning the observation space into two regions Z_0 and Z_1 such that whenever z lies in Z_0, we decide that H_0 was the correct hypothesis and whenever z lies in Z_1, we decide that H_1 is the correct hypothesis. The regions Z_0 and Z_1 are known as decision regions. The problem lies in choosing these regions such that we obtain the fewest errors in several repetitions of the experiment.

In order to quantify our judgment as to the best possible partitioning of the observation space, we have to first select a suitable criterion. While we will be discussing several such criteria in the sequel, let us at this stage assume that we choose as the correct hypothesis, the one that most likely gave rise to our observation z. If $P(H_i|z)$, $i = 0, 1$ denotes the probability that H_i was the true hypothesis given a particular value of the observation, we decide that the correct hypothesis is the one corresponding to the larger of the two probabilities. The decision rule will then be to choose H_0 if

$$P(H_0|z) > P(H_1|z) \tag{3.2.1}$$

and choose H_1 otherwise. We write the rule as

$$\frac{P(H_1|z)}{P(H_0|z)} \underset{H_0}{\overset{H_1}{\gtrless}} 1 \tag{3.2.2}$$

The criterion which we have used is known as the *maximum a posteriori probability* (MAP) criterion.

We can use Bayes' rule to write the criterion in a more useful form. We thus write

$$P(H_i|z) = \frac{p(z|H_i)P(H_i)}{p(z)} \qquad i = 0, 1 \tag{3.2.3}$$

where $P(H_i)$ denotes the probability that source H_i was active, so that

$$\frac{P(H_1|z)}{P(H_0|z)} = \frac{p(z|H_1)}{p(z|H_0)} \frac{P(H_1)}{P(H_0)}$$

and the test becomes

$$\frac{p(z|H_1)}{p(z|H_0)} \underset{H_0}{\overset{H_1}{\gtrless}} \frac{P(H_0)}{P(H_1)} \tag{3.2.4}$$

The ratio $\Lambda(z) = p(z|H_1)/p(z|H_0)$ is called the likelihood ratio and is of particular importance. The test therefore consists of comparing this ratio with a constant termed the threshold and is called a likelihood ratio test (LRT). We will see later that decision tests based on several other criteria also fall into the general class of likelihood ratio tests. But the thresholds will, in general, be different for the different tests.

Example 3.2.1 Consider a simple binary communication channel in which the source transmits over each T-second interval, either a signal pulse $y(t)$ of unit amplitude or no signal at all. Such a system is referred to as an on–off keying (OOK) system in digital communications. The communication channel adds noise $v(t)$ so that the received signal is either $y(t) + v(t)$ or $v(t)$ over each T-second interval. We observe the received signal at some instant during each signaling interval. The problem is to decide, on the basis of a single observation, whether or not the signal is present.

We will call the event that no signal was transmitted as the null hypothesis and the alternative hypothesis when a unit amplitude signal is transmitted as H_1. The received signal under the two hypotheses can then be written as

$$H_1 : z = 1 + v$$

$$H_0 : z = v$$

Let v be Gaussian, with zero mean and unit variance. The probability density of z under each hypothesis follows easily as

$$p(z|H_0) = \frac{1}{\sqrt{2\pi}} \exp\left[-\frac{z^2}{2}\right]$$

and

$$p(z|H_1) = \frac{1}{\sqrt{2\pi}} \exp\left[-\frac{(z-1)^2}{2}\right]$$

The likelihood ratio is then given by

$$\Lambda(z) = \frac{p(z|H_1)}{p(z|H_0)} = \exp\left(z - \frac{1}{2}\right)$$

The decision rule is

$$\exp\left(z - \frac{1}{2}\right) \underset{H_0}{\overset{H_1}{\gtrless}} \frac{P(H_0)}{P(H_1)} \tag{3.2.5}$$

It is often more convenient to work with the natural logarithm of the likelihood ratio. Since the logarithm is a monotonically increasing function, the inequality in the decision rule will still hold and we can write the rule in terms of the log-likelihood ratio as

$$L(z) = \ln \Lambda(z) = z - \frac{1}{2} \underset{H_0}{\overset{H_1}{\gtrless}} \ln \frac{P(H_0)}{P(H_1)}$$

or

$$z \underset{H_0}{\overset{H_1}{\gtrless}} \frac{1}{2} + \ln \frac{P(H_0)}{P(H_1)}$$

The decision rule therefore is to sample the received signal and compare it with a threshold. If the sample is greater than the threshold, it is decided that a "1" was sent; otherwise it is decided that a "0" was sent. Since the decision rule is completely determined by the sample z, it is called the "test statistic," for this problem. We will in general denote the test statistic by $l(z)$.

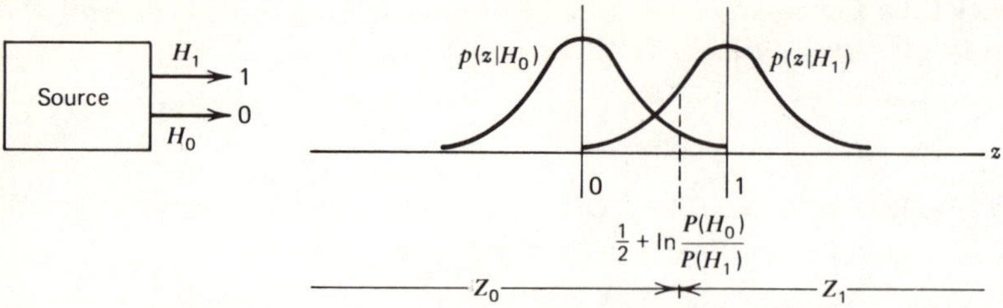

Fig. 3.2.1 Source, transition probabilities, and decision regions for Example 3.2.1.

The foregoing example illustrates the basic components of the decision problem. The first component is the *source*. The second component is the *probabilistic transition mechanism* which generates an observation z according to some probability law, depending on which hypothesis is true. The third component is the *observation space* Z which could be multidimensional. In the preceding example, the probabilities of interest are $p(z|H_0)$ and $p(z|H_1)$, and the observation space is the entire real line from $-\infty$ to $+\infty$. The decision rule then divides the observation space into the two regions Z_0 and Z_1, and assigns each point in the two regions to one of the two hypotheses. These concepts are illustrated in Fig. 3.2.1 for the example.

3.3 DECISION CRITERIA [2,4]

In the simple example considered in the previous section we divided the observation space into two regions in order to decide which hypothesis was true. It is clear that irrespective of where this point of division is chosen to be, we will occasionally make a wrong decision. Let us denote by D_i, our choice of H_i as the true hypothesis.

In making a decision in any binary hypotheses testing problem we have four possibilities to consider: (i) H_0 is the true hypothesis, decide D_0, (ii) H_1 is true, decide D_1, (iii) H_0 is true, decide D_1, and (iv) H_1 is true, decide D_0. The first two correspond to correct choices. The last two correspond to errors. We call (iii) an error of the first kind and (iv) an error of the second kind. In radar terminology, these are referred to as *false alarm* (decide target is present when it is not) and *miss* (decide no target is present when it is). We will assume that the prior probabilities $P(H_0)$ and $P(H_1)$ of H_0 and H_1 occurring are known.

The purpose of a decision criterion is to attach some relative importance to the four possible choices. The consequences of one decision may not be the same as the consequences of another. For example, in the radar context the consequences of a miss is quite different from that of a false alarm. To reflect these differences we assign a cost to each of the decisions. We thus define C_{ij} to be the cost associated with making a decision D_i when the true hypothesis is H_j. We would now like to determine our decision rule so that the average cost is minimized. This criterion is known as the *Bayes criterion*. If we let $P(D_i, H_j)$ denote the joint probability that we decide H_i and that the true hypothesis is H_j, the average cost can be written as

$$\bar{C} = C_{00}P(D_0, H_0) + C_{10}P(D_1, H_0) + C_{01}P(D_0, H_1) + C_{11}P(D_1, H_1) \tag{3.3.1}$$

An application of Bayes' rule yields

$$\begin{aligned}\bar{C} = {} & C_{00}P(D_0|H_0)P(H_0) + C_{10}P(D_1|H_0)P(H_0) \\ & + C_{01}P(D_0|H_1)P(H_1) + C_{11}P(D_1|H_1)P(H_1)\end{aligned} \tag{3.3.2}$$

where $P(D_i|H_j)$ is the probability of deciding H_i when H_j is the true hypothesis.

As discussed earlier, for the binary decision problem, the decision rule divides the observation space into two regions Z_0 and Z_1 by assigning each point in Z_0 to hypothesis H_0 and each point in Z_1 to H_1. The decision regions are shown in Fig. 3.3.1.

We can obtain the probabilities $P(D_i|H_j)$ $i, j = 0, 1$ by integrating the conditional density function of z given hypothesis H_j, over Z_i, the decision space corresponding to hypothesis H_i:

$$P(D_i|H_j) = \int_{Z_i} p(z|H_j)\, dz \tag{3.3.3}$$

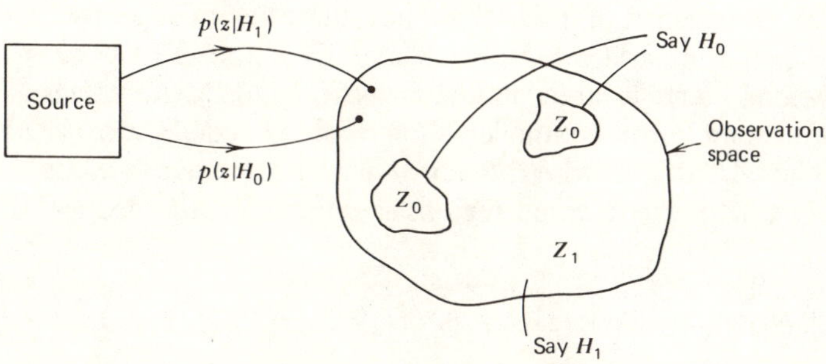

Fig. 3.3.1 Decision regions.

so that the average cost becomes

$$\bar{C} = C_{00}P(H_0)\int_{Z_0} p(z|H_0)\,dz + C_{10}P(H_0)\int_{Z_1} p(z|H_0)\,dz$$

$$+ C_{01}P(H_1)\int_{Z_0} p(z|H_1)\,dz + C_{11}P(H_1)\int_{Z_1} p(z|H_1)\,dz \qquad (3.3.4)$$

It is reasonable at this point to assume that the cost of making an incorrect decision is higher than the cost of a correct decision. We will therefore assume that

$$C_{10} > C_{00}$$

and

$$C_{01} > C_{11} \qquad (3.3.5)$$

Since $Z = Z_0 + Z_1 \underline{\Delta} Z_0 \cup Z_1$, and $Z_0 \cap Z_1 = \phi$, we can write

$$\int_{Z_1} p(z|H_i)\,dz = \int_{Z} p(z|H_i)\,dz - \int_{Z_0} p(z|H_i)\,dz$$

$$= 1 - \int_{Z_0} p(z|H_i)\,dz \qquad i = 0, 1 \qquad (3.3.6)$$

Consequently Eq. (3.3.4) becomes

$$\bar{C} = C_{10}P(H_0) + C_{11}P(H_1) + \int_{Z_0} \{[P(H_1)(C_{01} - C_{11})p(z|H_1)]$$

$$-[P(H_0)(C_{10} - C_{00})p(z|H_0)]\}\,dz \qquad (3.3.7)$$

The only variable quantity in the preceding integral is the region Z_0. The assumption in Eq. (3.3.5) implies that the two terms inside the brackets in the integral are both positive. Therefore all values of z such that the second term is larger should be assigned to Z_0. All values of z such that the second term is smaller should be excluded from Z_0, that is, assigned to Z_1. Points where the two terms are equal will be assigned arbitrarily to Z_1. The decision regions are, therefore, defined as follows:

If

$$P(H_1)(C_{01} - C_{11})p(z|H_1) \geqslant P(H_0)(C_{10} - C_{00})p(z|H_0) \qquad (3.3.8)$$

assign z to Z_1 and consequently decide that H_1 is true. Otherwise assign z

to Z_0 and say H_0 is true. In terms of the likelihood ratio

$$\Lambda(z) = \frac{p(z|H_1)}{p(z|H_0)} \tag{3.3.9}$$

and the threshold

$$\eta = \frac{P(H_0)(C_{10} - C_{00})}{P(H_1)(C_{01} - C_{11})} \tag{3.3.10}$$

the decision rule becomes

$$\Lambda(z) \underset{H_0}{\overset{H_1}{\gtrless}} \eta \tag{3.3.11}$$

and thus leads to a likelihood ratio test.

We note that the processing involved in computing $\Lambda(z)$ is independent of cost assignments and prior probabilities. In a practical implementation of the processor, we therefore leave the threshold as a variable quantity to accommodate changes in our cost assignments or prior probabilities.

In terms of the log-likelihood ratio, the decision rule becomes

$$\ln \Lambda(z) \underset{H_0}{\overset{H_1}{\gtrless}} \ln \eta \tag{3.3.12}$$

We note that if we choose $C_{10} - C_{00} = C_{01} - C_{11}$ in Eq. (3.3.10) the Bayes rule of Eq. (3.3.11) will be the same as was obtained using the MAP criterion.

We can give an alternate interpretation of the Bayes criterion as follows. We define the "conditional risk" or cost in making a decision in favor of hypothesis H_i as

$$C(H_i|z) = C_{i0}P(H_0|z) + C_{i1}P(H_1|z) \qquad i = 0, 1 \tag{3.3.13}$$

It can then easily be verified that the decision rule of Eq. (3.3.11) is equivalent to choosing that hypothesis for which the conditional risk is a minimum.

Example 3.3.1 We assume that under both hypotheses, the source output is a zero mean Gaussian signal. Under H_1, the variance of the signal is σ_1^2

and under H_0, the variance is σ_0^2. We thus have

$$p(z|H_i) = \frac{1}{\sqrt{2\pi}\,\sigma_i} \exp\left(-\frac{z^2}{2\sigma_i^2}\right) \qquad i = 0, 1$$

$$\Lambda(z) = \frac{\sigma_0}{\sigma_1} \exp\left[\frac{z^2}{2}\left(\frac{1}{\sigma_0^2} - \frac{1}{\sigma_1^2}\right)\right]$$

so that the decision rule is

$$\frac{\sigma_0}{\sigma_1} \exp\left[\frac{z^2}{2}\left(\frac{1}{\sigma_0^2} - \frac{1}{\sigma_1^2}\right)\right] \underset{H_0}{\overset{H_1}{\gtrless}} \eta$$

In terms of the log-likelihood ratio, we can write this as

$$\frac{z^2}{2}\left(\frac{1}{\sigma_0^2} - \frac{1}{\sigma_1^2}\right) + \ln\frac{\sigma_0}{\sigma_1} \underset{H_0}{\overset{H_1}{\gtrless}} \ln \eta$$

Let $\sigma_1^2 > \sigma_0^2$. We can use $l(z) = z^2$ as the test statistic and write the test in terms of $l(z)$ as

$$l(z) \underset{H_0}{\overset{H_1}{\gtrless}} \frac{2\sigma_0^2\sigma_1^2}{\sigma_1^2 - \sigma_0^2} \ln\left(\frac{\eta\sigma_1}{\sigma_0}\right) \triangleq \gamma$$

The performance of the detector can be determined by evaluating the errors associated with making the decisions. The quantities of interest are the probabilities of the two types of error introduced earlier. We recall that an error of the first kind (false alarm) corresponds to deciding H_1 when H_0 is true. We have from Eq. (3.3.3)

$$P_{\mathrm{I}} = P_F = P(D_1|H_0) = \int_{Z_1} p(z|H_0)\, dz \tag{3.3.14}$$

An error of the second kind (miss) corresponds to deciding H_0 when H_1 is true, so that

$$P_{\mathrm{II}} = P_M = P(D_0|H_1) = \int_{Z_0} p(z|H_1)\, dz \tag{3.3.15}$$

Similarly, the probability of detection P_D is given by

$$P_D = P(D_1|H_1) = \int_{Z_1} p(z|H_1)\, dz = 1 - P_M \tag{3.3.16}$$

The average probability of error in decision making is then given by

$$P_e = P_F \cdot P(H_0) + P_M \cdot P(H_1) \tag{3.3.17}$$

We note that the regions Z_0 and Z_1 consist of those values of z for which the likelihood ratio $\Lambda(z)$ is, respectively, less than or greater than the threshold η. We can therefore also write the various error probabilities in terms of $\Lambda(z)$ as

$$P_F = \int_{\eta}^{\infty} p(\Lambda(z)|H_0)\, d\Lambda \tag{3.3.18a}$$

$$P_M = \int_{-\infty}^{\eta} p(\Lambda(z)|H_1)\, d\Lambda \tag{3.3.18b}$$

and

$$P_D = \int_{\eta}^{\infty} p(\Lambda(z)|H_1)\, d\Lambda \tag{3.3.18c}$$

Finally, we can write the average risk $\bar{C}$ in Eq. (3.3.7) in terms of P_F and P_M as

$$\begin{aligned}\bar{C} = {} & C_{00}(1 - P_F) + C_{10}P_F \\ & + P(H_1)[(C_{11} - C_{00}) + (C_{01} - C_{11})P_M \\ & - (C_{10} - C_{00})P_F].\end{aligned} \tag{3.3.19}$$

Minimum Probability of Error Criterion

If we set $C_{00} = C_{11} = 0$ and $C_{10} = C_{01} = 1$, the expression for the risk in Eq. (3.3.5) becomes

$$\bar{C} = P(H_0)\int_{Z_1} p(z|H_0)\, dz + P(H_1)\int_{Z_0} p(z|H_1)\, dz \tag{3.3.20}$$

which is just the total probability of making an error. The criterion, therefore, yields the minimum probability of error. The decision rule becomes

$$\Lambda(z) \underset{H_0}{\overset{H_1}{\gtrless}} \frac{P(H_0)}{P(H_1)} \tag{3.3.21}$$

The receiver is then called an *ideal observer*. It can be seen that the decision rule is the same as for the MAP criterion.

Example 3.3.2 We assume that under the hypothesis H_1 we observe a constant signal of amplitude m corrupted by a zero mean Gaussian noise v, with variance σ^2; under the hypothesis H_0, we observe only the noise v. Given an observation z, we need to decide whether the level was m or zero.

We model the observation as follows.

$$H_1 : z = m + v, \quad p(z|H_1) = \frac{1}{\sqrt{2\pi}\,\sigma} \exp\left[-\frac{(z-m)^2}{2\sigma^2}\right]$$

$$H_0 : z = v, \qquad p(z|H_0) = \frac{1}{\sqrt{2\pi}\,\sigma} \exp\left[-\frac{z^2}{2\sigma^2}\right]$$

Assuming equally likely hypotheses, and minimum probability of error criterion, the likelihood ratio test is:

$$\ln\left[\frac{p(z|H_1)}{p(z|H_0)}\right] \underset{H_0}{\overset{H_1}{\gtrless}} 0$$

This can be simplified as

$$z \underset{H_0}{\overset{H_1}{\gtrless}} \frac{m}{2}$$

We note that the observation itself is the sufficient (test) statistic. In order to determine the performance, it is convenient to normalize our observation to have a standard deviation of unity. Let $\bar{z} = z/\sigma$ denote the normalized observation, and let $d = m/\sigma$, a design parameter. The decision rule then becomes

$$\bar{z} \underset{H_0}{\overset{H_1}{\gtrless}} \frac{d}{2}$$

Note that

$$p(\bar{z}|H_1) \quad \sim N(d, 1)$$

and

$$p(\bar{z}|H_0) \qquad \sim N(0, 1)$$

where $N(\mu, v)$ is a Gaussian random variable with mean μ and variance v.

The total probability of error P_e is given by

$$P_e = \frac{1}{2} \cdot P_{\mathrm{I}} + \frac{1}{2} \cdot P_{\mathrm{II}}$$

where $P_{\mathrm{I}} = \text{Prob.} \left\{ \bar{z} > \frac{d}{2} \text{ given that } \bar{z} \text{ corresponded to } H_0 \right\}$

$$= \int_{d/2}^{\infty} p(\bar{z}|H_0)\, d\bar{z} = \int_{d/2}^{\infty} \frac{1}{\sqrt{2\pi}} \exp\left(-\frac{\bar{z}^2}{2}\right) d\bar{z} = \operatorname{erfc}_*\left(\frac{d}{2}\right)^{\dagger}$$

and $P_{\mathrm{II}} = \text{Prob.} \left\{ \bar{z} < \frac{d}{2} \text{ given that } \bar{z} \text{ corresponded to } H_1 \right\}$

$$= \int_{-\infty}^{d/2} p(\bar{z}|H_1)\, dz = \int_{-\infty}^{d/2} \frac{1}{\sqrt{2\pi}} \exp\left(\frac{(\bar{z}-d)^2}{2}\right) d\bar{z} = \operatorname{erfc}_*\left(\frac{d}{2}\right)$$

Therefore,

$$P_e = \operatorname{erfc}_*\left(\frac{d}{2}\right)$$

We now briefly discuss other criteria which also lead to LRTs.

Minimax Criterion

We have seen that the Bayes criterion requires that, in addition to assigning costs to the various decisions, we must also assign prior probabilities to the two hypotheses. Frequently not enough is known about the hypotheses which will enable us to determine the prior probabilities. In such a case, the Bayes criterion cannot be applied. One approach to obtaining the decision rule is to use the Bayes solution corresponding to

†The error function is given by $\operatorname{erf}(x) = \frac{2}{\sqrt{\pi}} \int_{-\infty}^{x} e^{-y^2/2}\, dy$ so that $\operatorname{erfc}_*(x) = \frac{1}{2\sqrt{2}}(1 - \operatorname{erf}(x)) = \frac{1}{2\sqrt{2}} \operatorname{erfc}(x)$.

the value of $P(H_1)$ for which the average cost is a maximum. The criterion is thus referred to as a *minimax* criterion.

For a fixed assignment of costs, the threshold of the Bayes test varies with $P(H_1)$. So does the average cost, which can therefore be written as $\bar{C}(P(H_1))$. If we now plot the minimum average cost $\bar{C}_{\min}$ (Bayes cost) as a function of $P(H_1)$, we will obtain a curve somewhat like the one in Fig. 3.3.2. Typically, the curve will have a maximum at some value P_1^* of $P(H_1)$.

Suppose we now assume that $P(H_1)$ is a fixed value P_1 and design the Bayes test accordingly. The decision regions are then fixed by the threshold of the test and so are the error probabilities P_F and P_M. However, if the true value of $P(H_1)$ is some other value, the decision rule is no longer optimal and the average cost incurred will not be the Bayes cost but will be larger. In fact, as can be seen from Eq. (3.3.19), the variation in the average cost with $P(H_1)$ will be represented by a straight line. This straight line will of course be tangent to the $\bar{C}_{\min}$ curve at P_1, since the test is optimal for this value of $P(H_1)$. The actual cost incurred could thus be quite large depending on the true value of $P(H_1)$. However, if we were to design our test under the assumption that the prior probability was P_1^*, the tangent to the $\bar{C}_{\min}$ curve will be horizontal. We note that P_1^* represents the worst situation in that the Bayes cost is maximum at this point. We are then assured that irrespective of the true value of $P(H_1)$, the average cost will not exceed $\bar{C}_{\min}(P_1^*)$. This situation occurs when the partial derivative of $\bar{C}(P(H_1))$ with respect to $P(H_1)$ is zero and is given by

$$(C_{11} - C_{00}) + (C_{01} - C_{11})P_M - (C_{10} - C_{00})P_F = 0 \tag{3.3.22}$$

Equation (3.3.22) is referred to as the *minimax equation*.

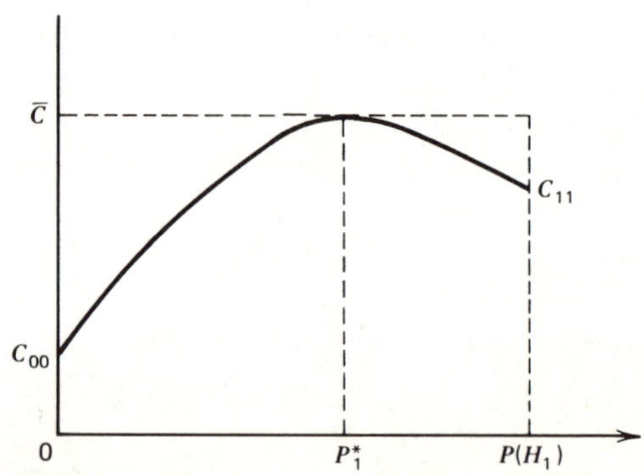

Fig. 3.3.2 Bayes cost as a function of P(H₁).

For the special case when $C_{00} = C_{11} = 0$, $C_{10} = C_{01} = 1$ the minimax equation reduces to

$$P_M = P_F \tag{3.3.23}$$

and the minimax cost is the average probability of error.

Example 3.3.3 We consider the same problem as in example 3.3.2 but assume that the prior probabilities $P(H_1)$ and $P(H_0)$ are not known. We would like to implement a minimax receiver in this case. For convenience, let us assume that $C_{ij} = 1 - \delta_{ij}$, $i, j = 0, 1$. Then, we make use of Eq. (3.3.23) to obtain the threshold.

We can easily show that for this example

$$P_F = \int_\gamma^\infty p(z|H_0)\, dz = \int_\gamma^\infty \frac{1}{\sqrt{2\pi}\,\sigma} \exp\left[-\frac{z^2}{2\sigma^2}\right] dz = \mathrm{erfc}_*(\gamma/\sigma)$$

and

$$P_M = \int_{-\infty}^\gamma p(z|H_1)\, dz = \int_{-\infty}^\gamma \frac{1}{\sqrt{2\pi}\,\sigma} \exp\left[-\frac{(z-m)^2}{2\sigma^2}\right] dz$$

$$= 1 - \mathrm{erfc}_*\left[\frac{(\gamma - m)}{\sigma}\right]$$

The optimum threshold, $\gamma = \gamma^*$, is obtained when $P_M = P_F$ according to Eq. (3.3.23). This results in the expression

$$\mathrm{erfc}_*\left(\frac{\gamma^*}{\sigma}\right) = 1 - \mathrm{erfc}_*\left[\frac{\gamma^* - m}{\sigma}\right]$$

from which

$$\gamma^* = \frac{m}{2}$$

The average probability of error is

$$P_e = P_F = \mathrm{erfc}_*\left(\frac{m}{2\sigma}\right) = \mathrm{erfc}_* \frac{d}{2}$$

which is the same result as we obtained in Example 3.3.2.

Neyman–Pearson Tests

In many situations, not only the prior probabilities, but also cost assignments are difficult to make. For example, in radar detection, the cost of failing to detect a target cannot be easily determined. A simple procedure to bypass this difficulty is to work with the conditional probabilities P_F and P_D. In general we would like to make P_F as small as possible and P_D as large as possible. However, these are usually conflicting objectives. It is usual therefore to determine a value of P_F which is acceptable and to seek a decision strategy that constrains P_F to this value while simultaneously maximizing P_D. Equivalently we can seek to minimize P_M. The test is then said to be a *Neyman–Pearson Test*. The value α chosen for P_F is called the *size* of the test and the maximum attainable value of P_D is called the *power* of the test, β.

We now obtain the decision rule by minimizing P_M subject to the constraint

$$P_F = \alpha$$

We thus construct the objective function (see Appendix B)

$$J = P_M + \lambda(P_F - \alpha) \tag{3.3.24}$$

where $\lambda \geqslant 0$ is a Lagrange multiplier. If we substitute for P_F and P_M from Eqs. (3.3.14) and (3.3.15), we get

$$J = \int_{Z_0} p(z|H_1)\,dz + \lambda\left[\int_{Z_1} p(z|H_0)\,dz - \alpha\right]$$

$$= \int_{Z_0} p(z|H_1)\,dz + \lambda\left[1 - \int_{Z_0} p(z|H_0)\,dz - \alpha\right]$$

$$= \lambda(1-\alpha) + \int_{Z_0} [p(z|H_1) - \lambda p(z|H_0)]\,dz \tag{3.3.25}$$

It easily follows that a likelihood ratio test (LRT) will minimize J. The decision rule becomes

$$\Lambda(z) \underset{H_0}{\overset{H_1}{\gtrless}} \lambda \tag{3.3.26}$$

The threshold of the test, η, is equal to the Lagrange multiplier λ and is chosen to satisfy the constraint on P_F. We thus have

$$\alpha = P_F = \int_{Z_1} p(z|H_0)\,dz = \int_{\eta}^{\infty} p(\Lambda(z)|H_0)\,dz \tag{3.3.27}$$

Example 3.3.4 We consider the problem of Example 3.2.1 and construct a Neyman–Pearson receiver for the case where P_F is constrained to be 0.1. From Example 3.2.1, the likelihood ratio is given by

$$\Lambda(z) = \exp\left(z - \frac{1}{2}\right)$$

so that the decision rule will be

$$z \underset{H_0}{\overset{H_1}{\gtrless}} \ln \eta + \frac{1}{2} \triangleq \gamma$$

To determine η, we set $P_F = 0.1$ which gives

$$\int_{\gamma}^{\infty} \frac{1}{\sqrt{2\pi}} \exp\left[\frac{-z^2}{2}\right] dz = 0.1$$

from which $\gamma = 1.29$. We therefore have

$$\eta = \exp\left(\gamma - \frac{1}{2}\right) = 2.2$$

The probability of detection is

$$P_D = P(D_1|H_1) = \int_{\gamma}^{\infty} \frac{1}{\sqrt{2\pi}} \exp\left[-\frac{(z-1)^2}{2}\right] dz = 0.614$$

It may be noted that this may be an unacceptable level of P_D. For larger values of P_F, the threshold γ becomes lower thus increasing P_D.

We have seen that all the three criteria discussed so far require the computation of a likelihood ratio. It is only the value of the threshold η with which this ratio is compared that varies with the criterion. The performance of the tests are best analyzed in terms of a graph of the detection probability P_D against the probability of false alarm P_F. Such a

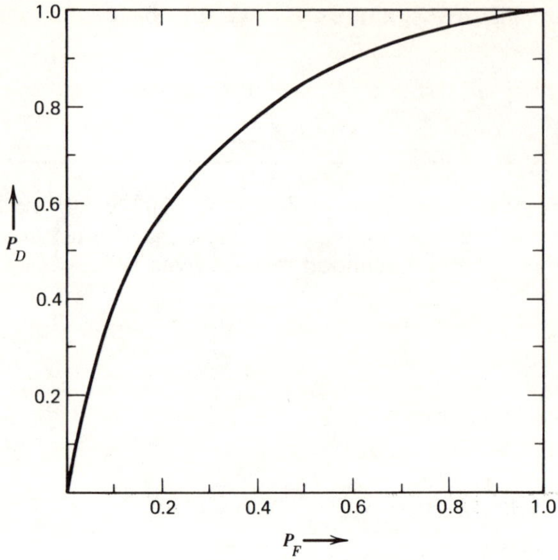

Fig. 3.3.3 Receiver operating characteristic for Example 3.3.4.

curve is called a receiver operating characteristic (ROC). It depends only on the probability density functions of the observation under the two hypotheses and not on any costs or prior probabilities. Figure 3.3.3 shows the ROC for Example 3.3.4.

For the Neyman–Pearson test, the slope of the ROC corresponding to any specified value of P_F represents the critical value of the likelihood ratio. This easily follows from the definitions of P_D and P_F. From Eq. (3.3.18c) we can write

$$\frac{dP_D}{d\eta} = \frac{d}{d\eta}\int_{\eta}^{\infty} p(\Lambda(z)|H_1)\, d\Lambda = -p(\eta|H_1) \tag{3.3.28}$$

However,

$$\begin{aligned} P_D(\eta) &= \Pr\{\Lambda(z) \geqslant \eta | H_1\} \\ &= \int_{\eta}^{\infty} p(\Lambda(z)|H_1)\, d\Lambda \\ &= \int_{\eta}^{\infty} \Lambda(z) p(\Lambda(z)|H_0)\, d\Lambda \end{aligned} \tag{3.3.29}$$

so that

$$\frac{dP_D(\eta)}{d\eta} = -\eta p(\eta|H_0) \tag{3.3.30}$$

Combining Eqs. (3.3.28) and (3.3.30) yields

$$\frac{p(\eta|H_1)}{p(\eta|H_0)} = \eta \tag{3.3.31}$$

from which it is clear that

$$\frac{dP_D}{dP_F} = \eta \tag{3.3.32}$$

For the Bayes criterion, the threshold η is determined by the costs and prior probabilities. The values of P_D and P_F corresponding to this threshold can be obtained by determining the point on the ROC at which the tangent has slope η.

It can easily be verified that the minimax equation represents a straight line in the $P_D - P_F$ plane. Thus, the probabilities P_D and P_F for this test can be determined from the point of intersection of the minimax line with the ROC. The slope of the tangent to the ROC at the point of intersection yields the threshold η. If the costs C_{ij} are such that either $C_{11} < C_{01} < C_{00} < C_{10}$ or $C_{00} < C_{10} < C_{11} < C_{01}$, the minimax line will not intersect the ROC in the region $0 < P_D, P_F < 1$. Thus the minimax solution does not exist in the usual sense. In the first case since C_{11} is less than C_{00}, the loss will always be greater when H_0 is true, so that the decision will be to choose H_0 always; whereas in the second case, the decision will be to choose H_1.

3.4 MULTIPLE MEASUREMENTS [2,6]

The discussions of the previous section can easily be extended to the more usual case in which the decision between the two hypotheses H_0 and H_1 is based on several observations. In this case, we have N observations, $z_1, z_2, \ldots, z_N$ which may represent successive measurements of the same physical parameter or the simultaneous measurement of several parameters or combinations thereof. The observations are described by the conditional density functions $p(z_1, z_2, \ldots, z_N|H_0)$ and $p(z_1, z_2, \ldots, z_N|H_1)$. We can derive the decision rules for this problem conveniently by considering the observations as a point in an N-dimensional space by defining the vector $\mathbf{z}$ as

$$\mathbf{z} = [z_1, z_2, \ldots, z_N]^T \tag{3.4.1}$$

The decision rule divides the observation space into two regions Z_0 and Z_1

such that we decide H_0 when the observation vector $\mathbf{z}$ lies in Z_0 and decide H_1 when $\mathbf{z}$ lies in Z_1. The surface separating the two regions is known as the decision surface. The particular criterion used determines the location of the decision surface. For applying the Bayes' criterion, we again must know the a priori probabilities $P(H_0)$ and $P(H_1)$ and the costs C_{ij}. As in Eq. (3.3.4), the average risk (cost) is given by

$$\bar{C} = P(H_0)\left[C_{00}\int_{Z_0} p(\mathbf{z}|H_0)\, d\mathbf{z} + C_{10}\int_{Z_1} p(\mathbf{z}|H_0)\, d\mathbf{z} \right] + P(H_1)\left[C_{01}\int_{Z_0} p(\mathbf{z}|H_1)\, d\mathbf{z} + C_{11}\int_{Z_1} p(\mathbf{z}|H_1)\, d\mathbf{z} \right] \tag{3.4.2}$$

where the integrals are N-fold integrals and $d\mathbf{z} = dz_1, dz_2, \ldots, dz_N$. We now choose our decision surface so as to minimize the average risk $\bar{C}$. Following a procedure similar to the one for the case of a single observation, we can easily show that the test requires the computation of the likelihood ratio

$$\Lambda(\mathbf{z}) = \frac{p(\mathbf{z}|H_1)}{p(\mathbf{z}|H_0)} = \frac{p(z_1, \ldots, z_N|H_1)}{p(z_1, \ldots, z_N|H_0)} \tag{3.4.3}$$

which is then compared with the threshold η in the usual manner, so that the decision rule is

$$\Lambda(\mathbf{z}) \underset{H_0}{\overset{H_1}{\gtrless}} \eta \tag{3.4.4}$$

If the costs are specified, but the prior probabilities of the two hypothesis are unknown, we may use the minimax criterion. For each value of $P(H_1)$, we compute the minimum average risk $\bar{C}_{\min}(P(H_1))$. We then determine the probability P_1^* at which $\bar{C}_{\min}$ is maximum. The minimax strategy will then correspond to the Bayes rule assuming $P(H_1) = P_1^*$ and the minimax cost is equal to $\bar{C}_{\min}(P_1^*)$. The error probabilities are given by

$$P_F = \int_{Z_1} p(\mathbf{z}|H_0)\, d\mathbf{z} = \int_\eta^\infty p(\Lambda(\mathbf{z})|H_0)\, d\Lambda \tag{3.4.5}$$

and

$$P_M = \int_{Z_0} p(\mathbf{z}|H_1)\, d\mathbf{z} = \int_0^\eta p(\Lambda(\mathbf{z})|H_1)\, d\Lambda \tag{3.4.6}$$

so that the minimax threshold can be obtained analogous to Eq. (3.3.22) as

$$C_{00}(1 - P_F) + C_{10}P_F = C_{01}P_M + C_{11}(1 - P_M) \tag{3.4.7}$$

In using the Neyman–Pearson criterion, we again specify an acceptable value for the false alarm probability P_F and seek to minimize the miss probability P_M. Use of a procedure similar to that for the single observation case then yields the optimum decision rule as

$$\Lambda(\mathbf{z}) \underset{H_0}{\overset{H_1}{\gtrless}} \eta \tag{3.4.8}$$

where η is determined from the specified value α of P_F

$$\int_{\eta}^{\infty} p(\Lambda(\mathbf{z})|H_0)\, d\mathbf{z} = \alpha \tag{3.4.9}$$

Again as discussed in the previous section, the receiver operating characteristic can be used to advantage to evaluate the performance of the various tests.

We illustrate the material in this section by means of two examples.

Example 3.4.1 We assume we have N independent observations z_i, $i = 1, 2, \ldots, N$ of the received signal. The observations under each hypothesis are Gaussian independent, identically distributed random variables. Under H_1, we assume that the variables have mean m and variance σ^2 while under H_0, they are zero mean with variance σ^2. We will design a decision rule assuming the minimum probability of error criterion.

For each i, we can write

$$p(z_i|H_1) = \frac{1}{\sqrt{2\pi}\,\sigma} \exp\left(-\frac{(z_i - m)^2}{2\sigma^2}\right)$$

$$p(z_i|H_0) = \frac{1}{\sqrt{2\pi}\,\sigma} \exp\left(-\frac{z_i^2}{2\sigma^2}\right)$$

Since the z_i are assumed independent, their joint density function under each hypothesis is

$$p(\mathbf{z}|H_j) = \prod_{j=1}^{N} p(z_i|H_j) \qquad j = 0, 1$$

In terms of the vectors

$$\mathbf{z} = [z_1, z_2, \ldots, z_N]^T$$

and

$$\mathbf{m} = [m, m, \ldots, m]^T$$

we can write the joint density functions as

$$p(\mathbf{z}|H_1) = \frac{1}{(2\pi\sigma^2)^{N/2}} \exp\left\{-\frac{1}{2\sigma^2}(\mathbf{z}-\mathbf{m})^T(\mathbf{z}-\mathbf{m})\right\}$$

and

$$p(\mathbf{z}|H_0) = \frac{1}{(2\pi\sigma^2)^{N/2}} \exp\left\{-\frac{1}{2\sigma^2}\mathbf{z}^T\mathbf{z}\right\}$$

so that the likelihood ratio is

$$\Lambda(\mathbf{z}) = \frac{p(\mathbf{z}|H_1)}{p(\mathbf{z}|H_0)} = \exp\left\{\frac{1}{\sigma^2}\left(\mathbf{m}^T\mathbf{z} - \frac{1}{2}\mathbf{m}^T\mathbf{m}\right)\right\}$$

The decision rule is

$$\exp\left\{\frac{1}{\sigma^2}\left(\mathbf{m}^T\mathbf{z} - \frac{1}{2}\mathbf{m}^T\mathbf{m}\right)\right\} \underset{H_0}{\overset{H_1}{\gtrless}} \eta$$

where the threshold η will be unity for the minimum probability of error criterion. Taking logarithms on both sides yields the more convenient form

$$\mathbf{m}^T\mathbf{z} \underset{H_0}{\overset{H_1}{\gtrless}} \frac{1}{2}\mathbf{m}^T\mathbf{m}$$

or explicitly

$$\sum_{i=1}^{N} z_i \underset{H_0}{\overset{H_1}{\gtrless}} \frac{Nm}{2}$$

which can be rearranged as

$$\frac{1}{N}\sum_{i=1}^{N} z_i \underset{H_0}{\overset{H_1}{\gtrless}} \frac{m}{2} \triangleq \gamma$$

The test statistic in this case is $\frac{1}{N}\sum_{i=1}^{N} z_i$ which is the sample mean $\bar{z}$ of the observations. The receiver operation therefore consists in determining the sample mean of the observations and comparing it with the threshold γ. If the sample mean exceeds the threshold, we decide that H_1 is the true hypothesis; otherwise, we decide H_0.

We can easily determine P_F and P_D for this case from

$$P_F = \int_{\gamma}^{\infty} p(\bar{z}|H_0)\, d\bar{z}$$

and

$$P_D = \int_{\gamma}^{\infty} p(\bar{z}|H_1)\, d\bar{z}$$

Now $\bar{z}$ is Gaussian under both hypotheses. Its mean and variance under H_1 are m and σ^2/N, respectively, whereas under H_0, it is zero mean but with the same variance σ^2/N. It therefore follows that

$$P_D = \operatorname{erfc}_*\left(\frac{\gamma - m}{\sigma/\sqrt{N}}\right) \quad \text{and} \quad P_F = \operatorname{erfc}_*\left(\frac{\gamma}{\sigma/\sqrt{N}}\right)$$

Example 3.4.2 We consider the same problem as in Example 3.4.1 but now assume that the observations under the two hypotheses have mean zero but different variances σ_1^2 and σ_0^2, respectively, with $\sigma_1^2 > \sigma_0^2$. It can easily be verified that the conditional densities are given by

$$p(\mathbf{z}|H_1) = \frac{1}{(2\pi\sigma_1^2)^{N/2}} \exp\left\{-\frac{1}{2\sigma_1^2}\mathbf{z}^T\mathbf{z}\right\}$$

and

$$p(\mathbf{z}|H_0) = \frac{1}{(2\pi\sigma_0^2)^{N/2}} \exp\left\{-\frac{1}{2\sigma_0^2}\mathbf{z}^T\mathbf{z}\right\}$$

The likelihood ratio is

$$\Lambda(\mathbf{z}) = \left(\frac{\sigma_0^2}{\sigma_1^2}\right)^{N/2} \exp\left\{\mathbf{z}^T\mathbf{z}\left(\frac{\sigma_1^2 - \sigma_0^2}{2\sigma_0^2\sigma_1^2}\right)\right\}$$

and the decision rule becomes

$$\left(\frac{\sigma_0^2}{\sigma_1^2}\right)^{N/2} \exp\left\{\mathbf{z}^T\mathbf{z}\left(\frac{\sigma_1^2 - \sigma_0^2}{2\sigma_0^2\sigma_1^2}\right)\right\} \underset{H_0}{\overset{H_1}{\gtrless}} 1$$

We can arrive at a simple form for the rule by taking logarithms on both sides and rearranging terms:

$$\mathbf{z}^T\mathbf{z} \underset{H_0}{\overset{H_1}{\gtrless}} \frac{2\sigma_0^2\sigma_1^2}{\sigma_1^2 - \sigma_0^2} N \ln\left(\frac{\sigma_1}{\sigma_0}\right) \triangleq \gamma$$

We see that the test statistic in this case is $l(\mathbf{z}) = \mathbf{z}^T\mathbf{z} = \sum_{i=1}^{N} z_i^2$. Under both hypotheses, it is the sum of the squares of N independent, identically distributed Gaussian random variables. Thus $l(\mathbf{z})$ is the χ^2 (chi square) statistics of these variables. We can therefore write down the density function for $l(\mathbf{z})$ under the two hypotheses as

$$p(l|H_i) = \begin{cases} \dfrac{l^{\left(\frac{N}{2}-1\right)}}{2^{N/2}\sigma_i^N\Gamma(N/2)} \exp\left(-l/2\sigma_i^2\right) & l \geqslant 0,\ i = 0, 1 \\ 0 & \text{Otherwise} \end{cases}$$

The expressions for P_D and P_F are given by

$$P_D = \int_\gamma^\infty \left[2^{N/2}\sigma_1^N\Gamma(N/2)\right]^{-1} l^{\left(\frac{N}{2}-1\right)} \exp\left(-l/2\sigma_1^2\right) dl$$

and

$$P_F = \int_\gamma^\infty \left[2^{N/2}\sigma_0^N\Gamma(N/2)\right]^{-1} l^{\left(\frac{N}{2}-1\right)} \exp\left(-l/2\sigma_0^2\right) dl$$

By making the change of variables $M = (\frac{N}{2} - 1)$ and $\gamma_i = \frac{\gamma}{2\sigma_i^2}$, $i = 0, 1$, we can express P_D and P_F in terms of the incomplete gamma function

$$I(u, M) \triangleq \int_0^{u\sqrt{M+1}} \frac{l^M}{M!} \exp(-l)\, dl$$

which has been tabulated [8]. We then obtain

$$P_D = 1 - I\left(\frac{\gamma_1}{\sqrt{M+1}}, M\right)$$

and

$$P_F = 1 - I\left(\frac{\gamma_0}{\sqrt{M+1}}, M\right)$$

In both the preceding examples, the observation space was N-dimensional. However the solution to the decision problem could be obtained explicitly in terms of the test (sufficient) statistic $l(\mathbf{z})$, which is a scalar quantity. We may think of this as choosing a new coordinate system for the observation space in which our decision is based on only one coordinate $l(\mathbf{z})$, while the other N-1 coordinates do not affect our decision and hence can be discarded. If we represent a point in the observation space by the vector $[l : \mathbf{y}^T]^T$ where $\mathbf{y}$ is an N-1 dimensional vector, we have

$$\Lambda(\mathbf{z}) = \Lambda(l, \mathbf{y}) = \frac{p(l, \mathbf{y}|H_1)}{p(l, \mathbf{y}|H_0)} = \frac{p(l|H_1)p(\mathbf{y}|l, H_1)}{p(l|H_0)p(\mathbf{y}|l, H_0)}$$

Since this is also equal to $\Lambda(l) = p(l|H_1)/p(l|H_0)$, we must have

$$p(\mathbf{y}|l, H_1) = p(\mathbf{y}|l, H_0)$$

3.5 MULTIPLE HYPOTHESES TESTING [2,6]

We have so far considered the problem of deciding between two hypotheses. In many situations the source has several outputs and we will have to decide which of the several hypotheses corresponding to these outputs is the correct one. With M outputs, there are a total of M alternatives that may occur during each experiment. We apply the Bayes criterion for this problem exactly as we did for the simple binary testing problem. We thus assume that the prior probabilities $P(H_0), P(H_1), \ldots, P(H_{M-1})$ are known. We associate a cost C_{ij} with choosing H_i when the correct hypothesis is H_j.

The average cost is then

$$\bar{C} = \sum_{i=0}^{M-1} \sum_{j=0}^{M-1} C_{ij} P(D_i|H_j) P(H_j) \tag{3.5.1}$$

where $P(D_i|H_j)$ is the probability of choosing hypothesis H_i when H_j is true. We again divide the observation space Z into several mutually exclusive subspaces $Z_0, Z_1, \ldots, Z_{M-1}$ by assigning the points in each subspace to the corresponding hypothesis. The objective is to find the decision surfaces so that the average cost $\bar{C}$ is minimized. We thus write

$$Z = Z_0 + Z_1 + \cdots + Z_{M-1} \underline{\Delta} Z_0 \cup Z_1 \cup \cdots \cup Z_{M-1} \tag{3.5.2}$$

In terms of these decision regions, the probability $P(D_i|H_j)$ can be written as

$$P(D_i|H_j) = \int_{Z_i} p(z|H_j)\, dz \tag{3.5.3}$$

Substitution of Eq. (3.5.3) in Eq. (3.5.1) yields

$$\begin{aligned} \bar{C} &= \sum_{i=0}^{M-1} \sum_{j=0}^{M-1} C_{ij} P(H_j) \int_{Z_i} p(z|H_j)\, dz \\ &= \sum_{i=0}^{M-1} C_{ii} P(H_i) \int_{Z_i} p(z|H_i)\, dz \\ &\quad + \sum_{i=0}^{M-1} \sum_{\substack{j=0 \\ j\neq i}}^{M-1} C_{ij} P(H_j) \int_{Z_i} p(z|H_j)\, dz \end{aligned} \tag{3.5.4}$$

We can put this in a more convenient form by noting that in the first summation Z_i can be written as $Z - \bigcup_{\substack{j=0 \\ j\neq i}}^{M-1} Z_j$. Then, since $\int_Z p(z|H_i)\, dz = 1$, we obtain for the average cost

$$\bar{C} = \sum_{i=0}^{M-1} C_{ii} P(H_i) + \sum_{i=0}^{M-1} \int_{Z_i} \sum_{\substack{j=0 \\ j\neq i}}^{M-1} P(H_j)(C_{ij} - C_{jj}) p(z|H_j)\, dz \tag{3.5.5}$$

As before, the first term represents the fixed cost and the integrals represent the variable cost that depends on our choice of the decision regions. Clearly we assign each observation z to the region in which the value of the integral is the smallest. Thus if we let

$$I_i(z) = \sum_{\substack{j=0 \\ j \neq i}}^{M-1} P(H_j)(C_{ij} - C_{jj})p(z|H_j) \tag{3.5.6}$$

our decision rule would be to choose that hypothesis which corresponds to the minimum value of $I_i(z)$ as the correct one.

We can rewrite the decision rule in terms of the likelihood ratios by defining

$$\Lambda_i(z) = \frac{p(z|H_i)}{p(z|H_0)}, \; i = 1 \ldots M-1 \tag{3.5.7}$$

and

$$J_i(z) = \frac{I_i(z)}{p(z|H_0)} = \sum_{\substack{j=1 \\ j \neq i}}^{M-1} P(H_j)(C_{ij} - C_{jj})\Lambda_j(z) \tag{3.5.8}$$

The decision rule is then to choose that hypothesis for which $J_i(z)$ is minimum. It can be seen that the decision rules correspond to hyperplanes in the $(M-1)$-dimensional space of the likelihood ratios $\Lambda_1(z) \ldots \Lambda_{M-1}(z)$.

Of particular interest is the case where $C_{ij} = 1$ for $i \neq j$ and $C_{ii} = 0$. The criterion in this case is the minimum probability of error. Then Eq. (3.5.6) reduces to

$$\begin{aligned} I_i(z) &= \sum_{\substack{j=0 \\ j \neq i}}^{M-1} P(H_j)p(z|H_j) = \sum_{\substack{j=0 \\ j \neq i}}^{M-1} P(H_j|z)p(z) \\ &= [1 - P(H_i|z)]\,p(z) \end{aligned} \tag{3.5.9}$$

For this particular assignment of costs, therefore, minimizing $I_i(z)$ is equivalent to maximizing $P(H_i|z)$. This is the posterior probability of hypothesis H_i given the observations z and the processor for the minimum probability of error is the maximum a posteriori probability computer.

An obvious extension is to the case when all the hypotheses are equally likely

$$P(H_1) = P(H_2) = \cdots = P$$

In this case Eq. (3.5.5) can be written

$$I_i(z) = \sum_{\substack{j=1 \\ j\neq i}}^{M-1} p(z|H_j)P = [1 - p(z|H_i)]P \tag{3.5.10}$$

The decision rule is then to choose the hypothesis H_i for which $p(z|H_i)$ is maximum. An example where this occurs is in a digital communication system using M-ary alphabets where each signal is transmitted with equal probability.

Example 3.5.1 Based on sampled observations of a signal, we are required to decide the correct hypothesis out of three possibilities ($M = 3$). Under each hypothesis H_i, the received signal is normally distributed with variance σ^2 and mean m_i. The means m_i are as follows:

$$m_1 = 1 \qquad m_2 = 2 \qquad \text{and} \qquad m_3 = -1$$

The probabilities of occurrence of the hypotheses are equal. Further, we assume minimum probability of error criterion (Bayes criterion with $C_{ij} = 1$ for $i \neq j$ and $C_{ij} = 0$ for $i = j$) to make our decision.

From our previous discussions, we see that the decision rule is to choose the hypothesis corresponding to the maximum of $p(\mathbf{z}|H_i)$, $i = 1, 2, 3$. If there are N independent sampled observations, we can write $p(\mathbf{z}|H_i)$ as

$$p(\mathbf{z}|H_i) = \frac{1}{(2\pi\sigma^2)^{N/2}} \exp\left[-\frac{1}{2}\sum_{j=1}^{N}\frac{(z_j - m_i)^2}{\sigma^2}\right], \qquad i = 1, 2, 3$$

The preceding density function can be rewritten as

$$p(\mathbf{z}|H_i) = \frac{1}{(2\pi\sigma^2)^{N/2}} \exp\left[-\frac{1}{2\sigma^2}\sum_{j=1}^{N} z_j^2\right]$$

$$\cdot \exp\left[\frac{1}{2\sigma^2}\sum_{j=1}^{N}(2z_j m_i - m_i^2)\right], \qquad i = 1, 2, 3$$

It is obvious that we choose that hypothesis H_i for which

$$\sum_{j=1}^{N}(2z_j m_i - m_i^2) = \left(\frac{2}{N}\sum_{j=1}^{N} z_j m_i\right) - m_i^2$$

is maximum. That is, we compute the following quantities

$$\left(\frac{2}{N}\sum_{j=1}^{N} z_j\right) - 1, \qquad \left(\frac{4}{N}\sum_{j=1}^{N} z_j\right) - 4, \qquad \text{and} \qquad \left(\frac{-2}{N}\sum_{j=1}^{N} z_j\right) - 1$$

based on the sampled observations, $\mathbf{z}$. If the second quantity exceeds the other two, then hypothesis H_2 is chosen, and so on. It is left as a simple exercise for the reader to show that the test statistic is the sample mean $\hat{m} = \frac{1}{N}\sum_{j=1}^{N} z_j$, of the received signal, and the decision regions are as follows:

$$\begin{aligned} \hat{m} &\geq 3/2, && \text{choose } H_2 \\ 0 \leq \hat{m} &< 3/2, && \text{choose } H_1 \\ \hat{m} &< 0, && \text{choose } H_3 \end{aligned}$$

3.6 COMPOSITE HYPOTHESIS TESTING [2,4]

In the previous sections, we have considered the case where the parameter characterizing a hypothesis assumed only a known single value. In many situations however, this parameter can take an unknown value on a range of values. For example, if the signal is a sine wave, its phase could be an unknown which lies between 0 and 2π. If the signal is a realization of a stochastic process, its mean or variance could lie on a range of values. Such hypotheses are called *composite hypotheses* as distinct from the simple hypothesis in which the characterizing parameter is fixed, and known. We confine our discussion in this section to the two hypotheses case for reasons of clarity. The extensions of the ideas presented to the multiple hypotheses case is quite straightforward.

We can view the set of parameters characterizing a hypothesis as a vector $\boldsymbol{\theta}$ in parameter space $\chi_{\boldsymbol{\theta}}$. The probability density governing the mapping from the parameter space to the observation space, $p(\mathbf{z}|\boldsymbol{\theta})$ is assumed known for all values of $\boldsymbol{\theta}$. Depending on how we characterize $\boldsymbol{\theta}$, we can proceed to derive rules for deciding between various hypotheses.

We may characterize $\boldsymbol{\theta}$ either as a random variable or as an unknown constant. If $\boldsymbol{\theta}$ is a random variable with a known probability density, the extension of the LRTs to the composite hypotheses case is quite straightforward. We first determine the conditional likelihood ratio $\Lambda(\mathbf{z}|\boldsymbol{\theta})$ assuming that $\boldsymbol{\theta}$ is given. The likelihood ratio can then be obtained as

$$\Lambda(\mathbf{z}) = \int_{\chi_{\boldsymbol{\theta}}} \Lambda(\mathbf{z}|\boldsymbol{\theta})p(\boldsymbol{\theta})\, d\boldsymbol{\theta} \tag{3.6.1}$$

The known density on $\boldsymbol{\theta}$ has enabled us to reduce the problem to a simple hypothesis-testing problem by integrating over $\boldsymbol{\theta}$.

When $\boldsymbol{\theta}$ is a random variable with an unknown density, the test procedure to be used is not clearly specified. In this case we might adopt the strategy of determining a least favorable density function, if one exists, and proceed accordingly. We may find the least favorable distribution by determining the average probability of choosing H_1 correctly, that is, $\mathrm{E}\{P(D_1|H_1)\}$ with respect to a particular distribution on $\boldsymbol{\theta}$ and then choosing the distribution to minimize this average probability. An alternate approach would be to try several densities based on any partial knowledge of $\boldsymbol{\theta}$ that may be available. In many cases the test structure turns out to be insensitive to the detailed behavior of the probability density.

If $\boldsymbol{\theta}$ is a nonrandom variable, a Bayes test is not meaningful since $\boldsymbol{\theta}$ has no probability density over which we can average $p(z|H_i, \boldsymbol{\theta})$. We therefore use a Neyman–Pearson test. For a given value of the parameter $\boldsymbol{\theta}$, we seek to maximize $P(D_1|H_1)$ subject to the constraint that $P(D_1|H_0)$ is a specified value. The probability $P(D_1|H_1)$ may depend on $\boldsymbol{\theta}$. If the test maximizes $P(D_1|H_1)$ independent of the value of $\boldsymbol{\theta}$, then the optimum test can be carried out without regard to the value of $\boldsymbol{\theta}$. We refer to such a test as a *uniformly most powerful* (UMP) test.

If a UMP test does not exist, we may try estimating the values of $\boldsymbol{\theta}$ under the two hypotheses H_1 and H_0 and use these estimates as the true values in a LRT. For nonrandom parameters, the usual estimate is the *maximum likelihood estimate*† obtained by maximizing the conditional density function $p(\mathbf{z}|\boldsymbol{\theta})$. We thus construct a "generalized likelihood ratio" and write the test as

$$\Lambda_g(z) = \frac{\max\limits_{\boldsymbol{\theta}_1} p(z|\boldsymbol{\theta}_1)}{\max\limits_{\boldsymbol{\theta}_0} p(z|\boldsymbol{\theta}_0)} \underset{H_0}{\overset{H_1}{\gtrless}} \eta \tag{3.6.2}$$

Example 3.6.1 The signals under the two hypotheses are Gaussian with variance σ^2. Under the null hypothesis, the mean is assumed to be zero, whereas under hypothesis H_1 the mean is a constant ($m_0 \leqslant m \leqslant m_1$). For a

†This estimate is defined in Chapter 5.

single observation the conditional densities are given by

$$p(z|H_1, m) = \frac{1}{\sqrt{2\pi}\,\sigma} \exp\left(-\frac{(z-m)^2}{2\sigma^2}\right) \qquad m_0 \leqslant m \leqslant m_1$$

and

$$p(z|H_0) = \frac{1}{\sqrt{2\pi}\,\sigma} \exp\left(-\frac{z^2}{2\sigma^2}\right)$$

For fixed m the likelihood ratio is

$$\Lambda(z) = \exp\left[\frac{mz - \dfrac{m^2}{2}}{\sigma^2}\right]$$

Using the log-likelihood ratio yields the decision rule as

$$L(z) = mz - \frac{m^2}{2} \underset{H_0}{\overset{H_1}{\gtrless}} \gamma$$

or

$$mz \underset{H_0}{\overset{H_1}{\gtrless}} \gamma^1 \triangleq \gamma + \frac{m^2}{2}$$

If $m_0 \geqslant 0$, then m can take on only nonnegative values and the test can be written as

$$z \underset{H_0}{\overset{H_1}{\gtrless}} \gamma^+ = \frac{\gamma}{m} + \frac{m}{2}$$

For a Neyman–Pearson test with P_F equal to some specified value α, we determine the threshold γ^+ from

$$P_F = \int_{\gamma^+}^{\infty} \frac{1}{\sqrt{2\pi}\,\sigma} \exp\left(\frac{-z^2}{2\sigma^2}\right) dz = \alpha$$

Note that the threshold γ^+ may be negative. If $m_1 \leqslant 0$, m takes on only nonpositive values and the test becomes

$$z \underset{H_0}{\overset{H_1}{\gtrless}} \gamma^- = \frac{\gamma}{m} + \frac{m}{2}$$

In this case, the threshold γ^- is determined from

$$P_F = \int_{-\infty}^{\gamma^-} \frac{1}{\sqrt{2\pi}\,\sigma} \exp\left(\frac{-z^2}{2\sigma^2}\right) dz = \alpha$$

For either of the two possibilities preceding, we see that the test (including the threshold) can be completely designed without any knowledge of the true value of m. Thus a UMP test exists in both cases. If however, m can take on both positive and negative values, a UMP test does not exist. In a situation such as this, if we were to determine the decision rule assuming that the mean is positive (negative) whereas it actually is negative (positive) very poor performance results. To demonstrate this, assume that the error of the first kind $P(D_1|H_0)$ has been fixed at some value α and the test has been designed as though the mean were positive. The situations resulting when the true mean is either positive or negative is shown in Fig. 3.6.1.

The resulting $P(D_1|H_1)$ as a function of the true value of the mean is shown in Fig. 3.6.2. Similar results hold when the test is designed assuming the mean is negative.

Note that when the test is designed assuming $m > 0$ ($m < 0$) for negative (positive) values of m, P_D is less than P_F since the threshold is on the wrong side.

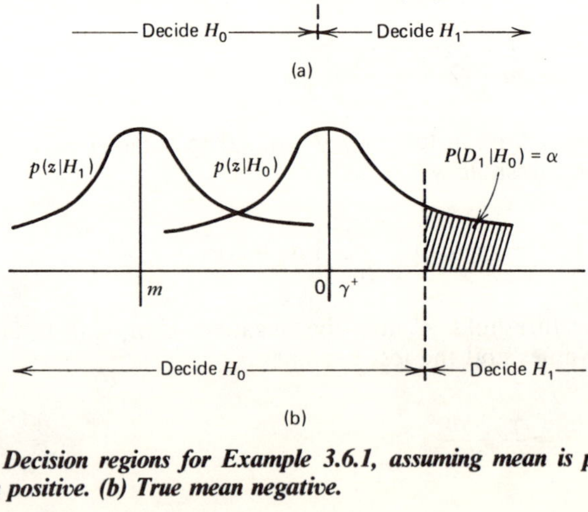

Fig. 3.6.1 Decision regions for Example 3.6.1, assuming mean is positive (a) True mean positive. (b) True mean negative.

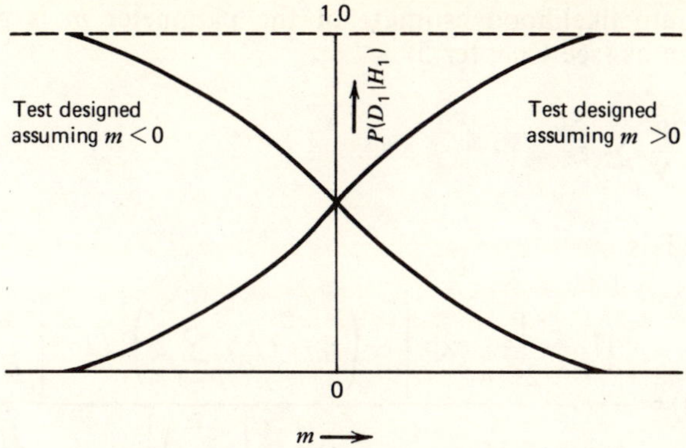

Fig. 3.6.2 Performance of tests for Example 3.6.1, $P(D_1|H_0)$ fixed.

Since a UMP test does not exist in the case when m can take on either positive or negative values, a reasonable choice in this case would be to use a two-sided test as shown in Fig. 3.6.3. The decision rule is to choose H_1 if $|z|>\gamma$.

This test obviously will not perform as well as the one-sided tests discussed earlier with correct assumptions but is much better in the case of incorrect assumptions.

Example 3.6.2 We will determine a generalized LRT test for the previous example. Assume that N independent observations are available. The relevant probability densities are

$$p(\mathbf{z}|m, H_1) = \prod_{i=1}^{N} \frac{1}{\sqrt{2\pi}\,\sigma} \exp\left[-\frac{(z_i - m)^2}{2\sigma^2}\right]$$

$$p(\mathbf{z}|H_0) = \prod_{i=1}^{N} \frac{1}{\sqrt{2\pi}\,\sigma} \exp\left[-\frac{z_i^2}{2\sigma^2}\right]$$

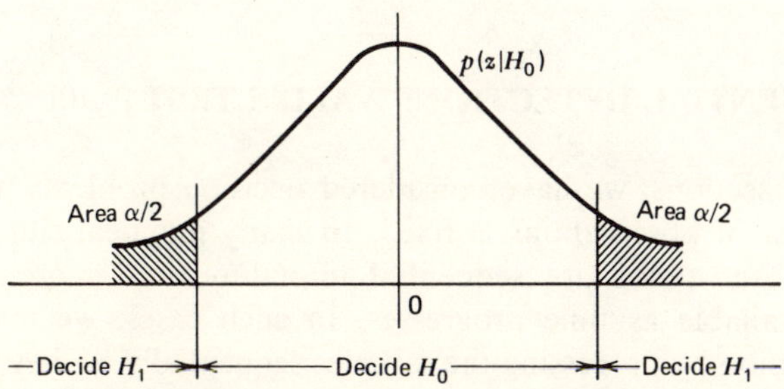

Fig. 3.6.3 Two-sided test for Example 3.6.1.

The maximum likelihood estimate of the parameter m is given by the sample mean as (see Chapter 5)

$$\hat{m} = \frac{1}{N} \sum_{i=1}^{N} z_i$$

and the LRT is

$$\Lambda_g(\mathbf{z}) = \frac{\prod_{i=1}^{N} \frac{1}{\sqrt{2\pi}\,\sigma} \exp\left[-\left(z_i - 1/N \sum_{i=1}^{N} z_i \right)^2 / 2\sigma^2 \right]}{\prod_{i=1}^{N} \frac{1}{\sqrt{2\pi}\,\sigma} \exp\left[-\left(\frac{z_i^2}{2\sigma^2} \right) \right]} \underset{H_0}{\overset{H_1}{\gtrless}} \gamma$$

Cancelling common terms and taking the logarithm yields

$$\frac{1}{2\sigma^2 N} \left(\sum_{i=1}^{N} z_i \right)^2 \underset{H_0}{\overset{H_1}{\gtrless}} \ln \gamma$$

An equivalent test is to choose

$$\left(\frac{1}{\sqrt{N}} \sum_{i=1}^{N} z_i \right)^2 \underset{H_0}{\overset{H_1}{\gtrless}} \gamma_1^2 \triangleq 2\sigma^2 \ln \gamma$$

That is

$$\left| \frac{1}{\sqrt{N}} \sum_{i=1}^{N} z_i \right| \underset{H_0}{\overset{H_1}{\gtrless}} \gamma_1$$

which is just the two-sided test discussed in the previous example.

3.7 SEQUENTIAL DETECTION: WALD'S TEST [9,10]

In previous sections, we have considered decision problems in which the total number of observations is fixed. In many practical situations, however, the observations are sequential in nature, and more information becomes available as time progresses. In such cases, we may desire to make a decision by processing the samples sequentially as they occur. Even though the Bayes procedure may be used in this case, the resulting test is fairly complicated. Instead, we focus our attention on a modified Neyman–Pearson type of test, referred to as Wald's sequential test or sequential probability ratio test (SPRT). These tests have found application in a number of areas such as pattern recognition, and radar [9,10]. For purposes of clarity, we again restrict our discussion to the simple binary hypothesis testing problem of Section 3.3.

We recall that in our discussion of the Neyman–Pearson rule in Section 3.3, we computed the likelihood ratio and compared it with a threshold η which was determined by the specified value of P_F. In the SPRT, at each stage of the decision procedure, the likelihood ratio is compared with two thresholds η_0 and η_1 which are determined by the specified values of the error probabilities P_F and P_M. If the likelihood ratio is greater than or equal to η_1 we decide in favor of H_1. If it is less than η_0, we decide that H_0 was the correct hypothesis. If its value lies between η_0 and η_1, we conclude that we do not have enough information to make a decision between the two hypotheses with the prescribed levels of P_F and P_M and hence take another observation.

The key problem in the design of SPRT is to determine the thresholds η_1 and η_0 in terms of a specified P_F and P_M. We develop the necessary relations in this section for determining these thresholds. We demonstrate that for a given level of performance (that is, for given values of P_F and P_M) this test will terminate, on the average, in fewer observation samples than the fixed sample size tests [9]. This can be claimed as the principal virtue of the method because of the costs involved in making additional measurements.

Let $z_1, z_2, \ldots, z_j, \ldots$ represent the observation samples and let $\mathbf{z}_N$ be the vector defined as

$$\mathbf{z}_N = [z_1, z_2, \ldots, z_N]^T$$

The likelihood ratio based on N samples is then given by

$$\Lambda(\mathbf{z}_N) = \frac{p(\mathbf{z}_N|H_1)}{p(\mathbf{z}_N|H_0)} \tag{3.7.1}$$

Thus, in general, in computing the likelihood ratio $\Lambda(\mathbf{z}_j)$ at each stage (that is for each value of j), we need to know the joint density functions of the samples $z_1, z_2, \ldots, z_j$ up to that stage. If however, these samples are independent, we can easily compute $\Lambda(\mathbf{z}_j)$ recursively as follows:

$$\Lambda(\mathbf{z}_j) = \frac{p(\mathbf{z}_j|H_1)}{p(\mathbf{z}_j|H_0)} = \prod_{i=1}^{j} \frac{p(z_i|H_1)}{p(z_i|H_0)} = \frac{p(z_j|H_1)}{p(z_j|H_0)} \prod_{i=1}^{j-1} \frac{p(z_i|H_1)}{p(z_i|H_0)}$$

That is

$$\Lambda(\mathbf{z}_j) = \Lambda(z_j)\Lambda(\mathbf{z}_{j-1}) \tag{3.7.2}$$

with the initial condition

$$\Lambda(\mathbf{z}_1) = \Lambda(z_1)$$

Let us assume that the error probabilities are specified as

$$P_F = \alpha \qquad \text{and} \qquad P_M = \beta \tag{3.7.3}$$

We now derive a pair of inequalities for the thresholds η_0 and η_1 in terms of α and β. We recall from Eqs. (3.3.14) and (3.3.16) that the false alarm and detection probabilities are given by

$$P_F = P(D_1|H_0) = \int_{Z_1} p(\mathbf{z}_j|H_0)\, d\mathbf{z}_j \tag{3.7.4}$$

and

$$P_D = P(D_1|H_1) = \int_{Z_1} p(\mathbf{z}_j|H_1)\, d\mathbf{z}_j$$

$$= \int_{Z_1} \Lambda(\mathbf{z}_j) p(\mathbf{z}_j|H_0)\, d\mathbf{z}_j \tag{3.7.5}$$

where Z_1 represents the decision region for hypothesis H_1. When H_1 is the true hypothesis and the decision is also H_1, we must have $\Lambda(\mathbf{z}_j) \geqslant \eta_1$, so that Eq. (3.7.5) can be rewritten as

$$P_D = P(D_1|H_1) \geqslant \eta_1 \int_{Z_1} p(\mathbf{z}_j|H_0)\, d\mathbf{z}_j \tag{3.7.6}$$

From Eq. (3.7.4), the integral on the right-hand side represents P_F which is assumed to have a value α. Since $P_D = 1 - P_M = 1 - \beta$, Eq. (3.7.6) yields the inequality

$$1 - \beta \geqslant \eta_1 \alpha$$

so that the threshold η_1 satisfies

$$\eta_1 \leqslant \frac{1-\beta}{\alpha} \tag{3.7.7}$$

Using a similar argument, it follows that

$$\eta_0 \geqslant \frac{\beta}{1-\alpha} \tag{3.7.8}$$

Often we use the log-likelihood ratio because of its computational advantages. Taking the logarithm of Eq. (3.7.2), we obtain the recursive relation for updating $L(\mathbf{z}_j)$ as

$$L(\mathbf{z}_j) = L(\mathbf{z}_{j-1}) + L(z_j) \tag{3.7.9}$$

The corresponding thresholds are given by $\ln \eta_1$ and $\ln \eta_0$. When the increments $L(z_j)$ in Eq. (3.7.9) are small, $L(\mathbf{z}_j)$ will at best exceed the thresholds only by a small quantity. In such a case, it is not unreasonable to treat the inequalities in Eqs. (3.7.7) and (3.7.8) as equalities to obtain the simple design rule:

$$\ln \eta_1 \simeq \ln\left(\frac{1-\beta}{\alpha}\right) \tag{3.7.10}$$

and

$$\ln \eta_0 \simeq \ln\left(\frac{\beta}{1-\alpha}\right) \tag{3.7.11}$$

We now calculate the average number of observations required to make a decision under each hypothesis. To do this let us assume that the test terminates in N samples, which implies that $L(\mathbf{z}_N)$ assumed either of the values $\ln \eta_0$ or $\ln \eta_1$. If the true hypothesis is H_0 and $L(\mathbf{z}_N) \geqslant \ln \eta_1$ we have false alarm with probability α. In this case, the probability that $L(\mathbf{z}_N)$ is greater than or equal to $\ln \eta_1$ will be $1-\alpha$. Similarly if the true hypothesis is H_1 and $L(\mathbf{z}_N) \leqslant \ln \eta_0$, we have miss with probability β, so that the probability that $L(\mathbf{z}_N)$ is less than or equal to $\ln \eta_1$ is $1-\beta$. The expected values of $L(\mathbf{z}_N)$ under the two hypotheses are therefore equal to

$$\mathrm{E}[L(\mathbf{z}_N)|H_1] = (1-\beta)\ln \eta_1 + \beta \ln \eta_0 \tag{3.7.12}$$

and

$$\mathrm{E}[L(\mathbf{z}_N)|H_0] = \alpha \ln \eta_1 + (1-\alpha)\ln \eta_0 \tag{3.7.13}$$

Let us define the binary variable k such that

$$k_n = \begin{cases} 1 \text{ if no decision was made up to } (n-1)\text{st stage} \\ 0 \text{ if a decision was made at an earlier stage.} \end{cases} \tag{3.7.14}$$

Since the test is assumed to terminate after N samples, we can use Eqs. (3.7.9) and (3.7.14) to write

$$L(\mathbf{z}_N) = \sum_{n=1}^{N} L(z_n) = \sum_{n=1}^{\infty} k_n L(z_n) \tag{3.7.15}$$

We note that k_n depends only on $z_1, z_2, \ldots, z_{n-1}$ and not on z_n and is hence independent of $L(z_n)$. Taking expectations on both sides of Eq. (3.7.15) under each hypothesis yields

$$\begin{aligned} \mathrm{E}\{L(\mathbf{z}_N)|H_i\} &= \mathrm{E}\left\{\sum_{n=1}^{\infty} k_n L(z_n)|H_i\right\} \\ &= \sum_{n=1}^{\infty} \mathrm{E}\{k_n\}\mathrm{E}\{L(z_n)|H_i\} \end{aligned} \tag{3.7.16}$$

If we assume that the observations are identically distributed under each hypothesis

$$\mathrm{E}\{L(\mathbf{z}_n)|H_1\} = \mathrm{E}\{L(z)|H_1\} \qquad \text{for all } n \tag{3.7.17}$$

we have

$$\mathrm{E}\{L(\mathbf{z}_N|H_i)\} = \mathrm{E}\{L(z)|H_i\} \sum_{n=1}^{\infty} \mathrm{E}\{k_n\} \tag{3.7.18}$$

It easily follows from our definition of k_n that

$$\mathrm{E}\{k_n\} = P(N \geqslant n) \tag{3.7.19}$$

so that

$$\sum_{n=1}^{\infty} \mathrm{E}\{k_n\} = \sum_{n=1}^{\infty} P(N \geqslant n) = \sum_{n=1}^{\infty} nP(N = n) = \mathrm{E}\{N\} \tag{3.7.20}$$

Substitution of Eqs. (3.7.18) and (3.7.20) in Eq. (3.7.12) yields the average number of samples when H_1 is the correct hypothesis as

$$\mathrm{E}[N|H_1] = \frac{(1-\beta)\ln\eta_1 + \beta\ln\eta_0}{\mathrm{E}[L(z)|H_1]} \tag{3.7.21}$$

Similarly Eq. (3.7.13) yields

$$\mathrm{E}[N|H_0] = \frac{\alpha\ln\eta_1 + (1-\alpha)\ln\eta_0}{\mathrm{E}[L(z)|H_0]} \tag{3.7.22}$$

We can easily show that the test terminates with probability one. That is

$$\lim_{n\to\infty} P(N \geqslant n) = 0 \tag{3.7.23}$$

We note that we take another sample if $\ln \eta_0 < L(z) < \ln \eta_1$. If we let $\delta = \ln \eta_1 - \ln \eta_0$, in general, $P(|L(z_i)| \leqslant \delta) = p < 1$. Thus the probability that the first n observations are all such that $|L(z_i)| < \delta$ is equal to p^n, since the observations are independent. This probability tends to zero as n tends to infinity, thus establishing Eq. (3.7.23).

It has also been shown by Wald and Wolfowitz [11] that for specified values of P_F and P_M, the SPRT minimizes the average number of observations $\mathrm{E}(N|H_1)$ and $\mathrm{E}(N|H_0)$. We omit the proof since it is somewhat involved.

Our development of the sequential method for decision making assumed the following: (i) The samples z_j are independent and identically distributed; (ii) the probabilities P_F and P_D were constant for the entire interval, resulting in constant threshold given by Eqs. (3.7.10) and (3.7.11). But neither of the assumptions is required in general. In the absence of independent samples, we cannot obtain the recursive relations of Eqs. (3.7.1) and (3.7.9). Consequently we must compute $\Lambda(\mathbf{z}_j)$ from the joint density functions $p(\mathbf{z}_j|H_i)$, $i = 0, 1$, and for each j. This, in general, is computationally difficult. Removal of assumption (ii) yields varying thresholds, $\eta_1(j)$ and $\eta_0(j)$, and it is no longer simple to calculate the average number of samples for decision making.

Although the SPRT terminates with probability 1, in many cases it is preferable to fix an upper limit N^*, on the total number of samples we can afford. This is called a truncated SPRT. At the end of N^* samples, a fixed sample size test is performed to obtain a decision. We conclude this section with an example.

Example 3.7.1 We consider the same problem as in Example 3.3.3, but assume that the samples are available in a sequence. From Example 3.3.3, we have the likelihood ratio at the jth stage as

$$\Lambda(\mathbf{z}_j) = \exp\left[\sum_{i=1}^{j} z_i - \frac{j}{2}\right]$$

and the log likelihood ratio is

$$L(\mathbf{z}_j) = \sum_{i=1}^{j} z_i - \frac{j}{2}$$

Let us assume that our acceptable level of performance is:

$$P_F = \alpha = 0.1 \qquad \text{and} \qquad P_M = \beta = 0.1$$

Using Eqs. (3.7.10) and (3.7.11), we obtain the two thresholds as

$$\ln \eta_1 = 2.197 \qquad \text{and} \qquad \ln \eta_0 = -2.197$$

The decision rule, therefore, is

$$L(\mathbf{z}_j) \geqslant 2.197, \text{ decide } H_1$$
$$L(\mathbf{z}_j) \leqslant -2.197, \text{ decide } H_0$$

and

$$-2.197 \leqslant L(\mathbf{z}_j) \leqslant 2.197, \text{ take another sample } z_{j+1}$$

The average number of samples for the test to terminate under each hypothesis is obtained using Eqs. (3.7.21) and (3.7.22). Since

$$\mathrm{E}[L(z)|H_1] = \mathrm{E}\left[\left(z - \frac{1}{2}\right)|H_1\right] = \frac{1}{2}$$

and

$$\mathrm{E}[L(z)|H_0] = \mathrm{E}\left[\left(z - \frac{1}{2}\right)|H_0\right] = -\frac{1}{2}$$

we obtain

$$\mathrm{E}[j|H_1] = 3.5 \text{ samples}$$

and

$$\mathrm{E}[j|H_0] = 3.5 \text{ samples}$$

The preceding results indicate that a sample size of 4 may yield the desired performance.

3.8 SUMMARY

In this chapter, we have developed the basic ideas of statistical hypothesis testing. The components of the problem were introduced in Section 3.2 and the basic decision rule for testing between two hypotheses was developed in Section 3.3. The test involves the computation of the likelihood ratio $\Lambda(z)$ and comparing it with a threshold in order to make a decision. The decision space is thus a one-dimensional space. The threshold of the test depends on the particular criterion chosen. The Bayes test requires the specification of costs and prior probabilities. In the absence of prior probabilities, we can use the minimax test which has its origins in the theory of games [7]. The Neyman–Pearson test can be used when assignment of neither costs nor prior probabilities can be made. We also determined the performance of the tests by computing the probabilities of the two types of error. The performance is most conveniently analyzed by consideration of the receiver operating characteristic which is a plot of detection probability P_D versus false alarm probability P_F.

These concepts were extended to the case of multiple observations in Section 3.4. While the observation space is multidimensional for this problem, the decision space is still one-dimensional. We also introduced the sufficient statistic which in a sense summarizes all the information needed for making a decision. This essentially involves choosing a new coordinate system for the observation space such that the decision can be made entirely in terms of one coordinate. The concept of a sufficient statistic plays a useful role in our discussions in the subsequent chapters.

Section 3.5 considered the M-ary hypothesis testing problem. The decision rule involves the computation of M-1 likelihood ratios from the density functions $p(z|H_i)$. The decision boundaries are hyperplanes in the $(M-1)$-dimensional space of the likelihood ratios. For the special case of minimum probability of error criterion, the test reduces to choosing the hypothesis corresponding to the largest of the posterior probabilities $P(H_i|z)$. While the optimum test is fairly straightforward to determine, the computation of the error probabilities is frequently difficult.

The composite hypothesis-testing problem, in which the parameters θ characterizing the hypotheses are not known, is considered in Section 3.6. The case in which the unknown parameter is random with known probability density is fairly straightforward since the likelihood ratio can be obtained by integrating the conditional likelihood ratio over the parameter space. If the density function for θ is not known, we may try integrating over several assumed densities and pick the least favorable one. Frequently, the test structure is insensitive to the choice of the density function. If θ is a nonrandom variable, we may consider applying the Neyman–Pearson test. This is useful if the test can be carried out without regard to the value of θ, that is, if a UMP test exists. Otherwise we may try estimating the values of θ under the two hypotheses and use these estimates as the true values in a likelihood ratio test. We concluded the chapter with a discussion of the sequential probability ratio test in Section 3.7.

The development of the discussions in this chapter have assumed that the density functions of the observations under each hypothesis is either known or can be determined. We have not attempted to deal with the problems in which the underlying density functions are not known. An approach to solving this class of problems is the *nonparametric* or *distribution-free* approach. The interested reader is referred to the literature for details of this approach (see, for example, [12]).

EXERCISES

3.1 Derive a likelihood ratio test for the following case. Under hypothesis H_1, the observation z is Gaussian with mean m_1 and standard deviation σ_1. Under hypothesis H_0, z is Gaussian with mean m_0 and standard deviation σ_0. Find the decision regions and the probability of error.

3.2 Consider the binary symmetric channel shown in Fig. E.3.1. ϵ is the crossover probability, i.e., probability that the channel output is 0(1) when the input is 1(0), and is small. Obtain a decision rule to minimize the total probability of error assuming equal prior probabilities. What is the probability of error? Assume $\epsilon < 1/2$.

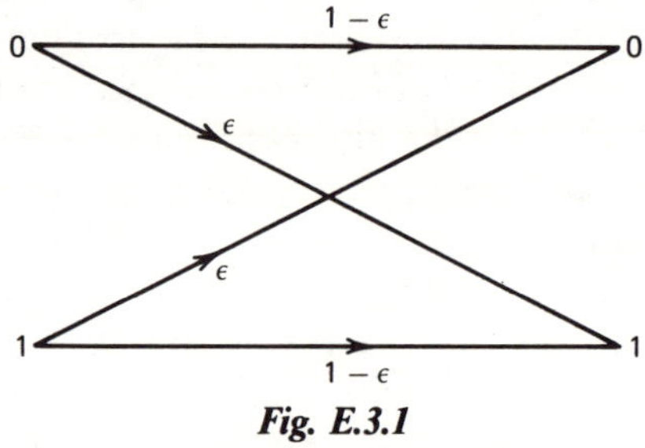

Fig. E.3.1

3.3 The probability density function of the observation z under the two hypotheses are as shown in Fig. E.3.2. Given the prior probabilities, $P(H_1) = 0.7$ and $P(H_0) = 0.3$ design the minimum probability of error test. Plot the ROC for this problem.

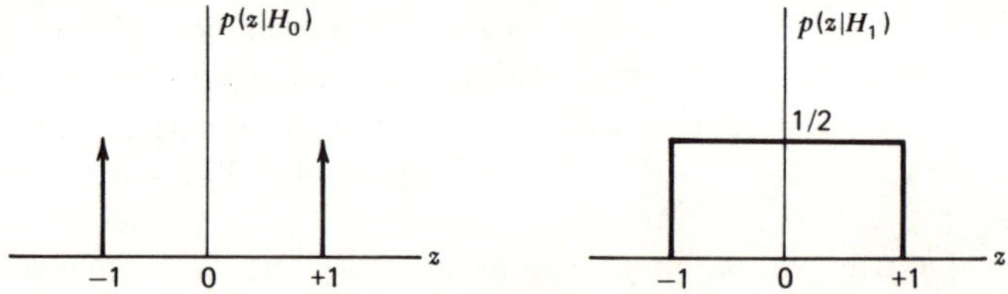

Fig. E.3.2

3.4 Consider the hypothesis testing problem in which

$$p(z|H_1) = \frac{1}{\sqrt{2\pi}} \exp\left(-\frac{z^2}{2}\right)$$

$$p(z|H_0) = \frac{1}{2} \exp\,(-|z|)$$

(a) Set up LRT and determine the decision regions. Plot the ROC.
(b) For $C_{00} = C_{11} = 0$ and $C_{01} = C_{10} = 1$, find P_F and P_D for the Bayes' test if $P(H_1) = 3/4$.
(c) For the same assignment of costs as in (b) find the performance of a minimax test.
(d) Set up the Neyman-Pearson test for $P_F = 0.2$.

3.5 The observed random variable z is Gaussian distributed on the three hypotheses,

$$p(z|H_i) = \frac{1}{\sqrt{2\pi}\,\sigma_i} \exp\left[-\frac{(z - m_i)^2}{2\sigma_i^2}\right] \qquad i = 0, 1, 2$$

where $m_0 = m_2 = 0$, $m_1 = 1$, and $\sigma_0 = \sigma_1 = 1$. Find the decision rules for minimum probability of error criterion assuming equal prior probabilities. What is the probability of error if $\sigma_2 = 2$?

3.6 Let y be a Gaussian random variable with mean m and variance σ^2. The observation under the two hypotheses is

$$H_1 : z = y^2$$

$$H_0 : z = e^y$$

Find the LRT to decide between the two hypotheses.

3.7 In a binary hypothesis-testing problem, the received signals under the two hypotheses are:

$$H_1 : z = x_1^2 + x_2^2$$

$$H_2 : z = x_1$$

where x_1 and x_2 are independent and identically Gaussian distributed with zero mean and unity variance. Obtain the optimum decision rule for Bayes' criterion.

3.8 Obtain a Bayes' criterion decision rule to descriminate between two hypotheses H_0 and H_1 based on the observation z. The distributions of z under the two hypotheses are as shown in Fig. E.3.3.

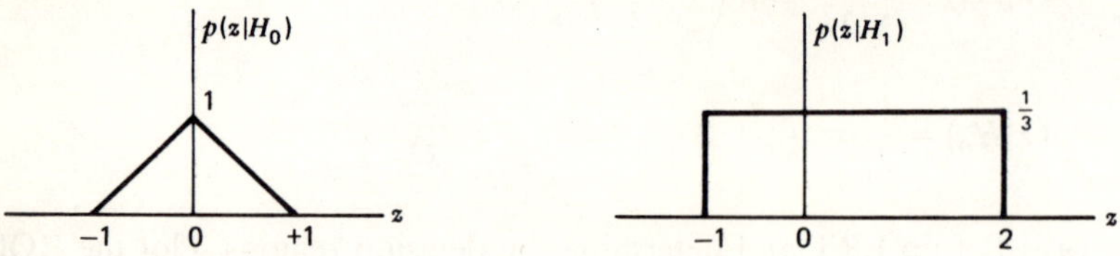

Fig. E.3.3

3.9 The observation z under both hypotheses H_0 and H_1 is Gaussian distributed, i.e.,

$$p(z|H_i) = \frac{1}{\sqrt{2\pi}\,\sigma} \exp\left(-\frac{(z-i)^2}{2\sigma^2}\right) \qquad i = 0, 1$$

The processor has a fixed square law detector, whose output y is such that $y = az^2$. Based on this modified measurement, obtain a decision rule for the minimum probability of error criterion. Assume $P(H_1) = P(H_0) = \frac{1}{2}$.

3.10 In many instances, the density functions of the observation under the two hypotheses are discrete. We can still derive a likelihood ratio test by permitting impulse functions in the densities. Let us assume that the observation under the two hypotheses has a Poisson distribution:

$$P(z = n|H_1) = \frac{m_1^n}{n!} \exp(-m_1) \qquad n = 0, 1, 2 \ldots$$

$$P(z = n|H_0) = \frac{m_0^n}{n!} \exp(-m_0) \qquad n = 0, 1, 2 \ldots$$

with $m_1 > m_0$

(a) Show that the LRT test is

$$z \underset{H_0}{\overset{H_1}{\gtrless}} \frac{\ln \eta + m_1 - m_0}{\ln m_1 - \ln m_0}$$

(b) Because z takes on only integer values, it is more convenient to write the decision rule as

$$z \underset{H_0}{\overset{H_1}{\gtrless}} \gamma^1 \qquad \gamma^1 = 0, 1, 2 \ldots$$

Show that the error probabilities are given by

$$P_F = 1 - \exp(-m_0) \sum_{n=0}^{\gamma^1 - 1} \frac{(m_0)^n}{n!}$$

and

$$P_M = \exp(-m_1) \sum_{n=0}^{\gamma^1 - 1} \frac{(m_1)^n}{n!}$$

Plot the ROC for this problem assuming $m_0 = 1$ and $m_1 = 2$.

3.11 In Exercise 3.10, P_F can assume only certain discrete values. Suppose we want to design a Neyman–Pearson rule for some intermediate value of P_F. It is clear that an ordinary LRT will not give this performance. We will have to use a *randomized decision* rule in which instead of assigning a point $z \in Z_1$ to H_1, we assign it to H_1 with a certain probability $P_a(z)$ and to H_0 with probability $1 - P_a(z)$. We do this by combining the results of two LRTs in a probabilistic manner.

Let us assume that for Exercise 3.10, we want $P_F = 1 - \frac{1}{2}\exp(-m_0)$.

(a) Show that a randomized rule combines two LRTs, one with a threshold $\gamma^1 = 0$ and the other with threshold $\gamma^1 = 1$. Find the test explicitly.

(b) Find P_D.

3.12 In a binary hypothesis testing problem, the observation z is Rayleigh distributed under both hypotheses with different parameters, i.e.,

$$p(z|H_i) = \frac{z}{\sigma_i^2} \exp\left(-\frac{z^2}{2\sigma_i^2}\right) \qquad z \geqslant 0 \qquad i = 0, 1$$

Obtain the likelihood ratio test for Bayes' criterion. Extend your results to N independent observations for the minimum probability of error criterion. Assume $P(H_1) = P(H_0) = \frac{1}{2}$. Derive the expressions for the resulting probability of error.

3.13 In a binary hypothesis-testing problem, the observed samples, z_i, are independent and identically distributed under each hypothesis. The distributions are

$$p(z_i|H_j) = \frac{1}{\sqrt{2\pi}\,\sigma_j} \exp\left[-\frac{(z_i - m_j)^2}{2\sigma_j^2}\right] \qquad \begin{array}{l} j = 0, 1 \\ i = 1 \ldots N \end{array}$$

where $\sigma_0 = 1$, $\sigma_1 = 2$, $m_0 = -1$, and $m_1 = 1$. Find the LRT for this problem. Express your results in terms of the quantities $l_1 = \sum_{i=1}^{N} z_i$ and $l_2 = \sum_{i=1}^{N} z_i^2$ and determine the decision regions in the $l_1 - l_2$ plane.

3.14 We are given k independent observations:

$$H_1 : z_k = v_k \qquad k = 1, 2, \ldots, K$$

$$H_0 : z_k = 1 + v_k \qquad k = 1, 2, \ldots, K$$

where v_k is a zero mean Gaussian random variable with variance σ^2.

(a) Compute the likelihood ratio and the threshold for the optimum receiver. Assume $C_{00} = C_{11} = 0$, $C_{01} = 2$, $C_{10} = 1$ and that $P(H_0) = .7$, $P(H_1) = .3$

(b) Find P_F and P_M

(c) Plot the ROC for $K = 1$, $\sigma^2 = 2$

(d) Find the value of K for which the risk is no more than one-half of the risk for $K = 1$.

3.15 Plot the ROC for the problem in Example 3.4.2. Assume (i) $N = 8$, $\sigma_1^2 = \sigma_0^2$; (ii) $N = 2$, $\sigma_1^2 = 4\sigma_0^2$. What do you conclude from your results?

3.16 Consider M-ary hypothesis testing with Bayes' criterion. Prove that the optimum decision criterion is to choose the hypothesis for which the conditional risk $C(H_i|z)$ is smallest, where

$$C(H_i|z) = \sum_{j=1}^{M-1} C_{ij} P(H_j|z), \qquad i = 0, 1, \ldots, M-1$$

3.17 The observation z is the output under one of the three hypotheses, H_i, $i = 0, 1, 2$. The probability density functions under the hypotheses are shown in Fig. E.3.4. What is the decision rule that minimizes the probability of error? Find the minimum probability of error. Assume equal prior probabilities.

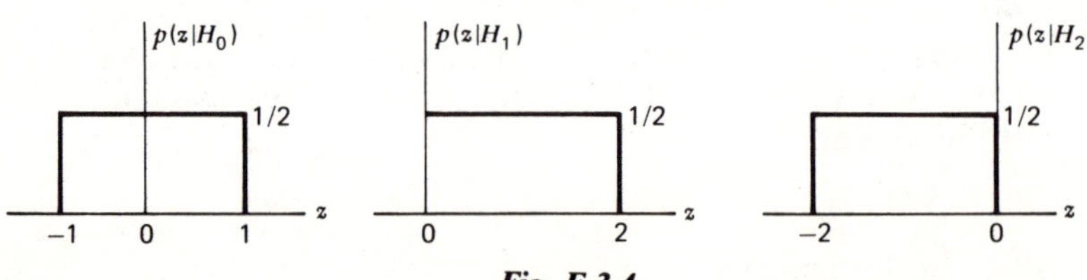

Fig. E.3.4

3.18 Design a likelihood ratio test to choose between the hypothesis H_1 and H_0, where

$$p(z|H_1) = \frac{1}{\sqrt{2\pi}\,\sigma} \exp\left[-\frac{z^2}{2\sigma^2}\right] \quad \text{and} \quad p(z|H_0) = \begin{cases} \frac{1}{2} & -1 \leqslant z \leqslant 1 \\ 0 & \text{otherwise} \end{cases}$$

but σ is unknown. Consider σ is (a) nonrandom, and lies in [0, 1], and (b) random and is uniformly distributed between 0 and 1.

3.19 In some radar problems, we have to establish the presence of an object in the background of unwanted reflections called clutter. Reflections from sea, land, ships, etc., can be characterized by a *log-normal* distribution

$$p_0(x|m, \sigma) = \frac{1}{\sqrt{2\pi}\, x\sigma} \exp\left[-\frac{(\ln x - m)^2}{2\sigma^2}\right]$$

$$x > 0, \sigma > 0, m > 0$$

or by the *Weibull* distribution

$$p_1(x|\alpha, \beta) = \left(\frac{x}{\beta}\right)^{\alpha-1} \exp\left[-\left(\frac{x}{\beta}\right)^{\alpha}\right] \qquad x \geqslant 0, \alpha > 0, \beta > 0$$

In practice, N independent measurements of the clutter reflections x_i, $i = 1, 2, \ldots, N$ are available. Based on these measurements, and without the knowledge of the nonrandom parameters m, σ, α, and β, we need to determine whether clutter is log-normally distributed or Weibull distributed. A convenient manner to implement such a test is to transform the measurements by taking the natural logarithm, i.e.,

$$z_i = \ln x_i$$

If x_is are log normally distributed, z_is are normally distributed with parameters m and σ:

$$p_0(z_i) = \frac{1}{\sqrt{2\pi}\,\sigma} \exp\left[-\frac{1}{2}\frac{(z_i - m)^2}{2\sigma^2}\right] \qquad i = 1, 2, \ldots, N$$

If x_is are Weibull distributed, z_is are distributed according to Type I extreme value distribution

$$p_1(z_i) = \frac{1}{b} \exp\left\{\left(\frac{z_i - a}{b}\right) - \exp\left(\frac{z_i - a}{b}\right)\right\}$$

$$a = \ln \beta \qquad b = 1/\alpha$$

In obtaining the generalized likelihood ratio using the transformed measurements, it is convenient to assume that the MLE of a and b are given by

$$\hat{a} = \hat{m} + \gamma \hat{b}$$
$$\hat{b} = \hat{\sigma}\sqrt{6}\,/\pi$$

where γ is the Euler constant, $\hat{m}$ and $\hat{\sigma}$ are the maximum likelihood estimates of m and σ. Show that the sufficient statistic is given by [13]

$$D(\mathbf{z}) = -\frac{1}{N}\sum_{i=1}^{N} \exp\left\{\frac{\pi(z_i - \hat{m})}{\hat{\sigma}\sqrt{6}}\right\}$$

3.20 In composite hypothesis testing, we can obtain an upper bound on the test performance by assuming that the receiver first measures the unknown parameters perfectly and then design the optimum likelihood ratio test. Such a bound is called a perfect measurement bound. The ROC of any test will be bounded by the ROC of this fictitious perfect measurement test.

Given that under H_0, the observation z is $N(0, \sigma_0^2)$ and under H_1, z is $N(0, \sigma_1^2)$ with $\sigma_1 > \sigma_0$, find the upper bound on the ROC for a P_F of 10^{-1} and determine if a UMP exists. Assume σ_1 is unknown.

3.21 Consider the following composite hypothesis testing problem. The observations are

$$H_1 : z_k = y_k + v_k \qquad k = 1, 2, \ldots, K$$
$$H_0 : z_k = v_k \qquad k = 1, 2, \ldots, K$$

with y_k and v_k independent identically distributed random variables with densities $N(0, \sigma_y^2)$ and $N(0, 1)$ where σ_y is unknown.

(a) Does a UMP test exist?

(b) If a UMP test does not exist, find a generalized LRT.

3.22 Derive a sequential ratio test when the observations are

$$H_0 : z_k = y_k^0 \qquad k = 1, 2 \ldots$$
$$H_1 : z_k = y_k^1 \qquad k = 1, 2 \ldots$$

where y_k^0 and y_k^1 are Gaussian, independent identically distributed variables with zero mean and variances σ_0^2 and σ_1^2, respectively. Assume $\sigma_1 > \sigma_0$ and that $P_F = 0.2$ and $P_M = 0.1$. Find the average number of samples for the test to terminate if $\sigma_0 = 1$, $\sigma_1 = 2$, and $P(H_0) = P(H_1) = 1/2$.

3.23 Consider the ternary hypothesis testing problem with

$$H_0 : z_k = v_k \qquad k = 1, 2, \ldots, K$$

$$H_1 : z_k = 1 + v_k \qquad k = 1, 2, \ldots, K$$

$$H_2 : z_k = -1 + v_k \qquad k = 1, , \ldots, K$$

Obtain a SPRT for this case if v_k is zero mean, white and Gaussian with variance σ^2.

REFERENCES

1. E. L. Lehmann, *Testing Statistical Hypotheses*, John Wiley and Sons, New York, 1950.
2. H. L. Van Trees, *Detection, Estimation, and Modulation Theory, Part I*, John Wiley and Sons, New York, 1968.
3. A. V. Balakrishnan, *Communication Theory*, McGraw-Hill Book Co. New York, 1968.
4. C. W. Helstrom, *Statistical Theory of Signal Detection*, Pergamon Press, Oxford, 1968.
5. I. Selin, *Detection Theory*, Princeton University Press, Princeton, N.J., 1965.
6. A. D. Whalen, *Detection of Signals in Noise*, Academic Press, New York, 1971.
7. D. Blackwell and M. A. Grishick, *Theory of Games and Statistical Decisions*, John Wiley and Sons, New York, 1954.
8. K. Pearson, *Tables of the Incomplete Γ-Function*, Cambridge University Press, Cambridge, 1934.
9. A. Wald, *Sequential Analysis*, John Wiley and Sons, New York, 1959.
10. K. S. Fu, *Sequential Methods in Pattern Recognition and Machine Learning*, Academic Press, New York, 1968.
11. A. Wald and J. Wolfowitz, Optimum Character of the Sequential Probability Ratio Test, *Ann. Math. Stat.*, **19**, 326–339 (1948).
12. J. D. Gibson and J. L. Melsa, *Introduction to Nonparametric Detection with Applications*, Academic Press, New York, 1975.
13. W. J. Szajnowsky, Discrimination Between Log Normal and Weibull Clutter, *IEEE Trans. Aerospace Electronic Sys.*, **AES-13**, No. 5, Sept., 480–485 (1977).

CHAPTER 4

DETECTION OF SIGNALS IN NOISE

4.1 INTRODUCTION

The preceding chapter introduced results required for the statistical detection of signals in noise. In this chapter we discuss the design of optimum receivers which will detect the presence of a signal or distinguish between different signals in the presence of noise. The term receiver is used to indicate the mathematical operation that is to be performed on the noisy observations. In order to design an optimum receiver we first specify a criterion of performance as in the previous chapter and then design a receiver that is the best with respect to the chosen criterion. It should be noted that a receiver designed to optimize a specific criterion may perform poorly with respect to other criteria, or even with respect to the chosen criterion if the underlying assumptions change.

In designing communication receivers, we primarily use the minimum probability of error criterion. As indicated in the previous chapter, for radar and sonar applications, the Neyman–Pearson criterion is usually preferred. In both cases, the procedure involves the computation of appropriate likelihood ratios.

We start with a discussion of the detection of known signals in additive noise. That is, the parameters characterizing the signal, such as amplitude, frequency, phase, and time of arrival are assumed known. While such an assumption may seem idealistic, in some practical systems these quantities are indeed known. For such situations at least, the ideal system represents a fairly good approximation. In any case, this discussion will serve to clarify the concepts used and will be extended in later sections to the problem of detection of signals with unknown parameters.

4.2 DETECTION OF KNOWN SIGNALS IN WHITE NOISE: THE CORRELATION RECEIVER [1–11]

Consider a simple binary communication system in which the transmitter sends out one of two signals $y_0(t)$ or $y_1(t)$ whose waveforms are completely known. The signal at the receiver is observed over an interval $[t_0, t_f]$ and is assumed to be corrupted by additive noise $v(t)$. We are required to design a receiver which operates on the received signal $z(t)$ and choose one of the two possible hypotheses:

$$H_0 : z(t) = y_0(t) + v(t) \qquad t_0 \leqslant t \leqslant t_f$$
$$H_1 : z(t) = y_1(t) + v(t) \qquad t_0 \leqslant t \leqslant t_f$$

We assume that the noise is Gaussian, white, with mean zero and spectral density $N_0/2$.

An obvious approach to solving this problem is to try to use our results from the previous chapter. We recall, however, that the results of Chapter 3 were obtained for a discrete set of observations. In order to apply those results to the present problem where the signal waveform is a continuous function of time, let us assume that we sample the signal so that N samples are available in the interval $[t_0, t_f]$ and later let N tend to infinity without limit. In this, we are assuming that the signals are all separable.

Let z_k, y_{ik} $(i = 0, 1)$ and v_k denote the sample values of $z(t)$, $y_i(t)$, and $v(t)$ at time $t = t_k$. The received samples under the two hypotheses in the interval $[t_0, t_f]$ are then given by

$$H_0 : z_k = y_{0k} + v_k \qquad 1 \leqslant k \leqslant N$$
$$H_1 : z_k = y_{1k} + v_k \qquad 1 \leqslant k \leqslant N \tag{4.2.1}$$

Let us first assume that the noise samples are independent and Gaussian with mean zero and variance σ^2. The observations under the two hypotheses are also independent, Gaussian, with

$$\mathrm{E}\{z_k | H_i\} = y_{ik}$$

and

$$\operatorname{var}\{z_k | H_i\} = \operatorname{var}\{v_k\} = \sigma^2 \qquad i = 0, 1 \tag{4.2.2}$$

Let us define the vectors $\mathbf{y}_0$, $\mathbf{y}_1$, and $\mathbf{z}$ as

$$\mathbf{y}_0 = [y_{01}, y_{02}, \ldots, y_{0N}]^T$$
$$\mathbf{y}_1 = [y_{11}, y_{12}, \ldots, y_{1N}]^T$$

and

$$\mathbf{z} = [z_1, z_2, \ldots, z_N]^T$$

It follows from Eq. (4.2.2) that the likelihood ratio is

$$\Lambda(\mathbf{z}) = \frac{p(\mathbf{z}|H_1)}{p(\mathbf{z}|H_0)} = \frac{\prod_{k=1}^{N} \frac{1}{\sqrt{2\pi}\,\sigma} \exp\left[-\frac{(z_k - y_{1k})^2}{2\sigma^2}\right]}{\prod_{k=1}^{N} \frac{1}{\sqrt{2\pi}\,\sigma} \exp\left[-\frac{(z_k - y_{0k})^2}{2\sigma^2}\right]}$$

$$= \exp\left\{-\frac{1}{2}\sum_{k=1}^{N}\left[\frac{2z_k y_{0k}}{\sigma^2} - \frac{2z_k y_{1k}}{\sigma^2} - \frac{(y_{0k}^2 - y_{1k}^2)}{\sigma^2}\right]\right\} \tag{4.2.3}$$

The decision rule is

$$\Lambda(\mathbf{z}) \underset{H_0}{\overset{H_1}{\gtrless}} \eta \tag{4.2.4}$$

Substituting Eq. (4.2.3) into Eq. (4.2.4), taking logarithms, and rearranging terms, yields the decision rule as

$$\sum_{k=1}^{N} z_k y_{1k} - \sum_{k=1}^{N} z_k y_{0k} \underset{H_0}{\overset{H_1}{\gtrless}} \sigma^2 \ln \eta + \frac{1}{2}\sum_{k=1}^{N}(y_{1k}^2 - y_{0k}^2) \tag{4.2.5}$$

In terms of the vectors $\mathbf{y}_0$, $\mathbf{y}_1$, and $\mathbf{z}$, the decision rule becomes

$$\mathbf{z}^T(\mathbf{y}_1 - \mathbf{y}_0) \underset{H_0}{\overset{H_1}{\gtrless}} \sigma^2 \ln \eta + \frac{1}{2}(\mathbf{y}_1^T\mathbf{y}_1 - \mathbf{y}_0^T\mathbf{y}_0) \tag{4.2.6}$$

The key step in implementing the rule is to obtain the quantity $\mathbf{z}^T(\mathbf{y}_1 - \mathbf{y}_0)$. The other terms in the rule do not depend on the observations and can be precomputed. We note that the sufficient statistic for this problem is $\mathbf{z}^T(\mathbf{y}_1 - \mathbf{y}_0) = \sum_{k=1}^{N} z_k(y_{1k} - y_{0k})$ which is a Gaussian random variable under both hypotheses.

In order to relate these results to our original problem of deciding between the two waveforms $y_0(t)$ and $y_1(t)$, let us assume that the noise process $v(t)$ is band limited so that

$$\Phi_v(\omega) = \begin{cases} N_0/2 & |\omega| < \Omega \\ 0 & \text{otherwise} \end{cases} \tag{4.2.7a}$$

with correlation function

$$\phi_v(\tau) = \frac{N_0\Omega}{2\pi} \frac{\sin \Omega\tau}{\Omega\tau} \tag{4.2.7b}$$

Thus, in order for the noise samples to be independent, we must sample $z(t)$ at intervals $\tau = k\pi/\Omega$, $k = 1, 2, 3$, and so on. The number of such samples in the interval $[t_0, t_f]$ will be $N = \Omega(t_f - t_0)/\pi$. The variance of the noise samples is obtained from Eq. (4.2.7b) as $\sigma^2 = N_0\Omega/2\pi$.

We may now seek to obtain the decision rule directly in terms of the original waveforms by letting the samples become dense. In the limit as N approaches infinity, $v(\cdot)$ becomes a true white process with the noise variance becoming an impulse of intensity $N_0/2$. We then have from Eq. (4.2.3)

$$\begin{aligned} \ln \Lambda[z(t)] &= \lim_{N\to\infty} \ln \Lambda(\mathbf{z}) \\ &= \frac{2}{N_0}\int_{t_0}^{t_f} z(t)y_1(t)\,dt - \frac{2}{N_0}\int_{t_0}^{t_f} z(t)y_0(t)\,dt \\ &\quad + \frac{1}{N_0}\int_{t_0}^{t_f}\left[y_0^2(t) - y_1^2(t)\right]dt \end{aligned} \tag{4.2.8}$$

so that the decision rule can be written as

$$\int_{t_0}^{t_f} z(t)y_1(t)\,dt - \int_{t_0}^{t_f} z(t)y_0(t)\,dt \underset{H_0}{\overset{H_1}{\gtrless}} \gamma \tag{4.2.9}$$

where the threshold is

$$\gamma = \frac{1}{2}N_0 \ln \eta + \frac{1}{2}\int_{t_0}^{t_f}\left[y_1^2(t) - y_0^2(t)\right]dt \tag{4.2.10}$$

The preceding derivation for obtaining the decision rule becomes involved when the observation noise is not white but is correlated (colored). In this case, the arguments used in going to the limit as N approaches infinity tend to become somewhat involved, since the correlation between samples must be taken into account. We present instead an alternative approach to the detection of signal waveforms observed in noise, which uses a representation of the noise in terms of a set of complete orthonormal functions. Basically the approach is to again convert the problem of deciding between two waveforms into one of deciding between two countable sets of random variables. However, these variables now are the coefficients of the expansion of the observed waveform in terms of the basis functions. If the basis functions are chosen appropriately, the coefficients will be uncorrelated random variables and the derivation of the decision rule will be straightforward. We will initially use a finite number K of these variables and later let K increase without limit.

Thus let $\{g_j(t)\}$, $j = 1, 2, \ldots$ denote a complete set of orthonormal functions over the interval $[t_0, t_f]$ such that

$$\int_{t_0}^{t_f} g_j(t) g_k(t)\, dt = \delta_{jk} \tag{4.2.11}$$

We defer until later, consideration of the choice of these functions. Let

$$v_K(t) = \sum_{j=1}^{K} v_j g_j(t) \tag{4.2.12}$$

represent the K-term approximation of $v(t)$ in terms of the functions $\{g_j(t)\}$, where

$$v_j = \int_{t_0}^{t_f} v(t) g_j(t)\, dt \qquad j = 1 \ldots K \tag{4.2.13}$$

Let us similarly define $z_K(t)$, $y_{1K}(t)$ and $y_{0K}(t)$ as the K-term approximations of $z(t)$, $y_1(t)$, and $y_0(t)$, respectively. Then

$$z(t) = \lim_{K\to\infty} z_K(t) = \lim_{K\to\infty} \sum_{j=1}^{K} z_j g_j(t) \tag{4.2.14}$$

where

$$z_j = \int_{t_0}^{t_f} z(t) g_j(t)\, dt \qquad j = 1 \ldots K \tag{4.2.15}$$

It is clear that z_j, $j = 1 \ldots K$ constitute a set of Gaussian random variables under each hypothesis.

The moments of these variables can be determined as follows. Under H_1 we have

$$\begin{aligned} z_j &= \int_{t_0}^{t_f} [y_1(t) + v(t)]\, g_j(t)\, dt \\ &= y_{1j} + v_j \end{aligned} \tag{4.2.16}$$

so that

$$\mathrm{E}\{z_j | H_1\} = y_{1j}$$

and

$$\operatorname{var}\{z_j|H_1\} = \mathrm{E}\{v_j^2\} \tag{4.2.17}$$

Now

$$\begin{aligned}\mathrm{E}\{v_j^2\} &= \mathrm{E}\left[\int_{t_0}^{t_f} v(t)g_j(t)\,dt \int_{t_0}^{t_f} v(\sigma)g_j(\sigma)\,dt\right] \\ &= \int_{t_0}^{t_f}\int_{t_0}^{t_f} \phi_v(t-\sigma)g_j(t)g_j(\sigma)\,dt\,d\sigma = \frac{N_0}{2}\end{aligned}$$

The last step follows from the fact that $v(t)$ is assumed to be white with spectral density $N_0/2$ and the orthogonality of the basis functions. Thus

$$\operatorname{var}\{z_j|H_1\} = \frac{N_0}{2} \tag{4.2.18}$$

Similarly, the moments under H_0 are given by

$$\mathrm{E}\{z_j|H_0\} = y_{0j}$$

and

$$\operatorname{var}\{z_j|H_0\} = \frac{N_0}{2} \tag{4.2.19}$$

It can easily be established that the variables z_j, $j = 1 \ldots K$ are independent under each hypothesis. We can therefore write the likelihood ratio for this problem as

$$\begin{aligned}\Lambda[z(t)] &= \lim_{K\to\infty} \Lambda[z_K(t)] = \lim_{K\to\infty} \frac{p[z_K(t)|H_1]}{p[z_K(t)|H_0]} \\ &= \lim_{K\to\infty} \frac{p(z_1, z_2, \ldots, z_K|H_1)}{p(z_1, z_2, \ldots, z_K|H_0)}\end{aligned} \tag{4.2.20}$$

Use of Eqs. (4.2.17) and (4.2.19) enables us to write the likelihood ratio $\Lambda[z_K(t)]$ as

$$\Lambda[z_K(t)] = \frac{\displaystyle\prod_{j=1}^{K} \frac{1}{\sqrt{\pi N_0}} \exp\left[-\frac{1}{2}\frac{(z_j - y_{1j})^2}{N_0/2}\right]}{\displaystyle\prod_{j=1}^{K} \frac{1}{\sqrt{\pi N_0}} \exp\left[-\frac{1}{2}\frac{(z_j - y_{0j})^2}{N_0/2}\right]} \tag{4.2.21}$$

Taking the logarithm and canceling common terms yields

$$\ln \Lambda[z_K(t)] = \frac{2}{N_0} \sum_{j=1}^{K} z_j(y_{1j} - y_{0j}) + \frac{1}{N_0} \sum_{j=1}^{K} (y_{0j}^2 - y_{1j}^2) \tag{4.2.22}$$

Now

$$\begin{aligned}\int_{t_0}^{t_f} z_K(t) y_{iK}(t)\, dt &= \int_{t_0}^{t_f} \left[\sum_{j=1}^{K} z_j g_j(t) \right] \left[\sum_{k=1}^{K} y_{ik} g_k(t) \right] dt \\ &= \sum_{j=1}^{K} \sum_{k=1}^{K} z_j y_{ik} \int_{t_0}^{t_f} g_j(t) g_k(t)\, dt \\ &= \sum_{j=1}^{K} z_j y_{ij} \qquad i = 0, 1\end{aligned} \tag{4.2.23}$$

Similarly

$$\sum_{j=1}^{K} y_{ij}^2 = \int_{t_0}^{t_f} y_{iK}^2(t)\, dt \qquad i = 0, 1 \tag{4.2.24}$$

Substitution of Eqs. (4.2.23) and (4.2.24) in Eq. (4.2.22) yields

$$\begin{aligned}\ln \Lambda[z_K(t)] &= \frac{2}{N_0} \int_{t_0}^{t_f} z_K(t)[y_{1K}(t) - y_{0K}(t)]\, dt \\ &\quad - \frac{1}{N_0} \int_{t_0}^{t_f} [y_{1K}^2(t) - y_{0K}^2(t)]\, dt\end{aligned} \tag{4.2.25}$$

The log-likelihood ratio $\ln \Lambda(t)$ obtained by taking the limit of Eq. (4.2.25) as K tends to infinity is

$$\ln \Lambda(t) = \frac{2}{N_0} \int_{t_0}^{t_f} z(t)[y_1(t) - y_0(t)]\, dt - \frac{1}{N_0} \int_{t_0}^{t_f} [y_1^2(t) - y_0^2(t)]\, dt \tag{4.2.26}$$

and the decision rule is

$$\ln \Lambda(t) \underset{H_0}{\overset{H_1}{\gtrless}} \ln \eta \tag{4.2.27}$$

Substitution of Eq. (4.2.26) in Eq. (4.2.27) yields the decision rule as

$$\int_{t_0}^{t_f} z(t)[y_1(t) - y_0(t)]\, dt \underset{H_0}{\overset{H_1}{\gtrless}} \frac{N_0}{2} \ln \eta + \frac{1}{2} \int_{t_0}^{t_f} [y_1^2(t) - y_0^2(t)]\, dt \tag{4.2.28}$$

which is the same as was obtained earlier (see Eqs. (4.2.9) and (4.2.10)). The structure of the receiver is shown in Fig. 4.2.1. The receiver is called a correlation receiver, since the received signal $z(t)$ is cross correlated with the signals $y_0(t)$ and $y_1(t)$. If the signals are of equal energy, the threshold is signal independent.

We now comment briefly on the choice of the basis functions $\{g_i(t)\}$. We recall from our discussion of series representations of stochastic signals in Chapter 2 that for white processes, *any* set of orthonormal functions can be used for decomposing the process. As we have seen, the coefficients of the expansion will be independent irrespective of the choice of basis functions. For colored (nonwhite) processes, our choice of basis functions is of course more restricted. We will discuss in the next section the use of the Karhunen–Loéve expansion for this case.

The nonuniqueness of the basis functions for the white observation noise problem enables us to give a simple derivation of the likelihood ratio test in terms of the sufficient statistic introduced in the previous chapter. There we saw that if we transformed the observation vector into two components, one of which was dependent on which hypothesis was true and the other independent of the hypothesis, the decision could be based entirely on the first component. The transformation involves choosing an appropriate set of coordinate or basis functions for the decision space. We

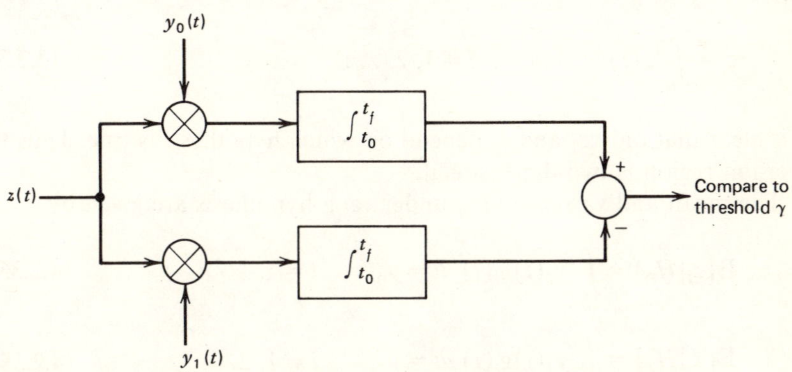

Fig. 4.2.1 Correlation receiver for binary decision problem.

will use a complete set of orthonormal functions chosen as follows. Let ϵ_1 and ϵ_0 denote the energy in the signals $y_1(t)$ and $y_0(t)$:

$$\epsilon_1 = \int_{t_0}^{t_f} y_1^2(t)\, dt \qquad \text{and} \qquad \epsilon_0 = \int_{t_0}^{t_f} y_0^2(t)\, dt \tag{4.2.29}$$

We choose as our first basis function

$$g_1(t) = \frac{y_1(t)}{\sqrt{\epsilon_1}} \qquad t_0 \leqslant t \leqslant t_f \tag{4.2.30}$$

The second function $g_2(t)$ is obtained from the signal $y_0(t)$ by subtracting the component correlated with $g_1(t)$ and normalizing. Thus $g_2(t)$ is chosen as

$$g_2(t) = \frac{1}{\sqrt{\epsilon_0 \epsilon_1 - \rho^2}} \left[\sqrt{\epsilon_1}\, y_0(t) - \frac{\rho}{\sqrt{\epsilon_1}} y_1(t) \right] \qquad t_0 \leqslant t \leqslant t_f \tag{4.2.31}$$

where the correlation coefficient ρ is given by

$$\rho = \int_{t_0}^{t_f} y_0(t) y_1(t)\, dt \tag{4.2.32}$$

The remaining orthonormal basis functions $g_i(t)$ can be arbitrary. In terms of the basis functions, we write the received waveform as

$$z(t) = \sum_{i=1}^{\infty} z_i g_i(t) \qquad t_0 \leqslant t \leqslant t_f \tag{4.2.33}$$

where

$$z_i = \int_{t_0}^{t_f} z(t) g_i(t)\, dt \qquad i = 1, 2, \ldots \tag{4.2.34}$$

It is clear that only z_1 and z_2 depend on which hypothesis is true. Thus the decision region is two-dimensional.

The mean and variance of z_i under each hypothesis are given by

$$\mathrm{E}\{z_i | H_0\} = \int_{t_0}^{t_f} y_0(t) g_i(t)\, dt = y_{0i} \qquad i = 1, 2 \tag{4.2.35a}$$

$$\mathrm{E}\{z_i | H_1\} = \int_{t_0}^{t_f} y_1(t) g_i(t)\, dt = y_{1i} \qquad i = 1, 2 \tag{4.2.35b}$$

and

$$\{\text{var } z_i | H_j\} = \int_{t_0}^{t_f}\int_{t_0}^{t_f} \phi_v(t-s) g_i(t) g_i(s)\, dt\, ds = \frac{N_0}{2}$$

$$j = 1, 2 \text{ and } i = 1, 2 \qquad \textbf{(4.2.36)}$$

Following an argument similar to that used in setting up Eq. (4.2.22) we can easily show that the decision rule becomes

$$\sum_{i=1}^{2} (2z_i - y_{1i})\frac{y_{1i}}{N_0} - \sum_{i=1}^{2} (2z_i - y_{0i})\frac{y_{0i}}{N_0} \underset{H_0}{\overset{H_1}{\gtrless}} \ln \eta \qquad \textbf{(4.2.37)}$$

or

$$\sum_{i=1}^{2} z_i(y_{1i} - y_{0i}) \underset{H_0}{\overset{H_1}{\gtrless}} \frac{N_0}{2}\ln \eta + \frac{1}{2}\left(\sum_{i=1}^{2} y_{1i}^2 - \sum_{i=1}^{2} y_{0i}^2\right) \qquad \textbf{(4.2.38)}$$

The implementation of the receiver is shown in Fig. 4.2.2.

The receiver is again a correlation receiver in which the observation $z(t)$ is correlated with the two basis functions $g_1(t)$ and $g_2(t)$. With the use of Eqs. (4.2.30), (4.2.31), and (4.2.35a, b) and a little algebraic manipulation, it can be shown that the structure of Fig. 4.2.2 is equivalent to that of Fig. 4.2.1.

It is clear from Fig. 4.2.2 that the structure can be reduced to a single correlation operation as shown in Fig. 4.2.3. This is equivalent to choosing

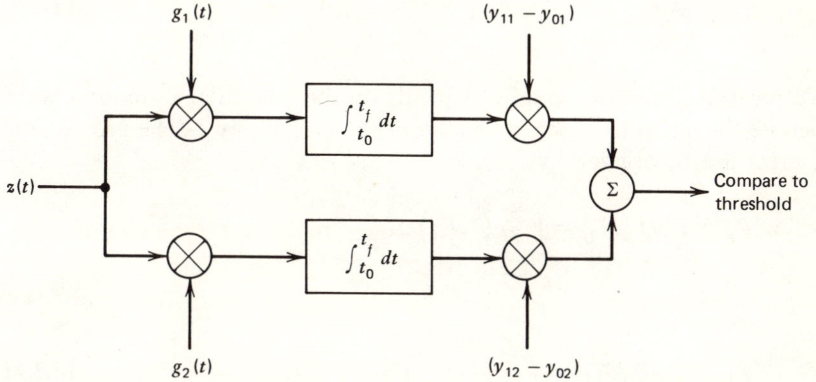

Fig. 4.2.2 Alternative implementation of correlation receiver for binary problem.

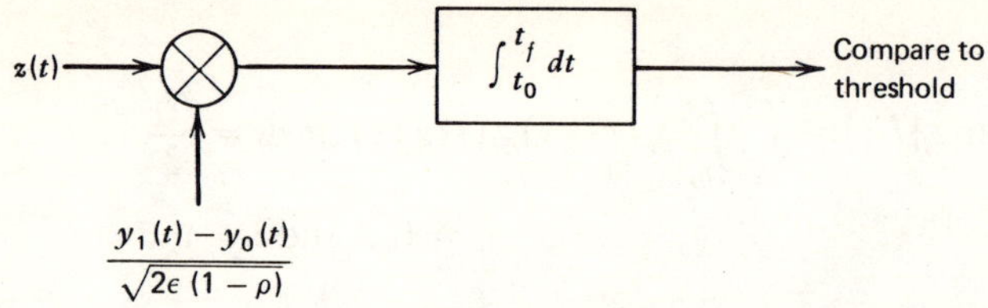

Fig. 4.2.3 Implementation of binary receiver with a single correlator.

as our first basis function

$$g_1(t) = \frac{y_1(t) - y_0(t)}{\sqrt{2\epsilon_1(1-\rho)}} \tag{4.2.39}$$

with the other basis functions being arbitrary.

If we again write $z(t)$ as in Eq. (4.2.35), only z_1 will depend on which hypothesis is true and the decision region becomes one-dimensional. Thus z_1 is the sufficient statistic l for this problem.

Receiver Performance

In order to determine the performance of the receiver, let us rewrite the decision rule in the following equivalent form:

$$\int_{t_0}^{t_f} z(t) y_1(t)\, dt - \int_{t_0}^{t_f} z(t) y_0(t)\, dt + \frac{1}{2}\int_{t_0}^{t_f}\left[y_0^2(t) - y_1^2(t)\right] dt$$

$$\underset{H_0}{\overset{H_1}{\gtrless}} \frac{N_0}{2} \ln \eta \triangleq \gamma \tag{4.2.40}$$

We recall that the threshold η depends on the particular criterion chosen. Denote the left side of the inequality in Eq. (4.2.40) by I. The probabilities of error are then given by

$$P_F = P(D_1|H_0) = \int_{\gamma}^{\infty} p(I|H_0)\, dI$$

and

$$P_M = P(D_0|H_1) = \int_{-\infty}^{\gamma} p(I|H_1)\, dI \tag{4.2.41}$$

We note that I is a Gaussian variable under both hypotheses. Under H_0, the received signal is $z(t) = y_0(t) + v(t)$, so that

$$I = \int_{t_0}^{t_f} [y_0(t) + v(t)] y_1(t)\, dt - \int_{t_0}^{t_f} [y_0(t) + v(t)] y_0(t)\, dt + \frac{1}{2}\int_{t_0}^{t_f} [y_0^2(t) - y_1^2(t)]\, dt \tag{4.2.42}$$

Then we have

$$\mathrm{E}\{I|H_0\} = -\frac{1}{2}\int_{t_0}^{t_f} [y_0(t) - y_1(t)]^2\, dt \tag{4.2.43}$$

where we have used the fact that $v(t)$ is zero mean.

The variance of I can also be calculated as

$$\begin{aligned}\mathrm{var}[I|H_0] &= \mathrm{E}\{[I - \mathrm{E}(I|H_0)]^2|H_0\}\\ &= \mathrm{E}\left\{\int_{t_0}^{t_f} v(t)[y_1(t) - y_0(t)]\, dt \times \int_{t_0}^{t_f} v(\tau)[y_1(\tau) - y_0(\tau)]\, d\tau\right\}\\ &= \int_{t_0}^{t_f}\int_{t_0}^{t_f} \mathrm{E}\{v(t)v(\tau)\}[y_1(t) - y_0(t)][y_1(\tau) - y_0(\tau)]\, dt d\tau\\ &= \frac{1}{2}N_0\int_{t_0}^{t_f} [y_1(t) - y_0(t)]^2\, dt\end{aligned} \tag{4.2.44}$$

since

$$\mathrm{E}\{v(t)v(\tau)\} = \frac{N_0}{2}\delta(t - \tau)$$

It may be similarly be shown that under H_1

$$\mathrm{E}\{I|H_1\} = \frac{1}{2}\int_{t_0}^{t_f} [y_1(t) - y_0(t)]^2\, dt \tag{4.2.45}$$

and

$$\mathrm{var}\{I|H_1\} = \frac{N_0}{2}\int_{t_0}^{t_f} [y_0(t) - y_1(t)]^2\, dt \tag{4.2.46}$$

As I is a Gaussian random variable, the means and variances of I under the two hypotheses are sufficient to write down the respective probability densities. We denote the average energy of the two signals by

$$\epsilon = \frac{1}{2}\int_{t_0}^{t_f}[y_0^2(t) + y_1^2(t)]\,dt = \frac{1}{2}(\epsilon_0 + \epsilon_1) \tag{4.2.47}$$

and define the (normalized) correlation coefficient

$$\bar{\rho} = \frac{1}{\epsilon}\int_{t_0}^{t_f} y_0(t)y_1(t)\,dt \tag{4.2.48}$$

We note that $|\bar{\rho}| \leqslant 1$. Then it can easily be verified that

$$\mathrm{E}\{I|H_0\} = -\epsilon(1-\bar{\rho})$$
$$\mathrm{E}\{I|H_1\} = \epsilon(1-\bar{\rho})$$

and

$$\mathrm{var}\{I|H_0\} = \mathrm{var}\{I|H_1\} = N_0\epsilon(1-\bar{\rho}) \tag{4.2.49}$$

Let γ^+ and γ^- be given by

$$\gamma^+ \underline{\Delta} \frac{\gamma + \epsilon(1-\bar{\rho})}{\sqrt{N_0\epsilon(1-\bar{\rho})}}; \qquad \gamma^- \underline{\Delta} \frac{\gamma - \epsilon(1-\bar{\rho})}{\sqrt{N_0\epsilon(1-\bar{\rho})}} \tag{4.2.50}$$

It easily follows that the error probabilities are given by

$$P_F = \int_{\gamma^+}^{\infty} \frac{1}{\sqrt{2\pi}} e^{-u^2/2}\,du = \mathrm{erfc}_*(\gamma^+) \tag{4.2.51}$$

and

$$P_M = \int_{-\infty}^{\gamma^-} \frac{1}{\sqrt{2\pi}} e^{-u^2/2}\,du = 1 - \mathrm{erfc}_*(\gamma^-) \tag{4.2.52}$$

The performance of the receiver thus depends only on the three parameters ϵ, $\bar{\rho}$, and N_0 and is independent of the particular waveform used. For equally likely hypotheses and the minimum probability of error criterion, γ equals zero. Thus $\gamma^+ = -\gamma^-$ and the two error probabilities become equal. The total probability of error is just P_F or P_M and decreases as $\epsilon(1-\bar{\rho})/N_0$ increases. For fixed N_0 therefore, the optimum system is one for which $\bar{\rho} = -1$ or $y_0(t) = -y_1(t)$. Such a system is known as an ideal binary communication system.

Example 4.2.1 We will determine the performance of three commonly used communication systems in which the binary signals are sine waves. We will assume that the signals are phase-synchronized at the transmitting and receiving ends so that we can normalize the phase to be zero. Such systems are referred to as "coherent" systems. We will use the minimum probability of error criterion in all cases.

In coherent phase shift keying (CPSK) the signals are

$$y_0(t) = A \sin \omega_c t \qquad \text{and} \quad y_1(t) = -A \sin \omega_c t \qquad t_0 \leqslant t \leqslant t_f$$

Since $\bar{\rho} = -1$, this is an ideal binary system and the probability of error is obtained as

$$P_e = \int_{\sqrt{2\epsilon/N_0}}^{\infty} \frac{1}{\sqrt{2\pi}} e^{-u^2/2}\, du = \text{erfc}_*\left(\sqrt{2\epsilon/N_0}\,\right)$$

In coherent frequency shift keying (CFSK) the binary signals are

$$y_0(t) = A \sin \omega_0 t \qquad \text{and} \quad y_1(t) = A \sin \omega_1 t \qquad t_0 \leqslant t \leqslant t_f$$

If the ratio ω_1/ω_0 is rational, the two signals are orthogonal and $\bar{\rho} = 0$. In this case, the error probability is given by

$$P_e = \int_{\sqrt{\epsilon/N_0}}^{\infty} \frac{1}{\sqrt{2\pi}} e^{-u^2/2}\, du = \text{erfc}_*\left(\sqrt{\epsilon/N_0}\,\right)$$

In on–off keying (OOK) the binary signals are

$$y_0(t) = 0 \qquad \text{and} \quad y_1(t) = A \sin \omega_1 t \qquad t_0 \leqslant t \leqslant t_f$$

Clearly $\bar{\rho} = 0$. Since $\epsilon_0 = 0$, we have $\epsilon = \epsilon_1/2$.
The error probability is

$$P_e = \int_{\sqrt{\epsilon_1/2N_0}}^{\infty} \frac{1}{\sqrt{2\pi}} e^{-u^2/2}\, du = \text{erfc}_*\left(\sqrt{\epsilon_1/2N_0}\,\right)$$

Matched Filter Implementation of Receiver

The implementation of the decision rule of Eq. (4.2.28) involves computation of the integrals

$$\int_{t_0}^{t_f} z(t) y_i(t)\, dt \qquad i = 0, 1 \qquad \textbf{(4.2.53)}$$

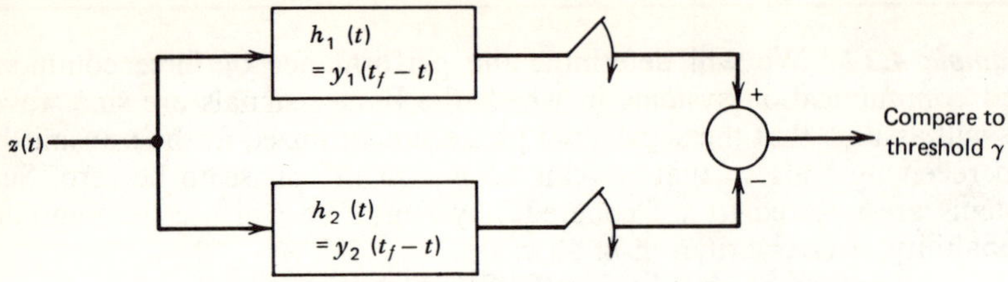

Fig. 4.2.4 Matched filter implementation of correlation receiver.

and requires one multiplier and integrator for determining each integral (see Fig. 4.2.1).

We can obtain an alternative implementation as follows. Suppose we apply $z(t)$ to a linear-time-invariant filter with impulse response $h_i(t)$. Then the output of the filter at time τ is given by

$$r(\tau) = \int_{t_0}^{t_f} h_i(\tau - t)z(t)\,dt \tag{4.2.54}$$

Comparison with Eq. (4.2.53) shows that if we let $h_i(\tau - t) = y_i(t)$ for $t_0 \leqslant t \leqslant t_f$ (or equivalently $h_i(t) = y_i(\tau - t)$ for $\tau - t_f \leqslant t \leqslant \tau - t_0$) the output $r(\tau)$ of the filter will be equal to the correlation between the signals $z(t)$ and $y_i(t)$. Such an implementation is referred to as a *matched filter*. The matched filter implementation is usually preferred in practice and is shown in Fig. 4.2.4. We note if $\tau < t_f$, the filter in Eq. (4.2.54) is noncausal (not physically realizable) since the output at time τ depends on values of the input over the interval $[\tau, t_f]$. If, however, τ is chosen to be equal to t_f the filter will be causal with impulse response

$$h_i(t) = y_i(t_f - t) \qquad \text{for} \quad 0 \leqslant t \leqslant t_f - t_0$$

M-ary Detection in White Gaussian Noise

The extension of the above results to the case where we have to choose between M possible signal waveforms is quite straightforward. We define the received signal under the hypotheses H_i ($i = 1 \ldots M$) as

$$H_i : z(t) = y_i(t) + v(t) \qquad t_0 \leqslant t \leqslant t_f \tag{4.2.55}$$

where $v(t)$ is zero mean, Gaussian white noise. We assume that the hypotheses are equally likely. We will restrict our discussion to the case when the signals are orthogonal with equal energy. That is

$$\rho_{ij} = \int_{t_0}^{t_f} y_i(t)y_j(t)\,dt = \begin{cases} \epsilon & i = j \\ 0 & i \neq j \end{cases} \tag{4.2.56}$$

The criterion to be used is the minimum probability of error.

Using arguments similar to those in the binary case, it can be shown that the decision rule is to choose the hypothesis for which $\int_{t_0}^{t_f}[z(t)-y_i(t)]^2\,dt$ is minimum. The integrand may be written as

$$z^2(t)+y_i^2(t)-2z(t)y_i(t)$$

Since the first two terms are independent of which hypothesis is true, the decision rule is to choose the hypothesis for which $\int_{t_0}^{t_f}z(t)y_i(t)\,dt$ is maximum. The receiver may be implemented as either a correlation receiver or as a matched filter.

To determine the performance of the receiver, we define the random variables

$$I_j=\int_{t_0}^{t_f}z(t)y_j(t)\,dt \qquad j=1\ldots M \tag{4.2.57}$$

Then

$$E\{I_j|H_i\}=\int_{t_0}^{t_f}[\,y_i(t)+v(t)\,]\,y_j(t)\,dt$$

$$=\begin{cases}0 & i\neq j\\ \epsilon & i=j\end{cases} \tag{4.2.58}$$

It can easily be verified that

$$E\{I_iI_j|H_k\}=0 \qquad \text{for} \quad i\neq j$$

and

$$E\{I_j^2|H_k\}=\begin{cases}\epsilon^2+(N_0\epsilon)/2 & j=k\\ (N_0\epsilon)/2 & j\neq k\end{cases} \tag{4.2.59}$$

Therefore the variance of I_j assuming hypothesis H_i is true is given by

$$\operatorname{var}\{I_j|H_i\}=\frac{N_0\epsilon}{2}=\sigma^2 \text{ for all } i \text{ and } j \tag{4.2.60}$$

Let us assume that H_i is the true hypothesis. The probability that no error is made is then equal to the probability that I_i is the maximum of the random variables $I_1, I_2, \ldots, I_M$. Suppose that I_i has a particular value x. Then the probability of no error is equal to the probability that I_j is less than x for $j\neq i$:

$$P(\text{no error}|I_i=x,\,H_i)=P(I_j<x:j=1\ldots M,\,j\neq i|H_i)$$

Since the variables I_j are Gaussian and uncorrelated, they are also independent so that we can write

$$P(\text{no error}|I_i = x, H_i) = \left[P(I_j < x|H_i)\right]^{M-1}$$

$$= \left[\int_{-\infty}^{x} \frac{1}{\sqrt{2\pi}\,\sigma} \exp\left(-\frac{u^2}{2\sigma^2}\right) du\right]^{M-1} \tag{4.2.61}$$

The probability of error is then given by

$$P_e = 1 - \int_{-\infty}^{\infty} P(\text{no error}|I_i = x, H_i) P(I_i = x|H_i)\, dx$$

$$= 1 - \int_{-\infty}^{\infty} \frac{1}{\sqrt{2\pi}\,\sigma} \exp\left[-\frac{(x-\epsilon)^2}{2\sigma^2}\right]$$

$$\times \left[\int_{-\infty}^{x} \frac{1}{\sqrt{2\pi}\,\sigma} \exp\left(-\frac{u^2}{2\sigma^2}\right) du\right]^{M-1} dx \tag{4.2.62}$$

This expression cannot in general be integrated analytically. However, values of P_e for certain values of ϵ and M have been numerically evaluated and are available in tables [12].

4.3 DETECTION OF KNOWN SIGNALS IN COLORED NOISE

In the previous section, we assumed that the observation noise $v(t)$ was Gaussian, white, with mean zero. We now extend our results to the case when the noise is colored with covariance function $\phi_v(t, u)$. We can obtain the solution in exactly the same manner as in the last section by expanding the observations in a set of orthonormal basis functions. The functions are again chosen so that the coefficients of the expansion are uncorrelated. Specifically, we will use the Karhunen–Loéve expansion in which the orthogonal functions are the eigenfunctions of the integral equation

$$\int_{t_0}^{t_f} \phi_v(t, u) g_j(u)\, du = \lambda_j g_j(t) \qquad t_0 \leqslant t \leqslant t_f \tag{4.3.1}$$

We can then write the expansion of $v(t)$ as

$$v(t) = \lim_{K\to\infty} \sum_{j=1}^{K} v_j g_j(t) \tag{4.3.2}$$

It follows from our discussion of the Karhunen–Loéve expansion in Chapter 2 that

$$E\{v_j\} = 0 \qquad \text{and} \qquad E\{v_j v_k\} = \lambda_j \delta_{jk} \tag{4.3.3}$$

We can similarly expand $z(t)$ in terms of $g_j(t)$ as

$$z(t) = \lim_{K \to \infty} \sum_{j=1}^{K} z_j g_j(t) \tag{4.3.4}$$

where

$$z_j = \int_{t_0}^{t_f} z(t) g_j(t)\, dt \tag{4.3.5}$$

and is hence Gaussian.

Under hypothesis H_1, we have

$$z_j = \int_{t_0}^{t_f} [y_1(t) + v(t)]\, g_j(t)\, dt$$

$$= y_{1j} + v_j \tag{4.3.6}$$

We can therefore write

$$E\{z_j | H_1\} = y_{1j} \qquad \text{and} \qquad \text{var}\{z_j | H_1\} = \lambda_j \tag{4.3.7}$$

where the last step follows from Eq. (4.3.3). Similarly, under hypothesis H_0, we have

$$E\{z_j | H_0\} = y_{0j} \qquad \text{and} \qquad \text{var}\{z_j | H_0\} = \lambda_j \tag{4.3.8}$$

It can easily be established by virtue of Eq. (4.3.3) that the variables z_j are independent under each hypothesis. Similar to Eq. (4.2.21), we can write the likelihood ratio for a K-term expansion as

$$\Lambda(z_k(t)) = \frac{\displaystyle\prod_{j=1}^{K} \frac{1}{\sqrt{2\pi\lambda_j}} \exp\left[-\frac{1}{2\lambda_j} (z_j - y_{1j})^2 \right]}{\displaystyle\prod_{j=1}^{K} \frac{1}{\sqrt{2\pi\lambda_j}} \exp\left[-\frac{1}{2\lambda_j} (z_j - y_{0j})^2 \right]} \tag{4.3.9}$$

Canceling common terms, taking the logarithm, and letting K tend to infinity yields the decision rule as

$$\ln[\Lambda(z(t))] = \sum_{j=1}^{\infty} \frac{z_j}{\lambda_j}(y_{1j} - y_{0j}) + \frac{1}{2}\sum_{j=1}^{\infty} \frac{1}{\lambda_j}(y_{0j}^2 - y_{1j}^2) \underset{H_0}{\overset{H_1}{\gtrless}} \ln \eta \tag{4.3.10}$$

We can rewrite the likelihood ratio in terms of $z(t)$ and the signals $y_1(t)$ and $y_0(t)$ by using Eqs. (4.3.5) and (4.3.6). We then obtain the decision rule as

$$\int_{t_0}^{t_f}\int_{t_0}^{t_f}[y_1(v) - y_0(v)]z(u)\left[\sum_{j=1}^{\infty} \frac{g_j(v)g_j(u)}{\lambda_j}\right] dv\, du + \frac{1}{2}\int_{t_0}^{t_f}\int_{t_0}^{t_f}[y_1(v)y_1(u) - y_0(v)y_0(u)] \times \left[\sum_{j=1}^{\infty} \frac{g_j(v)g_j(u)}{\lambda_j}\right] dv\, du \underset{H_0}{\overset{H_1}{\gtrless}} \ln \eta \tag{4.3.11}$$

The decision rule of Eq. (4.3.11) can be interpreted as a correlation detector as follows. Let us define the function $h(\cdot)$ by

$$h(u) = \int_{t_0}^{t_f}[y_1(v) - y_0(v)]\left[\sum_{j=1}^{\infty} \frac{1}{\lambda_j} g_j(v)g_j(u)\right] dv = \sum_{j=1}^{\infty} \frac{y_{1j} - y_{0j}}{\lambda_j} g_j(u) \tag{4.3.12}$$

The second term on the left side of Eq. (4.3.11) is an integral of known functions. Hence it can be combined with the $\ln \eta$ term on the right side to form a new threshold γ. Use of Eq. (4.3.12) then yields the decision rule as

$$\int_{t_0}^{t_f} z(u)h(u)\, du \underset{H_0}{\overset{H_1}{\gtrless}} \gamma \tag{4.3.13}$$

As in the case of white observation noise, we see that the receiver for colored observation noise can be interpreted as a correlation detector. We shall now show that the signal $h(\cdot)$ with which the observation $z(\cdot)$ is correlated satisfies an integral equation. To do so, let us multiply Eq.

(4.3.12) by $\phi_v(t, u)$ and integrate over the interval (t_0, t_f). We then get

$$\begin{aligned}\int_{t_0}^{t_f} \phi_v(t, u)h(u)\, du &= \int_{t_0}^{t_f} \phi_v(t, u)\left[\sum_{j=1}^{\infty} \frac{y_{1j} - y_{0j}}{\lambda_j} g_j(u)\right] du \\ &= \sum_{j=1}^{\infty} (y_{1j} - y_{0j})g_j(t) \\ &= y_1(t) - y_0(t) \end{aligned} \tag{4.3.14}$$

Thus $h(u)$ is obtained as the solution to the preceding integral equation.

Whitening Approach [13,14]

An alternative approach to obtaining the decision rule for the colored noise problem is to seek to convert it first into a problem in which the results of the previous section can be directly applied. We do this by "whitening" the observation noise. We thus do a preliminary processing of the observations by passing them through a linear time-varying filter with known impulse response. As can be expected the filter response will depend on the noise characteristics. At the output of the whitening filter, we will have a known signal in additive white noise and the problem thus reduces to the one considered in the previous section.

The output of the whitening filter can be written as

$$\begin{aligned} z_w(t) &= \int_{t_0}^{t_f} h_w(t, u)z(u)\, du \\ &= \int_{t_0}^{t_f} h_w(t, u)y_i(u)\, du + \int_{t_0}^{t_f} h_w(t, u)v(u)\, du \\ &= y_{wi}(t) + v_w(t) \qquad t \geqslant t_0 \qquad i = 0, 1 \end{aligned} \tag{4.3.15}$$

where $h_w(t, u)$ represents the impulse response of the whitening filter, $y_{wi}(t)$ is the output of this filter due to input $y_i(t)$, and $v_w(t)$ is the output due to input $v(t)$.

We want to choose $h_w(\cdot,\cdot)$ so that

$$\phi_{v_w}(t, u) = \mathrm{E}\{v_w(t)v_w(u)\} = \delta(t - u) \qquad t_0 \leqslant t, u \leqslant t_f$$

where we have arbitrarily specified a unity spectral height for the noise output. Assuming for the moment that such a filter has been found we can use Eq. (4.2.28) directly to determine the decision rule as

$$\int_{t_0}^{t_f} z_w(t)y_{w_1}(t)\, dt - \int_{t_0}^{t_f} z_w(t)y_{w_0}(t)\, dt \underset{H_0}{\overset{H_1}{\gtrless}} \gamma \tag{4.3.16}$$

where the threshold is

$$\gamma = \frac{1}{2}\ln \eta - \frac{1}{2}\int_{t_0}^{t_f}\left[y_{w0}^2(t) - y_{w1}^2(t)\right] dt \tag{4.3.17}$$

In terms of the original waveforms, the decision rule can be written as

$$\int_{t_0}^{t_f} dt \int_{t_0}^{t_f} h_w(t, u) z(u)\, du \int_{t_0}^{t_f} h_w(t, v) y_1(v)\, dv$$
$$- \int_{t_0}^{t_f} dt \int_{t_0}^{t_f} h_w(t, u) z(u)\, du \int_{t_0}^{t_f} h_w(t, v) y_0(v)\, dv \underset{H_0}{\overset{H_1}{\gtrless}} \gamma \tag{4.3.18}$$

The structure of the receiver is shown in Fig. 4.3.1.

All that remains is to determine the whitening filter $h_w(t, u)$. We can obtain a formal solution for $h_w(t, u)$ in terms of the eigenfunctions of $\phi_v(t, \sigma)$. In order to do so, let us expand $h_w(t, u)$ in terms of the orthogonal functions $g(t)$ as follows. We write

$$h_w(t, u) = \sum_{j=1}^{\infty} h_j g_j(t) g_j(u) \tag{4.3.19}$$

The noise output from the whitening filter is obtained from Eq. (4.3.12) as

$$v_w(t) = \int_{t_0}^{t_f} h_w(t, u) v(u)\, du \tag{4.3.20}$$

so that

$$\int_{t_0}^{t_f}\int_{t_0}^{t_f} h_w(t, u) h_w(s, v) \phi_v(u, v)\, du\, dv = \mathrm{E}\{v_w(t) v_w(s)\} = \delta(t - s) \tag{4.3.21}$$

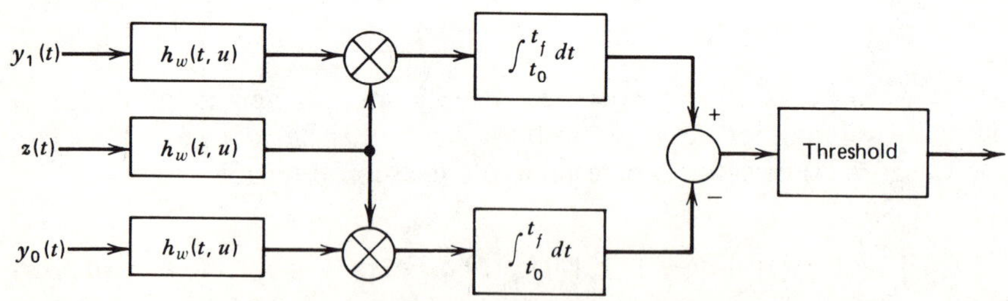

Fig. 4.3.1 Receiver structure for colored noise.

We now substitute for $h_w(\cdot, \cdot)$ from Eq. (4.3.19) in the left side of Eq. (4.3.21) and make use of Eq. (4.3.1) and the orthogonality of the eigenfunctions to obtain

$$\sum_{j=1}^{\infty} h_j^2 \lambda_j g_j(t) g_j(s) = \delta(t-s) \tag{4.3.22}$$

We recall from Chapter 2 that we can formally express the impulse function as a series

$$\delta(t-s) = \sum_{j=1}^{\infty} g_j(t) g_j(s) \tag{4.3.23}$$

Comparison of Eqs. (4.3.22) and (4.3.23) then yields

$$h_j = \frac{1}{\sqrt{\lambda_j}} \tag{4.3.24}$$

so that the whitening filter is given by

$$h_w(t, u) = \sum_{j=1}^{\infty} \frac{1}{\sqrt{\lambda_j}} g_j(t) g_j(u) \tag{4.3.25}$$

Finally, we can establish the equivalence of the decision rules of Eqs. (4.3.11) and (4.3.18) by substituting Eq. (4.3.25) in Eq. (4.3.18) and making use of the orthogonality of the eigenfunctions.

In going through the preceding formal derivation, we have ignored certain mathematical niceties. For example, if the correlation function $\phi_v(t, u)$ is not positive definite, the eigenfunctions may not form a complete orthonormal set. We can overcome this problem by assuming that the noise has a white component in addition to the correlated component. The thermal noise in the receiver can usually be used to account for this white component.

The determination of the whitening filter is, in general, fairly tedious. The expression for $h_w(\cdot, \cdot)$ in Eq. (4.3.25) is in the form of an infinite series and in general cannot be put in a closed form. Furthermore, the determination of the eigenfunctions involves solving the integral equation of (4.3.1). This in itself is usually a pretty formidable task. We discuss solution techniques for some special cases in Section 4.5.

One case of practical importance for which the whitening filter may be obtained in a straightforward fashion is that of stationary noise with rational spectrum and semi-infinite observation interval, that is $t_0 = -\infty$.

In this case, the whitening filter may be chosen to be time invariant. If $\Phi_v(s)$ is the power spectrum of the noise, we perform a spectral factorization and write (see Example 2.3.1)

$$\Phi_v(s) = \Phi_v^+(s)\Phi_v^-(s) \tag{4.3.26}$$

where $\Phi_v^+(\cdot)$ consists of all poles and zeroes in the left half-plane and $\Phi_v^-(\cdot)$ consists of all poles and zeros in the right half-plane. Then the transfer function of the whitening filter is given by

$$H_w(s) = \frac{1}{\Phi_v^+(s)} \tag{4.3.27}$$

Receiver Performance

We can determine the performance of the receiver of Eq. (4.3.11) or Eq. (4.3.16) from the results in Section 4.2 for the white observation noise problem by replacing $y_0(t)$ and $y_1(t)$ by $y_{w0}(t)$ and $y_{w1}(t)$, respectively, and setting $N_0 = 2$. For the minimum probability of error criterion and equal prior probabilities we can use Eqs. (4.2.43) and (4.2.44) to write the probability of error as

$$P_e = \int_d^\infty \frac{1}{\sqrt{2\pi}} e^{-z^2/2}\, dz \tag{4.3.28}$$

where

$$d^2 = \frac{1}{2}\int_{t_0}^{t_f} [y_{w0}(t) - y_{w1}(t)]^2\, dt \tag{4.3.29}$$

In terms of the original signals this becomes

$$d^2 = \frac{1}{2}\int_{t_0}^{t_f}\left[\int_{t_0}^{t_f} h_w(t, u)[y_0(u) - y_1(u)]\, du\right]^2 dt \tag{4.3.30}$$

The calculation of d^2 requires solving integral equations and is at best tedious. We also note that the performance of the detector is no longer independent of the signal shape.

It is clear from Eq. (4.3.28) that the probability of error decreases monotonically as d^2 increases. To optimize the performance of the communication system, therefore, the signals $y_0(t)$ and $y_1(t)$ must be chosen so as to maximize d^2 subject to the constraint

$$\frac{1}{2}\int_{t_0}^{t_f} [y_1^2(t) + y_0^2(t)]\, dt = \text{constant} = \epsilon \tag{4.3.31}$$

We thus form the objective function

$$J = d^2 - \mu\epsilon$$
$$= \int_{t_0}^{t_f}\left[\int_{t_0}^{t_f} h_w(t,u)[y_0(u) - y_1(u)]\,du\right]^2 dt$$
$$-\mu\int_{t_0}^{t_f}[y_1^2(t) + y_0^2(t)]\,dt \tag{4.3.32}$$

where μ is a Lagrange multiplier. It then follows by using the calculus of variations that the optimum signals $y_1(t)$ and $y_0(t)$ are given by the relations (see Appendix B)

$$\int_{t_0}^{t_f}\int_{t_0}^{t_f}[y_0(u) - y_1(u)]h_w(t,u)h_w(t,v)\,dt\,du$$
$$= 2\mu y_0(v) = -2\mu y_1(v) \tag{4.3.33}$$

We thus note that in the optimum system

$$y_1(v) = -y_0(v) \qquad t_0 \leqslant v \leqslant t_f \tag{4.3.34}$$

This is the same result found for the optimum binary system with white noise and corresponds to the case $\bar{\rho} = -1$ in Eq. (4.2.48). Substitution of Eqs. (4.3.33) and (4.3.34) in Eq. (4.3.29) yields the optimal value for d^2 as

$$d^2 = 2\mu\int_{t_0}^{t_f} y_0^2(t)\,dt = 2\mu\int_{t_0}^{t_f} y_1^2(t)\,dt \tag{4.3.35}$$

Since the signal energies are fixed, d^2 will be maximum (P_e will be minimum) when μ is maximum.

Signal Design

We now consider the problem of finding the optimum signals for transmission. Substitution of Eq. (4.3.34) in Eq. (4.3.33) yields

$$\int_{t_0}^{t_f}\int_{t_0}^{t_f} y_1(v)h_w(t,u)h_w(t,v)\,dt\,du = \mu y_1(v) \tag{4.3.36}$$

If we now multiply both sides of Eq. (4.3.36) by $\phi_v(\nu,\sigma)$, integrate over $t_0 \leqslant \nu \leqslant t_f$ and substitute Eq. (4.3.21) in the left side, we obtain

$$\int_{t_0}^{t_f}\phi_v(\nu,\sigma)y_1(\nu)\,d\nu = \frac{1}{\mu}y_1(\sigma) \tag{4.3.37}$$

We see that the optimum signal waveform is an eigenfunction of the covariance kernel $\phi_v(\nu, \sigma)$ corresponding to the eigenvalue $1/\mu$. Since d^2 is maximum when $1/\mu$ is minimum, we choose as the optimum signal $y_1(t)$, the eigenfunction corresponding to the *smallest* eigenvalue.

Finally, we note that if the noise is white, the eigenvalues are all equal (see Section 2.3) and it makes no difference which waveforms we choose as long as $y_1(t) = -y_0(t)$. This is in agreement with Section 4.2.

We also point out that perfect detection (probability of error is zero) occurs when $d^2 = \infty$. From Eq. (4.3.34) we see that this happens when the smallest eigenvalue of the kernel $\phi_v(u, \nu)$ is zero. In this case, for arbitrarily small time intervals and arbitrarily small energy levels, we achieve perfect detection. Such a situation is referred to as *singular detection*. Since this kind of performance cannot correspond to an actual physical situation, it is important that we make our mathematical model realistic enough to eliminate the possibility of singular detection. Again, a simple way of doing this is to assume that the observation noise has a white component.

4.4 DETECTION OF KNOWN SIGNALS IN NOISE: MAXIMUM SNR CRITERION

In communication problems, a measure of the efficiency of the system is obtained by the signal-to-noise ratio (SNR) which is the ratio of the signal power to the noise power. In fact, we have seen that the probability of error in the binary communication problem of the previous section was dependent on this ratio ϵ/N_0 and that the larger this ratio, the smaller the error probability. We may therefore use the signal-to-noise ratio itself as a criterion of the efficacy of the system and seek to design a receiver which maximizes this ratio [15].

The analysis which follows is applicable in many cases other than with the binary communication problem considered in the last section: for example, in radar systems where it is required to detect the presence of a signal echo in noise. The criterion is useful in applications where we are not interested in maintaining the fidelity of the transmitted pulse but are only interested in recognizing the presence of a signal in noise. In applications such as pulse-position-modulation in which the time of occurrence is of prime importance, it is necessary to retain the fidelity or sharpness of pulse rise time. In such systems, therefore, the signal-to-noise ratio is not a very meaningful measure of the performance of the system.

We consider the following problem. The received signal $z(t)$ consists of known signal $y(t)$ (defined over the interval (t_0, t_f)) and additive noise $v(t)$. The noise is assumed to be at least wide-sense stationary with zero mean and covariance $\phi_v(\tau)$. We seek to design a linear filter so as to maximize the output signal-to-noise ratio at some time T.

Let the impulse response of the filter be $h(t)$. Then the signal and noise components at the output at time T are

$$y_0(T) = \int_{t_0}^{t_f} h(T-\tau)y(\tau)\,d\tau \tag{4.4.1}$$

$$v_0(T) = \int_{t_0}^{t_f} h(T-\tau)v(\tau)\,d\tau \tag{4.4.2}$$

The noise power is given by

$$\begin{aligned} \mathrm{E}\{v_0^2(T)\} &= \mathrm{E}\left\{\int_{t_0}^{t_f} h(T-\tau)v(\tau)\,d\tau \int_{t_0}^{t_f} h(T-\sigma)v(\sigma)\,d\sigma\right\} \\ &= \int_{t_0}^{t_f}\int_{t_0}^{t_f} h(T-\tau)h(T-\sigma)\phi_v(\tau-\sigma)\,d\sigma\,d\tau \end{aligned} \tag{4.4.3}$$

The ratio of signal to noise power is then given by $y_0^2(T)/\mathrm{E}\{v_0^2(T)\}$. To make the maximization of this quantity meaningful, we assume that $y_0(T)$ is constrained to be a constant. The problem of maximizing the signal-to-noise ratio is then equivalent to that of minimizing the quantity $\mathrm{E}\{v_0^2(T)\}$ subject to the constraint $y_0(T)$ is a constant. We thus form the objective function (see Appendix B).

$$J = \mathrm{E}\{v_0^2(T)\} - 2\mu y_0(T) \tag{4.4.4}$$

where μ is a Lagrange multiplier. Substituting from Eq. (4.4.1) and Eq. (4.4.3) we can write Eq. (4.4.4) as

$$\begin{aligned} J &= \int_{t_0}^{t_f}\int_{t_0}^{t_f} h(T-\tau)h(T-\sigma)\phi_v(\tau-\sigma)\,d\sigma\,d\tau \\ &\quad -2\mu\int_{t_0}^{t_f} h(T-\tau)y(\tau)\,d\tau \\ &= \int_{t_0}^{t_f} h(T-\tau)\left[\int_{t_0}^{t_f} h(T-\sigma)\phi_v(\tau-\sigma)\,d\sigma - 2\mu y(\tau)\right]d\tau \end{aligned} \tag{4.4.5}$$

We now use the calculus of variations (Appendix B) to minimize J with respect to the filter function $h(t)$. The first variation in the cost function can easily be shown to be

$$\delta J = \int_{t_0}^{t_f}\int_{t_0}^{t_f}[\delta h(T-\tau)h(T-\sigma) + \delta h(T-\sigma)h(T-\tau)]\phi_v(\tau-\sigma)\,d\tau\,d\sigma - 2\mu\int_{t_0}^{t_f}\delta h(T-\tau)y(\tau)\,d\tau \tag{4.4.6}$$

which can be written as

$$\delta J = \int_{t_0}^{t_f} 2\delta h(T-\tau)\left[\left\{\int_{t_0}^{t_f} h(T-\sigma)\phi_v(\tau-\sigma)\,d\sigma\right\} - \mu y(\tau)\right]d\tau \tag{4.4.7}$$

For a minimum, δJ must be zero. Since $\delta h(\cdot)$ is arbitrary, it follows that the inner integral must be identically zero over the interval $t_0 \leqslant \tau \leqslant t_f$. Thus the optimum filter must satisfy the equation

$$\int_{t_0}^{t_f} h(T-\sigma)\phi_v(\tau-\sigma)\,d\sigma = \mu y(\tau) \qquad t_0 \leqslant \tau \leqslant t_f \tag{4.4.8}$$

Since the constant multiplier μ changes only the filter gain and does not affect the signal-to-noise ratio, we can arbitrarily set it equal to unity. Therefore the filter function should satisfy the integral equation

$$\int_{t_0}^{t_f} h(T-\sigma)\phi_v(\tau-\sigma)\,d\sigma = y(\tau) \qquad t_0 \leqslant \tau \leqslant t_f \tag{4.4.9}$$

Determination of the optimum filter therefore requires that we use information about the signal $y(t)$ over the entire observation interval. If $T < t_f$, the optimum system corresponds to a physically nonrealizable (noncausal) system: in order to make a decision at time T, we need information about the signal in the future, that is on the interval $[T, t_f]$. If the received signal needs to be processed in real time, we will have to make a decision at time T based only on the data available up to that time. Thus the receiver structure will not be truly optimum. However if the processing can be carried out in non–real time on stored or recorded data, we can arrive at a decision based on the signal-to-noise ratio at time T using the values of the received signal over the entire observation interval. The processing delay in this case is equal to $t_f - T$.

The maximum signal-to-noise ratio (SNR) obtained by substituting the optimum filter function in Eqs. (4.4.1) and (4.4.3) is given by

$$\mathrm{SNR}_{\max} = \frac{\left[\int_{t_0}^{t_f} h(T-\tau)y(\tau)\,d\tau\right]^2}{\int_{t_0}^{t_f} h(T-\tau)\left[\int_{t_0}^{t_f} h(T-\sigma)\phi_v(\tau-\sigma)\,d\sigma\right] d\tau}$$

$$= \int_{t_0}^{t_f} h(T-\tau)y(\tau)\,d\tau \tag{4.4.10}$$

where we have substituted Eq. (4.4.9) in the denominator.

To explicitly determine the optimum filter, we need to solve Eq. (4.4.9). This is a Fredholm integral equation of the first kind [16]. In general, the solution of such equations is not straightforward. We consider some solution techniques for integral equations in Section 4.5. A special case, in which the solution can be obtained easily is when the noise signal is white with autocorrelation $(N_0/2)\delta(\tau)$. For this case, we easily see that the solution to Eq. (4.4.9) is given by

$$h(T-\tau) = \frac{2}{N_0}y(\tau) \qquad t_0 \leqslant \tau \leqslant t_f \tag{4.4.11}$$

or equivalently

$$h(\tau) = \frac{2}{N_0}y(T-\tau) \qquad T-t_f \leqslant \tau \leqslant T-t_0 \tag{4.4.12}$$

which is precisely the matched filter discussed in Section 4.2. The maximum signal-to-noise ratio is obtained from Eq. (4.4.10) as

$$\mathrm{SNR}_{\max} = \int_{t_0}^{t_f} \frac{2}{N_0} y^2(\tau)\,d\tau = \frac{2\epsilon}{N_0} \tag{4.4.13}$$

where ϵ is the signal energy. We now consider the frequency domain representation of the filter in Eq. (4.4.12).

By taking the (bilateral) Laplace transform of Eq. (4.4.12), we obtain

$$H(s) = \frac{2}{N_0}Y(-s)e^{-sT} \tag{4.4.14}$$

If we assume that $y(t) = 0$ for $t > T$, then $y(T-\tau)$ in Eq. (4.4.12) is zero for $\tau < 0$. It therefore follows that $h(\tau)$ is a causal (physically realizable) function and that its transform, $H(s)$, has poles only in the left half-plane.

If $T < t_f$ but $y(t)$ is not zero for $t > T$, $y(T-\tau)$ is not zero for $\tau < 0$ and the optimum filter will not be physically realizable. As discussed earlier, we can overcome this problem by operating on stored data. If we want to restrict ourselves to causal filters, we can proceed as follows. Let $\{Y(-s)e^{-sT}\}_+$ and $\{Y(-s)e^{-sT}\}_-$ represent the transforms of the causal and anticausal parts of $y(T-\tau)$. The causal filter is then obtained as

$$H_c(s) = \frac{2}{N_0}\{Y(-s)e^{-sT}\}_+ \tag{4.4.15}$$

When the observation noise is not white, we may proceed in a manner similar to that in Section 4.3. That is, we first pass the observations through a whitening filter which whitens the observation noise and then design a filter matched to the signal output from the whitening filter. Specifically, if the impulse response of the whitening filter is $h_1(t, u)$, the output will be

$$\begin{aligned} z_1(t) &= \int_{t_0}^{t_f} h_1(t, u) z(u)\, du \\ &= \int_{t_0}^{t_f} h_1(t, u) y(u)\, du + \int_{t_0}^{t_f} h_1(t, u) v(u)\, du \\ &= y_1(t) + v_1(t) \end{aligned} \tag{4.4.16}$$

Since $v_1(t)$ is white, we can use the results of Eq. (4.4.12) to obtain the optimum matched filter $h_2(t)$. The filter $h_2(t)$ has to be matched to the signal $y_1(t)$ so that

$$h_2(t) = \frac{2}{N_0} y_1(T-t) \qquad T - t_f \leqslant t \leqslant T - t_0 \tag{4.4.17}$$

Again the basic problem is in determining the whitening filter. As discussed earlier, if $t_0 = -\infty$, this filter is easily determined in terms of the spectrum of the noise signal $v(t)$ as

$$H_1(s) = \frac{1}{\Phi_v^+(s)} \tag{4.4.18}$$

which represents a causal system. Transformation of Eq. (4.4.17) yields the transfer function of the filter $h_2(t)$ as

$$H_2(s) = \frac{2}{N_0}[Y_1(-s)e^{-sT}] \tag{4.4.19}$$

Since

$$Y_1(s) = H_1(s)Y(s)$$
$$= \frac{1}{\Phi_v^+(s)} Y(s) \tag{4.4.20}$$

it follows that

$$H_2(s) = \frac{2}{N_0}\left[\frac{Y(-s)e^{-sT}}{\Phi_v^-(s)}\right] \tag{4.4.21}$$

The realizable part of $H_2(s)$ is thus given by

$$H_{2c}(s) = \frac{2}{N_0}\left[\frac{Y(-s)e^{-sT}}{\Phi_v^-(s)}\right]_+ \tag{4.4.22}$$

The optimum causal filter obtained by cascading $H_1(s)$ and $H_{2c}(s)$ is thus given by

$$H(s) = H_1(s)H_{2c}(s)$$
$$= \frac{1}{\Phi_v^+(s)}\left\{\frac{Y(-s)e^{-sT}}{\Phi_v^-(s)}\right\}_+ \tag{4.4.23}$$

We conclude this section with an example.

Example 4.4.1 Consider the problem of designing a matched filter to detect the signal

$$y(t) = \begin{cases} e^{-t/2} - e^{-t} & t \geqslant 0 \\ 0 & t < 0 \end{cases}$$

The noise spectral density is

$$\Phi_v(s) = \frac{1}{1-s^2}$$

The signal-to-noise ratio is to be maximized at some specified time $T > 0$.
The noise spectrum may be factored as

$$\Phi_v(s) = \Phi_v^+(s)\Phi_v^-(s) = \frac{1}{1+s}\cdot\frac{1}{1-s}$$

so that the whitening filter has transfer function

$$H_1(s) = \frac{1}{\Phi_v^+(s)} = 1 + s$$

The signal output from the whitening filter is

$$Y_1(s) = Y(s)H_1(s) = \left[\frac{2}{1+2s} - \frac{1}{1+s}\right](1+s)$$

or

$$Y_1(s) = \frac{1}{1+2s}$$

The corresponding time function is

$$y_1(t) = \frac{1}{2}e^{-t/2} \qquad \text{for} \quad t \geqslant 0$$

and exists in white noise of spectral height unity. Since $y_1(t)$ is nonzero for all $t \geqslant 0$, the optimum matched filter in this case is unrealizable and is given by

$$H_2(s) = Y_1(-s)\, e^{-sT} = \frac{1}{1-2s} e^{-sT}$$

with impulse response

$$h_2(t) = \begin{cases} 1/2e^{(t-T)/2} & -\infty < t < T \\ 0 & t > T \end{cases}$$

We take as the realizable filter, the causal part of the preceding

$$h_{2c}(t) = \begin{cases} 1/2e^{(t-T)/2} & 0 \leqslant t \leqslant T \\ 0 & \text{elsewhere} \end{cases}$$

with transfer function

$$H_{2c}(s) = \int_0^T \frac{1}{2} e^{(t-T)/2} e^{-st}\, dt$$

$$= \frac{1}{1-2s}(e^{-sT} - e^{-T/2})$$

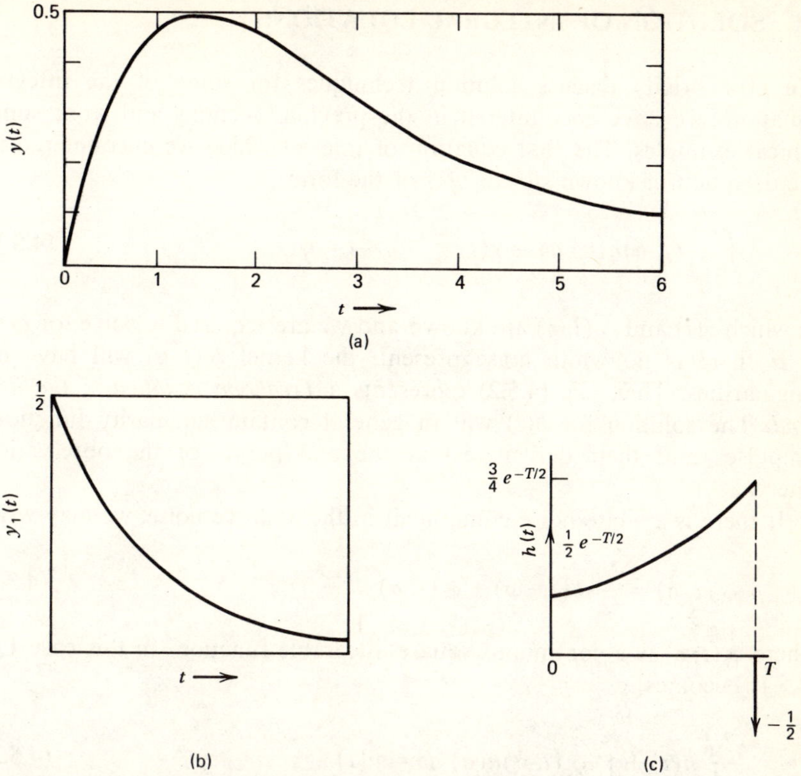

Fig. 4.4.1 Matched filter, Example 4.4.1. (a) Signal input. (b) Signal output from whitening filter. (c) Impulse response of optimum filter.

The overall matched filter is given by

$$H(s) = H_1(s)H_{2c}(s) = \frac{1+s}{1-2s}(e^{-sT} - e^{-T/2})$$
$$= \left[-1/2 + \frac{3/2}{1-2s}\right](e^{-sT} - e^{-T/2})$$

The impulse response of the optimal filter is given by

$$h(t) = -\frac{1}{2}\delta(t-T) + \frac{1}{2}e^{-T/2}\delta(t) + \frac{3}{4}e^{(t-T)/2}[\mathrm{u}(T-t) - \mathrm{u}(-t)]$$

where $\mathrm{u}(\cdot)$ denotes the unit step function. The various signals are illustrated in Fig. 4.4.1.

4.5 SOLUTION OF INTEGRAL EQUATIONS [16–20]

We now briefly discuss solution techniques for some of the integral equations we have encountered in the previous sections and work some typical examples. The first equation of interest which we encountered in the detection of known signals was of the form

$$\int_{t_0}^{t_f} \phi_v(t, \sigma)h(\sigma)\, d\sigma = y(t) \qquad t_0 \leqslant t \leqslant t_f \tag{4.5.1}$$

in which $y(t)$ and $\phi_v(t, \sigma)$ are known and we are required to solve for $h(t)$.

If there is no white noise present, the kernel $\phi_v(t, \sigma)$ will have no singularities. Then Eq. (4.5.2) represents a *Fredholm equation of the first kind.* The solution for $h(t)$ will in general contain singularity functions (impulses and their derivatives) at the end points of the observation interval.

If there is a white noise component in the additive noise, we may write

$$\phi_v(t, \sigma) = \frac{N_0}{2}\delta(t-\sigma) + \phi_c(t, \sigma) \tag{4.5.2}$$

where $\phi_c(t, \sigma)$ is a continuous square-integrable function. In this case Eq. (4.5.1) becomes

$$\frac{N_0}{2}h(t) + \int_{t_0}^{t_f} \phi_c(t, \sigma)h(\sigma)\, d\sigma = y(t) \tag{4.5.3}$$

This is called a Fredholm equation of the second kind.

The solutions to these equations are at best tedious and in many cases they do not appear to be solvable. When the kernel is of certain special types, however, straightforward procedures for solving Eq. (4.5.1) and (4.5.3) are available. One type occurs when the noise $v(t)$ is the steady state response of a linear time-invariant system excited with white noise. In this case, the noise output is stationary and the noise power spectrum is the ratio of two polynomials in s^2. Let us write the spectrum as

$$\Phi_v(s) = \int_{-\infty}^{\infty} \phi_v(\tau)e^{-s\tau}\, d\tau \triangleq \frac{N(s^2)}{D(s^2)} \tag{4.5.4}$$

In order for the noise to have finite power, the denominator $D(s^2)$ must be at least two degrees higher than the numerator. Kernels whose transforms satisfy Eq. (4.5.4) are called *rational* kernels.

The basic approach in solving an integral equation with rational kernel is to first find an equivalent differential equation. Because of the form of the kernel, the differential equation will be linear with constant coefficients and hence its solution can be easily determined. To obtain the differential equation, assume that

$$\Phi_v(s) = \frac{N(s^2)}{D(s^2)} \tag{4.5.5}$$

Multiplying both sides by $D(s^2)$ yields

$$D(s^2)\Phi_v(s) = N(s^2) \tag{4.5.6}$$

Recalling that multiplication by s in the frequency domain corresponds to differentiation with respect to t, we can write Eq. (4.5.6) as

$$D(p^2)\phi_v(t-u) = N(p^2)\delta(t-u) \tag{4.5.7}$$

where $p \triangleq d/dt$. Now the equation of interest is

$$y(t) = \int_{t_0}^{t_f} \phi_v(t-\sigma)h(\sigma)\,d\sigma \qquad t_0 \leqslant t \leqslant t_f \tag{4.5.8}$$

Operating on both sides by $D(p^2)$ we obtain

$$D(p^2)y(t) = \int_{t_0}^{t_f} D(p^2)\phi_v(t-\sigma)h(\sigma)\,d\sigma \qquad t_0 \leqslant t \leqslant t_f \tag{4.5.9}$$

Substitution of Eq. (4.5.7) in the right side of Eq. (4.5.9) then yields

$$D(p^2)y(t) = N(p^2)h(t) \qquad t_0 \leqslant t \leqslant t_f \tag{4.5.10}$$

The solution consists of three parts. The homogeneous solution is found by solving

$$N(p^2)h(t) = 0$$

and will consist of exponential terms. The particular solution is found by solving Eq. (4.5.10) directly without regard to the limits $[t_0, t_f]$ or equivalently by using transform techniques.

For example, transforming both sides of Eq. (4.5.10) yields

$$H(s) = \frac{D(s^2)Y(s)}{N(s^2)} = \frac{Y(s)}{\Phi_v(s)}$$

so that the particular solution is

$$h_p(t) = \frac{1}{2\pi j}\int_{c-j\infty}^{c+j\infty} \frac{Y(s)}{\Phi_v(s)} e^{st}\, ds \qquad t_0 \leqslant t \leqslant t_f$$

The total solution is found by adding the homogeneous and particular solutions to terms of the form

$$\sum_{i=0}^{p-q-1} a_i \delta^{(i)}(t - t_0) + b_i \delta^{(i)}(t - t_f)$$

where $\delta^{(i)}(t)$ is the ith derivative of the delta function and where p and q refer to the order of $D(s^2)$ and $N(s^2)$, respectively. The coefficients a_i and b_i are evaluated by substituting the assumed solution into the integral equation.

Example 4.5.1 We consider the problem of designing a correlation receiver for a known signal observed over the interval $[0, T]$ in the presence of noise with spectral density and autocorrelation function

$$\Phi_v(s) = \frac{2\alpha\beta}{-s^2 + \beta^2} \qquad \text{and} \quad \phi_v(\tau) = \alpha\, e^{-\beta|\tau|}$$

The relevant differential equation corresponding to Eq. (4.5.10) is easily seen to be

$$2\alpha\beta h(t) = \beta^2 y(t) - \ddot{y}(t) \qquad 0 \leqslant t \leqslant T$$

Thus, the equation reduces to a simple algebraic one for this example. We have

$$h(t) = \frac{1}{2\alpha\beta}\left[\, \beta^2 y(t) - \ddot{y}(t)\right]$$

Since $p - q - 1 = 0$, we modify the solution by including delta functions at the endpoints but not derivatives of the delta functions. The total solution is therefore of the form

$$h(t) = \frac{1}{2\alpha\beta}\left[\, \beta^2 y(t) - \ddot{y}(t)\right] + a\delta(t) + b\delta(t - T)$$

If we substitute this solution into the integral Eq. (4.5.1) we obtain

$$y(t)=\int_0^T \alpha e^{-\beta|t-\sigma|}\left\{\frac{1}{2\alpha\beta}\left[\beta^2 y(\sigma)-\ddot{y}(\sigma)\right] + a\delta(\sigma)+b\delta(\sigma-T)\right\}d\sigma \qquad 0 \leqslant t \leqslant T$$

Carrying out the integration on the right-hand side yields

$$y(t)=y(t)+e^{-\beta t}\left[a\alpha+\frac{\dot{y}(0)}{2\beta}-\frac{y(0)}{2}\right] + e^{\beta t}e^{-\beta T}\left[\alpha b-\frac{\dot{y}(T)}{2\beta}-\frac{y(T)}{2}\right] \qquad 0 \leqslant t \leqslant T$$

Since this equation must be satisfied for all t in the interval $[0, T]$ this implies that

$$a=\frac{1}{2\alpha\beta}\left[\beta y(0)-\dot{y}(0)\right] \qquad \text{and} \qquad b=\frac{1}{2\alpha\beta}\left[\beta y(T)+\dot{y}(T)\right]$$

The solution for $h(t)$ is thus

$$h(t)=\frac{1}{2\alpha\beta}\left\{\left[\beta y(0)-\dot{y}(0)\right]\delta(t)+\left[\beta y(T)+\dot{y}(T)\right]\delta(t-T) + \beta^2 y(t)-\ddot{y}(t)\right\} \qquad 0 \leqslant t \leqslant T$$

The output of the processor is

$$\int_0^T z(t)h(t)\,dt = az(0)+bz(T)+\frac{1}{2\alpha\beta}\int_0^T z(t)\left[\beta^2 y(t)-\ddot{y}(t)\right]dt$$

The solution to the Fredholm equation of the second kind proceeds in exactly the same way except that we do not need to add singularity functions at the end points.

For example, let us assume that there is also a white component in the noise spectrum so that

$$\Phi_v(s)=\frac{2\alpha\beta}{-s^2+\beta^2}+\frac{N_0}{2}=\frac{N_0/2(-s^2+\gamma^2)}{-s^2+\beta^2}=\frac{N(s^2)}{D(s^2)}$$

where $\gamma^2=\beta^2+4\alpha\beta/N_0$.

The optimum filter $h(t)$ is obtained by solving the integral equation

$$\int_0^T \phi_v(t-\sigma)h(\sigma)\,d\sigma = y(t)$$

which in this case reduces to (see Eq. 4.5.3)

$$\frac{N_0}{2}h(t) + \int_0^T \alpha e^{-\beta|t-\sigma|}h(\sigma)\,d\sigma = y(t)$$

The corresponding differential equation is obtained from Eq. (4.5.10) as

$$\frac{N_0}{2}\left[-\ddot{h}(t) + \gamma^2 h(t)\right] = -\ddot{y}(t) + \beta^2 y(t)$$

The complementary solution is easily seen to be

$$h_c(t) = \alpha_1 e^{\gamma t} + \alpha_2 e^{-\gamma t}$$

To obtain the particular solution, we note that

$$H(s) = \frac{2}{N_0}\left[\frac{-s^2+\beta^2}{-s^2+\gamma^2}\right]Y(s)$$

The particular solution can thus be obtained as

$$h_p(t) = \frac{1}{2\pi j}\int_{c-j\infty}^{c+j\infty}\frac{2}{N_0}\left[\frac{-s^2+\beta^2}{-s^2+\gamma^2}\right]Y(s)\,e^{st}\,ds$$

The total solution is obtained as

$$h(t) = h_p(t) + \alpha_1 e^{\gamma t} + \alpha_2 e^{-\gamma t} \qquad 0 \leqslant t \leqslant T$$

We conclude this section with a consideration of integral equations with separable kernels. A kernel is said to be separable if it can be written as

$$\phi_v(t, \sigma) = \sum_{i=1}^{N} \lambda_i g_i(t) g_i(\sigma) \qquad t_0 \leqslant t, u \leqslant t_f \tag{4.5.11}$$

where λ_i and $g_i(t)$ are the eigenvalues and eigenfunctions of the equation

$$\int_{t_0}^{t_f} \phi_v(t, \sigma) g_i(\sigma)\,d\sigma = \lambda_i g_i(t) \tag{4.5.12}$$

Note that N is assumed to be finite in Eq. (4.5.11), since if N were allowed to be infinite, all kernels could be considered separable. We can easily show in this case that

$$h(t)=\begin{cases}\dfrac{2}{N_0}\left[y(t)-\sum_{i=1}^{N}\dfrac{y_i\lambda_i}{N_0/2+\lambda_i}g_i(t)\right] & t_0\leqslant t\leqslant t_f\\ 0 & \text{elsewhere}\end{cases}$$

where

$$y(t)=\sum_{i=1}^{N}y_i g_i(t)$$

Recent results for the solution of integral equations have used a state variable model for the generation of the covariance kernel. The technique again is to obtain a set of differential equations from the integral equation which are then solved to obtain the required solution to the integral equation. The details are quite involved and the reader is referred to the literature [21–23].

4.6 DETECTION OF SIGNALS WITH UNKNOWN PARAMETERS

We have so far assumed in this chapter that the signals of concern were completely known. The only uncertainty was caused by additive noise. In many practical situations, however, this assumption is not valid. For example in a radar tracking problem, the reflected pulse acquires a random phase. In such cases the signals would be functions of an unknown parameter vector. Thus for the binary detection problem considered in this chapter, the received signals under the two hypotheses would be given by

$$\begin{aligned}H_1 &: z(t)=y_1(t,\boldsymbol{\theta}_1)+v(t) \qquad t_0\leqslant t\leqslant t_f\\ H_0 &: z(t)=y_0(t,\boldsymbol{\theta}_0)+v(t) \qquad t_0\leqslant t\leqslant t_f\end{aligned} \tag{4.6.1}$$

where $\boldsymbol{\theta}_0$ and $\boldsymbol{\theta}_1$ denote the unknown parameters. We assume that the signals $y_1(t, \boldsymbol{\theta}_1)$ and $y_0(t, \boldsymbol{\theta}_0)$ are conditionally deterministic. The problem is the waveform counterpart to the composite hypothesis testing problem in Chapter 3. Here we very briefly discuss the approach to solving such problems and defer till Chapter 7 some of the details.

The parameters $\boldsymbol{\theta}_i$ may be either random variables or nonrandom variables. In the former case, the associated probability density functions may or may not be known. The first case of known density functions is the one most frequently encountered in practice. The solution in this case is straightforward. We again expand the observations in a set of orthogonal basis functions and denote the coefficients of the K-term approximation by the K-dimensional vector $\mathbf{z}_K$. We then construct the likelihood ratio and let $K \to \infty$. To construct the likelihood ratio, we need the posterior density functions $p(\mathbf{z}_K|H_1)$ and $p(\mathbf{z}_K|H_0)$. If the prior densities for $\boldsymbol{\theta}_i$ are known, we can calculate these two densities as

$$p(\mathbf{z}_K|H_1) = \int_{\chi_{\boldsymbol{\theta}_1}} p(\mathbf{z}_K|\boldsymbol{\theta}_1, H_1)p(\boldsymbol{\theta}_1|H_1)\, d\boldsymbol{\theta}_1 \tag{4.6.2}$$

and

$$p(\mathbf{z}_K|H_0) = \int_{\chi_{\boldsymbol{\theta}_0}} p(\mathbf{z}_K|\boldsymbol{\theta}_0, H_0)p(\boldsymbol{\theta}_0|H_0)\, d\boldsymbol{\theta}_0 \tag{4.6.3}$$

where $\chi_{\boldsymbol{\theta}_i}$, $i = 0, 1$ denotes the space of the parameter $\boldsymbol{\theta}_i$.

The likelihood ratio is then given as

$$\Lambda[z(t)] = \lim_{K\to\infty} \frac{p(\mathbf{z}_K|H_1)}{p(\mathbf{z}_K|H_0)} \tag{4.6.4}$$

While the procedure to be used is conceptually straightforward, the actual computation of the integrals depends on the form of the functions to be integrated.

If $\boldsymbol{\theta}_0$ and $\boldsymbol{\theta}_1$ are nonrandom variables, we may again try estimating the values of θ_i under the two hypothesis H_1 and H_0 and use these estimates in the LRT. Using maximum likelihood estimates† of the parameters yields the generalized likelihood ratio

$$\Lambda_g[z(t)] = \lim_{K\to\infty} \frac{\max\limits_{\boldsymbol{\theta}_1} p(\mathbf{z}_K|\boldsymbol{\theta}_1)}{\max\limits_{\boldsymbol{\theta}_0} p(\mathbf{z}_K|\boldsymbol{\theta}_0)} \underset{H_0}{\overset{H_1}{\gtrless}} \gamma \tag{4.6.5}$$

which can be compared with a threshold γ to make a decision.

Example 4.6.1 Consider the on–off keying system of Example 3.2.1 in which a digital one is conveyed by transmitting a sinusoidal waveform of known amplitude and frequency and a digital zero by no transmission. The receiver observes the transmitted signal in the presence of additive, zero

†See Section 5.4.

mean, Gaussian, white noise $v(t)$ of spectral density $N_0/2$. In the case of transmission over fading channels, the attenuation in the received signal can be modeled as a random variable. Further, the transmission undergoes a random change in phase. The observations under the two hypotheses can thus be modeled as

$$H_1 : z(t) = \sqrt{2}\, E \cos(\omega_c t + \theta) + v(t) \qquad 0 \leqslant t \leqslant T$$
$$H_0 : z(t) = v(t) \qquad 0 \leqslant t \leqslant T$$

where E and θ are assumed to be random variables with known prior distributions. We will assume that there are an integral number of cycles corresponding to ω_c in the interval $[0, T]$.

To determine the decision rule for deciding between the two hypotheses, we first obtain the conditional likelihood ratio $\Lambda(z|E, \theta)$. Then the likelihood ratio can be obtained as

$$\Lambda(z(t)) = \int_{\chi_E} \int_{\chi_\theta} \Lambda(z(t)|E, \theta) p(E, \theta)\, dE\, d\theta$$

where χ_E and χ_θ are the ranges of E and θ, respectively. To derive the conditional likelihood ratio, we expand $z(t)$ in an orthonormal basis. We choose as our first basis function

$$g_1(t) = \sqrt{2} \cos \omega_c t$$

The other orthonormal functions can be chosen arbitrarily. It is clear that only the coefficient z_1 in the expansion will depend on which hypothesis is true. Hence

$$z_1 = \int_0^T z(t) g_1(t)\, dt$$

is a sufficient statistic. The conditional likelihood ratio can thus be written as

$$\Lambda(z(t)|E, \theta) = \frac{p(z_1|E, \theta, H_1)}{p(z_1|E, \theta, H_0)}$$

From our assumptions, it is clear that z_1 is Gaussian under both hypotheses with moments given by

$$E\{z_1|E, \theta, H_1\} = E \cos \theta \qquad \text{and} \qquad E\{z_1|E, \theta, H_0\} = 0$$

$$\text{var}\{z_1|E, \theta, H_1\} = \text{var}\{z_1|E, \theta, H_0\} = N_0$$

so that

$$\Lambda(z(t)|E,\theta)=\exp\left(\frac{2}{N_0}z_1E\cos\theta\right)\exp\left(-\frac{1}{N_0}E^2\cos^2\theta\right)$$

We now average this conditional likelihood function over the joint density function $p(E,\theta)$. A frequently used channel model is the Rayleigh fading channel in which E and θ are assumed to be independent with E having a Rayleigh density function and θ being uniformly distributed in $(0, 2\pi)$. Thus

$$p(E,\theta)=\begin{cases}\dfrac{1}{2\pi}\dfrac{2E}{\epsilon}\exp\left(\dfrac{-E^2}{\epsilon}\right) & E\geqslant 0,\ 0\leqslant\theta\leqslant 2\pi\\ 0 & \text{otherwise}\end{cases}$$

where ϵ is the average energy. In order to evaluate the likelihood function, we make the change of variables $l\triangleq E\cos\theta$. It then follows that

$$p(l)=\frac{1}{\sqrt{\pi\epsilon}}\exp\left[-\frac{l^2}{\epsilon}\right]\qquad -\infty<l<\infty$$

The likelihood ratio conditioned on the variable l is

$$\Lambda(z(t)|l)=\exp\left[\frac{2}{N_0}z_1l-\frac{1}{N_0}l^2\right]$$

so that

$$\begin{aligned}\Lambda(z(t)|l)&=\int_{-\infty}^{\infty}\Lambda(z(t)|l)p(l)\,dl\\&=\int_{-\infty}^{\infty}\frac{1}{\sqrt{\pi\epsilon}}\exp\left\{-l^2\left(\frac{\epsilon+N_0}{N_0}\right)+\frac{2}{N_0}z_1l\right\}dl\\&=\frac{N_0}{\epsilon+N_0}\exp\left[\frac{\epsilon/N_0}{\epsilon+N_0}z_1^2\right]\end{aligned}$$

The decision rule is

$$\frac{N_0}{\epsilon+N_0}\exp\left[\frac{\epsilon/N_0}{\epsilon+N_0}z_1^2\right]\underset{H_0}{\overset{H_1}{\gtrless}}\gamma$$

where γ is a threshold depending on the criterion chosen. Taking logarithms and rearranging yields the decision rule in terms of a new threshold

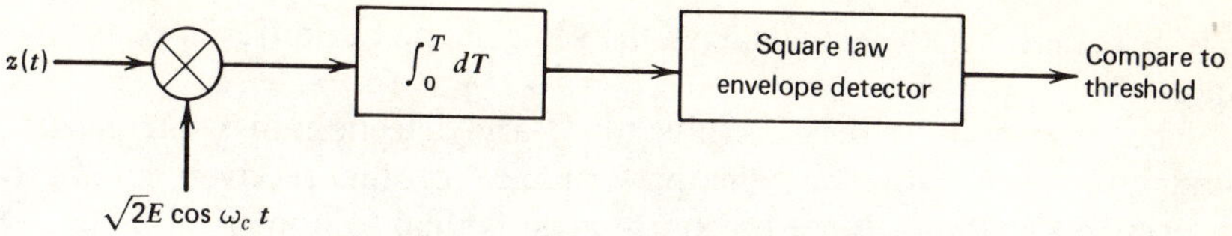

Fig. 4.6.1 Receiver structure for Example 4.6.1.

η as

$$z_1^2 \underset{H_0}{\overset{H_1}{\gtrless}} \eta$$

Thus the receiver can be implemented as a correlator followed by a square law envelope detector as shown in Fig. 4.6.1.

The performance of the receiver can be derived following the procedure of Section 4.2 and is left as an exercise for the reader.

Decision rules for the other communication schemes of Example 4.2.1 and their performance is discussed in the wider context of digital communications in Chapter 7. A similar situation also arises in detecting the presence of a fluctuating radar target and is discussed in Chapter 8.

4.7 SUMMARY

In this chapter, we have extended our results on hypothesis testing to the problem of detecting waveforms in the presence of noise. In Section 4.2, we considered the problem of deciding between two known signal waveforms when the observation noise is additive, white, Gaussian. Since the results of Chapter 3 were for discrete random variables, in order to apply those results to the present case, the observations were expanded in a set of orthonormal basis functions. Since the observations can now be expressed in terms of the coefficients of the expansion, the likelihood ratio can be computed easily. The decision rule consists in correlating the observations with each of the signal waveforms and comparing the difference with a threshold. The correlation operation can be replaced by a matched filter whose impulse response is matched to the signal waveform. The performance of the receiver is independent of signal shape and depends only on ϵ, the average energy of the two signals, ρ, the correlation between the signals and N_0, the spectral density of the observation noise. For the minimum probability of error criterion, the performance is optimum if

$\rho = -1$, that is if the signals have the same shape but differ in phase by π radians.

The extension of these results to M-ary detection in white noise is straightforward. However, the performance of the receiver cannot be determined analytically and recourse must be had to numerical tables.

In Section 4.3, we considered the case of colored observation noise with known covariance function $\phi_v(t, u)$. One approach to obtaining the decision rule in this case is to pass the observations through a whitening filter whose impulse response can be predetermined. The function of the whitening filter is to convert the colored observation noise to white noise so that the results of Section 4.2 become applicable. The determination of the whitening filter, however, is in most cases, difficult. A formal solution can be obtained in terms of the eigenvalues and eigenfunctions of an integral equation with kernel $\phi_v(t, u)$. The solution of the integral equation is in many cases a formidable task.

A case in which the solution is determined in a relatively straightforward fashion is when the noise is stationary, with rational spectrum and the observation interval is semi-infinite. The whitening filter is then obtained from a spectral factorization of the noise spectrum.

The evaluation of the receiver performance is again not easy. For minimum probability of error, the correlation ρ is again -1. The receiver performance is, however, not independent of signal shape in this case. In fact the optimum signal is the eigenfunction corresponding to the smallest eigenvalue.

A different criterion, namely maximum output signal-to-noise ratio was used in Section 4.4 to obtain the optimum linear time invariant filter for the detection of known signals in noise. The receiver is again a filter matched to the signal.

While Section 4.5 was concerned with solution techniques for some integral equations of interest, in Section 4.6 we discussed the detection of signals with unknown parameters. Such problems arise, for example, in communication systems in which there is fading over the channel or in analyzing radar returns for target ranging and tracking. Many problems in pattern classification also fall in this category. The problem is the waveform counterpart of the composite hypothesis testing problem of Chapter 3 and the techniques can be extended to this problem.

One topic that we have not discussed is the detection of stochastic signals in noise. In many applications, the observations, instead of consisting of a known signal observed in additive noise, must be modeled as a random signal in noise. Such instances arise for example, in multipath transmission over a number of randomly varying (fading) paths. At the receiver the effect of these multiple transmissions can be modeled as a narrow-band Gaussian signal. In radio astronomy or spectroscopy, the received signals will be very narrow-band random signals. While we can obtain the receiver structure for this case by following an approach similar to the known signal case, the structure of the receiver is difficult to determine explicitly [1,4,23–26].

We recall that in the case of the detection of a known signal in white noise, the receiver correlates the observations with the known signal to generate the sufficient statistic. For the detection of a random signal, we might intuitively expect that the sufficient statistic may be generated by correlating an *estimate* of the signal with the observation. While we will discuss the estimation of random waveforms in Chapter 6, we note that for the detection of Gaussian signals in Gaussian white noise, the receiver does indeed have an estimator–correlator structure and that the estimate to be used is the minimum-mean-square estimate [26,29].

Finally, the extension of the sequential detection procedure of Section 3 to the detection of known signals in noise is fairly straightforward and is left as an exercise to the reader.

EXERCISES

4.1 Design a receiver to choose between the hypotheses

$$H_1 : z(t) = y_1(t) + v(t)$$
$$H_0 : z(t) = y_0(t) + v(t)$$

if the signals $y_1(t)$ and $y_0(t)$ are as shown in Fig. E.4.1. Use minimum probability of error criterion. Assume that $v(t)$ is zero mean white, Gaussian with spectral density $N_0/2$. Plot the probability of error as a function of $2\varepsilon/N_0$. Assume equal prior probabilities for the two hypotheses.

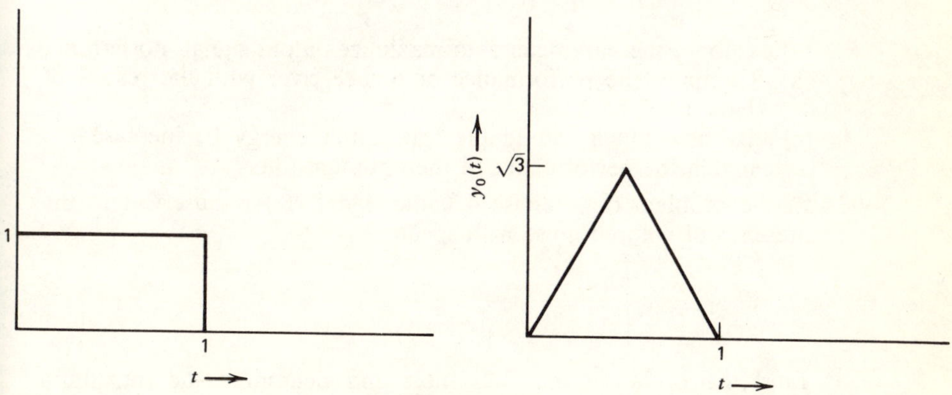

Fig. E4.1

4.2 In a ternary communication system, the transmitted signals over an interval $0 \leqslant t \leqslant T$ are as follows:

$$H_0 : y_0(t) = 0$$
$$H_1 : y_1(t) = A_1 \sin \omega_0 t$$
$$H_2 : y_2(t) = A_2 \sin \omega_0 t$$

The signals are observed in white additive noise of spectral density $N_0/2$

(a) Design a receiver which minimizes the probability of error.

(b) Find the probabilities of detection for the three hypotheses. Hence, find the probability of a correct decision. Assume equal prior probabilities for the hypotheses.

4.3 Design a filter that maximizes the output signal to noise ratio when the transmitted signal $y(t)$ is observed in additive white noise of spectral density $N_0/2$. The signal $y(t)$ is given by

$$y(t) = \begin{cases} e^{-t/2} - e^{3t/2} & 0 \leqslant t \leqslant T \\ 0 & \text{otherwise} \end{cases}$$

What is the maximum output signal-to-noise ratio? Assume

$$\int_0^T y^2(t)\, dt = 1$$

4.4 Instead of using the optimum filter, we want to use a simplified filter for the detection problem of Exercise 4.4. The filter has impulse response

$$h(t) = \begin{cases} e^{-\alpha t} & t \geqslant 0 \\ 0 & t < 0 \end{cases}$$

(a) Choose the parameter α to maximize output signal–noise ratio.

(b) Compare the performance of this receiver with the results of Exercise 4.3.

(c) By how much should the transmitter energy be increased to match the performance of the optimum filter?

4.5 In the problem of Exercise 4.3, the signal $y(t)$ is observed in the presence of colored noise with spectrum

$$\Phi_v(s) = N_0 \left[\frac{1 - s^2}{2 - s^2} \right]$$

Find the optimum matched filter and determine the maximum output signal-to-noise ratio.

4.6 Repeat Exercise 4.5 when the transmitted signal is

$$y(t)=\begin{cases} e^{-3t} & t \geqslant 0 \\ 0 & \text{otherwise} \end{cases}$$

and it is desired to maximize the output signal-to-noise ratio at some time $T > 0$. Determine the performance of the receiver as a function of T.

4.7 Consider the detection of a known signal in a mixture of white and colored noise. The observations are thus modeled as

$$H_1 : z(t) = y(t) + v_1(t) + v_2(t) \qquad t_0 \leqslant t \leqslant t_f$$
$$H_0 : z(t) = v_1(t) + v_2(t) \qquad t_0 \leqslant t \leqslant t_f$$

where $v_1(t)$ is zero mean, Gaussian white with spectral denity $N_0/2$ and $v_2(t)$ is zero mean, Gaussian with correlation function $\phi_{v_2}(\tau)$. Show that the test statistic can be chosen as

$$l = \int_{t_0}^{t_f} [z(t) - y(t)] h(t)\, dt$$

where $h(t)$ satisfies the integral equation

$$h(t) + \frac{2}{N_0} \int_{t_0}^{t_f} h(\tau)\phi_{v_2}(t-\tau)\, d\tau = \frac{1}{N_0} y(t) \qquad t_0 \leqslant t \leqslant t_f$$

4.8 Consider the integral equation

$$\int_0^T h(\tau)\phi_v(\tau - \sigma)\, d\tau = g(\sigma) \qquad 0 \leqslant t \leqslant T$$

where

$$\phi_v(\tau) = \alpha e^{-\alpha|\tau|}$$

Let $g(t)$ be such that

$$\dot{g}(0) = \alpha g(0) \qquad \text{and} \quad \dot{g}(T) = -\alpha g(T)$$

Find the solution to the integral equation.

4.9 Find the eigenvalues and eigenfunctions corresponding to the covariance kernel

$$\phi_x(t, u) = \sigma^2 \min(t, u)$$

by solving the integral equation

$$\lambda g(t) = \int_0^T \phi_x(t, u) g(u) \, du$$

Note that $x(t)$ is a Wiener process.

4.10 Derive Eq. (4.3.31) for optimum signal design.

4.11 In a binary communication system, the received signals are observed in additive noise with spectral density

$$\Phi_v(s) = \frac{a}{a - s^2} \qquad a > 0$$

Design the transmitted signals to yield minimum probability of error. What is P_e?

4.12 Assume that in Exercise 4.11, the noise has in addition a white component of spectral density $N_0/2$. Design the optimum signals and obtain P_e for this case.

4.13 Our discussions in Section 4.4 showed that for the detection of known signals in colored noise, the probability of error P_e was not independent of signal shape and that by appropriately choosing the transmitted waveforms, we could optimize P_e. In radar problems where a Neyman–Pearson criterion is used, we seek to maximize P_D for a given P_F level. Show how to choose the transmitted signals to maximize P_D. The problem of detection in colored noise in radar systems occurs when there is unwanted interference (clutter) present in the observations.

4.14 In certain applications involving band-limited signals, it is convenient to use the complex representation of signals discussed in Chapter 2. We consider the problem of detecting a known signal in the presence of band-limited, white Gaussian noise so that the observations under the two hypotheses are

$$H_1 : z(t) = y(t) + v(t)$$
$$H_0 : z(t) = v(t)$$

In terms of the respective complex envelopes, we can write

$$H_1 : \tilde{z}(t) = \tilde{y}(t) + \tilde{v}(t)$$
$$H_0 : \tilde{z}(t) = \tilde{v}(t)$$

where

$$z(t) = \sqrt{2}\, \mathrm{Re}[\tilde{z}(t) \exp(j\omega_0 t)]$$

and similar relations hold for $\tilde{y}(t)$ and $\tilde{v}(t)$. Here ω_0 is the center frequency of the band-limited signals. Show that a sufficient statistic for the decision problem is

$$\tilde{y}_1 = \int_0^T \tilde{z}(t)\tilde{y}(t)\, dt$$

Obtain the decision rule and determine P_e.

4.15 Extend the results of Exercise 4.14 to the colored observation noise case.

4.16 In many digital communication systems, orthogonal signals are preferred because of various reasons. Consider a binary system in which

$$\int_0^T y_1(t)y_0(t)\, dt = 0$$

and

$$\int_0^T y_i^2(t)\, dt = \varepsilon \qquad i = 0, 1$$

The additive noise is colored with spectral density $\Phi_v(s)$. How will you design the signals for minimizing P_e?

4.17 For the digital communication problem discussed in Example 4.6.1, obtain the probability of error expression assuming equal prior probabilities.

4.18 In Example 4.6.1, assume that the phase θ is exactly known. Derive the optimum receiver and its performance.

4.19 In Example 4.6.1, we assumed a Rayleigh fading channel. This is also a valid model in certain radar situations where the aspect of the target is changing. Derive the P_F and P_D expressions and sketch the ROC.

4.20 In Example 4.6.1, we assumed additive Gaussian white noise. Suppose that the noise is colored with spectral density

$$\Phi_v(s) = \frac{-s^2}{1-s^2}$$

Can you derive the optimum receiver? What is its performance?

4.21 In radar applications the transmitted waveform is often a segment of a sinusoidal signal of known frequency. The frequency of the return signal from the target, however, is shifted by an amount called the Doppler frequency which is approximately equal to $\omega_d = 2v\omega_0/c$ where v is the radial velocity of the target and c is the velocity of propagation of the electromagnetic wave. For targets of unknown velocity, ω_d can be modeled as a random variable with known density function. The received observations corresponding to whether a target is present or not will thus be

$$H_1 : z(t) = \sqrt{2}\ E \sin(\omega_d t + \theta) + v(t) \qquad 0 \leqslant t \leqslant T$$

$$H_0 : z(t) = v(t) \qquad 0 \leqslant t \leqslant T$$

where E and θ are assumed known. Determine the likelihood ratio test for this problem under the assumption that $v(t)$ is zero mean, Gaussian, white.

4.22 The range (distance from transmitter) of a radar target is determined from the delay between the transmitted and return signals. For targets of unknown range, the delay will be a random variable. The decision problem in this case can be modeled as

$$H_1 : z(t) = y(t - \tau) + v(t) \qquad 0 \leqslant t \leqslant T$$

$$H_0 : z(t) = v(t) \qquad 0 \leqslant t \leqslant T$$

where $y(t)$ is a known waveform and τ is the unknown delay. Derive the likelihood ratio test for this problem under the assumption that $v(t)$ is zero mean, white, Gaussian, and τ is uniformly distributed in $[\tau_0, \tau_1]$.

4.23 [30] This problem is concerned with the detection of a target from observing it in a number of antenna beams. The received signal in different beams appears with different amplitudes that depend on the target angle relative to the antenna system. When the target is present, the complex envelopes of the received waveforms are given by

$$\tilde{y}_i(t) = A \exp(j\Psi_i) G_i(\theta, \phi)\tilde{s}(t) + \tilde{v}_i(t) \qquad i = 1, 2, 3$$

where $A \geqslant 0$ is the signal amplitude, Ψ_i is the unknown carrier phase of the signal in the ith beam, and $\tilde{s}(t)$ is the known complex envelope of the transmitted signal. The noise $\tilde{v}_i(t)$ is complex, zero mean, white, Gaussian, with

$$\mathrm{E}[\tilde{v}_i(t)\tilde{v}_i(t')] = 4N_0\delta(t - t')$$

and

$$\mathrm{E}[\tilde{v}_i(t)\tilde{v}_j(t')] = 0$$

$G_i(\theta, \phi)$, a real quantity, denotes the beam pattern of the ith beam measured with respect to the angular coordinates θ and ϕ, say azimuth and elevation. Note that the beams are incoherent because Ψ_i differs from beam to beam. Assume that we do not know the nonrandom real parameters A, Ψ_i, θ, ϕ.

If the target is not present, the received waveform is

$$\tilde{y}_i(t) = \tilde{v}_i(t) \qquad i = 1, 2, 3$$

Obtain the generalized likelihood ratio for signal detection. Show that the LRT reduces to computing

$$l = |\tilde{y}_1|^2 + |\tilde{y}_2|^2 + |\tilde{y}_3|^2$$

where

$$\tilde{y}_i = \int_0^T \tilde{y}_i(t)\tilde{s}^*(t)\, dt$$

and T is the observation interval.
[Hint: Maximize the relevant quantity with respect to one parameter at a time for A and Ψ_i. For maximization with respect to θ,ϕ, assume that a proportionality exists between $|G_i(\theta, \phi)|$ and $|\tilde{y}_i|$, $i = 1, 2, 3$, and that such maximization is possible.]

4.24 [30] In the previous problem assume that all Ψ_is are equal. Now we have a coherent reception problem. Assume that $G_i(\theta, \phi)$ is proportional to $\mathrm{Re}[\tilde{y}_i e^{-j\Psi}]$. Show that the LRT reduces to computing

$$l = \frac{1}{2}\sum_{i=1}^{3} |\tilde{y}_i|^2 + \frac{1}{2}\left|\sum_{i=1}^{3} \tilde{y}_i^2\right|$$

4.25 [31] This problem is basically an extension of the previous problem to the case where we have to establish the presence of an interfering target. Also we assume that we are processing the sampled observations. The sampled signal from the three beams may be written for the target plus noise case as

$$\tilde{y}_i = A_s \exp(j\Psi_s) G_i(\theta_s, \phi_s) + \tilde{v}_i \qquad i = 1, 2, 3$$

where A_s, Ψ_s are target signal related amplitude and carrier phase parameters. $G_i(\theta_s, \phi_s)$ is the beam pattern of the ith beam with respect to target angular location (θ_s, ϕ_s). The sampled noise $\tilde{v}_i$ is Gaussian, zero mean, white in which the real and imaginary parts have equal variances σ^2. Further $\tilde{v}_i$ is uncorrelated with $\tilde{v}_j$, $i \neq j$. The received waveform when both an interfering target and the target of interest are present may be written as

$$\tilde{y}_i = A_s \exp(j\Psi_s)G_i(\theta_s, \phi_s) + A_I \exp(j\Psi_I)G_i(\theta_I, \phi_I) + \tilde{v}_i \qquad i = 1, 2, 3$$

where the quantites with subscript I refer to the interfering target. Obtain the generalized likelihood ratio assuming that the nonrandom real parameters A_s, Ψ_s, θ_s, ϕ_s, and A_I, Ψ_I, θ_I, ϕ_I are not known. Show that the LRT reduces to computing

$$l = \sum_{i=1}^{3} |\tilde{y}_i|^2 - \left| \sum_{i=1}^{3} \tilde{y}_i^2 \right|$$

and comparing it with a threshold. What differences do you see between this and the previous problem.? [Note: Refer to [31] for a detailed exposition of the problem including performance analysis.]

REFERENCES

1. H. L. Van Trees, *Detection, Estimation, and Modulation Theory, Part I,* John Wiley and Sons, New York, 1968.
2. A. D. Whalen, *Detection of Signals in Noise,* Academic Press, New York, 1971.
3. J. B. Thomas, *An Introduction to Statistical Communication Theory,* John Wiley and Sons, New York, 1969.
4. C. W. Helstrom, *Statistical Theory of Signal Detection,* Pergamon Press, Oxford, England, 1968.
5. J. M. Wozencraft and I. M. Jacobs, *Principles of Communication Engineering,* John Wiley and Sons, New York, 1965.
6. I. Selin, *Detection Theory,* Princeton University Press, Princeton, 1965.
7. W. W. Peterson, T. G. Birdsall, and W. C. Fox, The Theory of Signal Detectability, *IRE Trans. Info. Theory,* **IT-4**, September, 171 (1954).
8. D. Middleton and D. Van Meter, Detection and Extraction of Signals in Noise from the Point of View of Statistical Decision Theory, *J. SIAM,* **3**, 192 (1955); **4**, 86 (1956).
9. T. Kailath, Correlation Detection of Signals Perturbed by a Random Channel, *Trans. IRE Info. Theory,* **IT-6**, No. 3, July, 361–366 (1960).
10. P. M. Woodward, *Probability and Information Theory, with Radar Applications,* McGraw-Hill Book Co., New York, 1953.
11. U. Grenander, Stochastic Processes and Statistical Inference, *Arkiv. Mat.*, **1**, 295 (1950).
12. R. H. Urbano, Analysis and Tabulation of the *M* Positions Experiment Integral and Related Error Function Integrals, *Tech. Rep. No. AFCRC TR-55-100,* Apr. (1955), AFCRC, Bedford, Mass.

13. H. W. Bode and C. E. Shannon, A Simplified Derivation of Linear Least-Square Smoothing and Prediction Theory, *Proc. IRE,* **38**, April, 417–425 (1950).
14. L. A. Zadeh and J. R. Ragazzini, An Extension of Wiener's Theory of Prediction, *J. Appl. Phys.*, **4**, July, 645–655 (1950).
15. D. O. North, Analysis of the Factors which Determine Signal/Noise Discrimination in Radar, *Report PTR-6C,* June (1943) RCA Laboratories, Princeton, N.J.; also in *Proc. IEEE,* **51**, July (1963).
16. R. Courant and D. Hilbert, *Methods of Mathematical Physics, Vol. I,* English Translation, Interscience, New York, 1953.
17. F. Smithies, *Integral Equations,* Cambridge University Press, London, 1965.
18. L. A. Zadeh and J. R. Ragazzini, Optimum Filters for the Detection of Signals in Noise, *Proc. IRE,* **40**, 1233 (1952).
19. K. S. Miller and L. A. Zadeh, Solution of an Integral Equation Occurring in the Expansion of Second-Order Stationary Random Functions, *IRE Trans. Info. Theory,* **IT-2,** 72 (1956).
20. D. Youla, The Solution of a Homogeneous Wiener–Hopf Integral Equation Occurring in the Expansion of Second-Order Stationary Random Functions, *IRE Trans. Info. Theory,* **IT-3,** 187 (1957).
21. A. Baggeroer, A State Variable Approach to the Solution of Fredholm Integral Equations, *IEEE Trans. Info. Theory*, **IT-15,** No. 5, Sept., 556–569 (1969).
22. H. L. Van Trees, *Detection Estimation and Modulation Theory,* Part II, John Wiley and Sons, New York, 1971.
23. D. Middleton, On the Detection of Stochastic Signals in Additive Normal Noise, *IRE Trans. Info Theory,* **IT-3,** No. 2, June, 86 (1957).
24. T. Kailath, Solution of an Integral Equation Occurring in Multipath Communication Problems, *IRE Trans. Info. Theory,* **IT-6,** No. 3, June, 412 (1960).
25. J. Hájek, On Linear Statistical Problems in Stochastic Processes, *Czech. Math. J.,* **12**, No. 87, 404 (1962).
26. T. T. Kadota and L. A. Shepp, On the Best Finite Set of Linear Observables for Discriminating Two Gaussian Signals, *IEEE Trans. Info. Theory,* **IT-13,** No. 2, April, 278–284 (1967).
27. T. E. Duncan, *Probability Densities for Diffusion Processes with Applications to Nonlinear Filtering Theory and Detection Theory,* Ph.D. Dissertation, Stanford University, Stanford, Ca. 1967.
28. T. Kailath, A General Likelihood-Ratio Formula for Random Signals, *IEEE Trans. Info. Theory,* **IT-15,** No. 3, May, 350–361 (1969).
29. T. Kailath, Likelihood Ratios for Gaussian Processes, *IEEE Trans. Info. Theory,* **IT-16,** No. 3, May, 276–288 (1970).
30. E. M. Hofstetter and D. F. DeLong, Detection and Parameter Estimation in an Amplitude Comparison Monopulse Radar, *IEEE Trans. Info. Theory,* **IT-15,** No. 1, Jan., 22–30 (1969).
31. R. J. McAulay and T. P. McGarty, Maximum Likelihood Detection of Unresolved Radar Targets and Multipaths, *IEEE Trans. Aerospace Electronic Syst.,* **AES-10,** No. 6, Nov., 821–829 (1974).

CHAPTER 5

ESTIMATION THEORY

5.1 INTRODUCTION

In the preceding chapters, we were concerned primarily with the detection of signals. We wished to know whether a signal was present or not and were not explicitly concerned with determining the signal waveform or parameters associated with it. In many instances, however, we need to know these parameters. For example, in radar tracking, the time at which the return pulse arrives at the transmitter is of interest in determining the range of the target. In a gun control problem, it is necessary not only to estimate the present position and heading of the target, but also to predict its future position.

As in the previous chapters we assume that we have available a received waveform which consists of the transmitted signal (which may be distorted in the transmission medium) and local receiver noise. Based on either a finite number of samples or a continuous observation of the received signal, we wish to estimate either the transmitted signal or parameters associated with it. We start with a discussion of the estimation of parameters in this chapter and defer the estimation of waveforms until Chapter 6.

5.2 ESTIMATION OF PARAMETERS

We start our discussion by considering a simple example.

Example 5.2.1 A generator is available in which the voltage is some value between 0 and 10 volts. We wish to determine this value based on a measurement that is corrupted by additive noise. The observed variable is thus

$$z = \theta + v$$

where v is modeled as an independent zero mean Gaussian random variable. The problem is to observe z and estimate θ.

The probability density governing the observation process is

$$p(z|\theta) = \frac{1}{\sqrt{2\pi}\,\sigma_v} \exp\left(-\frac{(z-\theta)^2}{2\sigma_v^2}\right)$$

There are several ways in which an estimate may be found. If we treat θ as a random variable, then, as we did in the chapter on detection, we may choose as our basis for determining the estimate, the maximum a posteriori (MAP) criterion. The approach in this case would be to form the posterior density function $p(\theta|z)$ and determine the value of θ which maximizes this density.

On the other hand, if we treat θ as a constant but unknown value, we may seek to determine θ by maximizing the density function $p(z|\theta)$. Such an estimate is called a maximum likelihood estimate (the estimate can be interpreted as that value of θ which most probably gave rise to the observed quantity z).

The example brings out the structure of the estimation problem in the following manner. The first component is the source whose output is a parameter which can be regarded as a point in parameter space. The parameter may be either a random variable or a nonrandom, unknown quantity. The second component is the probabilistic transition mechanism that governs the effect of the parameter on the observation. The third component is the observation space. The estimation rule then provides a mapping of the observation space into an estimate. This is shown in Fig. 5.2.1.

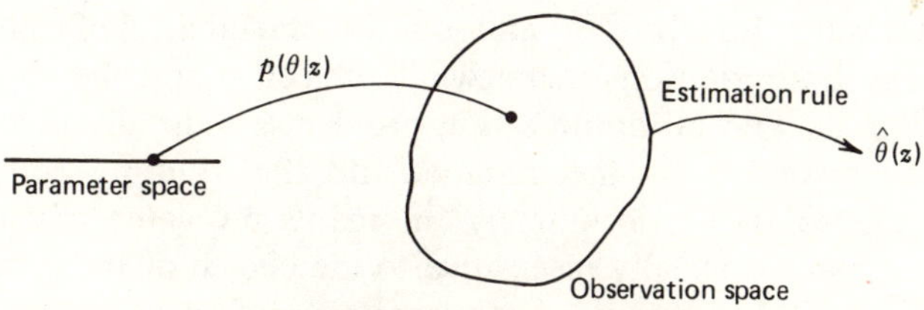

Fig. 5.2.1 Estimation model.

We note that the first three components are similar to the decision problem. We develop suitable estimation rules in this chapter for both random and nonrandom variables and discuss some of the properties of such estimators. We start with the random parameter case.

5.3 RANDOM PARAMETERS: BAYES ESTIMATES [1–6]

In the detection problem, we saw that we had to specify two quantities, namely the set of costs C_{ij} and the a priori probabilities $P(H_i)$. For the estimation problem, if we designate the estimate of a parameter θ by $\hat{\theta}(z)$, we have to assign a nonnegative cost to all pairs $[\theta, \hat{\theta}(z)]$ over the range of interest. For most problems of interest, it is usual to choose the cost $C(\theta, \hat{\theta}(z))$ as a function of the parameter estimation error $\tilde{\theta}(z)$ defined as

$$\tilde{\theta}(z) = \theta - \hat{\theta}(z) \tag{5.3.1}$$

so that we can write the cost as $C(\tilde{\theta}(z))$.

In the preceding discussion we have denoted the estimate $\hat{\theta}$ and the error $\tilde{\theta}$ as functions of z since they are explicitly dependent on the observations. Normally, in the sequel we will suppress this dependence and write $\hat{\theta}$ and $\tilde{\theta}$ except where clarity requires it.

Some typical cost functions are shown in Fig. 5.3.1. They correspond to the cases

$$C(\theta, \hat{\theta}) = (\theta - \hat{\theta})^2 : \text{Squared error}$$

$$C(\theta, \hat{\theta}) = |\theta - \hat{\theta}| : \text{Absolute value of error}$$

$$C(\theta, \hat{\theta}) = \begin{cases} 1 & |\theta - \hat{\theta}| \geqslant \dfrac{\Delta}{2} \\ 0 & |\theta - \hat{\theta}| < \dfrac{\Delta}{2} \end{cases} : \text{Uniform cost} \tag{5.3.2}$$

In a given situation, we choose a cost function on the basis of the use to which the results will be put and on the tractability of the resulting problem. The first criterion is somewhat subjective and the choice of the cost function, as with all optimization problems, is usually a compromise between the goals set by the designer and the resulting complexity in obtaining the solution. Fortunately, in many problems of interest the optimum estimate is usually insensitive to the choice of the criterion. The same estimate will be optimum for a large class of cost functions.

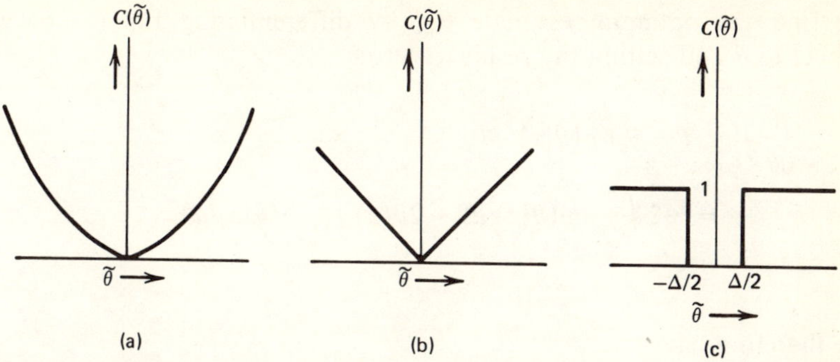

Fig. 5.3.1 Cost functions. (a) Squared error. (b) Absolute value. (c) Uniform cost.

Corresponding to the a priori probabilities in the detection problem, we have an a priori density $p(\theta)$ for the random parameter estimation problem. The density function $p(\theta)$ is assumed to be known.

Once we have specified the cost function and the a priori probability, we may write an expression for the average cost or risk as

$$\bar{C} = \int_{-\infty}^{\infty} \int_{-\infty}^{\infty} C(\theta - \hat{\theta}(z)) p(z, \theta)\, dz\, d\theta \tag{5.3.3}$$

We want to choose $\hat{\theta}$ so as to minimize the average cost. We do this by making use of Bayes' rule to write $\bar{C}$ in terms of two integrals only one of which depends on $\hat{\theta}$ as

$$\bar{C} = \int_{-\infty}^{\infty} \left[\int_{-\infty}^{\infty} C(\theta - \hat{\theta}(z)) p(\theta|z)\, d\theta \right] p(z)\, dz \tag{5.3.4}$$

It is clear that in the preceding equation the inner integral and $p(z)$ are nonnegative. Therefore we can minimize $\bar{C}$ by minimizing the inner integral with respect to $\hat{\theta}$. Thus the optimal estimate is obtained by minimizing the conditional risk

$$\bar{C}(\hat{\theta}|z) = \int_{-\infty}^{\infty} C(\theta - \hat{\theta}(z)) p(\theta|z)\, d\theta \tag{5.3.5}$$

Let us now determine the estimates for the typical cost functions of Fig. 5.3.1. For the squared error cost, if we denote the conditional risk as $\bar{C}_{\rm ms}$ Eq. (5.3.5) becomes

$$\bar{C}_{\rm ms}(\hat{\theta}|z) = \int_{-\infty}^{\infty} (\theta - \hat{\theta})^2 p(\theta|z)\, d\theta \tag{5.3.6}$$

We find the optimum estimate $\hat{\theta}_{ms}$ by differentiating Eq. (5.3.6) with respect to $\hat{\theta}$ and setting the result to zero:

$$\begin{aligned}\frac{d}{d\hat{\theta}}\int_{-\infty}^{\infty}(\theta-\hat{\theta})^2 p(\theta|z)\,d\theta \\ &= -2\int_{-\infty}^{\infty}\theta p(\theta|z)\,d\theta + 2\hat{\theta}(z)\int_{-\infty}^{\infty}p(\theta|z)\,d\theta \\ &= 0\end{aligned} \tag{5.3.7}$$

We therefore have

$$\hat{\theta}_{ms}(z) = \int_{-\infty}^{\infty}\theta p(\theta|z)\,d\theta \tag{5.3.8}$$

where we have used the fact that $\int_{-\infty}^{\infty}p(\theta|z)\,d\theta$ equals one.

This represents a unique minimum since the second derivative is positive (why?). The term on the right side of Eq. (5.3.8) represents the mean $E(\theta|z)$ of the a posteriori density. The mean square estimate is therefore sometimes referred to as the conditional mean estimate.

To find the Bayes estimate for the absolute value cost function we write Eq. (5.3.5) as

$$\begin{aligned}\bar{C}_{abs}(\hat{\theta}|z) &= \int_{-\infty}^{\infty}|\tilde{\theta}(z)|p(\theta|z)\,d\theta \\ &= \int_{-\infty}^{\hat{\theta}(z)}(-\theta+\hat{\theta}(z))p(\theta|z)\,d\theta + \int_{\hat{\theta}(z)}^{\infty}(-\hat{\theta}(z)+\theta)p(\theta|z)\,d\theta\end{aligned} \tag{5.3.9}$$

Differentiating both sides with respect to $\hat{\theta}(z)$ and setting the result to zero gives

$$\int_{-\infty}^{\hat{\theta}_{abs}(z)}p(\theta|z)\,d\theta = \int_{\hat{\theta}_{abs}(z)}^{\infty}p(\theta|z)\,d\theta \tag{5.3.10}$$

Thus the estimate $\hat{\theta}_{abs}(z)$ is the median of the conditional density $p(\theta|z)$. For the uniform cost function, the conditional risk becomes

$$\bar{C}_{unf}(\hat{\theta}|z) = 1 - \int_{\hat{\theta}_{unf}-\Delta/2}^{\hat{\theta}_{unf}+\Delta/2}p(\theta|z)\,d\theta \tag{5.3.11}$$

To minimize the conditional risk, we need to maximize the integral on the right hand side. For small Δ we see that the best choice for $\hat{\theta}$ is the value of θ at which the a posteriori density has its maximum. Thus the estimate corresponds to the maximum a posteriori (MAP) estimate discussed in Section 5.2 and is given by the mode of the a posteriori density.

For most densities of interest, we can find $\hat{\theta}_{\text{map}}$ as a solution to the equation

$$\left.\frac{\partial \ln p(\theta|z)}{\partial \theta}\right|_{\theta=\hat{\theta}_{\text{map}}(z)} = 0 \tag{5.3.12}$$

Equation (5.3.12) is referred to as the MAP equation. We may rewrite Eq. (5.3.12) to separate the role of the observation z and the a priori knowledge. We thus write

$$p(\theta|z) = \frac{p(z|\theta)p(\theta)}{p(z)} \tag{5.3.13}$$

Taking logarithms on both sides yields

$$\ln p(\theta|z) = \ln p(z|\theta) + \ln p(\theta) - \ln p(z)$$

Differentiating with respect to θ yields the MAP equation

$$\left.\frac{\partial \ln p(z|\theta)}{\partial \theta}\right|_{\theta=\hat{\theta}(z)} + \left.\frac{\partial \ln p(\theta)}{\partial \theta}\right|_{\theta=\hat{\theta}(z)} = 0 \tag{5.3.14}$$

The first term gives the dependence on z and the second corresponds to the a priori knowledge.

We shall be mostly concerned with mean square and maximum a posteriori estimates in the sequel.

Example 5.3.1 We consider the problem of estimating a parameter θ observed in additive noise. The observations are given by

$$z_i = \theta + v_i \qquad i = 1 \ldots N$$

We assume that the v_i are independent, identically distributed Gaussian random variables with zero mean and variance σ_v^2. We also assume that θ is Gaussian, zero mean, and with variance σ_θ^2.

It is clear from our previous discussions that the first step in determining the estimate is to find the relevant probability density function $p(\theta|\mathbf{z})$, where $\mathbf{z}$ denotes the set of observations $\{z_1, z_2, \ldots, z_N\}$. Now

$$p(\theta|\mathbf{z}) = \frac{p(\mathbf{z}|\theta)p(\theta)}{p(\mathbf{z})} = \frac{p(\mathbf{z}|\theta)p(\theta)}{\int_{\chi_\theta} p(\mathbf{z}|\theta)p(\theta)\, d\theta}$$

Because of the form of the observation equations we have

$$p(\mathbf{z}|\theta) = \prod_{i=1}^{N} p(z_i|\theta) = \prod_{i=1}^{N} \frac{1}{\sqrt{2\pi}\,\sigma_v} \exp\left(-\frac{(z_i-\theta)^2}{2\sigma_v^2}\right)$$

Also

$$p(\theta) = \frac{1}{\sqrt{2\pi}\,\sigma_\theta} \exp\left(-\frac{\theta^2}{2\sigma_\theta^2}\right)$$

The computation of $p(\mathbf{z})$ can tend to be algebraically tedious. We note instead that $p(\mathbf{z})$ is just a normalizing constant which causes

$$\int_{-\infty}^{\infty} p(\theta|\mathbf{z})\, d\theta = 1$$

Substituting for $p(\mathbf{z}|\theta)$ and $p(\theta)$ thus yields

$$p(\theta|\mathbf{z}) = \frac{\left(\prod_{i=1}^{N} \frac{1}{\sqrt{2\pi}\,\sigma_v}\right)\frac{1}{\sqrt{2\pi}\,\sigma_\theta}}{p(\mathbf{z})} \exp\left\{-\frac{1}{2}\left[\sum_{i=1}^{N}\left(\frac{z_i-\theta}{\sigma_v}\right)^2 + \frac{\theta^2}{\sigma_\theta^2}\right]\right\}$$

Let

$$\sigma_m^2 = \frac{\sigma_\theta^2\sigma_v^2}{N\sigma_\theta^2 + \sigma_v^2}$$

We can complete the square in the exponent by multiplying with $\exp\left[\pm\frac{1}{2}\left(\frac{\sigma_m^2}{\sigma_v^2}\sum_{i=1}^{N} z_i\right)^2\right]$. We can then write $p(\theta|\mathbf{z})$ as

$$p(\theta|\mathbf{z}) = k(z)\exp\left\{-\frac{1}{2\sigma_m^2}\left[\theta - \frac{\sigma_m^2}{\sigma_v^2}\left(\sum_{i=1}^{N} z_i\right)\right]^2\right\}$$

where $k(z)$ denotes all the terms not explicitly involving θ. We see that $p(\theta|\mathbf{z})$ is just a Gaussian density. Thus the mean, mode, and the median of the density are the same, implying that

$$\hat{\theta}_{\text{ms}}(\mathbf{z}) = \hat{\theta}_{\text{map}}(\mathbf{z}) = \hat{\theta}_{\text{abs}}(\mathbf{z})$$

The estimate is given by

$$\hat{\theta}_{\text{ms}} = \frac{\sigma_m^2}{\sigma_v^2}\left(\sum_{i=1}^{N} z_i\right)$$

$$= \frac{\sigma_\theta^2}{\sigma_\theta^2 + \dfrac{\sigma_v^2}{N}}\left(\frac{1}{N}\sum_{i=1}^{N} z_i\right)$$

We note that the observations enter into the a posteriori density only through their sum. Thus

$$l(\mathbf{z}) = \sum_{i=1}^{N} z_i$$

is a sufficient statistic for the estimation problem.

Example 5.3.2 Suppose now our observations are of the form

$$z_i = y(\theta) + v_i$$

where $y(\cdot)$ is a nonlinear function of θ. As before, we will have

$$p(\theta|\mathbf{z}) = \frac{\left(\displaystyle\prod_{i=1}^{N} \frac{1}{\sqrt{2\pi}\,\sigma_v}\right)\dfrac{1}{\sqrt{2\pi}\,\sigma_\theta}}{p(\mathbf{z})} \exp\left\{-\frac{1}{2}\left[\sum_{i=1}^{N} \frac{[z_i - y(\theta)]^2}{\sigma_v^2} + \frac{\theta^2}{\sigma_\theta^2}\right]\right\}$$

The determination of the conditional mean in this case will be quite complicated since the posterior density is no longer Gaussian. In this case $\hat{\theta}_{\text{map}}$ will not in general be the same as $\hat{\theta}_{\text{ms}}$.

Example 5.3.3 As a final example in this section, we consider a binary communication problem. We assume that the signals corresponding to the two hypotheses H_1 and H_0 are $+1$ and -1, respectively. We are required to estimate the value of the signal based on a single observation. We thus have

$$z = y + v$$

where v is assumed $N(0, \sigma_v^2)$. If we assume the prior probabilities for the two hypotheses to be the same, we have

$$p(y) = 0.5\delta(y-1) + 0.5\delta(y+1)$$

Since

$$p(z|y) = \frac{1}{\sqrt{2\pi}\,\sigma_v} \exp\left[-\frac{(z-y)^2}{2\sigma_v^2}\right]$$

we can compute $p(z)$ as

$$\begin{aligned}
p(z) &= \int_{-\infty}^{\infty} p(z|y)p(y)\,dy \\
&= \int_{-\infty}^{\infty} \frac{1}{\sqrt{2\pi}\,\sigma_v} \exp\left[-\frac{(z-y)^2}{2\sigma_v^2}\right]\{0.5\delta(y-1) + 0.5\delta(y+1)\}\,dy \\
&= \frac{1}{2\sqrt{2\pi}\,\sigma_v}\left\{\exp\left(-\frac{(z-1)^2}{2\sigma_v^2}\right) + \exp\left(-\frac{(z+1)^2}{2\sigma_v^2}\right)\right\}
\end{aligned}$$

The a posteriori density is given by

$$\begin{aligned}
p(y|z) &= \frac{p(z|y)p(y)}{p(z)} \\
&= \frac{\exp\left[-\dfrac{(z-y)^2}{2\sigma_v^2}\right]\{\delta(y-1) + \delta(y+1)\}}{\exp\left(-\dfrac{(z-1)^2}{2\sigma_v^2}\right) + \exp\left(-\dfrac{(z+1)^2}{2\sigma_v^2}\right)}
\end{aligned}$$

Obviously, to maximize $p(y|z)$, we have to choose $\hat{y} = \pm 1$ as otherwise the density is zero. Since minimizing $|z-y|$ maximizes the density, we choose $\hat{y}$ to be the value closest to z. We therefore have

$$\hat{y}_{\text{map}} = \operatorname{sgn} z$$

where

$$\operatorname{sgn} z = \begin{cases} +1 & \text{if } z \geqslant 0 \\ -1 & \text{if } z < 0 \end{cases}$$

To find the mean square estimate, we find the mean of the posterior density. Thus

$$\hat{y}_{ms} = \int_{-\infty}^{\infty} y p(y|z)\, dy$$

$$= \int_{-\infty}^{\infty} y \frac{\exp\left[-\dfrac{(z-y)^2}{2\sigma_v^2}\right]\{\delta(y-1)+\delta(y+1)\}}{\exp\left(-\dfrac{(z-1)^2}{2\sigma_v^2}\right)+\exp\left(-\dfrac{(z+1)^2}{2\sigma_v^2}\right)}\, dy$$

$$= \frac{\exp\left(\dfrac{z}{\sigma_v^2}\right)-\exp\left(-\dfrac{z}{\sigma_v^2}\right)}{\exp\left(\dfrac{z}{\sigma_v^2}\right)+\exp\left(\dfrac{z}{\sigma_v^2}\right)} = \tanh\frac{z}{\sigma_v^2}$$

Thus the MAP and minimum mean square estimates are not equivalent.

Invariance of Estimators [8–10]

We have seen from the preceding examples that different cost criteria yield different estimates. If an estimate could be found which was optimum for several cost functions, this would be desirable because of the subjective judgments used in choosing $C(\tilde{\theta})$. As we shall establish in the sequel, the minimum mean square estimate possesses this property under certain conditions.

We will assume that the cost function we are considering is a symmetric, convex-upward function and that the a posteriori density is symmetric about its conditional mean. Hence we have

$$C(\tilde{\theta}) = C(-\tilde{\theta}) \qquad \text{(Symmetry)} \tag{5.3.15a}$$

$$C(b\theta_1 + (1-b)\theta_2) \leqslant bC(\theta_1) + (1-b)C(\theta_2) \qquad \text{(Convexity)}$$
$$\text{for any } b \in (0, 1) \text{ and for all } \theta_1 \text{ and } \theta_2 \tag{5.3.15b}$$

$$p(\theta - \hat{\theta}_{ms}|z) = p(\hat{\theta}_{ms} - \theta|z) \qquad \text{(Symmetry about conditional mean)} \tag{5.3.15c}$$

We obtain the estimate for this cost function by minimizing the conditional risk

$$\bar{C}(\hat{\theta}|z) = \mathrm{E}\{C(\theta - \hat{\theta})|z\} \tag{5.3.16}$$

Let $\delta = \theta - \hat{\theta}_{ms}$. Then we can write

$$\begin{aligned}\bar{C}(\hat{\theta}|z) &= \int_{-\infty}^{\infty} C(\delta + \hat{\theta}_{ms} - \hat{\theta})p(\delta|z)\,d\delta \\ &= \int_{-\infty}^{\infty} C(-\delta - \hat{\theta}_{ms} + \hat{\theta})p(\delta|z)\,d\delta \qquad \text{by (5.3.15a)} \\ &= \int_{-\infty}^{\infty} C(\delta - \hat{\theta}_{ms} + \hat{\theta})p(\delta|z)\,d\delta \qquad \text{by (5.3.15c)}\end{aligned} \tag{5.3.17}$$

We can therefore, by using the first and third expressions on the right side of Eq. (5.3.17), write for the conditional risk

$$\begin{aligned}\bar{C}(\hat{\theta}|z) &= \int_{-\infty}^{\infty} \left[\tfrac{1}{2}C(\delta + \hat{\theta}_{ms} - \hat{\theta}) + \tfrac{1}{2}C(\delta - \hat{\theta}_{ms} + \hat{\theta})\right]p(\delta|z)\,d\delta \\ &\geqslant \int_{-\infty}^{\infty} C\left(\tfrac{1}{2}(\delta + \hat{\theta}_{ms} - \hat{\theta}) + \tfrac{1}{2}(\delta - \hat{\theta}_{ms} + \hat{\theta})\right)p(\delta|z)\,d\delta \\ &\qquad\qquad \text{by Eq. (5.3.15b)} \\ &= \bar{C}[\delta|z]\end{aligned} \tag{5.3.18}$$

Equality will be achieved if $\hat{\theta} = \hat{\theta}_{ms}$.

We have thus established that the minimum mean-square estimate $\hat{\theta}_{ms}$ is also the optimal estimate for all cost functions satisfying Eq. (5.3.15).

For nonconvex cost functions we require the further condition that

$$\lim_{x\to\infty} C(x)p(x|z) = 0 \tag{5.3.19}$$

The proof is similar to the one for the case of convex cost functions.

5.4 MAXIMUM LIKELIHOOD ESTIMATES

In our discussion of the maximum a posteriori estimate, we found that we had to solve the equation

$$\left.\frac{\partial p(\theta|z)}{\partial \theta}\right|_{\theta = \hat{\theta}_{map}} = 0 \tag{5.4.1}$$

Since

$$p(\theta|z) = \frac{p(z|\theta)p(\theta)}{p(z)} \tag{5.4.2}$$

we required the prior density $p(\theta)$ to be available. In terms of the logarithms of the density functions, the MAP equation was given by

$$\left[\frac{\partial \ln p(z|\theta)}{\partial \theta} + \frac{\partial \ln p(\theta)}{\partial \theta}\right]_{\theta=\hat{\theta}_{\text{map}}} = 0 \tag{5.4.3}$$

If the density $p(\theta)$ were not available, we could try to determine the estimate by considering only the first term on the left side of Eq. (5.4.3). The estimate that we obtain is called a maximum likelihood estimate and is obtained as a solution to the likelihood equation

$$\left.\frac{\partial \ln p(z|\theta)}{\partial \theta}\right|_{\theta=\hat{\theta}_{\text{ml}}} = 0 \tag{5.4.4}$$

If the parameter we are seeking to estimate is a nonrandom but unknown constant, we would not be able to define a prior probability density for the parameter and hence could not use a Bayes estimator. In such cases the maximum likelihood estimate (MLE) is applicable. We recall that the MLE is obtained by maximizing the density function $p(\theta|z)$ and can be interpreted as that value of θ which most probably gave rise to the observation z. This estimate can also be used when the parameter is random but has unknown prior density.

Example 5.4.1 We determine a maximum likelihood estimator for the problem of Example 5.3.1 where we have

$$p(\mathbf{z}|\theta) = \prod_{i=1}^{N} \frac{1}{\sqrt{2\pi}\,\sigma_v} \exp\left(-\frac{(z_i-\theta)^2}{2\sigma_v^2}\right)$$

Taking logarithms and differentiating we have

$$\frac{\partial \ln p(\mathbf{z}|\theta)}{\partial \theta} = \frac{1}{\sigma_v^2}\left(\sum_{i=1}^{N} z_i - N\theta\right)$$

Thus

$$\hat{\theta}_{\text{ml}} = \frac{1}{N}\sum_{i=1}^{N} z_i$$

The estimate is the sample mean of the observations.

Example 5.4.2 We consider the problem of Example 5.3.2. The likelihood equation for this problem is

$$\frac{\partial \ln p(\mathbf{z}|\theta)}{\partial \theta}\bigg|_{\theta=\hat{\theta}_{\text{ml}}} = \frac{1}{\sigma_v^2}\sum_{i=1}^{N}[z_i - y(\theta)]\frac{\partial y(\theta)}{\partial \theta}\bigg|_{\theta=\hat{\theta}_{\text{ml}}} = 0$$

or

$$\left[\frac{\partial y(\theta)}{\partial \theta}\frac{1}{\sigma_v^2}\right]\left[\frac{1}{N}\sum_{i=1}^{N} z_i - y(\theta)\right]_{\theta=\hat{\theta}_{\text{ml}}} = 0$$

If $y^{-1}(\cdot)$ exists and if $\frac{1}{N}\sum_{i=1}^{N} z_i$ is in the range of $y(\cdot)$, we can solve the preceding to obtain

$$\hat{\theta}_{\text{ml}} = y^{-1}\left[\frac{1}{N}\sum_{i=1}^{N} z_i\right]$$

We note that the MLE commutes over nonlinear operations. That is, if the estimate of θ is $\hat{\theta}_{\text{ml}}$, the estimate of $y(\theta)$ is $y(\hat{\theta}_{\text{ml}})$. This is not true for MMSE or MAP estimation.

5.5 PROPERTIES OF ESTIMATORS

We discuss in this section some desirable properties of estimators. As noted earlier, the estimate is a function of the observations. As the observations change, the value of the estimate changes. Since the observations are random variables, so is the estimate. It is obviously desirable that for repeated observations, the estimates group near the true value. For nonrandom parameters, we say that an estimate is (*conditionally*) *unbiased* if the mean value of the estimate is equal to the true value

$$\mathrm{E}\{\hat{\theta}\} = \int_{-\infty}^{\infty} \hat{\theta} p(z|\theta)\, dz = \theta \tag{5.5.1}$$

The difference between the mean $\mathrm{E}\{\hat{\theta}\}$ and the true value is called the *bias* in the estimate. For random parameters, we say that the estimate is (*unconditionally*) *unbiased* if the mean of the estimate is equal to the mean of the parameter being estimated

$$E\{\hat{\theta}\} = \iint_{-\infty}^{\infty} \hat{\theta} p(z, \theta)\, dz\, d\theta = \int_{-\infty}^{\infty} \hat{\theta}(z) p(z)\, dz = E\{\theta\} \tag{5.5.2}$$

In this case, the difference between the two means represents the bias in the estimate.

As the number of observations used in forming an estimate increases, we would like the estimate to tend towards its true value. Thus if we let $\hat{\theta}^N$ denote the estimate based on N observations, we say that the estimate is consistent if

$$\lim_{N \to \infty} P(|\theta - \hat{\theta}^N| < \epsilon) = 1 \tag{5.5.3}$$

for arbitrarily small $\epsilon > 0$.

Finally, an estimator is said to be a minimum variance estimator if the variance of the estimation error is less than that of any other estimator.

Bounds on Estimates: Nonrandom Parameters

In determining the effectiveness of an estimator, we can compute the bias and the variance. This is often difficult to do. Rather, we seek an indirect approach to the problem by determining a lower bound for the variance of *any* unbiased estimator and then compare the variance of $\hat{\theta}(z)$ computed using any procedure, with this bound. While there are several such bounds available in the literature, we concern ourselves with the Cramér–Rao bound [10,11]. We derive the bound for the nonrandom parameter case. The derivation for the random parameter case follows along essentially the same lines.

Since $\hat{\theta}$ is assumed to be an unbiased estimate, we have

$$E[\hat{\theta} - \theta] = \int_{-\infty}^{\infty} [\hat{\theta} - \theta] p(z|\theta)\, dz = 0 \tag{5.5.4}$$

Differentiate both sides with respect to θ to get

$$\begin{aligned} \frac{d}{d\theta} \int_{-\infty}^{\infty} [\hat{\theta} - \theta] p(z|\theta)\, dz & \\ &= \int_{-\infty}^{\infty} \frac{\partial}{\partial \theta} \{[\hat{\theta} - \theta] p(z|\theta)\}\, dz \\ &= -\int_{-\infty}^{\infty} p(z|\theta)\, dz + \int_{-\infty}^{\infty} \frac{\partial p(z|\theta)}{\partial \theta} [\hat{\theta} - \theta]\, dz = 0 \end{aligned} \tag{5.5.5}$$

assuming that the conditions permitting an interchange of the orders of integration and differentiation have been satisfied. The first integral equals -1. We also note that for any function $f(x)$, $df(x)/dx$ may be expressed as $(d \ln f(x)/dx)f(x)$. Thus Eq. (5.5.5) may be written as

$$\int_{-\infty}^{\infty} \frac{\partial \ln p(z|\theta)}{\partial\theta} p(z|\theta)[\hat{\theta}-\theta]\, dz = 1 \tag{5.5.6}$$

Rewriting, we have

$$\int_{-\infty}^{\infty} \left[\frac{\partial \ln p(z|\theta)}{\partial\theta} \sqrt{p(z|\theta)} \right] \left[\sqrt{p(z|\theta)}\, (\hat{\theta}-\theta) \right] dz = 1 \tag{5.5.7}$$

Using Schwarz's inequality† we find

$$\int_{-\infty}^{\infty} \left[\frac{\partial \ln p(z|\theta)}{\partial\theta} \right]^2 p(z|\theta)\, dz \int_{-\infty}^{\infty} (\hat{\theta}-\theta)^2 p(z|\theta)\, dz \geqslant 1 \tag{5.5.8}$$

with equality holding if and only if

$$\frac{\partial \ln p(z|\theta)}{\partial\theta} = [\hat{\theta}-\theta]k(\theta) \tag{5.5.9}$$

for all z and θ. Here $k(\cdot)$ is not a function of z or $\hat{\theta}$ but may be a function of θ. From Eq. (5.5.8) we have the inequality

$$\operatorname{var}\,[\hat{\theta}-\theta|\theta] = \mathrm{E}\{(\hat{\theta}-\theta)^2|\theta\} \geqslant \left\{ \mathrm{E}\left\{ \left[\frac{\partial \ln p(z|\theta)}{\partial\theta} \right]^2 \middle| \theta \right\} \right\}^{-1} \tag{5.5.10}$$

The right side of Eq. (5.5.10) represents a lower bound on the variance of any unbiased estimate of θ and is the Cramér–Rao bound referred to earlier. Any estimate that satisfies this bound with an equality is called an *efficient* estimate. We can derive an alternative expression for the right side

†The form of the Schwarz inequality used here is

$$\left[\int g(x)h(x)\, dx \right]^2 \leqslant \int g^2(x)\, dx \int h^2(x)\, dx$$

with equality holding when $g(x)$ and $h(x)$ are linearly related.

of Eq. (5.5.10) by noting that

$$\int_{-\infty}^{\infty} p(z|\theta)\, dz = 1 \tag{5.5.11}$$

Differentiating both sides with respect to θ, we have

$$\int_{-\infty}^{\infty} \frac{\partial p(z|\theta)}{\partial \theta}\, dz = \int_{-\infty}^{\infty} \frac{\partial \ln p(z|\theta)}{\partial \theta} p(z|\theta)\, dz = 0 \tag{5.5.12}$$

Differentiating again with respect to θ yields

$$\int_{-\infty}^{\infty} \frac{\partial^2 \ln p(z|\theta)}{\partial \theta^2} p(z|\theta)\, dz + \int_{-\infty}^{\infty} \left[\frac{\partial \ln p(z|\theta)}{\partial \theta} \right]^2 p(z|\theta)\, dz = 0 \tag{5.5.13}$$

or

$$\mathrm{E}\left[\frac{\partial^2 \ln p(z|\theta)}{\partial \theta^2} \middle| \theta \right] = -\mathrm{E}\left[\left(\frac{\partial \ln p(z|\theta)}{\partial \theta} \right)^2 \middle| \theta \right] \tag{5.5.14}$$

As noted earlier the maximum likelihood technique is used most often for estimation of nonrandom parameters. We can derive some useful properties of these estimators by using the Cramér–Rao bound.

Let us assume that Eq. (5.5.9) is satisfied and therefore an efficient estimate $\hat{\theta}$ exists. Suppose we set $\theta = \hat{\theta}_{\mathrm{ml}}$ on both sides of Eq. (5.5.9). Setting $\theta = \hat{\theta}_{\mathrm{ml}}$ in the left side gives us the likelihood equation

$$\left. \frac{\partial \ln p(z|\theta)}{\partial \theta} \right|_{\theta = \hat{\theta}_{\mathrm{ml}}} = 0 \tag{5.5.15}$$

Since the right side must also be zero we get

$$\left[\hat{\theta} - \hat{\theta}_{\mathrm{ml}} \right] k(\theta) = 0 \tag{5.5.16}$$

This is satisfied if

$$\hat{\theta} = \hat{\theta}_{\mathrm{ml}} \tag{5.5.17}$$

Thus if an efficient estimate exists, it is $\hat{\theta}_{\mathrm{ml}}$ and can be obtained as a solution to the likelihood equation.

If an efficient estimate does not exist, we will not know how good $\hat{\theta}_{ml}$ is. In fact, we will not know how closely the variance of any estimate will approach the bound.

Under reasonably general conditions it may be established that the ML estimate is consistent, asymptotically efficient, and asymptotically Gaussian [10].

Example 5.5.1 We consider the problem of Example 5.4.1 where

$$\frac{\partial \ln p(\mathbf{z}|\theta)}{\partial \theta} = \frac{N}{\sigma_v^2}\left(\frac{1}{N}\sum_{i=1}^{N} z_i - \theta\right)$$

so that

$$\hat{\theta}_{ml} = \frac{1}{N}\sum_{i=1}^{N} z_i$$

and

$$\mathrm{E}\{\hat{\theta}_{ml}\} = \frac{1}{N}\sum_{i=1}^{N} z_i = \frac{1}{N}\sum_{i=1}^{N} \theta = \theta$$

Thus $\hat{\theta}_{ml}$ is unbiased.

Since $\partial \ln p(\mathbf{z}|\theta)/\partial\theta$ has the form required by Eq. (5.5.9) an efficient estimate exists and is given by $\hat{\theta}_{ml}$. The variance of the estimate is given by the Cramér–Rao bound. We have

$$\frac{\partial^2 \ln p(\mathbf{z}|\theta)}{\partial \theta^2} = -\frac{N}{\sigma_v^2}$$

Using Eq. (5.5.14) and the fact that the estimate is efficient yields

$$\mathrm{var}\left[\hat{\theta}_{ml} - \theta\right] = \frac{\sigma_v^2}{N}$$

Example 5.5.2 [12] We consider the problem of estimating certain unknown (nonrandom) parameters of a signal of known form observed in noise. Such problems commonly arise in the estimation of modulated waveforms, and are discussed in some detail in Chapter 7. The observations are assumed to be of the form

$$z(t) = y(t, \theta) + v(t) \qquad t_0 \leqslant t \leqslant t_f \tag{5.5.18}$$

where θ is an unknown parameter. In the first instance we assume that the noise is zero mean, Gaussian, white with spectral height $N_0/2$ and subsequently extend our results to the colored noise case.

The first step in determining the ML estimate of θ is to obtain the likelihood function (conditional density) $p(z|\theta)$. In order to do so, we use an argument similar to that in Section 4.2. If we use the sampling approach discussed there, we sample $z(t)$ so that we have N samples in the interval (t_0, t_f).

Let z_k, y_k, and v_k denote the sample values at instant t_k. That is, $z_k = z(t_k)$, $y_k = y(t_k, \theta)$, and $v_k = v(t_k)$. As in Section 4.2, we initially assume that $v(t)$ is lowpass with (double-sided) bandwith 2Ω and later let Ω tend to infinity. If we let $t_k = k\pi/\Omega$ so that $N = \Omega(t_f - t_0)/\pi$ the noise samples v_k will be independent. The received samples can be written as

$$z_k = y_k + v_k = y(t_k, \theta) + v_k \tag{5.5.19}$$

It follows that

$$\mathrm{E}\{z_k|\theta\} = y_k$$

and

$$\mathrm{var}\,\{z_k|\theta\} = \mathrm{var}\,\{v_k\} = \frac{N_0\Omega}{2\pi} = \sigma^2 \tag{5.5.20}$$

Let $\mathbf{z}_N = [z \ldots z_N]^T$.

Since the samples are independent, we have

$$\begin{aligned} p(\mathbf{z}_N|\theta) &= \prod_{k=1}^{N} p(z_k|\theta) \\ &= \prod_{k=1}^{N} \frac{1}{\sqrt{2\pi}\,\sigma} \exp\left[-\frac{(z_k - y_k)^2}{2\sigma^2}\right] \\ &= \left(\frac{1}{\sqrt{2\pi}\,\sigma}\right)^N \exp\left\{-\frac{1}{2\sigma^2}\sum_{k=1}^{N}(z_k - y_k)^2\right\} \end{aligned} \tag{5.5.21}$$

It easily follows that

$$\frac{\partial \ln p(\mathbf{z}_N|\theta)}{\partial\theta} = \frac{1}{\sigma^2}\sum_{k=1}^{N}(z_k - y_k)\frac{\partial y_k}{\partial\theta} \tag{5.5.22}$$

Letting $N \to \infty$, we obtain the maximum likelihood estimate of θ as the solution to the equation

$$\int_{t_0}^{t_f} [z(t) - y(t, \theta)] \frac{\partial y(t, \theta)}{\partial \theta} \, dt \bigg|_{\theta = \hat{\theta}_{\text{ml}}} = 0 \tag{5.5.23}$$

We can also use an expansion in a set of orthonormal basis functions $\{g_k(t)\}$ to derive this result. If we write the K-term approximation of $z(t)$ over the interval $[t_0, t_f]$ as

$$z_K(t) = \sum_{k=1}^{K} z_k g_k(t) \tag{5.5.24}$$

we have

$$z_k = y_k(\theta) + v_k$$

where now y_k and v_k are the coefficients of the expansion of $y(t, \theta)$ and $v(t)$, respectively.

It follows from the assumptions that

$$\mathrm{E}\{z_k\} = y_k(\theta)$$

and

$$\operatorname{var}\{z_k\} = \operatorname{var}\{v_k\} = \frac{N_0}{2} \tag{5.5.25}$$

Since the coefficients are independent, we have

$$\begin{aligned} p(z_K(t)|\theta) &= \prod_{k=1}^{K} p(z_k) \\ &= \prod_{k=1}^{K} \frac{1}{\sqrt{\pi N_0}} \exp\left[-\frac{(z_k - y_k(\theta))^2}{N_0} \right] \\ &= \left(\frac{1}{\sqrt{\pi N_0}} \right)^K \exp\left\{ -\frac{1}{N_0} \sum_{k=1}^{K} (z_k - y_k(\theta))^2 \right\} \end{aligned} \tag{5.5.26}$$

We now obtain the estimate of θ based on the K-term approximation by setting the partial derivative of $\ln p(z_k(t)|\theta)$ with respect to θ equal to zero to get

$$\sum_{k=1}^{K} (z_k - y_k(\theta)) \frac{\partial y_k(\theta)}{\partial \theta} = 0 \tag{5.5.27}$$

In the limit as $K \to \infty$, it can be shown that Eq. (5.5.27) yields the same result as Eq. (5.5.23). We note that Eq. (5.5.23) represents a necessary but not sufficient condition on $\hat{\theta}_{ml}$.

In particular, if θ is an amplitude parameter so that

$$y(t, \theta) = \theta y(t)$$

Eq. (5.5.23) reduces to

$$\int_{t_0}^{t_f} [z(t) - \theta y(t)] y(t)\, dt \bigg|_{\theta = \hat{\theta}_{ml}} = 0 \tag{5.5.28}$$

and the ML estimate is

$$\hat{\theta}_{ml} = \frac{\int_{t_0}^{t_f} z(t) y(t)\, dt}{\int_{t_0}^{t_f} y^2(t)\, dt} \tag{5.5.29}$$

If we normalize the signal $y(t)$ so that

$$\int_{t_0}^{t_f} y^2(t)\, dt = 1$$

we obtain for the estimate

$$\hat{\theta}_{ml} = \int_{t_0}^{t_f} z(t) y(t)\, dt \tag{5.5.30}$$

which corresponds to the correlation receiver operation discussed earlier. Since

$$\begin{aligned} E\{\hat{\theta}_{ml}\} &= E\left\{ \int_{t_0}^{t_f} [\theta y(t) + v(t)] y(t)\, dt \right\} \\ &= \theta \end{aligned} \tag{5.5.31}$$

the estimate is unbiased. From Eq. (5.5.22), it easily follows that

$$\begin{aligned} \frac{\partial \ln p(z|\theta)}{\partial \theta} &= \frac{2}{N_0} \int_{t_0}^{t_f} [z(t) - \theta y(t)] y(t)\, dt \\ &= \frac{2}{N_0} [\hat{\theta}_{ml} - \theta] \end{aligned} \tag{5.5.32}$$

and is of the form of Eq. (5.5.9). Thus $\hat{\theta}_{ml}$ is an efficient estimate and its variance is given by Eq. (5.5.10) as $N_0/2$.

If the unknown parameter is a phase parameter, then

$$y(t, \theta) = A \sin(\omega_0 t + \theta)$$

With A and ω_0 known, the estimate is obtained from Eq. (5.5.23) by solving

$$\int_{t_0}^{t_f} \left[z(t) - A \sin\left(\omega_0 t + \hat{\theta}\right) \right] \cos\left(\omega_0 t + \hat{\theta}\right) dt = 0 \tag{5.5.33}$$

If we assume that $\omega_0(t_f - t_0) = k\pi$ for some integer k, the second integral is zero and the estimate is obtained as a solution of

$$\int_{t_0}^{t_f} z(t) \cos\left(\omega_0 t + \hat{\theta}\right) dt = 0 \tag{5.5.34}$$

Use of trigonometric identities then enables us to write

$$\hat{\theta}_{\text{ml}} = \tan^{-1} \left\{ \frac{\int_{t_0}^{t_f} z(t) \cos \omega_0 t \, dt}{\int_{t_0}^{t_f} z(t) \sin \omega_0 t \, dt} \right\} \tag{5.5.35}$$

If the measurement noise is colored, we can use the results of Section 4.4 to derive the estimator equations. We first pass the observations through a whitening filter with impulse response $h_w(\cdot, \cdot)$. The output of the whitening filter is then given by

$$\begin{aligned} z_w(t) &= \int_{t_0}^{t_f} h_w(t, u) z(u) \, du \\ &= \int_{t_0}^{t_f} h_w(t, u) y(u, \theta) \, du + \int_{t_0}^{t_f} h_w(t, u) v(u) \, du \\ &= y_w(t, \theta) + v_w(t) \qquad t_0 \leqslant t \leqslant t_f \end{aligned} \tag{5.5.36}$$

where $v_w(t)$ is white with spectral height arbitrarily chosen to be unity.

The estimator equations follow from Eq. (5.5.23) as

$$\int_{t_0}^{t_f} [z_w(t) - y_w(t, \theta)] \frac{\partial y_w(t, \theta)}{\partial \theta} dt = 0 \tag{5.5.37}$$

In terms of the original waveforms Eq. (5.5.37) can be written as

$$\int_{t_0}^{t_f} \left\{ \int_{t_0}^{t_f} h_w(t, u) z(u) \, du - \int_{t_0}^{t_f} h_w(t, u) y(u, \theta) \, du \right\} \cdot \left\{ \frac{\partial \int_{t_0}^{t_f} h_w(t, v) y(v, \theta) \, dv}{\partial \theta} \, dt \right\} = 0 \tag{5.5.38}$$

Equation (5.5.38) can be written as

$$\int_{t_0}^{t_f}\left\{\int_{t_0}^{t_f}h_w(t,u)z(u)\,du-\int_{t_0}^{t_f}h_w(t,u)y(u,\theta)\,du\right\}$$
$$\cdot\left\{\int_{t_0}^{t_f}h_w(t,v)\frac{\partial y(v,\theta)}{\partial\theta}\,dv\right\}dt=0 \qquad \textbf{(5.5.39)}$$

The ML estimate of θ can be obtained by solving Eq. (5.5.39). The determination of $h_w(\cdot,\cdot)$ was discussed in Sec. 4.5.

Random Parameter Estimation: Bounds and MAP Estimates

Analogous to our derivation for the nonrandom parameter case, we can establish bounds on the variance of any estimate of a random parameter. Here, in addition to requiring that $\partial p(z,\theta)/\partial\theta$ and $\partial^2 p(z,\theta)/\partial\theta^2$ both be absolutely integrable with respect to z and θ, we will also assume that the following condition holds:

$$\lim_{\theta\to\pm\infty} p(\theta)\int_{-\infty}^{\infty}[\hat{\theta}-\theta]p(z|\theta)\,dz=0 \qquad \textbf{(5.5.40)}$$

We can then establish along the same lines as in the nonrandom parameter case that if $\hat{\theta}$ is any unbiased estimate of θ, then

$$\mathrm{E}\{(\hat{\theta}-\theta)^2\}\geqslant\left(\mathrm{E}\left\{\left[\frac{\partial\ln p(z,\theta)}{\partial\theta}\right]^2\right\}\right)^{-1}$$
$$=\left\{-\mathrm{E}\left[\frac{\partial^2\ln p(z,\theta)}{\partial\theta}\right]\right\}^{-1} \qquad \textbf{(5.5.41)}$$

with equality if and only if

$$\frac{\partial\ln p(z,\theta)}{\partial\theta}=k[\hat{\theta}-\theta] \qquad \textbf{(5.5.42)}$$

for all z and θ. We note that in Eq. (5.5.42) k cannot be a function of $\hat{\theta}$, z, or θ.

Arguing as in Eqs. (5.5.15)–(5.5.17) we see that if Eq. (5.5.42) is satisfied, the MAP estimate will be efficient. Because the MSE estimate cannot have a larger variance, this implies that $\hat{\theta}_{\mathrm{ms}}=\hat{\theta}_{\mathrm{map}}$ whenever an efficient estimate exists. In such a case it is easier to solve the MAP equation to determine the estimate than to compute the conditional mean. The variance of the estimate can be determined from the bound given in Eq. (5.5.41).

Example 5.5.3 We consider the same problem as in Example 5.5.2, but assume that θ is a Gaussian random variable with zero mean and variance σ_θ^2. We consider only the case when the observation noise is Gaussian, white. The extension to the colored noise problem is obvious.

To find the MAP estimate of θ, we note that

$$p(z, \theta) = p(z|\theta)p(\theta)$$

so that to obtain the MAP estimate we need to add $d \ln p(\theta)/d\theta$ to the left side of Eq. (5.5.23). We then get

$$\hat{\theta}_{\text{map}} = \frac{2}{N_0}\sigma_\theta^2 \int_{t_0}^{t_f} [z(t) - y(t, \theta)] \frac{\partial y(t, \theta)}{\partial \theta} dt \bigg|_{\theta = \hat{\theta}_{\text{map}}} \tag{5.5.43}$$

Again, Eq. (5.5.43) is a necessary, but not sufficient condition on $\hat{\theta}_{\text{map}}$.

Simultaneous Estimation of Multiple Parameters

The discussion so far has been concerned with the estimation of a single parameter. We can easily extend our results to the case of estimation of several parameters. For the parameter vector

$$\boldsymbol{\theta} = [\theta_1, \theta_2, \ldots, \theta_n]^T$$

we define the parameter estimation error as

$$\tilde{\boldsymbol{\theta}}(\mathbf{z}) = \begin{bmatrix} \theta_1 - \hat{\theta}_1(\mathbf{z}) \\ \theta_2 - \hat{\theta}_2(\mathbf{z}) \\ \vdots \\ \theta_n - \hat{\theta}_n(\mathbf{z}) \end{bmatrix} \tag{5.5.44}$$

where the symbol ˆ refers to estimated values. The mean square estimate is then determined by minimizing the average risk

$$\bar{C}_{\text{ms}} = \int_{-\infty}^{\infty} \int_{-\infty}^{\infty} C(\tilde{\boldsymbol{\theta}}(\mathbf{z}))p(\mathbf{z}, \boldsymbol{\theta})\, d\mathbf{z}\, d\boldsymbol{\theta} \tag{5.5.45}$$

where

$$C(\tilde{\boldsymbol{\theta}}(\mathbf{z})) = \sum_{i=1}^{n} \tilde{\theta}_i^2(\mathbf{z}) = \tilde{\boldsymbol{\theta}}^T(\mathbf{z})\, \tilde{\boldsymbol{\theta}}(\mathbf{z}) \tag{5.5.46}$$

Rewriting Eq. (5.3.55) in terms of the conditional density $p(\boldsymbol{\theta}|\mathbf{z})$ gives

$$\overline{C}_{\text{ms}} = \int_{-\infty}^{\infty} p(\mathbf{z})\, d\mathbf{z} \int_{-\infty}^{\infty} \left[\sum_{i=1}^{n} (\theta_i - \hat{\theta}_i(\mathbf{z}))^2 \right] p(\boldsymbol{\theta}|\mathbf{z})\, d\boldsymbol{\theta} \tag{5.5.47}$$

As before we can minimize the inner integral. Since the terms are positive, we minimize them separately to get the estimate of the ith element as

$$(\hat{\theta}_i)_{\text{ms}}(\mathbf{z}) = \int_{-\infty}^{\infty} \theta_i\, p(\boldsymbol{\theta}|\mathbf{z})\, d\boldsymbol{\theta} \tag{5.5.48}$$

or in vector notation,

$$\hat{\boldsymbol{\theta}}_{\text{ms}} = \int_{-\infty}^{\infty} \boldsymbol{\theta}\, p(\boldsymbol{\theta}|\mathbf{z})\, d\boldsymbol{\theta} \tag{5.5.49}$$

For MAP estimation, we must solve the set of simultaneous equations

$$\left. \frac{\partial \ln p(\boldsymbol{\theta}|\mathbf{z})}{\partial \theta_i} \right|_{\boldsymbol{\theta} = \hat{\boldsymbol{\theta}}_{\text{map}}} = 0 \qquad i = 1, 2, \ldots, n \tag{5.5.50}$$

We can write Eq. (5.5.50) compactly in terms of the partial derivative operator $\nabla_{\boldsymbol{\theta}}$ as

$$\nabla_{\boldsymbol{\theta}}[\ln p(\boldsymbol{\theta}|\mathbf{z})]_{\boldsymbol{\theta} = \hat{\boldsymbol{\theta}}_{\text{map}}} = 0 \tag{5.5.51}$$

For ML estimates we solve the set of equations

$$\nabla_{\boldsymbol{\theta}}[\ln p(\mathbf{z}|\boldsymbol{\theta})]_{\boldsymbol{\theta} = \hat{\boldsymbol{\theta}}_{\text{ml}}} = 0 \tag{5.5.52}$$

We can obtain lower bounds on the variance of the estimation error for the multiple parameter case similar to the ones derived earlier. For nonrandom variables this bound is given by

$$\text{var}\, \tilde{\theta}_i(\mathbf{z}) \geqslant \psi_{ii} \tag{5.5.53}$$

where ψ_{ii} is the iith element of the $n \times n$ matrix $\psi = \mathbf{J}^{-1}$ with $\mathbf{J}$ given by

$$\begin{aligned} J_{ij} &= \mathrm{E}\left\{ \frac{\partial \ln p(\mathbf{z}|\boldsymbol{\theta})}{\partial \theta_i} \frac{\partial \ln p(\mathbf{z}|\boldsymbol{\theta})}{\partial \theta_j} \right\} \\ &= -\mathrm{E}\left\{ \frac{\partial^2 \ln p(\mathbf{z}|\boldsymbol{\theta})}{\partial \theta_i \partial \theta_j} \right\} \end{aligned} \tag{5.5.54}$$

The matrix $\mathbf{J}$ is commonly referred to as the Fisher information matrix. Equality in Eq. (5.5.53) holds if and only if

$$\hat{\theta}_i(\mathbf{z}) - \theta_i = \sum_{j=1}^{n} k_{ij}(\boldsymbol{\theta}) \frac{\partial \ln p(\mathbf{z}|\boldsymbol{\theta})}{\partial \theta_j} \tag{5.5.55}$$

for θ_i and $\mathbf{z}$.

For random parameters, we replace the conditional density function in the right sides of Eqs. (5.5.54) to (5.5.55) by the joint density function $p(\boldsymbol{\theta}, \mathbf{z})$.

We conclude this section by considering an example.

Example 5.5.3 We consider the problem of estimating a random vector parameter $\boldsymbol{\theta}$ observed in white Gaussian noise. The observations are of the form

$$z_i = \sum_{j=1}^{n} h_{ij}\theta_j + v_i \qquad i = 1 \ldots m \tag{5.5.56}$$

with v_i being a set of independent variables. In vector notation, the above can be written as

$$\mathbf{z} = \mathbf{H}\boldsymbol{\theta} + \mathbf{v} \tag{5.5.57}$$

where $\mathbf{z}$ and $\mathbf{v}$ are m-dimensional column vectors, $\boldsymbol{\theta}$ is n-dimensional and the observation matrix $\mathbf{H}$ is $m \times n$. Let us assume that the prior density for $\boldsymbol{\theta}$ is Gaussian, $N(\boldsymbol{\mu}_\theta, \mathbf{V}_\theta)$, and that for $\mathbf{v}$ is $N(\mathbf{0}, \mathbf{V}_\mathbf{v})$. Since the variables v_i are independent, the matrix $\mathbf{V}_\mathbf{v}$ will be diagonal. The relevant densities for this problem will then be Gaussian with

$$p(\mathbf{z}|\boldsymbol{\theta}) \sim N(\mathbf{H}\boldsymbol{\theta}, \mathbf{V}_\mathbf{v}) \tag{5.5.58a}$$

$$p(\mathbf{z}) \sim N\big(\mathbf{H}\boldsymbol{\mu}_\theta, \mathbf{H}\mathbf{V}_\theta\mathbf{H}^T + \mathbf{V}_\mathbf{v}\big) \tag{5.5.58b}$$

Explicitly these densities can be written as

$$p(\mathbf{z}|\boldsymbol{\theta}) = \frac{1}{(2\pi)^{m/2}|\mathbf{V}_\mathbf{v}|^{1/2}} \exp\left\{-\frac{1}{2}(\mathbf{z} - \mathbf{H}\boldsymbol{\theta})^T \mathbf{V}_\mathbf{v}^{-1}(\mathbf{z} - \mathbf{H}\boldsymbol{\theta})\right\} \tag{5.5.59a}$$

$$p(\mathbf{z}) = \frac{1}{(2\pi)^{m/2}|\mathbf{H}\mathbf{V}_\theta\mathbf{H}^T + \mathbf{V}_\mathbf{v}|^{1/2}} \times \exp\left\{-\frac{1}{2}(\mathbf{z} - \mathbf{H}\boldsymbol{\mu}_\theta)^T \big[\mathbf{H}\mathbf{V}_\theta\mathbf{H}^T + \mathbf{V}_\mathbf{v}\big]^{-1}(\mathbf{z} - \mathbf{H}\boldsymbol{\mu}_\theta)\right\} \tag{5.5.59b}$$

$$p(\boldsymbol{\theta}) = \frac{1}{(2\pi)^{n/2}|\mathbf{V}_\theta|^{1/2}} \exp\left\{-\frac{1}{2}(\boldsymbol{\theta} - \boldsymbol{\mu}_\theta)^T \mathbf{V}_\theta^{-1}(\boldsymbol{\theta} - \boldsymbol{\mu}_\theta)\right\} \tag{5.5.59c}$$

Use of the matrix lemma and and some algebraic manipulations enables us to write the a posteriori density as

$$p(\boldsymbol{\theta}|\mathbf{z}) = \frac{p(\mathbf{z}|\boldsymbol{\theta})p(\boldsymbol{\theta})}{p(\mathbf{z})}$$
$$= \frac{|\mathbf{H}\mathbf{V}_\theta\mathbf{H}^T + \mathbf{V}_\mathrm{v}|^{1/2}}{(2\pi)^{n/2}|\mathbf{V}_\theta|^{1/2}|\mathbf{V}_\mathrm{v}|^{1/2}} \exp\left\{ -\frac{1}{2}(\boldsymbol{\theta} - \boldsymbol{\xi})^T\boldsymbol{\Sigma}^{-1}(\boldsymbol{\theta} - \boldsymbol{\xi})\right\} \quad (5.5.60)$$

where

$$\boldsymbol{\Sigma}^{-1} = \mathbf{V}_\theta^{-1} + \mathbf{H}^T\mathbf{V}_\mathrm{v}^{-1}\mathbf{H} \quad (5.5.61)$$

and

$$\boldsymbol{\xi} = \boldsymbol{\Sigma}(\mathbf{H}^T\mathbf{V}_\mathrm{v}^{-1}\mathbf{z} + \mathbf{V}_\theta^{-1}\boldsymbol{\mu}_\theta) \quad (5.5.62)$$

Since this density is in a Gaussian form, we have

$$\hat{\boldsymbol{\theta}}_{\mathrm{map}} = \hat{\boldsymbol{\theta}}_{\mathrm{ms}} = \boldsymbol{\xi} = \boldsymbol{\Sigma}(\mathbf{H}^T\mathbf{V}_\mathrm{v}^{-1}\mathbf{z} + \mathbf{V}_\theta^{-1}\boldsymbol{\mu}_\theta)$$
$$= (\mathbf{V}_\theta^{-1} + \mathbf{H}^T\mathbf{V}_\mathrm{v}^{-1}\mathbf{H})^{-1}(\mathbf{H}^T\mathbf{V}_\mathrm{v}^{-1}\mathbf{z} + \mathbf{V}_\theta^{-1}\boldsymbol{\mu}_\theta) \quad (5.5.63)$$

The error variance is given by

$$\mathbf{V}_{\tilde{\theta}} = \boldsymbol{\Sigma} = (\mathbf{V}_\theta^{-1} + \mathbf{H}^T\mathbf{V}_\mathrm{v}^{-1}\mathbf{H}^{-1})^{-1}$$

which can be written using matrix identities (see Appendix C) as

$$\mathbf{V}_{\tilde{\theta}} = \mathbf{V}_\theta - \mathbf{V}_\theta\mathbf{H}^T(\mathbf{V}_\mathrm{v} + \mathbf{H}\mathbf{V}_\theta\mathbf{H}^T)^{-1}\mathbf{H}\mathbf{V}_\theta \quad (5.5.64)$$

The estimates above were obtained using a measurement vector of dimension n. If more measurements are available, we may construct a new estimate by repeating the procedure above and using the new observation vector. However, it seems reasonable to obtain the updated estimate as a function of the new observation and the old estimate since the first n measurements have already been used in an optimal fashion.

To see how this can be done, let us denote the estimate based on n measurements as $\hat{\boldsymbol{\theta}}(n)$. For simplicity, we will assume $\boldsymbol{\mu}_\theta = 0$. Let $\mathbf{z}(n)$ denote the measurement vector and $\mathbf{H}(n)$ the corresponding observation matrix. The estimate based on $(n + 1)$ measurements is, from our previous derivation, given by

$$\hat{\boldsymbol{\theta}}_{n+1} = \boldsymbol{\Sigma}_{n+1}[\mathbf{H}^T(n+1)\mathbf{V}_\mathrm{v}^{-1}(n+1)\mathbf{z}(n+1)] \quad (5.5.65)$$

where

$$\Sigma_{n+1}^{-1} = \mathbf{V}_\theta^{-1} + \mathbf{H}^T(n+1)\mathbf{V}_v^{-1}(n+1)\mathbf{H}(n+1) \tag{5.5.66}$$

$$\mathbf{H}(n+1) = \begin{bmatrix} \mathbf{H}(n) \\ \hdashline \mathbf{h}_{n+1} \end{bmatrix}, \quad \mathbf{z}(n+1) = \begin{bmatrix} \mathbf{z}(n) \\ \hdashline z_{n+1} \end{bmatrix}$$

$$\mathbf{V}_v(n+1) = \begin{bmatrix} \mathbf{V}_v(n) & \mathbf{0} \\ \mathbf{0} & \text{cov } v_{n+1} \end{bmatrix} \tag{5.5.67}$$

and $\mathbf{h}_{n+1} = [h_{1,n+1} h_{2,n+1} \dots h_{n+1,n+1}]$. Substituting in the expression for Σ_{n+1} yields

$$\begin{aligned} \Sigma_{n+1}^{-1} &= \mathbf{V}_\theta^{-1} + \mathbf{H}^T(n)\mathbf{V}_v^{-1}(n)\mathbf{H}(n) + \mathbf{h}_{n+1}^T(\text{cov } v_{n+1})^{-1}\mathbf{h}_{n+1} \\ &= \Sigma_n^{-1} + \mathbf{h}_{n+1}^T(\text{cov } v_{n+1})^{-1}\mathbf{h}_{n+1} \end{aligned} \tag{5.5.68}$$

Again using matrix identities, the preceding can be written as

$$\Sigma_{n+1} = \Sigma_n - \Sigma_n \mathbf{h}_{n+1}^T\left[\mathbf{h}_{n+1}\Sigma_n\mathbf{h}_{n+1}^T + \text{cov } v_{n+1}\right]^{-1}\mathbf{h}_{n+1}\Sigma_n \tag{5.5.69}$$

Thus the estimate $\hat{\theta}_{n+1}$ can be written as

$$\begin{aligned} \hat{\theta}_{n+1} &= \Sigma_{n+1}\left[\mathbf{H}^T(n)\mathbf{V}_v^{-1}(n)\mathbf{z}(n) + \mathbf{h}_{n+1}^T(\text{cov } v_{n+1})^{-1}z_{n+1}\right] \\ &= \Sigma_n\mathbf{H}^T(n)\mathbf{V}_v^{-1}(n)\mathbf{z}(n) + \Sigma_n\mathbf{h}_{n+1}^T(\text{cov } v_{n+1})^{-1}z_{n+1} \\ &\quad - \Sigma_n\mathbf{h}_{n+1}^T(\mathbf{h}_{n+1}\Sigma_n\mathbf{h}_{n+1}^T + \text{cov } v_{n+1})^{-1}\mathbf{h}_{n+1} \\ &\quad \times \Sigma_n\left[\mathbf{H}^T(n)\mathbf{V}_v^{-1}(n)\mathbf{z}(n) + \mathbf{h}_{n+1}^T(\text{cov } v_{n+1})^{-1}z_{n+1}\right] \\ &= \hat{\theta}_n + \Sigma_n\mathbf{h}_{n+1}^T\left[\mathbf{h}_{n+1}\Sigma_n\mathbf{h}_{n+1}^T + \text{cov } v_{n+1}\right]^{-1}\left[z_{n+1} - \mathbf{h}_{n+1}\hat{\theta}_n\right] \end{aligned} \tag{5.5.70}$$

Thus the present estimate is equal to the old estimate plus a linear correction term based on the new measurement. The computations involved in processing the data sequentially as in the preceding are much less than in obtaining the estimate directly from processing the measurements together in a batch.

5.6 LINEAR MEAN SQUARE ESTIMATION

In our discussion of the estimation of a random parameter θ from observations of a related random variable z, we saw that the minimum mean square estimate was given by the mean of the posterior density $p(\theta|z)$:

$$\hat{\theta}_{\text{ms}} = \mathrm{E}\{\theta|z\} = \int_{-\infty}^{\infty} \theta p(\theta|z)\, d\theta \tag{5.6.1}$$

As discussed earlier, the computation of this conditional mean becomes quite involved in many problems. In general the conditional mean will be a nonlinear function of the observations z. Thus rather than seeking the mean square estimate of θ, we look for the best linear mean square estimate of θ, that is, the best estimate of θ expressible as a linear function of the observation z. If we write the linear minimum mean square estimate of θ as

$$\hat{\theta}_{\text{lms}} = az + b \tag{5.6.2}$$

the problem reduces to one of finding the best constants a and b so as to minimize

$$\begin{aligned} \bar{C}_{\text{lms}} &= \mathrm{E}\{(\theta - \hat{\theta}_{\text{lms}})^2\} \\ &= \iint_{-\infty}^{\infty} (\theta - \hat{\theta}_{\text{lms}})^2 p(z, \theta)\, dz\, d\theta \end{aligned} \tag{5.6.3}$$

As in Section 5.3, we can easily show that $\hat{\theta}_{\text{lms}}$ can be obtained by minimizing

$$\int_{-\infty}^{\infty} (\theta - \hat{\theta}_{\text{lms}})^2 p(\theta|z)\, d\theta \tag{5.6.4}$$

If the conditional mean were expressible as a linear function of the observations z, then the linear mean square estimate would also be the minimum mean square estimate, $\hat{\theta}_{\text{lms}} = \hat{\theta}_{\text{ms}}$.

It is instructive to determine under what conditions this can happen [13]. Assume

$$\mathrm{E}\{\theta|z\} = az + b \tag{5.6.5}$$

We start with the characteristic equation associated with the density function $p(z, \theta)$

$$C_{z\theta}(\omega_1, \omega_2) = \iint_{-\infty}^{\infty} p(z, \theta) e^{j(\omega_1 z + \omega_2 \theta)}\, dz\, d\theta \tag{5.6.6}$$

Differentiate both sides with respect to $j\omega_2$ and set $\omega_2 = 0$ to get

$$\left.\frac{\partial}{\partial j\omega_2} C_{z\theta}(\omega_1, \omega_2)\right|_{\omega_2=0} = \iint_{-\infty}^{\infty} \theta p(z, \theta) e^{j\omega_1 z}\, dz\, d\theta$$

$$= \int_{-\infty}^{\infty} \left[\int_{-\infty}^{\infty} \theta p(\theta|z)\, d\theta \right] e^{j\omega_1 z} p(z)\, dz \qquad \textbf{(5.6.7)}$$

Substituting for the inner integral on the right-hand side of Eq. (5.6.7) from Eq. (5.6.5) yields

$$\left.\frac{\partial}{\partial j\omega_2} C_{z\theta}(\omega_1, \omega_2)\right|_{\omega_2=0} = \int_{-\infty}^{\infty} (az + b) e^{j\omega_1 z} p(z)\, dz$$

$$= a \frac{\partial}{\partial j\omega_1} \int_{-\infty}^{\infty} e^{j\omega_1 z} p(z)\, dz + b \int_{-\infty}^{\infty} e^{j\omega_1 z} p(z)\, dz$$

$$= \left(a \frac{\partial}{\partial j\omega_1} + b \right) C_z(\omega_1) \qquad \textbf{(5.6.8)}$$

It can also be shown by retracing our steps that Eq. (5.6.8) is a sufficient condition for the conditional mean $E\{\theta|z\}$ to be of the form of Eq. (5.6.5). It can easily be verified that Eq. (5.6.8) is satisfied if $p(z, \theta)$ is Gaussian. It follows that if the joint density function is Gaussian, the conditional mean is linear in the observations. Thus the minimum mean square estimate is linear.

To return to our original problem, we will now seek the best linear estimate $\hat{\theta}_{\text{lms}}$ of the parameter θ by choosing the constants a and b of Eq. (5.6.2) to minimize the risk function of Eq. (5.6.3). Thus substituting for $\hat{\theta}_{\text{lms}}$ in Eq. (5.6.3) yields

$$\overline{C}_{\text{lms}} = \iint_{-\infty}^{\infty} [\theta - az - b]^2 p(z, \theta)\, dz\, d\theta \qquad \textbf{(5.6.9)}$$

Setting the partial derivatives with respect to a and b equal to zero yields the two equations

$$\iint_{-\infty}^{\infty} z[\theta - az - b] p(z, \theta)\, dz\, d\theta = 0 \qquad \textbf{(5.6.10)}$$

and

$$\iint_{-\infty}^{\infty} [\theta - az - b] p(z, \theta)\, dz\, d\theta = 0 \qquad \textbf{(5.6.11)}$$

or equivalently

$$aE\{z^2\} + bE\{z\} = E\{\theta z\} \tag{5.6.12}$$

and

$$aE\{z\} + b = E\{\theta\} \tag{5.6.13}$$

Solving for a and b then gives

$$a = \frac{E(z\theta) - E(z)E(\theta)}{E(z^2) - E^2(z)} \tag{5.6.14}$$

$$b = E\{\theta\} - aE\{z\} \tag{5.6.15}$$

In terms of the correlation coefficient ρ

$$\rho = \frac{E\{[z - E(z)][\theta - E\{\theta\}]\}}{\sqrt{E\{(z - E(z))^2\}E\{(\theta - E(\theta))^2\}}} \tag{5.6.16}$$

we can write

$$a = \frac{\rho\sigma_\theta}{\sigma_z} \tag{5.6.17}$$

The optimal cost obtained by substituting these values of a and b in Eq. (5.6.9) can easily be shown to be

$$\overline{C}_{\text{lms}} = \sigma_\theta^2(1 - \rho^2) \tag{5.6.18}$$

If $\boldsymbol{\theta}$ is a vector parameter, the optimal estimate can easily be shown to be

$$\hat{\boldsymbol{\theta}}_{\text{lms}} = \mathbf{Az} + \mathbf{b} \tag{5.6.19a}$$

where the matrix $\mathbf{A}$ and the vector $\mathbf{b}$ are given by

$$\begin{aligned}\mathbf{A} &= [E(\mathbf{zz}^T) - E(\mathbf{z})E(\mathbf{z}^T)]^{-1}[E(\boldsymbol{\theta}\mathbf{z}^T) - E(\boldsymbol{\theta})E(\mathbf{z}^T)] \\ &= \mathbf{V}_{\mathbf{z}}^{-1}\mathbf{V}_{\boldsymbol{\theta}\mathbf{z}}\end{aligned} \tag{5.6.19b}$$

and

$$\mathbf{b} = E\{\boldsymbol{\theta}\} - \mathbf{A}E\{\mathbf{z}\} \tag{5.6.19c}$$

Example 5.6.1 We consider the problem of estimating a parameter θ on the basis of a single observation z. Let z and θ be jointly normal with means μ_z and μ_θ, respectively. Let the elements of the covariance matrix

$$\mathrm{E}\left\{\begin{bmatrix} z - \mu_z \\ \theta - \mu_\theta \end{bmatrix}\begin{bmatrix} z - \mu_z & \theta - \mu_\theta \end{bmatrix}\right\}$$

be given by v_{ij}, $i, j = 1, 2$.

The joint characteristic function of z and θ is given by

$$C_{z\theta}(\omega_1, \omega_2) = \exp\left\{j(\mu_z\omega_1 + \mu_\theta\omega_2) - \tfrac{1}{2}\left(v_{11}\omega_1^2 + 2v_{12}\omega_1\omega_2 + v_{22}\omega_2^2\right)\right\}$$

Thus

$$\frac{\partial}{\partial j\omega_2} C_{z\theta}(\omega_1, \omega_2)\bigg|_{\omega_2=0} = (\mu_\theta + v_{12}j\omega_1)\exp\left(j\mu_z\omega_1 - \tfrac{1}{2}v_{11}\omega_1^2\right)$$

Also since

$$C_z(\omega_1) = \exp\left(j\omega_1\mu_z - \tfrac{1}{2}v_{11}\omega_1^2\right)$$

we have

$$\frac{\partial}{\partial j\omega_1} C_z(\omega_1) = (\mu_z + v_{11}j\omega_1)\exp\left(j\omega_1\mu_z - \tfrac{1}{2}v_{11}\omega_1^2\right)$$

and

$$\left(a\frac{\partial}{\partial j\omega_1} + b\right)C_z(\omega_1) = (a\mu_z + b + av_{11}j\omega_1)\exp\left(j\omega_1\mu_z - \tfrac{1}{2}v_{11}\omega_1^2\right)$$

Thus Eq. (5.6.8) will be satisfied with

$$a\mu_z + b = \mu_\theta$$

and

$$av_{11} = v_{12}$$

Solving for a and b yields

$$a = \frac{v_{12}}{v_{11}}$$

and

$$b = \mu_\theta - a\mu_z$$

so that

$$\hat{\theta}_{\text{ms}} = \text{E}\{\theta|z\} = az + b$$

Thus the linear mean square estimate is also the minimum mean square estimate in this case. We note that a and b obtained previously are the same as in Eqs. (5.6.14) and (5.6.15).

It can easily be shown by direct calculation that

$$\text{E}\{\tilde{\theta}z\} = \text{E}\{\theta z\} - a\text{E}\{z^2\} - b\text{E}\{z\} = 0$$

so that the error in the estimate is orthogonal to the observation z.

Example 5.6.2 We can use our result on mean square estimation of jointly Gaussian random variables of the previous example to derive an expression for the conditional density $p(\theta|z)$.

Let us assume for simplicity that $\mu_z = \mu_\theta = 0$. Since $p(\theta|z)$ will be Gaussian, we need determine only its mean and variance. From Example 5.6.1 we know that

$$\text{E}\{\theta|z\} = az + b$$

with

$$a = \frac{\rho\sigma_\theta}{\sigma_z} \quad \text{and} \quad b = 0$$

The variance is given by

$$\begin{aligned} \text{var}\,\{\theta|z\} &= \text{E}\{(\theta - az - b)^2|z\} = \bar{C}_{\text{ms}} \\ &= \sigma_\theta^2(1-\rho^2) \end{aligned}$$

so that

$$p(\theta|z) = \frac{1}{\sqrt{2\pi\sigma_\theta^2(1-\rho^2)}}\exp\left\{-\frac{(\theta - az)^2}{2\sigma_\theta^2(1-\rho^2)}\right\}$$

Orthogonality Principle [6,7,14]

It was shown in Example 5.6.1 that the error in the estimate was orthogonal to the observation z. This result is valid for general linear mean square estimation also and is of significance in setting up the equations needed to obtain the estimate. We will now derive this result for the more general problem of estimating a vector parameter $\boldsymbol{\theta}$ from the observation vector $\mathbf{z}$.

Let us assume that $E\{\mathbf{z}\} = E\{\boldsymbol{\theta}\} = 0$. Then we can write the linear mean square estimate as

$$\hat{\boldsymbol{\theta}}_{\text{lms}} = \mathbf{A}z \tag{5.6.20}$$

We will now show that the constant matrix $\mathbf{A}$ that minimizes the mean square error

$$\begin{aligned} \overline{C}_{\text{lms}} &= E\{[\boldsymbol{\theta} - \mathbf{Az}]^T[\boldsymbol{\theta} - \mathbf{Az}]\} \\ &= E\{\|\boldsymbol{\theta} - \mathbf{Az}\|^2\} \end{aligned} \tag{5.6.21}$$

is such that the error

$$\tilde{\boldsymbol{\theta}} = \boldsymbol{\theta} - \mathbf{Az} \tag{5.6.22}$$

is orthogonal to $\mathbf{z}$. That is

$$E\{(\boldsymbol{\theta} - \mathbf{Az})^T\mathbf{z}\} = 0 \tag{5.6.23}$$

To see this, suppose $\hat{\boldsymbol{\theta}} = \mathbf{A}_1\mathbf{z}$, represents any other linear estimate. Then

$$\begin{aligned} E\{\|\boldsymbol{\theta} - \mathbf{A}_1\mathbf{z}\|^2\} &= E\{\|\boldsymbol{\theta} - \mathbf{Az} + (\mathbf{A} - \mathbf{A}_1)\mathbf{z}\|^2\} \\ &= E\{\|\boldsymbol{\theta} - \mathbf{Az}\|^2\} + (\mathbf{A} - \mathbf{A}_1)^T E\{\mathbf{z}^T(\boldsymbol{\theta} - \mathbf{Az})\} \\ &\quad + E\{(\boldsymbol{\theta} - \mathbf{Az})^T\mathbf{z}\}(\mathbf{A} - \mathbf{A}_1) \\ &\quad + E\{\|(\mathbf{A} - \mathbf{A}_1)\mathbf{z}\|^2\} \\ &\geqslant E\{\|\boldsymbol{\theta} - \mathbf{Az}\|^2\} \end{aligned} \tag{5.6.24}$$

since the last term is nonnegative and the second and third terms are zero from Eq. (5.6.23).

The orthogonality principle can be interpreted in terms of projections on a linear space defined as follows. Let $\mathbf{Z}$ be the linear space formed from a finite collection of random vectors $\mathbf{z}_1, \mathbf{z}_2, \ldots, \mathbf{z}_n$ such that $E\{\mathbf{z}_k\mathbf{z}_k^T\} < \infty$ for each i. Thus $\mathbf{Z}$ is the space of random vectors with finite second moments. We define the inner product of two vectors $\mathbf{x}$ and $\mathbf{y}$ on this space as

$$\langle \mathbf{x}, \mathbf{y} \rangle = \mathrm{E}\{\mathbf{x}^T\mathbf{y}\} \tag{5.6.25}$$

with associated norm

$$\|\mathbf{x}\| = \left[\mathrm{E}\{\mathbf{x}^T\mathbf{x}\}\right]^{1/2} \tag{5.6.26}$$

We say that the two vectors $\mathbf{x}$ and $\mathbf{y}$ are orthogonal if

$$\mathrm{E}\{\mathbf{x}^T\mathbf{y}\} = 0$$

If $\mathbf{Z}_1$ and $\mathbf{Z}_2$ are two complementary and orthogonal subspaces of $\mathbf{Z}$, then every vector $\mathbf{x} \in \mathbf{Z}$ can be decomposed as $\mathbf{x} = \mathbf{x}_1 + \mathbf{x}_2$ with $\mathbf{x}_1 \in \mathbf{Z}_1$ and $\mathbf{x}_2 \in \mathbf{Z}_2$ so that $\mathbf{x}_1$ and $\mathbf{x}_2$ are orthogonal. The operator P such that

$$P\mathbf{x} = \mathbf{x}_1$$

is then the projection operator onto the subspace $\mathbf{Z}_1$. The projection theorem for linear spaces says that if $\mathbf{x}$ is any vector in $\mathbf{Z}$ and $\mathbf{x}'$ is a vector in a subspace $\mathbf{Z}'$ of $\mathbf{Z}$, then the norm of the vector $\mathbf{x} - \mathbf{x}'$ has a minimum value if and only if $\mathbf{x}'$ is the projection of $\mathbf{x}$ on $\mathbf{Z}$ [15].

Let us now return to the problem of estimating the parameter vector $\boldsymbol{\theta}$. Suppose we have N measurements of the vector $\mathbf{z}$. Then, the best mean square estimate of $\boldsymbol{\theta}$ is

$$\hat{\boldsymbol{\theta}}_{\mathrm{ms}} = \mathrm{E}\{\boldsymbol{\theta} | \mathbf{z}_1, \ldots, \mathbf{z}_N\} \tag{5.6.27}$$

If we restrict ourselves to linear estimates, it follows from Eq. (5.6.23) and the discussions above that the best linear estimate $\hat{\boldsymbol{\theta}}_{\mathrm{lms}}$ is the projection of the vector $\boldsymbol{\theta}$ on to the space generated by the vectors $\mathbf{z}_1, \mathbf{z}_2, \ldots, \mathbf{z}_N$. If we write

$$\hat{\boldsymbol{\theta}}_{\mathrm{lms}} = \sum_{i=1}^{N} \alpha_i \mathbf{z}_i \tag{5.6.28}$$

then the weight factors α_i are obtained from the orthogonality principle by solving the set of equations

$$\mathrm{E}\left\{\left(\boldsymbol{\theta} - \sum_{i=1}^{N} \alpha_i \mathbf{z}_i\right)\mathbf{z}_j^T\right\} = 0 \qquad j = 1, 2, \ldots, N \tag{5.6.29}$$

If the condition of Eq. (5.6.8) is satisfied, then $\hat{\boldsymbol{\theta}}_{\mathrm{ms}} = \mathrm{E}\{\boldsymbol{\theta}|\mathbf{z}\}$ will be an element of the space generated by $\mathbf{z}_1 \ldots \mathbf{z}_N$, and in this case $\hat{\boldsymbol{\theta}}_{\mathrm{ms}} = \hat{\boldsymbol{\theta}}_{\mathrm{lms}}$.

When the conditional expectation is not an element of this space, the projection of $\boldsymbol{\theta}$ on this space gives the best linear mean square estimate. We denote this by

$$\hat{\boldsymbol{\theta}}_{\text{lms}} = \hat{\mathrm{E}}\{\boldsymbol{\theta}|\mathbf{z}_1 \ldots \mathbf{z}_N\} \tag{5.6.30}$$

where $\hat{\mathrm{E}}$ is called the wide-sense conditional expectation of $\boldsymbol{\theta}$ relative to the set $\{\mathbf{z}_1 \ldots \mathbf{z}_N\}$. Since the wide-sense conditional expectation is equivalent to a projection operator [14], we can show that this expectation follows the same combinatorial rules as conventional conditional expectations. For example for any scalars c_1 and c_2

$$\hat{\mathrm{E}}\{c_1\boldsymbol{\theta}_1 + c_2\boldsymbol{\theta}_2|\mathbf{z}\} = c_1\hat{\mathrm{E}}\{\boldsymbol{\theta}_1|\mathbf{z}\} + c_2\hat{\mathrm{E}}\{\boldsymbol{\theta}_2|\mathbf{z}\} \tag{5.6.31}$$

Other relations may similarly be stated. These rules must be understood as equations valid with probability 1 and not as identities.

We conclude this chapter on parameter estimation with a discussion of reproducing densities, which is especially of use in pattern recognition problems.

5.7 REPRODUCING DENSITIES

We have seen in our discussion of the estimation of parameters that we need to specify the prior probability density function $p(\theta)$. The estimation procedure involves determining the a posteriori density function

$$p(\theta|z) = \frac{p(z|\theta)p(\theta)}{p(z)} \tag{5.7.1}$$

The estimation problem becomes particularly simple if the functional forms for $p(\theta)$ and $p(\theta|z)$ are the same with only the associated parameters being different. We call such a density a reproducing density or a conjugate prior density (with respect to the transition density $p(z|\theta)$) [16,17]. Since the choice of the a priori density is somewhat arbitrary in most cases, it is frequently convenient to choose a reproducing density.

We will use the concept of reproducing densities to obtain estimates of parameters associated with probability density functions. We are given the functional form of the probability density function underlying a certain experiment so that the density functions are completely known except for some parameters. On the basis of measurements of the outcomes of the experiment, we are required to estimate the parameters by obtaining a reproducing density for the parameter. We generate this reproducing density [16] by choosing an a priori set Z_{n_0} of n_0 (fictitious) observations and a nonnegative function $g(\theta)$ and setting

$$p(\theta)=\frac{p(Z_{n_0}|\theta)g(\theta)}{\int p(Z_{n_0}|\theta)g(\theta)d\theta} \tag{5.7.2}$$

where $g(\theta)$ is any nonnegative function of θ such that the denominator in Eq. (5.7.2) exists. Usually $g(\theta)$ can be chosen as unity. We illustrate the technique by considering two examples.

Example 5.7.1 Let z be a scalar observation with

$$z=\begin{cases}1 & \text{with probability } \theta\\ 0 & \text{with probability } 1-\theta\end{cases}$$

We make n observations of the random variable z. Based on our observations e are required to estimate θ.To find a simple reproducing density, we consider a specific a priori sequence Z_{n_0} consisting of α_0 ones and $n_0-\alpha_0$ zeros, for n_0 and α_0 any two positive integers, $\alpha_0<n_0$. Then by Eq. (5.7.2) with $g(\theta)=1$,

$$p(\theta)=\frac{p(Z_{n_0}|\theta)}{\int p(Z_{n_0}|\theta)\,d\theta}=\begin{cases}\dfrac{\theta^{\alpha_0}(1-\theta)^{n_0-\alpha_0}}{\int_0^1\theta^{\alpha_0}(1-\theta)^{n_0-\alpha_0}\,d\theta} & 0<\theta<1\\ 0 & \text{elsewhere}\end{cases}$$

$$=\begin{cases}\dfrac{\Gamma(n_0+2)}{\Gamma(n_0+1)\Gamma(n_0-\alpha_0+1)}\theta^{\alpha_0}(1-\theta)^{n_0-\alpha_0} & 0<\theta<1\\ 0 & \text{elsewhere}\end{cases}$$

which is the beta density with parameters α_0 and n_0. Let the sequence Z_n consisting of n observations have α ones and $n-\alpha$ zeros. Then

$$p(\theta|Z_n)=\frac{p(Z_n|\theta)p(\theta)}{p(Z_n)}=\frac{p(Z_n|\theta)p(\theta)}{\int p(Z_n|\theta)p(\theta)\,d\theta}$$

$$=\begin{cases}\dfrac{\theta^{\alpha}(1-\theta)^{n-\alpha}\theta^{n_0}(1-\theta)^{n_0-\alpha_0}}{\int\theta^{(\alpha+\alpha_0)}(1-\theta)^{(n+n_0)-(\alpha+\alpha_0)}\,d\theta} & 0<\theta<1\\ 0 & \text{elsewhere}\end{cases}$$

$$=\begin{cases}\dfrac{\Gamma(n+n_0+2)\theta^{(\alpha+\alpha_0)}(1-\theta)^{(n+n_0)-(\alpha+\alpha_0)}}{\Gamma(\alpha+\alpha_0+1)\Gamma(n+n_0-(\alpha+\alpha_0))+1} & 0<\theta<1\\ 0 & \text{elsewhere}\end{cases}$$

which is also the beta density with parameters $\alpha + \alpha_0$ and $n + n_0$. The mean square estimate of θ is given by

$$\hat{\theta}_{\text{ms}}(n) = \mathrm{E}\{\theta | Z_n\} = \frac{\alpha + \alpha_0 + 1}{n + n_0 + 2}$$

To put the estimate in a sequential form, we note that if $z_n = 1$, then with $(n - 1)$ observations, there would have been $\alpha - 1$ ones and $n - \alpha$ zeros. Thus

$$\hat{\theta}_{\text{ms}}(n-1) = \frac{(\alpha - 1) + \alpha_0 + 1}{(n-1) + n_0 + 2} = \frac{\alpha + \alpha_0}{n + n_0 + 1}$$

If $z_n = 0$, then the sequence Z_{n-1} would consist of α ones and $n - 1 - \alpha$ zeros and

$$\hat{\theta}_{\text{ms}}(n-1) = \frac{\alpha + \alpha_0 + 1}{n + n_0 + 1}$$

We can combine the two to write

$$\hat{\theta}_{\text{ms}}(n-1) = \frac{\alpha + \alpha_0 + 1 - z_n}{n + n_0 + 1}$$

Thus

$$\begin{aligned}\hat{\theta}_{\text{ms}}(n) &= \frac{z_n}{n + n_0 + 2} + \frac{\alpha + \alpha_0 + 1 - z_n}{n + n_0 + 2} \\ &= \frac{z_n}{n + n_0 + 2} + \frac{n + n_0 + 1}{n + n_0 + 2}\hat{\theta}_{\text{ms}}(n-1)\end{aligned}$$

with

$$\hat{\theta}_{\text{ms}}(0) = \frac{\alpha_0 + 1}{n_0 + 2}$$

Example 5.7.2 The number of events over an interval τ in an experiment obey a Poisson law with mean value θ

$$\Pr(n \text{ events} | \theta) = \begin{cases} \dfrac{(\theta\tau)^n e^{-\theta\tau}}{n!} & \theta \geqslant 0 \\ 0 & \text{otherwise} \end{cases}$$

We want to observe the number of events and estimate the parameter θ. To find a reproducing density, we assume a specific a priori sample Z_{n_0} of n_0 events in time τ_0. Then

$$p(\theta) = \frac{p(Z_{n_0}|\theta)}{\int p(Z_{n_0}|\theta)\,d\theta} = \begin{cases} \dfrac{\dfrac{(\theta\tau_0)^{n_0}e^{-\theta\tau_0}}{n_0!}}{\displaystyle\int_0^\infty \frac{(\theta\tau_0)^{n_0}e^{-\theta\tau_0}}{n_0!}\,d\theta} & \theta \geqslant 0 \\ 0 & \text{otherwise} \end{cases}$$

But

$$\int_0^\infty (\theta\tau_0)^{n_0}e^{-\theta\tau_0}\,d\theta = \frac{1}{\tau_0}\Gamma(n_0+1)$$

so that a reproducing density is

$$p(\theta) = \begin{cases} \dfrac{\tau_0^{n_0+1}}{(n_0+1)}\theta^{n_0}e^{-\theta\tau_0} & \theta \geqslant 0 \\ 0 & \text{otherwise} \end{cases}$$

which is the gamma density with parameters τ_0, n_0. Now if we observed n_1 events in time τ_1

$$p(\theta|Z_{n_1}) = \frac{p(Z_{n_1}|\theta)p(\theta)}{\int [\text{num}]\,d\theta}$$

$$= \begin{cases} \dfrac{\dfrac{(\theta\tau_1)^{n_1}e^{-\theta\tau_1}}{n_1!}\,\dfrac{\tau_0^{n_0+1}}{\Gamma(n_0+1)}\theta^{n_0}e^{-\theta\tau_0}}{\int [\text{num}]\,d\theta} & \theta \geqslant 0 \\ 0 & \text{otherwise} \end{cases}$$

$$= \begin{cases} \dfrac{(\tau_0+\tau_1)^{n_0+n_1}}{(n_0+n_1)}\theta^{(n_0+n_1)}e^{-\theta(\tau_0+\tau_1)} & \theta \geqslant 0 \\ 0 & \text{otherwise} \end{cases}$$

where [num] indicates the expression in the numerator in the corresponding term. This is a gamma density with parameters $\tau^1 = \tau_0 + \tau_1$ and $n^1 = n_0 + n_1$. The mean square estimate $\hat{\theta}_{\text{ms}}$ is given by

$$\hat{\theta}_{\text{ms}} = \mathrm{E}\{\theta|Z_{n_1}\} = \frac{n^1+1}{\tau^1} = \frac{1+n_0+n_1}{\tau_0+\tau_1}$$

If we conduct N successive trials of the experiment and observe n_1 events in τ_1, n_2 events in $\tau_2 \ldots n_N$ events in τ_N, we can easily show using the preceding procedure that $p(\theta|Z_{n_1}, Z_{n_2}, \ldots, Z_{n_N})$ will also be a gamma density with parameters

$$\tau^N = \sum_{i=0}^{N} \tau_i, \; n^N = \sum_{i=0}^{N} n_i$$

and

$$\hat{\theta}_{\text{ms}} = \frac{\sum_{i=1}^{N} n_i + 1}{\sum_{i=1}^{N} \tau_i}$$

We list here a few other commonly encountered distributions, the reproducing densities for estimating unknown parameters associated with these, and the mean square estimates of the parameters:

1. N-Variate Gaussian density, mean $\mathbf{m}$ and covariance matrix $\mathbf{V}$.
   (a) If the unknown parameter is the mean $\mathbf{m}$, the appropriate prior density is Gaussian with mean $\mathbf{m}_0$ and covariance matrix $\mathbf{V}_0$:

$$p(\mathbf{m}) = \frac{1}{(2\pi)^{N/2}|\mathbf{V}_0|^{1/2}} \exp\left[-\frac{1}{2}(\mathbf{m}-\mathbf{m}_0)^T \mathbf{V}_0^{-1}(\mathbf{m}-\mathbf{m}_0)\right]$$

   The estimate after n observations is

$$\hat{\mathbf{m}}(n) = \frac{n_0}{n+n_0}\mathbf{m}_0 + \frac{n}{n+n_0}\left[\frac{1}{n}\sum_{j=1}^{n} \mathbf{z}_j\right]$$

   where n_0 is an arbitrary number.
   (b) If the unknown parameter is $\mathbf{V}$, we assume for the prior density of $\mathbf{V}^{-1}$, the Wishart density of the form

$$p(\mathbf{V}^{-1}) = \begin{cases} k(n_0)\left|\dfrac{n_0}{2}\mathbf{V}_0\right|^{(n_0-1)/2} |\mathbf{V}^{-1}|^{(n_0-N-2)/2} \exp\left[-\dfrac{1}{2}\text{tr}\, n_0 \mathbf{V}_0 \mathbf{V}^{-1}\right] & \text{if } \mathbf{V}^{-1} \text{ is positive definite} \\ 0 & \text{otherwise} \end{cases}$$

where n_0 is a scalar and

$$k(n_0) = \left[(\pi)^{N(N-1)/4} \prod_{\alpha=1}^{N} \Gamma\left(\frac{n_0-\alpha}{2}\right) \right]^{-1}$$

The mean square estimate of $\mathbf{V}$ is

$$\hat{\mathbf{V}}(n) = \frac{n_0}{n+n_0}\mathbf{V}_0 + \frac{n}{n+n_0}\left[\frac{1}{n}\sum_{j=1}^{n} (\mathbf{z}_j - \mathbf{m})(\mathbf{z}_j - \mathbf{m})^T \right]$$

(c) If both $\mathbf{m}$ and $\mathbf{V}$ are unknown, we choose a Gaussian form for the prior density of $\mathbf{m}$ as in (a) preceding and a Wishart form as in (b) for the prior density of $\mathbf{V}$. The mean square estimate of $\mathbf{m}$ is the same as in (a). The estimate of $\mathbf{V}$ is now given by

$$\hat{\mathbf{V}}(n) = \frac{1}{n+n_0}\left[(n-1)\mathbf{S}_n + \frac{1}{n}\left(\sum_{j=1}^{n}\mathbf{z}_j\right)\left(\sum_{j=1}^{n}\mathbf{z}_j\right)^T \right.$$
$$\left. + \left(n_0\mathbf{V}_0 + n_0\mathbf{m}_0\mathbf{m}_0^T\right) - (n+n_0)\hat{\mathbf{m}}(n)\hat{\mathbf{m}}^T(n) \right]$$

where

$$\mathbf{S}_n = \frac{1}{n-1}\sum_{j=1}^{n}\left[\mathbf{z}_j - \frac{1}{n}\sum_{i=1}^{n}\mathbf{z}_i\right]\left[\mathbf{z}_j - \frac{1}{n}\sum_{i=1}^{n}\mathbf{z}_i\right]^T$$

2. Rayleigh density, parameter σ^2:
We assume a gamma density with parameters τ_0 and n_0 as the prior density for $\theta = 1/\sigma^2$

$$p(\theta) = \begin{cases} \dfrac{(\tau_0)^{n_0+1}}{\Gamma(n_0+1)}\theta^{n_0}e^{-\theta\tau_0} & \theta \geqslant 0 \\ 0 & \text{otherwise} \end{cases}$$

The mean square estimate of θ is

$$\hat{\theta}(n) = \frac{(n_0+n+1)}{\left[\tau_0 + \sum_{j=1}^{n} z_j^2/2\right]}$$

3. Exponential density, parameter θ: The prior density for θ is again a gamma density with parameters τ_0 and n_0. The estimate is

$$\hat{\theta}(n) = \frac{(n_0+n+1)}{\left[\tau_0 + \sum_{j=1}^{n} z_j\right]}$$

5.8 SUMMARY

In this chapter, we have considered the estimation of parameters from noisy observations. We started with the problem of Bayesian estimation of random parameters in Section 5.3. The density function of interest in the estimation problem is the posterior density $p(\theta|z)$. Two criteria of particular interest are the minimum mean square error (MMSE) criterion and the maximum a posteriori (MAP) criterion. The MMSE estimate is the mean of the a posteriori density (conditional mean), while the MAP estimate is the value of θ at which the posterior density is a maximum. It was also shown that for a large class of cost functions, the conditional mean is the optimum estimate whenever $p(\theta|z)$ is unimodal and symmetric about its mean.

The maximum likelihood estimate (MLE) of the parameter θ was considered in Section 5.4. The MLE is useful when θ is either a nonrandom parameter or is random with unknown prior density. The MLE is obtained as the value of θ which maximizes the conditional density $p(z|\theta)$.

Section 5.5. discusses the properties of estimators. Two measures of the performance of an estimator are the bias and the variance. In general, the bias should be zero and the variance as small as possible. Since the variance is usually difficult to determine, we established the Cramér–Rao inequality to provide a bound on the variance of any unbiased estimate. If any estimate exists which meets this bound, then it will be equal to the ML estimate. Similar bounds were also stated for random parameter estimation. In this case, if an efficient estimate exists, it can be determined using the MAP technique.

The extension to multiple parameters is straightforward and involves extensions to multiple dimensions of the results obtained for the scalar case.

The problem of MMSE estimation involves determining the conditional density $p(\theta|z)$. In many applications of interest, the determination of this density is not an easy task. It then becomes appropriate to use a suboptimal estimate which can be easily determined. Section 5.6 considers the problem of determining the best estimate in the class of linear estimates for this problem. The conditions under which the linear mean square (LMS) estimate is also the MMSE estimate are presented in terms of the characteristic function associated with the joint density function $p(z, \theta)$. Since this condition is satisfied if $p(z, \theta)$ is Gaussian, it follows that for this case the LMS estimate is the optimum estimate. By defining a linear space formed from a collection of random vectors of finite variance, the LMS estimate can be interpreted as the projection on the subspace of observation vectors. An important consequence of this result is the orthogonality principle which states that the LMS estimate is such that the estimation error is orthogonal to the observations.

We concluded the chapter with a brief discussion of reproducing densities. These are particularly useful in the estimation of the parameters of probability density functions.

The discussions in this chapter also bring out the similarities between the detection and estimation problems. Both are based on the determination of the likelihood function (a posteriori density) and in manipulations of this function. In many applications, these manipulations will be the same whether the problem is one of detection or estimation.

EXERCISES

5.1 A random parameter λ is observed through another random variable z with

$$p(z|\lambda) = \lambda e^{-\lambda z} \qquad z \geqslant 0, \lambda > 0$$
$$= 0 \qquad \lambda < 0$$

Assuming the prior density of λ to be

$$p(\lambda) = \frac{l^n}{\Gamma(n)} e^{-\lambda l} \lambda^{n-1} \qquad \lambda \geqslant 0$$
$$= 0 \qquad \lambda < 0$$

obtain $\hat{\lambda}_{map}$ and $\hat{\lambda}_{ms}$. Find $E\{[\lambda - \hat{\lambda}_{ms}]^2\}$.

5.2 In Example 5.3.1, denote the estimate of θ with N observations as $\hat{\theta}(N)$, and the corresponding variance as $\sigma^2(N)$. Now another sample z_{N+1} is observed. Express $\hat{\theta}(N+1)$ in terms of $\hat{\theta}(N)$, z_{N+1} and $\sigma^2(N)$. This provides a sequential estimation algorithm for $\hat{\theta}(N)$.

5.3 We are given a sequence of independent observations $z_1, z_2, \ldots, z_N$ with mean m and variance σ^2.

(a) Is the sample mean

$$\mu = \frac{1}{N} \sum_{i=1}^{N} z_i$$

an unbiased estimate of m? What is the variance of μ?

(b) We may seek to estimate the variance as

$$V = \frac{1}{N} \sum_{i=1}^{N} [z_i - \mu]^2$$

Is this an unbiased estimate of σ^2? Can you find its variance?

5.4 We have N independent observations, z_i, $i = 1 \ldots N$ of a Gaussian variable with mean m and variance σ^2, which is unknown. Obtain $\hat{\sigma}^2_{\text{ml}}$. Is this estimate unbiased? Is this efficient? What is the variance of this estimate?

5.5 Consider Exercise 5.4 with both mean m and variance σ^2 unknown. Obtain the maximum likelihood estimates.

5.6 The log-normal distribution is useful in characterizing the density functions of the radar cross section (RCS) of large metal objects of irregular shape, sea clutter, etc. This density function is given by

$$p(z|m, \sigma) = \frac{1}{\sqrt{2\pi}\, z\sigma} \exp\left\{-\frac{1}{2\sigma^2}\ln^2(z/m)\right\} \quad z > 0,\ m > 0,\ \sigma > 0$$

where m is the median of z and σ is the standard deviation of $\ln(z/m)$. An important quantity of interest is the ratio of the mean to median of the variable z:

$$\rho = \frac{\mathrm{E}\{z\}}{m} = \exp\left(\frac{\sigma^2}{2}\right).$$

Assume that we have N independent observations of the variable z. Show that the maximum likelihood estimates of the parameters m and ρ are

$$\hat{m} = \left[\prod_{i=1}^{N} z_i\right]^{1/N}$$

and

$$\hat{\rho} = \left[\prod_{i=1}^{N} \left(\frac{z_i}{\hat{m}}\right)^{\ln(z_i/\hat{m})}\right]^{1/2N}$$

Obtain the variance of the estimates as the number of observations becomes very large. [Note that the ML estimate commutes over nonlinear operations. See reference [18] for details and some numerical experiments.]

5.7 We observe a random parameter x, Gaussian distributed with zero mean and variance σ_x^2 through two independent observation channels.

$$z_1 = x + v_1$$
$$z_2 = x + v_2$$

where

$$p(v_i) = \frac{1}{\sqrt{2\pi}\,\sigma_i} \exp\left[-\frac{v_i^2}{2\sigma_i^2}\right]$$

Obtain $\hat{x}_{\text{ms}}$, and $\hat{x}_{\text{map}}$ as a function of z_1 and z_2. What is the mean square error?

5.8 In Exercise 5.7 consider x to be nonrandom real. Obtain $\hat{x}_{\text{ml}}$. Is this an efficient estimate? What is the variance of the estimate?

5.9 Obtain a MAP and MMSE estimate of x_1 from the observation

$$z = x_1 + x_2$$

where x_1 and x_2 are independent and are Rayleigh distributed with parameters σ_1^2 and σ_2^2.

5.10 Consider the problem of pulse amplitude modulation communication. A signal $x(t)$ is sampled at regular intervals to give a sequence of amplitudes $x_1, x_2, \ldots, x_n$. The x_is modulate a carrier and are received in a T-second interval in additive Gaussian white noise with zero mean and spectral height $N_0/2$. The received signal in the jth T-second interval is

$$z(t) = x_j \sin \omega_c t + v(t) \qquad jT \leqslant t \leqslant (j+1)T$$

How can you estimate x_j in the sense of maximum likelihood?

5.11 We observe a random parameter x in multiplicative and additive noise as

$$z = v_1 x + v_2$$

where

$$p(x) = \frac{1}{\sqrt{2\pi}\,\sigma_x} \exp\left[-\frac{(x-m_x)^2}{2\sigma_x^2}\right]$$

$$p(v_1) = \frac{1}{\sqrt{2\pi}\,\sigma_1} \exp\left[-\frac{(v_1-m_1)^2}{2\sigma_1^2}\right]$$

and

$$p(v_2) = \frac{1}{\sqrt{2\pi}\,\sigma_2} \exp\left[-\frac{(v_2-m_2)^2}{2\sigma_2^2}\right]$$

Obtain the linear MS estimate of x based on the observation z. Extend your results to N independent observations.

5.12 Repeat Exercise 5.10 when the observation model is

$$z(t) = x_j \sin(\omega_c t + \theta) + v(t) \qquad jT < t \leqslant (j+1)T$$

when θ is unknown and is uniformly distributed between 0 and 2π radians.

5.13 In a pulse frequency modulation (PFM) system, the samples $x_1, x_2, \ldots, x_N$ of a message $x(t)$ change the frequency of the carrier, and are observed with additive noise. The observed signal in a T-second interval is

$$z(t) = \sqrt{\frac{2\epsilon}{T}} \sin(\omega_c t + \beta x_j) + v(t)$$

where β is the *modulation index*, and $v(t)$ is Gaussian white, with zero mean and spectral height $N_0/2$. Assume that x_j is uniformly distributed between $-M$ and M. Obtain the estimator equations, and suggest a practical implementation.

5.14 If $\mathbf{x}$, $\mathbf{y}$, and $\mathbf{z}$ are zero-mean Gaussian vectors such that $E\{\mathbf{y}^T\mathbf{z}\} = 0$ show that

$$E\{\mathbf{x}|\mathbf{y}, \mathbf{z}\} = E\{\mathbf{x}|\mathbf{y}\} + E\{\mathbf{x}|\mathbf{z}\}$$

5.15 Consider independent samples $x_1, x_2, \ldots, x_n$ from a Gaussian distribution

$$p(x) = \frac{1}{\sqrt{2\pi}\,\sigma} \exp\left(-\frac{(x-m)^2}{2\sigma^2}\right)$$

We have to estimate the unknown mean m, and variance σ^2 assuming a proper reproducing density function for m and σ^2. Set up the estimation algorithms in a sequential manner.

REFERENCES

1. A. M. Mood and F. A. Graybill, *Introduction to the Theory of Statistics,* 2nd Ed., McGraw-Hill Book Co., New York, 1963.
2. R. A. Fisher, Theory of Satistical Estimation, *Proc. Cambridge Phil. Soc.,* **22,** 700–725 (1925).
3. D. Middleton, *Introduction to Statistical Communication Theory,* McGraw-Hill Book Co., New York, 1963.

4. H. L. Van Trees, *Detection, Estimation and Modulation Theory, Part I,* John Wiley and Sons, New York, 1968.
5. R. Deutsch, *Estimation Theory,* Prentice Hall, Englewood Cliffs, N. J,, 1965.
6. A. P. Sage and J. L. Melsa, *Estimation Theory with Applications to Communications and Control,* McGraw-Hill Book Co., New York, 1971.
7. T. W. Anderson, The Integral of a Symmetric Unimodal Function over a Symmetric Convex Set and Some Probability Inequalities, *Proc. Am. Math. Soc.,* **6**, 170–176 (1955).
8. S. Sherman, Non-Mean Square Error Criteria, *I.R.E. Trans. Info. Theory,* **IT-4,** No. 3, 125–126 (1958).
9. A. J. Viterbi, *Principles of Coherent Communication,* McGraw-Hill Book Co., New York, 1966.
10. H. Cramér, *Mathematical Methods of Statistics,* Princeton University Press, Princeton, N. J., 1946.
11. C. R. Rao, Information and the Accuracy Attainable in the Estimation of Statistical Parameters, *Bull. Calcutta Math. Soc.,* **37**, 81–91 (1945).
12. P. Swerling, Parameter Estimation for Waveforms in Additive Gaussian Noise, *J. SIAM,* **7**, No. 2, 152–166 (1959).
13. A. V. Balakrishnan, On a Characterization of Processes for which Optimal Mean-Square Systems Are of a Specified Form, *IRE Trans. Info. Theory,* **IT-6,** Sept., 491–500 (1960).
14. J. L. Doob, *Stochastic Processes,* John Wiley and Sons, New York, 1950.
15. K. Hoffman and R. Kunze, *Linear Algebra,* Prentice Hall, Englewood Cliffs, N. J., 1961.
16. E. A. Patrick, *Fundamentals of Pattern Recognition,* Prentice-Hall, Englewood Cliffs, N. J., 1972.
17. J. D. Spragins, *Reproducing Distributions for Machine Learning, Stanford Electronics Labs. Tech. Rept. 6103-7,* Stanford, Ca., Nov. 1963.
18. W. J. Szajnowski, Estimators of Log Normal Distribution Parameters, *IEEE Trans. Aerospace Electronic* Syst., **AES-13,** No. 5, Sept. 533–536 (1977).

CHAPTER 6

ESTIMATION OF WAVEFORMS

6.1 INTRODUCTION

In the previous chapter, we developed estimation techniques for both random and nonrandom parameters. Many physical situations, for example in analog communications, radar tracking, and recognition of patterns with time varying characteristics, involve the estimation of waveforms. These waveforms to be estimated can be modeled as realizations of stochastic processes, and are usually observed in noise. When unknown deterministic signals are observed in noise, essentially the same estimation techniques can be applied. If a signal which is known except for certain nonrandom parameters is observed in a noisy background, the MLE techniques of Chapter 5 can be used.

In this chapter we consider the mean square estimation of stochastic processes observed in additive noise. We define the three basic estimation problems: prediction, filtering, and smoothing. We first develop the orthogonal projection property of linear estimators of waveforms. The classical theory of Weiner filtering for discrete and continuous time stationary stochastic processes is presented with examples in Section 6.3. In Section 6.4, recursive estimation schemes for estimation of nonstationary processes are considered. The discrete and continuous time Kalman filters are derived using system theory concepts. For nonlinear signal and observation models, approximate estimation schemes are presented in Section 6.5. We conclude with a consideration of the MAP estimate of certain nonlinear observation models.

6.2 LINEAR MMSE ESTIMATION OF WAVEFORMS: PRELIMINARIES [1–7]

We are given two related stochastic processes $y(t)$ and $z(t)$. Our problem is to estimate various parameters of the first process (signal) in terms of certain values $z(\xi)$ of the second process. We will assume without loss of generality that the processes $y(t)$ and $z(t)$ have zero mean. We shall refer to the quantity to be estimated as $g(t)$. For example

$$g(t) = y(t) \quad \text{or} \quad \dot{y}(t) \quad \text{or} \quad y(t+\alpha)$$

The estimation is to be based on the values of $z(t)$ at certain instants $t = \xi$ forming a set I on the time axis. This set could consist of a number of discrete points or an interval on the time axis. Our problem is to find a suitable transformation of the data that can be used as a best estimate $\hat{g}(t)$ of $g(t)$.

We now consider several specific problems.

(i) If $g(t) = y(t+\alpha)$ with $\alpha > 0$, we have the *prediction* problem. If $z(t) = y(t)$ and if the observation interval I consists of the single point t, we are required to predict the value of $y(\cdot)$, α units in the future in terms of its value at present time. If the observation interval is $(-\infty, t)$, our data consists of the entire past history of $y(\cdot)$.

(ii) If $g(t) = y(t)$, we have the *filtering* problem.

(iii) If the observation interval is $[t_0, t_f]$ and we are required to estimate $y(t)$ for any $t \in [t_0, t_f]$, the problem is one of *smoothing* or interpolation. Of particular interest is the *fixed-lag* smoothing problem where we are required to estimate $y(t+\alpha)$ for $\alpha < 0$ based on observations over the interval $[t_0, t]$.

If we want to use the mean square cost function in our estimation problem, we form the estimation error $\tilde{g}(t)$ defined as

$$\tilde{g}(t) = g(t) - \hat{g}(t) \tag{6.2.1}$$

The risk is

$$\bar{C} = \mathrm{E}\{[g(t) - \hat{g}(t)]^2\} \tag{6.2.2}$$

If I is a set of discrete points, we are required to find the estimate $\hat{g}(t)$ based on the set of random variables $z(\xi)$, $\xi \in I$. From our discussions in

Chapter 5, it readily follows that for fixed t, the estimate $\hat{g}(t)$ is given by the conditional mean

$$\hat{g}(t) = \mathrm{E}\{ g(t)|z(\xi), \xi \in I \} \tag{6.2.3}$$

However, if I is an interval, say $[t_0, t_f]$, $z(\xi)$ represents a sample function of a stochastic process and the conditioning in Eq. (6.2.3) has to be interpreted a little more carefully. In this case the conditioning is really on the σ-algebra generated by $z(\cdot)$. For a detailed discussion see, for example, [6]. We can overcome this problem by using an approach similar to the one used in Chapter 4. Thus if we use a K-term approximation of $z(\xi)$ in terms of a complete orthonormal set, the conditioning can be expressed in terms of a set of random variables, namely, the coefficients of the expansion. We can then use a suitable limiting process to obtain $\hat{g}(t)$. We leave the details to the interested reader.

The determination of the conditional mean in Eq. (6.2.3) is not straightforward, since it requires the joint distribution of two variables $g(t)$ and $z(\xi)$ for each t and ξ of interest. Rather, just as we did in the case of random parameters, we confine ourselves to linear estimates. In this case our estimate is of the form

$$\hat{g}(t) = L[z(\xi), \xi \in I] \tag{6.2.4}$$

where L is a linear operator.

For example, if we want to predict $y(t+\alpha)$ in terms of $y(t)$ and its derivative $\dot{y}(t)$, we write for the estimate

$$\hat{y}(t+\alpha) = ay(t) + b\dot{y}(t) \tag{6.2.5}$$

The problem then will be to choose the constants a and b so as to minimize the mean square error

$$\mathrm{E}\{[y(t+\alpha) - ay(t) - b\dot{y}(t)]^2\} \tag{6.2.6}$$

In general our problem is to find a linear operator

$$L[z(\xi), \xi \in I]$$

operating on the data $z(\xi)$ so as to minimize the mean square error

$$\overline{C}_{\mathrm{lms}} = \mathrm{E}\{[g(t) - L(z(\xi))]^2\} \tag{6.2.7}$$

The principle of orthogonality applied to this problem then implies that if an optimum solution exists, this solution is found by choosing L such that the estimation error

$$\tilde{g}(t) = g(t) - L[z(\xi)]$$

is orthogonal to $z(\xi)$ for every $\xi \in I$.

That is, if

$$E\{[g(t) - L(z(\xi))]z(\xi_i)\} = 0 \quad \text{for each } \xi_i \in I \tag{6.2.8}$$

then $\bar{C}_{\text{lms}}$ is minimum and is given by

$$\bar{C}_{\text{lms}} = E\{[g(t) - L(z(\xi))]g(t)\} \tag{6.2.9}$$

The proof is similar to the one used in Chapter 5. Suppose L is such that it satisfies Eq. (6.2.8). Let L_1 be any other linear operator operating on the set of data $z(\xi)$, $\xi \in I$. We can write

$$\tilde{g}(t) = g(t) - L_1[z(\xi)] = g(t) - L[z(\xi)] + L[z(\xi)] - L_1[z(\xi)] \tag{6.2.10}$$

Since the difference of two linear operators is linear, we write

$$L[z(\xi)] - L_1[z(\xi)] = L_2[z(\xi)] \tag{6.2.11}$$

Then

$$\begin{aligned} E\{[g(t) - L_1[z(\xi)]]^2 &= E\{[g(t) - L[z(\xi)] + L_2[z(\xi)]]^2\} \\ &= E\{[g(t) - L[z(\xi)]]^2\} + 2E\{[g(t) \\ &\quad - L[z(\xi)]]L_2[z(\xi)]\} + E\{L_2[z(\xi)]^2\} \end{aligned} \tag{6.2.12}$$

The difference $g(t) - L[z(\xi)]$ is by assumption orthogonal to the data. It must therefore be orthogonal to a linear combination of $z(\xi)$. Thus the second term on the right side of Eq. (6.2.12) is zero. Since the last term is nonnegative we have

$$E\{[g(t) - L_1[z(\xi)]]^2\} \geqslant E\{[g(t) - L[z(\xi)]]^2\} \tag{6.2.13}$$

thus establishing the orthogonality principle. We note that equality holds in Eq. (6.2.13) only when $E\{[L_2[z(\xi)]]^2\} = 0$ implying that $L_2[z(\xi)] = 0$ (that is $L_1 = L$) with probability 1. The expression for $\bar{C}_{\text{lms}}$ follows from the definition and Eq. (6.2.8).

As in the case of random parameters, if $g(t)$ and $z(t)$ are jointly normal (with zero mean) then

$$E\{ g(t)|z(\xi), \xi \in I \}$$

is a linear combination of the values of $z(\xi)$. In this case the linear mean square estimate of $g(t)$ is also the minimum mean square estimate.

We now consider a few examples.

Example 6.2.1 Consider the problem of predicting a signal based on present measured values. Thus we want to estimate $y(t+\alpha)$ in terms of $y(t)$ as

$$\hat{y}(t+\alpha) = ay(t)$$

We assume that $y(t)$ is a stationary process.

From the orthogonality principle, it follows that the constant a must be such that the error $y(t+\alpha) - ay(t)$ is orthogonal to the data $y(t)$:

$$E\{[y(t+\alpha) - ay(t)]y(t)\} = 0$$

This yields

$$a = \frac{\phi_y(\alpha)}{\phi_y(0)}$$

where $\phi_y(\alpha)$ denotes the auto correlation function of $y(t)$. The mean square error is given by Eq. (6.2.9) as

$$\begin{aligned}\overline{C}_{\text{lms}} &= E\{[y(t+\alpha) - ay(t)]y(t+\alpha)\} \\ &= \phi_y(0) - a\phi_y(\alpha) \\ &= \phi_y(0) - \frac{\phi_y^2(\alpha)}{\phi_y(0)}\end{aligned}$$

Example 6.2.2 Suppose we want to estimate $y(t+\alpha)$ in terms of $y(t)$ and its derivative $\dot{y}(t)$:

$$\hat{y}(t+\alpha) = ay(t) + b\dot{y}(t)$$

The error must be orthogonal to $y(t)$ and $\dot{y}(t)$ so that

$$E\{[y(t+\alpha) - ay(t) - b\dot{y}(t)]y(t)\} = 0$$

and

$$E\{[y(t+\alpha) - ay(t) - b\dot{y}(t)]\dot{y}(t)\} = 0$$

We note that

$$\phi_{y\dot{y}}(\tau), = -\dot{\phi}_y(\tau), \ \phi_{\dot{y}\dot{y}}(\tau) = -\ddot{\phi}_y(\tau)$$

and that $\dot{\phi}_y(0) = 0$ for $\dot{y}(t)$ to exist.

We thus have

$$a = \frac{\phi_y(\alpha)}{\phi_y(0)} \qquad b = \frac{\dot{\phi}_y(\alpha)}{\ddot{\phi}_y(0)}$$

The mean square error is

$$\bar{C}_{\text{lms}} = E\{[y(t+\alpha) - ay(t) - b\dot{y}(t)]y(t+\alpha)\}$$

$$= \phi_y(0) - a\phi_y(\alpha) + b\dot{\phi}_y(\alpha)$$

Example 6.2.3 As a final example we consider an interpolation problem in which we are given the values of a signal at the end points of an interval and are required to estimate the signal at any instant in the interval. Thus given $y(0)$ and $y(T)$ we write for the estimate

$$\hat{y}(t) = ay(0) + by(T) \qquad \text{for } t \in [0, T]$$

Since the error is orthogonal to the data, we have

$$E\{[y(t) - ay(0) - by(T)]y(0)\} = 0$$

and

$$E\{[y(t) - ay(0) - by(T)]y(T)\} = 0$$

Hence

$$a\phi_y(0) + b\phi_y(T) = \phi_y(t)$$

and

$$a\phi_y(T) + b\phi_y(0) = \phi_y(T-t)$$

Solving these two equations for a and b yields

$$a = \frac{\phi_y(0)\phi_y(t) - \phi_y(T)\phi_y(T-t)}{\phi_y^2(0) - \phi_y^2(T)}$$

and

$$b = \frac{\phi_y(0)\phi_y(T-t) - \phi_y(t)\phi_y(T)}{\phi_y^2(0) - \phi_y^2(T)}$$

We note that for $t = 0$, $a = 1$, $b = 0$; for $t = T$, $a = 0$, and $b = 1$.

6.3 ESTIMATION OF STATIONARY PROCESSES: THE WIENER FILTER [8]

In the examples in the preceding section, the data consisted of a finite number of random variables. We now consider the problem of estimating a stochastic process $g(t)$ based on observations of a related process $z(t)$ by a linear combination of known values

$$\{z(\xi) : t_0 \leqslant \xi \leqslant t_f\}$$

of $z(t)$, so that the observation interval consists of the closed interval $[t_0, t_f]$. We assume that the processes $g(t)$ and $z(t)$ have zero mean. We write for our estimate

$$\begin{aligned} \hat{g}(t) &= \lim_{\Delta\xi \to 0} \sum_{i=1}^{N} h(t, \xi_i) z(\xi_i) \Delta\xi \\ &= \int_{t_0}^{t_f} h(t, \xi) z(\xi)\, d\xi \end{aligned} \tag{6.3.1}$$

We note that $g(t)$ represents the output of a linear time varying system whose input is $z(t)$.

Since the integral in Eq. (6.3.1) is the limit of a sum, the problem is one of estimating $g(t)$ by a linear combination of the random variables $z(\xi_i)$. The weights $h(t, \xi_i)\Delta\xi$ associated with these variables are to be chosen so as to minimize the mean square error. From the orthogonality principle, it follows that these weights have to be chosen such that the error in the estimate is orthogonal to the data; that is,

$$\mathrm{E}\left\{\left[g(t) - \lim_{\Delta\xi \to 0} \sum_{i=1}^{N} h(t, \xi_i) z(\xi_i) \Delta\xi\right] z(\tau)\right\} = 0, \qquad t_0 \leqslant \tau \leqslant t_f \tag{6.3.2}$$

In terms of the integral, the preceding can be written as

$$\mathrm{E}\left\{\left[g(t) - \int_{t_0}^{t_f} h(t, \xi) z(\xi)\, d\xi\right] z(\tau)\right\} = 0, \qquad t_0 \leqslant \tau \leqslant t_f \tag{6.3.3}$$

In terms of correlation functions, this becomes

$$\phi_{gz}(t, \tau) = \int_{t_0}^{t_f} h(t, \xi) \phi_z(\xi, \tau)\, d\xi, \qquad t_0 \leqslant \tau \leqslant t_f \tag{6.3.4}$$

which is an integral equation of the type we have encountered earlier in Chapter 4. This is referred to in the literature as the Wiener–Hopf integral equation [1–8]. The mean square error can be written from Eq. (6.2.9) as

$$\begin{aligned} \overline{C}_{\mathrm{lms}} &= \mathrm{E}\left\{\left[g(t) - \int_{t_0}^{t_f} h(t, \xi) z(\xi)\, d\xi\right] g(t)\right\} \\ &= \phi_g(t, t) - \int_{t_0}^{t_f} h(t, \xi) \phi_{zg}(\xi, t)\, d\xi \end{aligned} \tag{6.3.5}$$

We now consider the solution of Eq. (6.3.4) for some special cases.

Estimation of Signals Observed in Additive Noise

Of particular importance is the case where the observations $z(t)$ consist of a signal observed in additive noise:

$$z(t) = y(t) + v(t), \qquad t_0 \leqslant t \leqslant t_f \tag{6.3.6}$$

We desire to estimate $g(t)$ in the interval $t_0 \leqslant t \leqslant t_f$. This was first considered by Wiener in connection with problems in tracking and control [8]. As discussed in the previous section, if we choose $g(t) = y(t + \alpha)$, $\alpha > 0$, we have a prediction problem. If $g(t) = y(t)$ we have a filtering problem and if $g(t) = y(t + \alpha)$, $\alpha < 0$, the problem is one of interpolation or smoothing.

Let us now consider the problem of solving Eq. (6.3.4). From our discussions in Chapter 4, we know that the solution is rather difficult except in the special case when the observations $z(t)$ and the desired estimate $\hat{g}(t)$ are jointly stationary. For this to happen, the observation interval must be semi-infinite, that is, $t_0 = -\infty$ and the system (filter) must be time invariant. If we restrict ourselves to causal systems, then the filter is permitted to use only past data in constructing the estimate. Under these conditions Eq. (6.3.1) becomes

$$\hat{g}(t) = \int_{-\infty}^{t} h(t-\xi) z(\xi)\, d\xi \tag{6.3.7}$$

Application of the orthogonality principle yields the equation which the optimal filter must satisfy:

$$\phi_{gz}(t-\tau) = \int_{-\infty}^{t} h(t-\xi)\phi_z(\xi-\tau)\, d\xi, \qquad -\infty < \tau \leqslant t \tag{6.3.8}$$

We solve this equation in a manner similar to our solution in Chapter 4 [9]. Thus we note that if $z(t)$ were a white process so that

$$\phi_z(\tau) = \delta(\tau) \tag{6.3.9}$$

the equation that the optimum filter satisfies reduces to

$$\phi_{gz}(t-\tau) = \int_{-\infty}^{t} h(t-\xi)\delta(\xi-\tau)\, d\xi = h(t-\tau) \quad \text{for} \quad -\infty < \tau \leqslant t \tag{6.3.10}$$

or in terms of transfer functions

$$H(s) = \left[\Phi_{gz}(s)\right]_{+} \tag{6.3.11}$$

We see that the solution of the integral equation is straightforward.

When the observation $z(t)$ is not white, we can consider passing it through a whitening filter $H_w(s)$. The input to the optimum filter is now the whitened observation $\nu(t)$ and the filter is designed so that its output is the desired mean square estimate $\hat{g}(t)$. This is illustrated in Fig. 6.3.1.

The structure of the filter is similar to the one we encountered in our solution to the matched filter problem (see Fig. 4.3.1). The filter $H_w(s)$ in

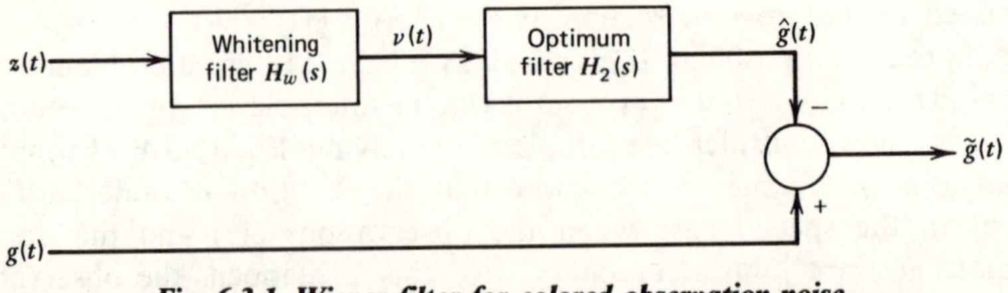

Fig. 6.3.1 Wiener filter for colored observation noise.

the estimation problem is, however, designed to whiten the observation $z(t)$, whereas in the detection problem, only the observation noise was whitened.

The optimum filter $H_2(s)$ operates on the whitened observations $\nu(t)$ and is obtained from Eq. (6.3.11) by replacing $z(t)$ with $\nu(t)$. We thus get

$$H_2(s) = \left[\Phi_{g\nu}(s)\right]_+ \tag{6.3.12}$$

The whitening filter is given by

$$H_w(s) = \frac{1}{\Phi_z^+(s)} \tag{6.3.13}$$

It only remains to determine $\Phi_{g\nu}(s)$. We note that

$$\begin{aligned} \Phi_{g\nu}(s) &= H_w(-s)\Phi_{gz}(s) \\ &= \frac{\Phi_{gz}(s)}{\Phi_z^-(s)} \end{aligned} \tag{6.3.14}$$

so that the overall filter is

$$H(s) = H_w(s)H_2(s) = \frac{1}{\Phi_z^+(s)}\left[\frac{\Phi_{gz}(s)}{\Phi_z^-(s)}\right]_+ \tag{6.3.15}$$

The filter so obtained is usually referred to as the *Wiener filter*. The error associated with the Wiener filter can be determined as follows. From Eq. (6.3.5), we have, with $t_0 = -\infty$, and for stationary signals

$$\begin{aligned} \overline{C}_{\text{lms}} &= \phi_g(0) - \int_{-\infty}^{t} h(t-\xi)\phi_{gz}(t-\xi)\, d\xi \\ &= \phi_g(0) - \int_0^{\infty} h(\lambda)\phi_{gz}(\lambda)\, d\lambda \end{aligned} \tag{6.3.16}$$

A problem of special interest is the estimation of $y(t+\alpha)$. In this case

$$\begin{aligned} \phi_{gz}(\tau) &= \mathrm{E}\{g(t+\tau)z(t)\} \\ &= \mathrm{E}\{y(t+\tau+\alpha)z(t)\} \\ &= \phi_{yz}(\tau+\alpha) \end{aligned} \tag{6.3.17}$$

Thus

$$\Phi_{gz}(s) = \Phi_{yz}(s)e^{\alpha s} \tag{6.3.18}$$

and the optimal filter obtained from Eq. (6.3.15) is

$$H(s)=\frac{1}{\Phi_z^+(s)}\left[\frac{e^{\alpha s}\Phi_{yz}(s)}{\Phi_z^-(s)}\right]_+ \tag{6.3.19}$$

The mean square error in this case is, from Eq. (6.3.16),

$$\overline{C}_{\text{lms}}=\phi_y(0)-\int_0^\infty h(\lambda)\phi_{yz}(\lambda+\alpha)\,d\lambda \tag{6.3.20}$$

Let us denote the integral on the right side of Eq. (6.3.20) as

$$F(\alpha)=\int_0^\infty h(\lambda)\phi_{yz}(\lambda+\alpha)\,d\lambda \tag{6.3.21}$$

Since $h(\lambda)=0$ for $\lambda<0$, we can replace the lower limit by $-\infty$. Let

$$k(t)=\mathcal{L}^{-1}\left[\frac{\Phi_{yz}(s)}{\Phi_z^-(s)}\right] \tag{6.3.22}$$

where $\mathcal{L}^{-1}$ denotes the inverse Laplace transform. Then

$$\mathcal{L}^{-1}\left\{\left[\frac{\Phi_{yz}(s)e^{\alpha s}}{\Phi_z^-(s)}\right]_+\right\}=\begin{cases} k(t+\alpha) & t\geqslant 0\\ 0 & t<0\end{cases} \tag{6.3.23}$$

We can substitute Eq. (6.3.23) in Eq. (6.3.19) and write

$$H(s)=\frac{1}{\Phi_z^+(s)}\int_0^\infty k(t+\alpha)e^{-st}\,dt \tag{6.3.24}$$

so that

$$h(\lambda)=\frac{1}{2\pi j}\int_{c-j\infty}^{c+j\infty}\frac{e^{s\lambda}\,ds}{\Phi_z^+(s)}\int_0^\infty k(t+\alpha)e^{-st}\,dt \tag{6.3.25}$$

Substitution of Eq. (6.3.25) in Eq. (6.3.21) yields

$$F(\alpha)=\int_0^\infty k(t+\alpha)\frac{1}{2\pi j}\int_{c-j\infty}^{c+j\infty}\frac{e^{-s(t+\alpha)}}{\Phi_z^+(s)}$$
$$\times\int_{-\infty}^{\infty}\phi_{yz}(\lambda+\alpha)e^{s(\lambda+\alpha)}\,d\lambda\,ds\,dt \tag{6.3.26}$$

We note that the innermost integral is $\Phi_{yz}(-s)$ so that Eq. (6.3.26) becomes

$$F(\alpha) = \int_0^\infty k(t+\alpha)\left[\frac{1}{2\pi j}\int_{c-j\infty}^{c+j\infty} K(-s)e^{-s(t+\alpha)}\,ds\right]dt \tag{6.3.27}$$

where

$$K(s) = \mathcal{L}[k(t)] = \frac{\Phi_{yz}(s)}{\Phi_z^-(s)} \tag{6.3.28}$$

Finally we have

$$\begin{aligned} F(\alpha) &= \int_0^\infty [k(t+\alpha)]^2\,dt \\ &= \int_\alpha^\infty k^2(t)\,dt \end{aligned} \tag{6.3.29}$$

We see that $F(\alpha)$ is a monotone decreasing function of α.

Substituting for $F(\alpha)$ in Eq. (6.3.21) yields the mean square error

$$\overline{C}_{\text{lms}} = \phi_y(0) - \int_\alpha^\infty k^2(t)\,dt \tag{6.3.30}$$

The error is maximum for $\alpha = \infty$ (prediction time is infinite) and is minimum for $\alpha = -\infty$ (smoothing time or lag is infinite). In fact, for the case of $\alpha = -\infty$, the mean square error obtained is called the irreducible error. This irreducible error is also obtained for the filtering problem ($\alpha = 0$), but with the observation interval being infinity, $(-\infty, \infty)$. In such a case, the Eq. (6.3.8) has the upper limit ∞, instead of t. The solution for $H(s)$ can be obtained by taking the bilateral Laplace transform, but the filter will be physically unrealizable (noncausal).

We illustrate the material in this section by considering four examples.

Example 6.3.1 Let the input signal be stationary, zero mean, with autocorrelation function

$$\phi_y(\tau) = \frac{7}{12}e^{-|\tau|/2}$$

and power spectral density

$$\Phi_y(s) = \frac{7/3}{-4s^2+1}$$

Assume that the noise is zero mean with spectral density

$$\Phi_v(s) = \frac{5/3}{-s^2+1}$$

and is uncorrelated with the signal. We seek to estimate $y(t+\alpha)$.

In this case, we note that since the signal and noise are uncorrelated

$$\Phi_{yz}(s) = \Phi_y(s) = \frac{7/3}{-4s^2+1}$$

and

$$\Phi_z(s) = \Phi_y(s) + \Phi_v(s) = \frac{7/3}{-4s^2+1} + \frac{5/3}{-s^2+1} = \frac{-9s^2+4}{(-4s^2+1)(-s^2+1)}$$

$$= \frac{(3s+2)}{(2s+1)(s+1)} \cdot \frac{(-3s+2)}{(-2s+1)(-s+1)}$$

The optimal filter as given by Eq. (6.3.15) is, therefore,

$$H(s) = \frac{(2s+1)(s+1)}{3s+2}\left[e^{\alpha s} \cdot \frac{7/3}{-4s^2+1} \cdot \frac{(-2s+1)(-s+1)}{(-3s+2)}\right]_+$$

$$= \frac{(2s+1)(s+1)}{3s+2}\left[e^{\alpha s}\left\{\frac{1}{2s+1} + \frac{1/3}{-3s+2}\right\}\right]_+$$

Let

$$K(s) = \frac{1}{2s+1} + \frac{1/3}{-3s+2}$$

Then

$$k(t) = \begin{cases} 1/2e^{-t/2} & t \geqslant 0 \\ 1/9e^{2t/3} & t < 0 \end{cases}$$

The causal part of $k(t+\alpha)$ is given by

$$k(t+\alpha) = 1/2e^{-(t+\alpha)/2} \qquad \alpha > 0$$

$$k(t+\alpha) = \begin{cases} 1/9e^{2(t+\alpha)/3} & 0 \leqslant t < |\alpha| \\ 1/2e^{-(t+\alpha)/2} & t \geqslant |\alpha| \end{cases} \qquad \alpha < 0$$

so that

$$[K(s)e^{\alpha s}]_+ = \begin{cases} \dfrac{e^{-\alpha/2}}{2s+1} & \alpha > 0 \\ \dfrac{1}{2s+1} & \alpha = 0 \\ \dfrac{1}{3}\dfrac{e^{2\alpha/3} - e^{s\alpha}}{3s-2} + \dfrac{1}{2}\dfrac{e^{s\alpha}}{s+1/2} & \alpha < 0 \end{cases}$$

The optimum filter in each case is given by

$$H(s) = \frac{(2s+1)(s+1)}{3s+2}[K(s)e^{\alpha s}]_+$$

To determine the mean square error, we compute

$$F(\alpha) = \int_\alpha^\infty k^2(t)\, dt$$

We have

$$F(\alpha) = \begin{cases} \frac{1}{4}e^{-\alpha} & \alpha > 0 \\ \frac{1}{4} & \alpha = 0 \\ \frac{1}{108}(1 - e^{4\alpha/3}) + \frac{1}{4} & \alpha < 0 \end{cases}$$

and the mean square error is

$$\overline{C}_{\text{lms}} = \phi_y(0) - F(\alpha) = 7/12 - F(\alpha)$$

These various functions are sketched in Fig. 6.3.2. The minimum error that can be obtained when $\alpha = -\infty$ is 35/108.

Example 6.3.2 Consider the problem of estimating a stationary signal $y(t)$ with rational power spectrum, in the presence of white noise of spectral height $N_0/2$. If the signal and noise are uncorrelated, we can write

$$\Phi_z(s) = \Phi_y(s) + \Phi_v(s) = \Phi_y(s) + N_0/2$$

From Eq. (6.3.15) the optimum filter is given by

$$H(s) = \frac{1}{\Phi_z^+(s)}\left[\frac{\Phi_{gz}(s)}{\Phi_z^-(s)}\right]_+$$

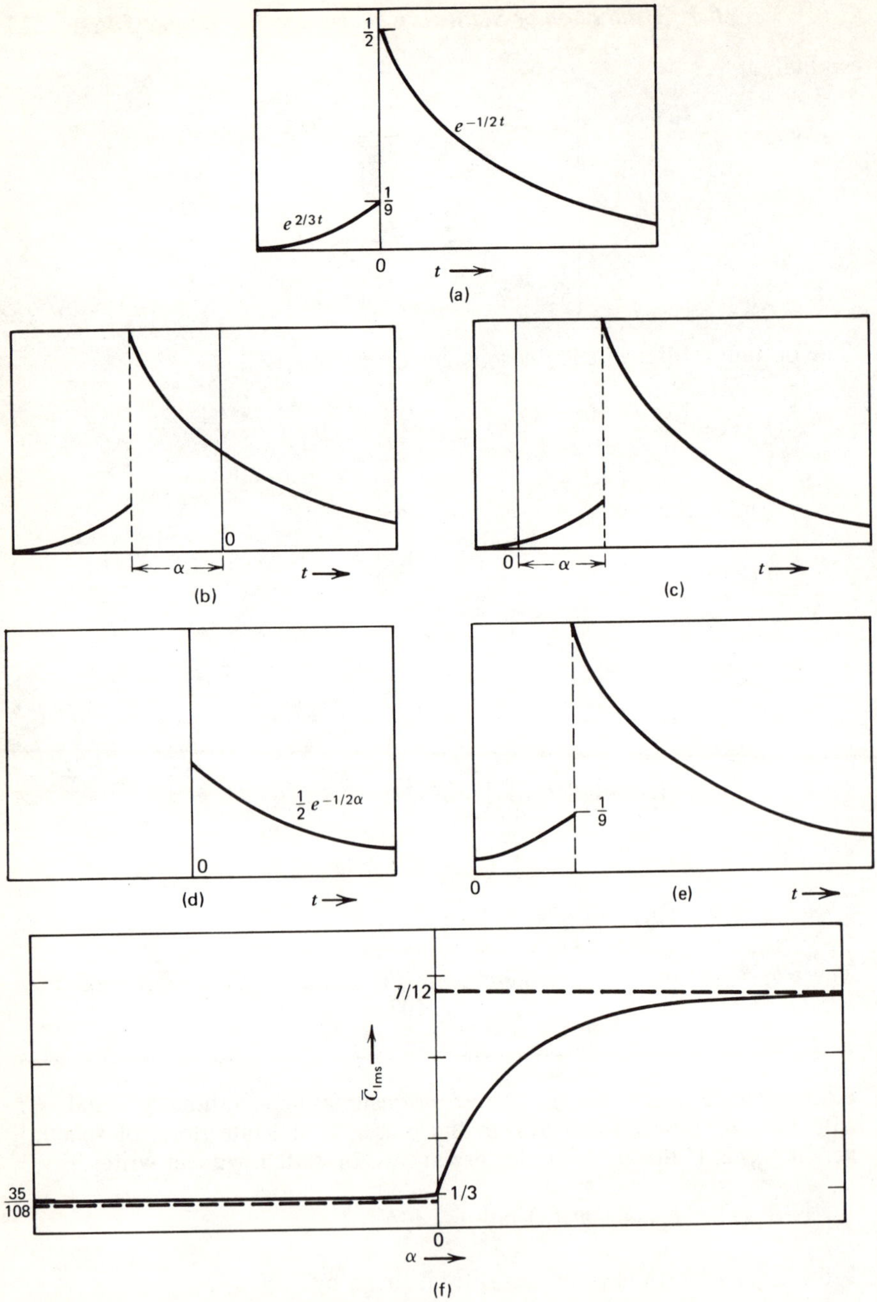

Fig. 6.3.2 Computation of error for Example 6.3.1. (a) k(t); (b) k(t + a) for a > 0; (c) k(t + a) for a < 0; (d) causal part of (b); (e) causal part of (c); (f) cost versus a.

Because $g(t) = y(t)$ in this case, we have

$$\Phi_{gz}(s) = \Phi_{yz}(s) = \Phi_y(s)$$

By adding and subtracting $\Phi_v(s)$ to the numerator of the term in braces, we can write the expression for $H(s)$ as

$$\begin{aligned} H(s) &= \frac{1}{\Phi_z^+(s)}\left[\Phi_z^+(s) - \frac{\Phi_v(s)}{\Phi_z^-(s)}\right]_+ \\ &= 1 - \frac{1}{\Phi_z^+(s)}\left[\frac{N_0/2}{\Phi_z^-(s)}\right]_+ \end{aligned}$$

From the form of $\Phi_z(s)$ we note that it is a rational function with equal order numerator and denominator. The leading coefficient in the numerator will be $N_0/2$. Thus $\Phi_z^-(s)$ will also be a rational function with equal order numerator and denominator with leading coefficient $\sqrt{N_0/2}$. Correspondingly, $(N_0/2)/\Phi_z^-(s)$ will have leading coefficient $\sqrt{N_0/2}$ and can be written as

$$\frac{N_0/2}{\Phi_z^-(s)} = \sqrt{\frac{N_0}{2}} + \Phi_1(s)$$

where $\Phi_1(s)$ has numerator degree at least one less than denominator. Since $\Phi_1(s)$ has all its poles in the right half-plane

$$\left[\frac{N_0/2}{\Phi_z^-(s)}\right]_+ = \sqrt{\frac{N_0}{2}}$$

and the optimum filter is

$$H(s) = 1 - \frac{\sqrt{\frac{N_0}{2}}}{\Phi_z^+(s)}$$

For example, if $\Phi_y(s) = \dfrac{1}{1-s^2}$,

$$\Phi_z(s) = \frac{1}{1-s^2} + \frac{N_0}{2} = \frac{1 + N_0/2 - N_0/2s^2}{1-s^2}$$

and

$$\Phi_z^+(s) = \frac{\sqrt{1 + N_0/2} + \sqrt{N_0/2}\, s}{1 + s}$$

so that the optimum filter is

$$H(s) = 1 - \frac{\sqrt{N_0/2}\,(1 + s)}{\sqrt{1 + N_0/2} + \sqrt{N_0/2}\, s} = \frac{\sqrt{1 + 2N_0} - 1}{s + \sqrt{1 + 2/N_0}}$$

Example 6.3.3 We consider the problem of Example 6.3.2 but with $N_0/2 = 1$. The estimate is then given by

$$\hat{Y}(s) = H(s)Z(s) = \left[1 - \frac{1}{\Phi_z^+(s)}\right] Z(s)$$

Since $(1/\Phi_z^+(s))Z(s)$ represents the output of the whitening filter, we have

$$\hat{Y}(s) = Z(s) - \nu(s)$$

so that the whitened observation is given by

$$\nu(t) = z(t) - \hat{y}(t)$$

By rewriting the equation as

$$z(t) = \hat{y}(t) + \nu(t)$$

we see that the observation has essentially been decomposed into two components, one of which is white. Furthermore, our discussions in this section show that the solution to the estimation problem can be obtained completely in terms of this white process.

Example 6.3.4 Consider the feedback control system shown in Fig. 6.3.3 which consists of a fixed plant $G_p(s)$ and a compensator $G_c(s)$. The input $z(t)$ to the plant consists of a stationary signal $y(t)$ corrupted by additive noise $v(t)$. It is desired to design the compensator $G_c(s)$ so that the system output $c(t)$ follows the signal input $y(t)$ in a mean square sense. It can easily be verified that the input–output transfer function $T(s)$ is given by

$$T(s) = \frac{C(s)}{Z(s)} = \frac{G_c(s)G_p(s)}{1 + G_c(s)G_p(s)}$$

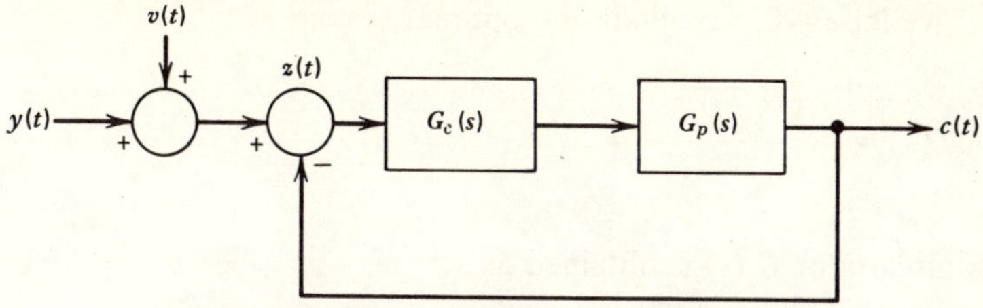

Fig. 6.3.3 Block diagram for control example.

If $G_p(s)$ has no poles or zeros in the right half-plane, we can solve the problem by posing it as a Wiener filtering problem. We then design the system (filter) $T(s)$ such that $c(t)$ is the mean square estimate of $y(t)$ by using Eq. (6.3.15). Once $T(s)$ is determined, $G_c(s)$ can easily be obtained.

For example, let the signal $y(t)$ be obtained by passing white noise of spectral height unity through an integrator. Let $v(t)$ be white with spectral height $N_0/2$. Let $G_p(s) = \dfrac{1}{s(s+1)}$.

We note that the two-sided Laplace transform of $\phi_y(\tau)$ does not exist in this case (why?). We overcome this problem by replacing the integrator by a system with transfer function $1/(s+\varepsilon)$ and let ε tend to zero later. We can then write for the power spectrum of $y(t)$

$$\Phi_y(s) = \frac{1}{\varepsilon^2 - s^2}$$

Since $v(t)$ is white, we can use the results of Example 6.3.2 to obtain the optimum system $T(s)$. We have

$$\begin{aligned}\Phi_z(s) &= \Phi_y(s) + \Phi_v(s) = \frac{1}{\varepsilon^2 - s^2} + \frac{N_0}{2}\\ &= \frac{N_0}{2}\left[\frac{\sqrt{(2/N_0)+\varepsilon^2} + s}{\varepsilon + s}\right]\left[\frac{\sqrt{(2/N_0)+\varepsilon^2} - s}{\varepsilon - s}\right]\end{aligned}$$

and

$$\Phi_z^+(s) = \frac{s + \sqrt{\varepsilon^2 + (2/N_0)}}{s+\varepsilon}\sqrt{\frac{N_0}{2}}$$

so that

$$T(s) = 1 - \frac{s+\varepsilon}{s + \sqrt{\varepsilon^2 + (2/N_0)}} = \frac{\sqrt{\varepsilon^2 + (2/N_0)} - \varepsilon}{s + \sqrt{\varepsilon^2 + (2/N_0)}}$$

Now if we let $\varepsilon \to 0$, we obtain the optimal system as

$$T(s) = \frac{\sqrt{2/N_0}}{s + \sqrt{2/N_0}}$$

The compensator $G_c(s)$ is obtained as

$$G_c(s) = \frac{1}{G_p(s)} \frac{T(s)}{1 - T(s)}$$
$$= \sqrt{2/N_0}\,(s+1)$$

To obtain a practical realization of this compensator, it is usual to add poles in the interior of the left-half s-plane so that the degree of denominator is at least as large as that of the numerator. The poles are chosen so that they will not significantly alter the response of the system. In case the plant has poles or zeros in the right half-plane, the design of the overall system must be modified to take into account certain practical constraints. This is because the system may become unstable due to the presence of right half-plane poles. The problem must be solved as a constrained optimization problem. The reader is referred to [10] for further details.

We conclude this section with a brief discussion of the Wiener filter for discrete-time systems.

Wiener Filter for Discrete Systems

Our observation interval in this case consists of a set of discrete time instants t_k for $k \in [k_0, k_f]$. We observe the signal $z(k)$ at these instants and are required to estimate the signal $g(k)$ based on this set of observations. We will assume that the observation intervals are equally spaced and that

$$t_{k+1} - t_k = T \qquad \text{for} \quad k \in [k_0, k_f]$$

Since we are seeking a linear minimum mean square estimate, we write

$$\hat{g}(k) = \sum_{j=k_0}^{k_f} h(k,j) z(j) \tag{6.3.31}$$

Again the weights $h(k,j)$ are such that the error in the estimate is orthogonal to the observations. We thus have

$$\mathrm{E}\left\{\left[g(k)-\sum_{j=k_0}^{k_f} h(k,j)z(j)\right]z(l)\right\}=0 \quad \text{for} \quad l\in[k_0,k_f] \tag{6.3.32}$$

and therefore

$$\phi_{gz}(k,l)=\sum_{j=k_0}^{k_f} h(k,j)\phi_z(j,l) \quad \text{for} \quad l\in[k_0,k_f] \tag{6.3.33}$$

Equation (6.3.33) is the discrete analog of Eq. (6.3.4). The solution of the equation is again difficult in general. We defer until the next section the solution of this equation for a particular class of nonstationary signals.

In this section, we obtain a solution under the assumption that the processes are all stationary and that the observation interval is semi-infinite. If we restrict ourselves to causal and time invariant systems, Eq. (6.3.33) reduces to

$$\phi_{gz}(k-l)=\sum_{j=-\infty}^{k} h(k-j)\phi_z(j-l) \qquad l\in[-\infty,k] \tag{6.3.34}$$

Making the change of variables $k-l=n$ and $k-j=m$ leads to

$$\phi_{gz}(n)=\sum_{m=0}^{\infty} h(m)\phi_z(n-m) \qquad n\in[0,\infty] \tag{6.3.35}$$

which is the discrete analog of Eq. (6.3.8). We can solve Eq. (6.3.35) in a similar fashion to the solution of Eq. (6.3.8). We thus again note that if $z(\cdot)$ were a white sequence such that

$$\phi_z(n-m)=\delta_{nm} \tag{6.3.36}$$

the solution to Eq. (6.3.35) is easily obtained as

$$h(n)=\phi_{gz}(n) \qquad n\in[0,\infty] \tag{6.3.37}$$

with corresponding transfer function

$$H(z)=[\Phi_{gz}(z)]_+ \tag{6.3.38}$$

where $[\cdot]_+$ denotes the Z-transform of the causal part of the autocorrelation sequence $\phi_{gz}(n)$. If we restrict ourselves to signals with rational spectra, this is obtained from $\Phi_{gz}(z)$ by grouping together all terms corresponding to poles and zeros within the unit circle.

If $z(\cdot)$ is not a white sequence, we can again convert it into a white sequence by passing $z(\cdot)$ through a whitening filter. The transfer function of the whitening filter is given by

$$H_w(z) = \frac{1}{\Phi_z^+(z)} \tag{6.3.39}$$

The filter $H_2(z)$ operates on the output of the whitening filter $\nu(\cdot)$ to yield an estimate of the desired signal $g(\cdot)$. We obtain the transfer function of this filter from Eq. (6.3.38) by replacing $z(\cdot)$ with $\nu(\cdot)$ so that

$$H_2(z) = \left[\Phi_{g\nu}(z)\right]_+ \tag{6.3.40}$$

Finally, since

$$\begin{aligned}\Phi_{g\nu}(z) &= H_w(z^{-1})\Phi_{gz}(z)\\ &= \frac{\Phi_{gz}(z)}{\Phi_z^-(z)}\end{aligned} \tag{6.3.41}$$

the overall filter obtained by cascading $H_w(z)$ and $H_2(z)$ is

$$H(z) = \frac{1}{\Phi_z^+(z)}\left[\frac{\Phi_{gz}(z)}{\Phi_z^-(z)}\right]_+ \tag{6.3.42}$$

Estimation of Vector Processes

So far we have restricted ourselves to the estimation of scalar processes. The extension to vector processes is quite straightforward. In this case we are given values of the observed process $\mathbf{z}(t)$ which is assumed to be an n-vector, and are required to estimate the related m-vector process $\mathbf{g}(t)$ by a linear combination of the data. Thus if $g_i(t)$ is the ith component of the vector $\mathbf{g}(t)$, we write for its estimate

$$\hat{g}_i(t) = \sum_{j=1}^{n} \int_{t_0}^{t_f} H_{ij}(t, \xi) z_j(\xi)\, d\xi \qquad i = 1 \ldots m \tag{6.3.43}$$

In vector-matrix notation, this can be written as

$$\hat{\mathbf{g}}(t) = \int_{t_0}^{t_f} \mathbf{H}(t, \xi)\mathbf{z}(\xi)\, d\xi \tag{6.3.44}$$

We could choose as our estimate either the minimum mean square estimate or the minimum variance estimate. In the former case our risk function is

$$\begin{aligned} \bar{C}_{\text{lms}} &= \mathrm{E}\left\{ [\mathbf{g}(t) - \hat{\mathbf{g}}(t)]^T [\mathbf{g}(t) - \hat{\mathbf{g}}(t)] \right\} \\ &= \mathrm{E}\left\{ \sum_{i=1}^{m} [\tilde{g}_i(t)]^2 \right\} \end{aligned} \tag{6.3.45}$$

In the latter case we seek to minimize the error covariance. Since this is now a matrix, this implies that we minimize the associated quadratic form

$$\begin{aligned} \bar{C}_{\text{lmv}} &= \mathrm{E}\left\{ \mathbf{a}^T [\mathbf{g}(t) - \hat{\mathbf{g}}(t)][\mathbf{g}(t) - \hat{\mathbf{g}}(t)]^T \mathbf{a} \right\} \\ &= \mathbf{a}^T \operatorname{cov} \tilde{\mathbf{g}}(t)\mathbf{a} \end{aligned} \tag{6.3.46}$$

where $\mathbf{a}$ is any arbitrary vector.

In either case, it can be shown by using the orthogonality principle that the matrix $\mathbf{H}(t, \xi)$ must be chosen such that

$$\boldsymbol{\phi}_{\mathbf{gz}}(t, \tau) = \int_{t_0}^{t_f} \mathbf{H}(t, \xi)\boldsymbol{\phi}_{\mathbf{z}}(\xi, \tau)\, d\tau \qquad t_0 \leqslant \tau \leqslant t_f \tag{6.3.47}$$

Again the solution of this equation is not an easy task. We might expect that if we restrict ourselves to stationary processes and semi-infinite observation intervals, we may be able to get a solution as in the scalar case. In this case Eq. (6.3.47) becomes

$$\boldsymbol{\phi}_{\mathbf{gz}}(t - \tau) = \int_{-\infty}^{t} \mathbf{H}(t - \xi)\boldsymbol{\phi}_{\mathbf{z}}(\xi - \tau)\, d\xi \qquad t_0 \leqslant \tau \leqslant t_f \tag{6.3.48}$$

We may again seek to obtain the overall filter as a cascade of a whitening filter and an optimal filter that operates on the whitened observations to yield an estimate of the desired signal. We recall that in the scalar problem, this involved a spectral factorization of $\Phi_z(s)$ into terms containing poles and zeros only in the left half-plane. We might seek a similar factorization for the present problem also. Unfortunately, in this case $\boldsymbol{\Phi}_{\mathbf{z}}(s)$ is a matrix and the spectral factorization is not trivial. Here we are seeking a matrix

$\mathbf{\Phi}^+(s)$ such that

$$\mathbf{\Phi}_{\mathbf{z}}(s) = \Phi^+(s)\left[\mathbf{\Phi}^+(-s)\right]^T \tag{6.3.49}$$

such that $\mathbf{\Phi}^+(s)$ is analytic (has no poles) in the right half-plane and $[\mathbf{\Phi}^+(s)]^{-1}$ is also analytic in the right half-plane. We will denote $[\mathbf{\Phi}^+(-s)]^T$ as $\mathbf{\Phi}^-(s)$ for convenience.

Assuming such a matrix has been found, the transfer matrix of the whitening filter is given by

$$H_w(s) = \left[\mathbf{\Phi}^+(s)\right]^{-1} \tag{6.3.50}$$

The transfer function of the optimal filter follows analogous to our derivation in the scalar case as

$$H_2(s) = \left\{\mathbf{\Phi}_{\mathbf{gz}}(s)\left[\mathbf{\Phi}^-(-s)\right]^{-1}\right\}_+ \tag{6.3.51}$$

and the overall filter obtained as a cascade of $H_w(s)$ and $H_2(s)$ is

$$\mathbf{H}(s) = \left\{\mathbf{\Phi}_{\mathbf{gz}}(s)\left[\mathbf{\Phi}^-(s)\right]^{-1}\right\}_+\left[\mathbf{\Phi}^+(s)\right]^{-1} \tag{6.3.52}$$

The critical part of this solution as explained earlier, is in the spectral factorization of Eq. (6.3.49). While several techniques are available for accomplishing this, the techniques are usually cumbersome [11,12]. We discuss one such technique which involves the solution to a nonlinear algebraic equation, in Section 6.5.

6.4 ESTIMATION OF NONSTATIONARY PROCESSES: THE KALMAN FILTER [13–15]

We now turn to the solution of the integral Eq. (6.3.4) or (6.3.33) for nonstationary processes. We consider measurement models of the type of Eq. (6.3.6) and assume that the observation noise is white. Since the approach that we take for the solution of the estimation problem is no more complicated for vector processes than for scalar processes, we assume that our processes are in general vectors. The observations are thus given by

$$\mathbf{z}(t) = \mathbf{y}(t) + \mathbf{v}(t) \qquad t_0 \leqslant t \leqslant t_f \tag{6.4.1}$$

where $\mathbf{z}(t)$, $\mathbf{y}(t)$, and $\mathbf{v}(t)$ are $m \times 1$ vectors and $\mathbf{v}(t)$ has covariance

$$E\{\mathbf{v}(t)\mathbf{v}^T(s)\} = \mathbf{V}_{\mathbf{v}}(t)\delta(t-s) \tag{6.4.2}$$

where $\mathbf{V}_{\mathbf{v}}(t)$ is now an $m \times m$ matrix.

The integral equation for the estimation problem is then given by Eq. (6.3.47) as

$$\boldsymbol{\phi}_{\mathbf{gz}}(t, \tau) = \int_{t_0}^{t_f} \mathbf{H}(t, \xi)\boldsymbol{\phi}_{\mathbf{z}}(\xi, \tau)\, d\xi, \qquad t_0 \leqslant \tau \leqslant t_f \tag{6.4.3}$$

We have seen earlier that the solution to the integral equation is straightforward if $\mathbf{z}(\cdot)$ is a white process. In the event $\mathbf{z}(t)$ is not white, we have seen in our earlier discussions that, for the stationary problem, $\mathbf{z}(t)$ can be decomposed into two components one of which is white, and has mean zero. In the general case under discussion here also, it can be shown that $\mathbf{z}(t)$ can be decomposed into two components, one of which can be determined from past values of $\mathbf{z}(\cdot)$ and the other, a white process $\boldsymbol{\nu}(t)$, containing the new information in $\mathbf{z}(t)$. This result is known as the Wold decomposition [1,6]. The process $\boldsymbol{\nu}(t)$ is referred to as the *innovations* process associated with the observations $\mathbf{z}(t)$ [16] and is obtained by a suitable causal linear transformation of the process $\mathbf{z}(t)$. It follows that the estimate $\hat{\mathbf{g}}(t)$ can be obtained completely in terms of $\boldsymbol{\nu}(t)$ as

$$\hat{\mathbf{g}}(t) = \int_{t_0}^{t_f} \mathbf{H}(t, \xi)\boldsymbol{\nu}(\xi)\, d\xi \tag{6.4.4}$$

where, by the principle of orthogonality, the matrix $\mathbf{H}(t, \xi)$ satisfies

$$E\left\{\left[\mathbf{g}(t) - \int_{t_0}^{t_f} \mathbf{H}(t, \xi)\boldsymbol{\nu}(\xi)\, d\xi\right]\boldsymbol{\nu}^T(\tau)\right\} = 0 \qquad t_0 \leqslant \tau \leqslant t_f \tag{6.4.5}$$

or

$$E\{\mathbf{g}(t)\boldsymbol{\nu}^T(\tau)\} = \int_{t_0}^{t_f} \mathbf{H}(t, \xi) E\{\boldsymbol{\nu}(\xi)\boldsymbol{\nu}^T(\tau)\}\, d\xi \qquad t_0 \leqslant \tau \leqslant t_f \tag{6.4.6}$$

Let the covariance matrix of $\boldsymbol{\nu}(t)$ be given by

$$\operatorname{cov} \boldsymbol{\nu}(t) = E\{\boldsymbol{\nu}(t)\boldsymbol{\nu}^T(\tau)\} = \mathbf{V}_{\boldsymbol{\nu}}(t)\delta(t-\tau) \tag{6.4.7}$$

Substitution in Eq. (6.4.6) then yields

$$E\{\mathbf{g}(t)\boldsymbol{\nu}^T(\tau)\} = \mathbf{H}(t, \tau)\mathbf{V}_{\boldsymbol{\nu}}(\tau) \tag{6.4.8}$$

Solving for $\mathbf{H}(t, \tau)$, we obtain

$$\mathbf{H}(t, \tau) = \mathrm{E}\{\mathbf{g}(t)\boldsymbol{\nu}^T(\tau)\}\mathbf{V}_\nu^{-1}(\tau) \tag{6.4.9}$$

Using the preceding in Eq. (6.4.4) yields

$$\hat{\mathbf{g}}(t) = \int_{t_0}^{t_f} \mathrm{E}\{\mathbf{g}(t)\boldsymbol{\nu}^T(\tau)\}\mathbf{V}_\nu^{-1}(\tau)\boldsymbol{\nu}(\tau)\, d\tau \tag{6.4.10}$$

It is worthwhile, at this point, to derive analogous equations for the corresponding discrete-time problem also. We seek an estimate of the process $\mathbf{g}(k)$ based on observations of the form

$$\mathbf{z}(k) = \mathbf{y}(k) + \mathbf{v}(k) \qquad k_0 \leqslant k \leqslant k_f \tag{6.4.11}$$

where $\mathbf{v}(k)$ is an independent random process with covariance matrix

$$\mathrm{E}\{\mathbf{v}(k)\mathbf{v}^T(j)\} = \mathbf{V}_\mathbf{v}(k)\delta_{kj} \tag{6.4.12}$$

If we convert the observations $\mathbf{z}(k)$ into an independent (white) process $\boldsymbol{\nu}(k)$, the estimate can be written as

$$\hat{\mathbf{g}}(k) = \sum_{j=k_0}^{k_f} \mathbf{H}(k,j)\boldsymbol{\nu}(j) \tag{6.4.13}$$

The weights $\mathbf{H}(k,j)$ are such that the estimation error is orthogonal to the data set $\{\boldsymbol{\nu}(j)\}$. We thus have

$$\mathrm{E}\left\{\left[\mathbf{g}(k) - \sum_{j=k_0}^{k_f} \mathbf{H}(k,j)\boldsymbol{\nu}(j)\right]\boldsymbol{\nu}^T(l)\right\} = 0 \qquad k_0 \leqslant l \leqslant k_f \tag{6.4.14}$$

or equivalently

$$\mathrm{E}\{\mathbf{g}(k)\boldsymbol{\nu}^T(l)\} = \sum_{j=k_0}^{k_f} \mathbf{H}(k,j)\mathrm{E}\{\boldsymbol{\nu}(j)\boldsymbol{\nu}^T(l)\} \qquad k_0 \leqslant l \leqslant k_f \tag{6.4.15}$$

Since $\boldsymbol{\nu}(k)$ is white, its covariance is

$$\mathrm{E}\{\boldsymbol{\nu}(k)\boldsymbol{\nu}^T(l)\} = \mathbf{V}_\nu(k)\delta_{kl} \tag{6.4.16}$$

Substituting Eq. (6.4.16) into Eq. (6.4.15) and solving for $\mathbf{H}(\cdot, \cdot)$ yields

$$\mathbf{H}(k, l) = \mathrm{E}\{\mathbf{g}(k)\boldsymbol{\nu}^T(l)\}\mathbf{V}_\nu^{-1}(l) \tag{6.4.17}$$

and the required estimate is

$$\hat{\mathbf{g}}(k) = \sum_{j=k_0}^{k_f} \mathrm{E}\{\mathbf{g}(k)\boldsymbol{\nu}^T(j)\}\mathbf{V}_\nu^{-1}(j)\boldsymbol{\nu}(j) \tag{6.4.18}$$

It only remains to find the transformation that yields the innovations process $\boldsymbol{\nu}(t)$ ($\boldsymbol{\nu}(k)$ in the discrete case) and its associated covariance. For stationary processes, at least for the scalar case, this transformation is easy to find. For nonstationary processes, however, this is not so easy. For such processes, even if we find the innovations process and obtain a solution as in Eq. (6.4.10) or (6.4.18) a major difficulty remains in the practical implementation of the estimator. The solution is specified in terms of the impulse response matrix of a time-varying system. To translate the impulse response into physical hardware is again not an easy task. We therefore seek an alternative solution to the problem which can be easily implemented using either analog or digital hardware. We do this by deriving a differential equation (a difference equation for the discrete time problem) which the estimate must satisfy. The solution to the estimation problem is thus given not in explicit analytical terms but rather in an algorithmic form; that is, we specify a set of algorithms, which, when applied to the data gives us the desired estimate. This approach to the problem was first suggested by Kalman and the resulting estimator is usually referred to as the Kalman filter [13,14].

For reasons which will soon become apparent, we will start our derivation of the Kalman filter algorithms with the discrete-time problem.

The Optimum Linear Discrete Filter

We now consider the determination of the estimate in Eq. (6.4.18). Since the estimate is based on measurements up to k_f, we denote the estimate as $\hat{\mathbf{g}}(k|k_f)$. In order to proceed with our discussion, at this point we need to make some specific assumptions about the signal $\mathbf{y}(k)$. In particular we assume that the signal $\mathbf{y}(k)$ is obtained as the output of a linear discrete system driven by white noise. We use the state-variable model†

$$\begin{aligned}\mathbf{x}(k+1) &= \boldsymbol{\phi}(k+1, k)\mathbf{x}(k) + \boldsymbol{\Gamma}(k)\mathbf{w}(k)\\ \mathbf{y}(k) &= \mathbf{C}(k)\mathbf{x}(k)\end{aligned} \tag{6.4.19}$$

where the input $\mathbf{w}(k)$ is zero-mean white with covariance matrix

†The matrix $\boldsymbol{\phi}(\cdot, \cdot)$ represents the state-transition matrix of the discrete-time system and should not be confused with the autocorrelation matrix.

$$\text{cov } \mathbf{w}(k) = \mathrm{E}\{\mathbf{w}(k)\mathbf{w}^T(j)\} = \mathbf{V_w}(k)\delta_{kj} \tag{6.4.20}$$

The observation equation is

$$\mathbf{z}(k) = \mathbf{y}(k) + \mathbf{v}(k) \qquad k_0 \leqslant k \leqslant k_f \tag{6.4.21}$$

where $\mathbf{v}(k)$ is zero mean white with covariance matrix

$$\text{cov } \mathbf{v}(k) = \mathrm{E}\{\mathbf{v}(k)\mathbf{v}^T(j)\} = \mathbf{V_v}(k)\delta_{kj} \tag{6.4.22}$$

and is uncorrelated with $\mathbf{w}(k)$.

We further assume that $\mathbf{x}(k_0)$ is a random variable with known mean $\boldsymbol{\mu}_\mathbf{x}(k_0)$ and covariance $\mathbf{V_x}(k_0)$ uncorrelated with $\mathbf{w}(k)$ and $\mathbf{v}(k)$ for all k. Thus

$$\mathrm{E}\{\mathbf{w}(k)\mathbf{v}^T(j)\} = 0 \quad \text{all } k, j$$
$$\mathrm{E}\{\mathbf{x}(k_0)\mathbf{w}^T(j)\} = 0 \quad \text{all } j$$

and

$$\mathrm{E}\{\mathbf{x}(k_0)\mathbf{v}^T(j)\} = 0 \quad \text{all } j \tag{6.4.23}$$

The problem is hence slightly different from the one considered in the preceding discussions in that $\mathbf{x}(k)$ (and hence $\mathbf{y}(k)$) is no longer zero mean.

We will seek to obtain an estimate of the state of the model $\mathbf{x}(k)$ based on the noisy observations. Let us denote by $\mathbf{Z}(j)$ the set of observations $\{\mathbf{z}(k_0), \mathbf{z}(k_0+1), \ldots, \mathbf{z}(j)\}$:

$$\mathbf{Z}(j) = \{\mathbf{z}(i), i \in [k_0, j]\} \tag{6.4.24}$$

The estimate we are seeking is

$$\hat{\mathbf{x}}(k|k_f) = \hat{\mathrm{E}}\{\mathbf{x}(k)|\mathbf{Z}(k_f)\} \tag{6.4.25}$$

where $\hat{\mathrm{E}}$ denotes the wide-sense conditional expectation as explained earlier.

The estimate of the signal $\mathbf{y}(k)$ can be obtained as

$$\begin{aligned}\hat{\mathbf{y}}(k|k_f) &= \hat{\mathrm{E}}\{\mathbf{y}(k)|\mathbf{Z}(k_f)\} \\ &= \mathrm{E}\{\mathbf{C}(k)\mathbf{x}(k)|\mathbf{Z}(k_f)\} \\ &= \mathbf{C}(k)\hat{\mathrm{E}}\{\mathbf{x}(k)|\mathbf{Z}(k_f)\}\end{aligned} \tag{6.4.26}$$

where the last expression follows since $\hat{E}$ commutes over linear operations. We may therefore write

$$\hat{\mathbf{y}}(k|k_f) = \mathbf{C}(k)\hat{\mathbf{x}}(k|k_f) \tag{6.4.27}$$

In terms of the innovations sequence $\boldsymbol{\nu}(k)$, the estimate $\hat{\mathbf{x}}(k|k_f)$ can be written from Eq. (6.4.18) as

$$\hat{\mathbf{x}}(k|k_f) = \sum_{j=k_0}^{k_f} E\{\mathbf{x}(k)\boldsymbol{\nu}^T(j)\}\mathbf{V}_\nu^{-1}(j)\boldsymbol{\nu}(j) \tag{6.4.28}$$

If $k_f = k$, the problem corresponds to one of filtering a signal in the presence of noise and the estimate $\hat{\mathbf{x}}(k|k)$ is referred to as the filtered estimate. For $k > k_f$, we have the prediction problem and the corresponding estimate is the predicted estimate. Of particular importance is the one-stage predicted estimate $\hat{\mathbf{x}}(k|k-1)$. If $k < k_f$, we have the interpolation problem and the estimate is called a smoothed estimate.

We now determine the innovations sequence $\boldsymbol{\nu}(k)$ associated with the observation model Eq. (6.4.21). We have seen in Example 6.3.3 that for the continuous-time stationary problem in which the signal and observation noise were uncorrelated, the innovations were given by

$$\nu(t) = z(t) - \hat{y}(t)$$

so that $\hat{y}(t)$ corresponds to that component of $z(t)$ which is determined by past values of $z(\cdot)$. For the discrete-time problem under consideration, it follows that the innovations sequence should be

$$\begin{aligned}\boldsymbol{\nu}(k) &= \mathbf{z}(k) - \hat{\mathbf{y}}(k|k-1) \\ &= \tilde{\mathbf{y}}(k|k-1) + \mathbf{v}(k)\end{aligned} \tag{6.4.29}$$

where $\tilde{\mathbf{y}}(k|k-1)$ represents the estimation error. For the estimate $\hat{\mathbf{y}}(k|k-1)$ to be unbiased, $\tilde{\mathbf{y}}(k|k-1)$ must have mean zero. We will see later, how this can be achieved. It then follows that since $\mathbf{v}(k)$ is a zero-mean sequence, $\boldsymbol{\nu}(k)$ will also be a zero-mean sequence. We now establish that $\boldsymbol{\nu}(k)$ as defined previously is a white sequence.

Let $k > l$. Then

$$\begin{aligned}E\{\boldsymbol{\nu}(k)\boldsymbol{\nu}^T(l)\} &= E\{[\tilde{\mathbf{y}}(k|k-1) + \mathbf{v}(k)]\boldsymbol{\nu}^T(l)\} \\ &= E\{\mathbf{v}(k)\boldsymbol{\nu}^T(l)\}\end{aligned}$$

(since $\tilde{\mathbf{y}}(k|k-1)$ is orthogonal to $\boldsymbol{\nu}(l)$ for $l \leqslant k-1$)

$$= \mathrm{E}\{\mathbf{v}(k)[\tilde{\mathbf{y}}(l|l-1) + \mathbf{v}(l)]^T\}$$
$$= 0 \tag{6.4.30}$$

The last result follows by causality ($\mathbf{v}(k)$ is in the future as far as $\tilde{\mathbf{y}}(l|l-1)$ is concerned and hence cannot be correlated with it) and the fact that $\mathbf{v}(\cdot)$ is a white sequence. The result holds for $k < l$, by interchanging k and l in the preceding.

For $k = l$ we have

$$\begin{aligned}\mathrm{E}\{\boldsymbol{\nu}(k)\boldsymbol{\nu}^T(k)\} &= \mathrm{E}\{[\tilde{\mathbf{y}}(k|k-1) + \mathbf{v}(k)][\tilde{\mathbf{y}}(k|k-1) + \mathbf{v}(k)]^T\}\\ &= \mathbf{V}_{\tilde{\mathbf{y}}}(k|k-1) + \mathbf{V}_{\mathbf{v}}(k)\end{aligned} \tag{6.4.31}$$

where

$$\begin{aligned}\mathbf{V}_{\tilde{\mathbf{y}}}(k|k-1) &= \mathrm{E}\{\tilde{\mathbf{y}}(k|k-1)\tilde{\mathbf{y}}^T(k|k-1)\}\\ &= \mathbf{C}(k)\mathbf{V}_{\tilde{\mathbf{x}}}(k|k-1)\mathbf{C}^T(k)\end{aligned} \tag{6.4.32}$$

Thus $\boldsymbol{\nu}(k)$ is a zero-mean, white sequence, with covariance kernel

$$\mathbf{V}_{\boldsymbol{\nu}}(k) = \mathbf{C}(k)\mathbf{V}_{\tilde{\mathbf{x}}}(k|k-1)\mathbf{C}^T(k) + \mathbf{V}_{\mathbf{v}}(k) \tag{6.4.33}$$

One-Stage Predictor Algorithms

Since the computation of the innovations sequence involves determining the estimate $\hat{\mathbf{y}}(k|k-1)$, we first set up algorithms for obtaining the one-stage predicted estimate $\hat{\mathbf{x}}(k|k-1)$. If we set $\hat{\mathbf{g}}(k) = \hat{\mathbf{x}}(k|k-1)$ in Eq. (6.4.18), we obtain

$$\hat{\mathbf{x}}(k|k-1) = \sum_{j=k_0}^{k-1} \mathrm{E}\{\mathbf{x}(k)\boldsymbol{\nu}^T(j)\}\mathbf{V}_{\boldsymbol{\nu}}^{-1}(j)\boldsymbol{\nu}(j) \tag{6.4.34}$$

Indexing upwards by one gives

$$\begin{aligned}\hat{\mathbf{x}}(k+1|k) &= \sum_{j=k_0}^{k} \mathrm{E}\{\mathbf{x}(k+1)\boldsymbol{\nu}^T(j)\}\mathbf{V}_{\boldsymbol{\nu}}^{-1}(j)\boldsymbol{\nu}(j)\\ &= \sum_{j=k_0}^{k} \mathrm{E}\{[\boldsymbol{\phi}(k+1, k)\mathbf{x}(k) + \boldsymbol{\Gamma}(k)\mathbf{w}(k)]\boldsymbol{\nu}^T(j)\}\mathbf{V}_{\boldsymbol{\nu}}^{-1}(j)\boldsymbol{\nu}(j)\end{aligned} \tag{6.4.35}$$

We note that in Eq. (6.4.35), $\boldsymbol{\nu}(j)$ depends on values of $\mathbf{x}(l)$ only for $l = k_0, k_0 + 1, \ldots, j-1$. Thus since $\mathbf{w}(k)$ is white

$$E\{\mathbf{w}(k)\boldsymbol{\nu}^T(j)\} = 0, \qquad j \leqslant k \tag{6.4.36}$$

We can therefore write Eq. (6.4.35) as

$$\begin{aligned}\hat{\mathbf{x}}(k+1|k) &= \boldsymbol{\phi}(k+1,k)\sum_{j=k_0}^{k} E\{\mathbf{x}(k)\boldsymbol{\nu}^T(j)\}\mathbf{V}_{\nu}^{-1}(j)\boldsymbol{\nu}(j)\\ &= \boldsymbol{\phi}(k+1,k)E\{\mathbf{x}(k)\boldsymbol{\nu}^T(k)\}\mathbf{V}_{\nu}^{-1}(k)\boldsymbol{\nu}(k)\\ &\quad + \boldsymbol{\phi}(k+1,k)\sum_{j=k_0}^{k-1} E\{\mathbf{x}(k)\boldsymbol{\nu}^T(j)\}\mathbf{V}_{\nu}^{-1}(j)\boldsymbol{\nu}(j)\end{aligned} \tag{6.4.37}$$

Define $\mathbf{K}(k+1,k)$ by

$$\mathbf{K}(k+1,k) \triangleq \boldsymbol{\phi}(k+1,k)E\{\mathbf{x}(k)\boldsymbol{\nu}^T(k)\}\mathbf{V}_{\nu}^{-1}(k) \tag{6.4.38}$$

Substitution of Eqs. (6.4.38) and (6.4.34) in Eq. (6.4.37) then yields

$$\hat{\mathbf{x}}(k+1|k) = \boldsymbol{\phi}(k+1,k)\hat{\mathbf{x}}(k|k-1) + \mathbf{K}(k+1,k)\boldsymbol{\nu}(k) \tag{6.4.39}$$

or

$$\begin{aligned}\hat{\mathbf{x}}(k+1|k) &= \boldsymbol{\phi}(k+1,k)\hat{\mathbf{x}}(k|k-1)\\ &\quad + \mathbf{K}(k+1,k)[\mathbf{z}(k) - \mathbf{C}(k)\hat{\mathbf{x}}(k|k-1)]\end{aligned} \tag{6.4.40}$$

where we have used Eq. (6.4.29) and the fact that

$$\hat{\mathbf{y}}(k|k-1) = \mathbf{C}(k)\hat{\mathbf{x}}(k|k-1)$$

The matrix $\mathbf{K}(k+1,k)$ is referred to as the Kalman gain for one-stage prediction.

The form of the solution in Eq. (6.4.40) is computationally very useful. The result is a sequential algorithm for determining $\hat{\mathbf{x}}(k+1|k)$ based on $\hat{\mathbf{x}}(k|k-1)$ and the new observation $\mathbf{z}(k)$. The new estimate is formed by predicting forward from the old estimate and then correcting it by the innovation term which contains the new information. A block diagram of the estimator including the message model is shown in Fig. 6.4.1.

All that is needed to complete the specification of the estimator are the Kalman gain $\mathbf{K}(k+1,k)$ and the error variance $\mathbf{V}_{\tilde{\mathbf{y}}}(k|k-1)$ needed to specify the covariance kernel $\mathbf{V}_{\nu}(k)$.

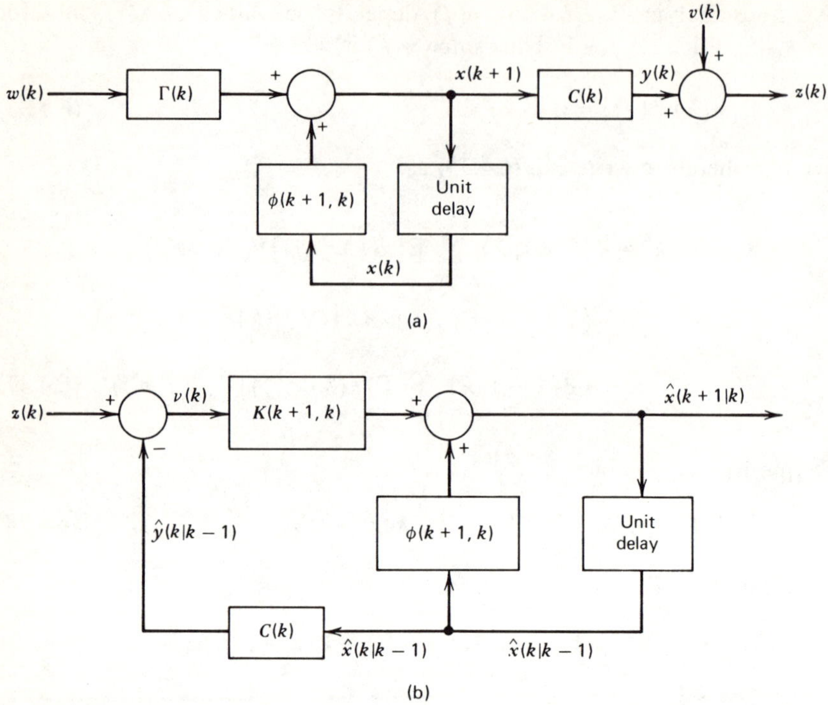

Fig. 6.4.1 Block diagram for one-stage prediction for the discrete Kalman Filter. (a) Message model. (b) One-stage predictor.

Substitution of Eq. (6.4.29) in Eq. (6.4.38) yields

$$\begin{aligned}\mathbf{K}(k+1,k) &= \boldsymbol{\phi}(k+1,k)\mathrm{E}\{\mathbf{x}(k)[\mathbf{C}(k)\tilde{\mathbf{x}}(k|k-1)+\mathbf{v}(k)]^T\}\mathbf{V}_\nu^{-1}(k)\\ &= \boldsymbol{\phi}(k+1,k)\mathrm{E}\{\mathbf{x}(k)\tilde{\mathbf{x}}^T(k|k-1)\}\mathbf{C}^T(k)\mathbf{V}_\nu^{-1}(k) \quad \textbf{(6.4.41)}\end{aligned}$$

since $\mathbf{x}(k)$ and $\mathbf{v}(k)$ are uncorrelated. We can write

$$\mathbf{x}(k) = \hat{\mathbf{x}}(k|k-1) + \tilde{\mathbf{x}}(k|k-1) \quad \textbf{(6.4.42)}$$

Since $\tilde{\mathbf{x}}(k|k-1)$ is orthogonal to $\boldsymbol{\nu}(j)$ for $j \leqslant k-1$ it is also orthogonal to $\hat{\mathbf{x}}(k|k-1)$ (since $\hat{\mathbf{x}}(k|k-1)$ is a linear combination of $\boldsymbol{\nu}(j)$ for $j \leqslant k-1$). Equation (6.4.41) can therefore be written as

$$\mathbf{K}(k+1,k) = \boldsymbol{\phi}(k+1,k)\mathbf{V}_{\tilde{\mathbf{x}}}(k|k-1)\mathbf{C}^T(k)\mathbf{V}_\nu^{-1}(k) \quad \textbf{(6.4.43)}$$

To obtain an expression for the error variance $\mathbf{V}_{\tilde{\mathbf{x}}}(k|k-1)$, we subtract Eq.

(6.4.40) from Eq. (6.4.19) and use Eq. (6.4.29) to get

$$\begin{aligned}\tilde{\mathbf{x}}(k+1|k) = {} & \boldsymbol{\phi}(k+1,k)\tilde{\mathbf{x}}(k|k-1) + \boldsymbol{\Gamma}(k)\mathbf{w}(k) \\ & -\mathbf{K}(k+1,k)[\mathbf{C}(k)\tilde{\mathbf{x}}(k|k-1) + \mathbf{v}(k)]\end{aligned} \tag{6.4.44}$$

Thus we have

$$\begin{aligned}\tilde{\mathbf{x}}(k+1|k) = {} & [\boldsymbol{\phi}(k+1,k) - \mathbf{K}(k+1,k)\mathbf{C}(k)]\tilde{\mathbf{x}}(k|k-1) \\ & +\boldsymbol{\Gamma}(k)\mathbf{w}(k) - \mathbf{K}(k+1,k)\mathbf{v}(k)\end{aligned} \tag{6.4.45}$$

Since $\mathbf{w}(\cdot)$ and $\mathbf{v}(\cdot)$ are white and uncorrelated with each other, the difference equation for the error variance may be written from Eq. (2.10.30) as

$$\begin{aligned}\mathbf{V}_{\tilde{\mathbf{x}}}(k+1|k) = {} & [\boldsymbol{\phi}(k+1,k) - \mathbf{K}(k+1,k)\mathbf{C}(k)] \\ & \times \mathbf{V}_{\tilde{\mathbf{x}}}(k|k-1)[\boldsymbol{\phi}(k+1,k) - \mathbf{K}(k+1,k)\mathbf{C}(k)]^T \\ & +\boldsymbol{\Gamma}(k)\mathbf{V}_{\mathbf{w}}(k)\boldsymbol{\Gamma}^T(k) \\ & +\mathbf{K}(k+1,k)\mathbf{V}_{\mathbf{v}}(k)\mathbf{K}^T(k+1,k)\end{aligned} \tag{6.4.46}$$

Substituting for $\mathbf{K}(k+1,k)$ from Eq. (6.4.43) and simplifying

$$\begin{aligned}\mathbf{V}_{\tilde{\mathbf{x}}}(k+1|k) = {} & \boldsymbol{\phi}(k+1,k)\mathbf{V}_{\tilde{\mathbf{x}}}(k|k-1)\boldsymbol{\phi}^T(k+1,k) + \boldsymbol{\Gamma}(k)\mathbf{V}_{\mathbf{w}}(k)\boldsymbol{\Gamma}^T(k) \\ & -\boldsymbol{\phi}(k+1,k)\mathbf{V}_{\tilde{\mathbf{x}}}(k|k-1)\mathbf{C}^T(k) \\ & \times[\mathbf{C}(k)\mathbf{V}_{\tilde{\mathbf{x}}}(k|k-1)\mathbf{C}^T(k) + \mathbf{V}_{\mathbf{v}}(k)]^{-1} \\ & \cdot\mathbf{C}(k)\mathbf{V}_{\tilde{\mathbf{x}}}(k|k-1)\boldsymbol{\phi}^T(k+1,k)\end{aligned} \tag{6.4.47}$$

To specify the initial conditions for the Eqs. (6.4.40) and (6.4.47), we note from Eq. (6.4.44) that

$$E\{\tilde{\mathbf{x}}(k+1|k)\} = [\boldsymbol{\phi}(k+1,k) - \mathbf{K}(k+1,k)\mathbf{C}(k)]E\{\tilde{\mathbf{x}}(k|k-1)\} \tag{6.4.48}$$

For an unbiased estimate, we must have

$$E\{\mathbf{x}(k) - \hat{\mathbf{x}}(k|k-1)\} = E\{\tilde{\mathbf{x}}(k|k-1)\} = 0 \tag{6.4.49}$$

Since Eq. (6.4.48) is a homogeneous difference equation, if we choose $E\{\tilde{\mathbf{x}}(k_0)\} = 0$, then $E\{\tilde{\mathbf{x}}(k+1|k)\} = 0$ for all $k \geqslant k_0$. We therefore have

$$\hat{\mathbf{x}}(k_0) = E\{\mathbf{x}(k_0)\} = \boldsymbol{\mu}_{\mathbf{x}}(k_0) \tag{6.4.50}$$

In this case

$$\mathbf{V}_{\tilde{\mathbf{x}}}(k_0) = \mathrm{E}\big[[\mathbf{x}(k_0) - \hat{\mathbf{x}}(k_0)][\mathbf{x}(k_0) - \hat{\mathbf{x}}(k_0)]^T\big]$$
$$= \mathbf{V}_{\mathbf{x}}(k_0) \qquad \textbf{(6.4.51)}$$

The algorithms are summarized in Table 6.4.1a. As can be seen from Eq. (6.4.47), the error variance $\mathbf{V}_{\tilde{\mathbf{x}}}(k|k-1)$ and hence the gain $\mathbf{K}(k+1, k)$ do not depend on the observations and may be precomputed and stored. From a practical viewpoint, it is easier to process these equations also sequentially along with the estimator equation. The error variance is directly available and can be used to measure the performance of the filter.

Discrete Filtering Algorithms

We now consider the problem of obtaining the filtered estimate $\hat{\mathbf{x}}(k|k)$. If we set $k_f = k$ in Eq. (6.4.28) we have

$$\hat{\mathbf{x}}(k|k) = \sum_{j=k_0}^{k} \mathrm{E}\{\mathbf{x}(k)\boldsymbol{\nu}^T(j)\}\mathbf{V}_{\nu}^{-1}(j)\boldsymbol{\nu}(j)$$
$$= \sum_{j=k_0}^{k-1} \mathrm{E}\{\mathbf{x}(k)\boldsymbol{\nu}^T(j)\}\mathbf{V}_{\nu}^{-1}(j)\boldsymbol{\nu}(j) + \mathrm{E}\{\mathbf{x}(k)\boldsymbol{\nu}^T(k)\}\mathbf{V}_{\nu}^{-1}(k)\boldsymbol{\nu}(k)$$
$$\textbf{(6.4.52)}$$

TABLE 6.4.1 ONE-STAGE PREDICTION AND FILTERING ALGORITHMS FOR THE DISCRETE KALMAN FILTER

Message Model:

$$\mathbf{x}(k+1) = \boldsymbol{\phi}(k+1, k)\mathbf{x}(k) + \boldsymbol{\Gamma}(k)\mathbf{w}(k)$$
$$\mathbf{y}(k) = \mathbf{C}(k)\mathbf{x}(k) \qquad \textbf{(6.4.19)}$$

Observation Model:

$$\mathbf{z}(k) = \mathbf{y}(k) + \mathbf{v}(k) \qquad \textbf{(6.4.21)}$$

Prior Statistics:

$$\mathrm{E}\{\mathbf{w}(k)\} = \mathrm{E}\{\mathbf{v}(k)\} = 0,\ \mathrm{E}\{\mathbf{x}(k_0)\} = \boldsymbol{\mu}_{\mathbf{x}}(k_0)$$
$$\mathrm{E}\{\mathbf{w}(k)\mathbf{w}^T(l)\} = \mathbf{V}_{\mathbf{w}}(k)\delta_{kl},\ \mathrm{E}\{\mathbf{v}(k)\mathbf{v}^T(l)\} = \mathbf{V}_{\mathbf{v}}(k)\delta_{kl}$$
$$\mathrm{E}\{\mathbf{w}(k)\mathbf{v}^T(l)\} = \mathrm{E}\{\mathbf{w}(k)\mathbf{x}^T(k_0)\} = \mathrm{E}\{\mathbf{v}(l)\mathbf{x}^T(k_0)\} = 0$$
$$\mathrm{var}\ \mathbf{x}(k_0) = \mathbf{V}_{\mathbf{x}}(k_0)$$

TABLE 6.4.1a PREDICTOR ALGORITHMS

Predictor Algorithm:

$$\hat{\mathbf{x}}(k+1|k) = \boldsymbol{\phi}(k+1, k)\hat{\mathbf{x}}(k|k-1) + \mathbf{K}(k+1, k)\boldsymbol{\nu}(k) \tag{6.4.39}$$

$$\boldsymbol{\nu}(k) = \mathbf{z}(k) - \mathbf{C}(k)\hat{\mathbf{x}}(k|k-1) \tag{6.4.29}$$

$$\hat{\mathbf{x}}(k_0) = \boldsymbol{\mu}_{\mathbf{x}}(k_0) \tag{6.4.50}$$

Gain Algorithm:

$$\mathbf{K}(k+1, k) = \boldsymbol{\phi}(k+1, k)\mathbf{V}_{\tilde{\mathbf{x}}}(k|k-1)\mathbf{C}^T(k)\mathbf{V}_{\boldsymbol{\nu}}^{-1}(k) \tag{6.4.43}$$

Variance Equations:

$$\mathbf{V}_{\boldsymbol{\nu}}(k) = \mathbf{C}(k)\mathbf{V}_{\tilde{\mathbf{x}}}(k|k-1)\mathbf{C}^T(k) + \mathbf{V}_{\mathbf{v}}(k) \tag{6.4.33}$$

$$\begin{aligned}\mathbf{V}_{\tilde{\mathbf{x}}}(k+1|k) = {} & \boldsymbol{\phi}(k+1, k)\mathbf{V}_{\tilde{\mathbf{x}}}(k|k-1)\boldsymbol{\phi}^T(k+1, k) + \boldsymbol{\Gamma}(k)\mathbf{V}_{\mathbf{w}}(k)\boldsymbol{\Gamma}^T(k) \\ & - \boldsymbol{\phi}(k+1, k)\mathbf{V}_{\tilde{\mathbf{x}}}(k|k-1)\mathbf{C}^T(k)\mathbf{V}_{\boldsymbol{\nu}}^{-1}(k)\mathbf{C}(k) \\ & \cdot \mathbf{V}_{\tilde{\mathbf{x}}}(k|k-1)\boldsymbol{\phi}^T(k+1, k)\end{aligned} \tag{6.4.47}$$

$$\mathbf{V}_{\tilde{\mathbf{x}}}(k_0) = \mathbf{V}_{\mathbf{x}}(k_0) \tag{6.4.51}$$

TABLE 6.4.1b FILTERING ALGORITHMS

Filtering Algorithm:

$$\hat{\mathbf{x}}(k|k) = \boldsymbol{\phi}(k, k-1)\hat{\mathbf{x}}(k-1|k-1) + \mathbf{K}(k)\boldsymbol{\nu}(k) \tag{6.4.57}$$

$$\hat{\mathbf{x}}(k+1|k) = \boldsymbol{\phi}(k+1, k)\hat{\mathbf{x}}(k|k) \tag{6.4.56}$$

Gain Algorithm:

$$\mathbf{K}(k) = \mathbf{V}_{\tilde{\mathbf{x}}}(k|k-1)\mathbf{C}^T(k)\mathbf{V}_{\boldsymbol{\nu}}^{-1}(k) \tag{6.4.61}$$

Variance Equations:

$$\mathbf{V}_{\tilde{\mathbf{x}}}(k|k) = [\mathbf{I} - \mathbf{K}(k)\mathbf{C}(k)]\mathbf{V}_{\tilde{\mathbf{x}}}(k|k-1) \tag{6.4.62}$$

$$\mathbf{V}_{\tilde{\mathbf{x}}}(k+1|k) = \boldsymbol{\phi}(k+1, k)\mathbf{V}_{\tilde{\mathbf{x}}}(k|k)\boldsymbol{\phi}^T(k+1, k) + \boldsymbol{\Gamma}(k)\mathbf{V}_{\mathbf{w}}(k)\boldsymbol{\Gamma}^T(k) \tag{6.4.63}$$

We recognize from Eq. (6.4.34) that the first term on the right-hand side in the equation preceding is the one-stage predicted estimate $\hat{\mathbf{x}}(k|k-1)$. We can therefore write for the filtered estimate

$$\hat{\mathbf{x}}(k|k) = \hat{\mathbf{x}}(k|k-1) + \mathbf{K}(k)\boldsymbol{\nu}(k) \tag{6.4.53}$$

where we have defined the gain $\mathbf{K}(k)$ as

$$\mathbf{K}(k) = \mathrm{E}\{\mathbf{x}(k)\boldsymbol{\nu}^T(k)\}\mathbf{V}_{\nu}^{-1}(k) \tag{6.4.54}$$

Comparison with Eq. (6.4.38) shows that

$$\mathbf{K}(k+1, k) = \boldsymbol{\phi}(k+1, k)\mathbf{K}(k) \tag{6.4.55}$$

Since $\hat{\mathbf{x}}(k+1|k)$ is the wide-sense conditional expectation of $\mathbf{x}(k)$ given the data sequence $\mathbf{Z}(k)$, we have

$$\begin{aligned}\hat{\mathbf{x}}(k|k-1) &= \hat{\mathrm{E}}[\mathbf{x}(k)|\mathbf{Z}(k)] \\ &= \hat{\mathrm{E}}[\boldsymbol{\phi}(k, k-1)\mathbf{x}(k) + \boldsymbol{\Gamma}(k-1)\mathbf{w}(k-1)|\mathbf{Z}(k-1)] \\ &= \boldsymbol{\phi}(k, k-1)\hat{\mathrm{E}}\{\mathbf{x}(k-1)|\mathbf{Z}(k-1)\} \\ &= \boldsymbol{\phi}(k, k-1)\hat{\mathbf{x}}(k-1|k-1)\end{aligned} \tag{6.4.56}$$

Substitution in Eq. (6.4.53) yields a direct sequential algorithm for $\hat{\mathbf{x}}(k|k)$ as

$$\hat{\mathbf{x}}(k|k) = \boldsymbol{\phi}(k, k-1)\hat{\mathbf{x}}(k-1|k-1) + \mathbf{K}(k)\boldsymbol{\nu}(k) \tag{6.4.57}$$

where

$$\begin{aligned}\boldsymbol{\nu}(k) &= \mathbf{z}(k) - \mathbf{C}(k)\hat{\mathbf{x}}(k|k-1) \\ &= \mathbf{z}(k) - \mathbf{C}(k)\boldsymbol{\phi}(k, k-1)\hat{\mathbf{x}}(k-1|k-1)\end{aligned} \tag{6.4.58}$$

To obtain an expression for the filtered estimate error variance we proceed as follows. From Eq. (6.4.57), we have

$$\begin{aligned}\tilde{\mathbf{x}}(k|k) &= \mathbf{x}(k) - \hat{\mathbf{x}}(k|k) \\ &= \mathbf{x}(k) - \hat{\mathbf{x}}(k|k-1) - \mathbf{K}(k)\boldsymbol{\nu}(k) \\ &= \tilde{\mathbf{x}}(k|k-1) - \mathbf{K}(k)[\mathbf{C}(k)\tilde{\mathbf{x}}(k|k-1) + \mathbf{v}(k)] \\ &= [\mathbf{I} - \mathbf{K}(k)\mathbf{C}(k)]\tilde{\mathbf{x}}(k|k-1) - \mathbf{K}(k)\mathbf{v}(k)\end{aligned} \tag{6.4.59}$$

Thus the error variance is given by

$$\mathbf{V}_{\tilde{\mathbf{x}}}(k|k) = [\mathbf{I} - \mathbf{K}(k)\mathbf{C}(k)]\mathbf{V}_{\tilde{\mathbf{x}}}(k|k-1)[\mathbf{I} - \mathbf{K}(k)\mathbf{C}(k)]^T + \mathbf{K}(k)\mathbf{V}_{\mathbf{v}}(k)\mathbf{K}^T(k) \tag{6.4.60}$$

which upon substituting for $\mathbf{K}(k)$ as

$$\mathbf{K}(k) = \mathbf{V}_{\tilde{\mathbf{x}}}(k|k-1)\mathbf{C}^T(k)\mathbf{V}_{\nu}^{-1}(k) \tag{6.4.61}$$

can be simplified to yield

$$\mathbf{V}_{\tilde{\mathbf{x}}}(k|k) = [\mathbf{I} - \mathbf{K}(k)\mathbf{C}(k)]\mathbf{V}_{\tilde{\mathbf{x}}}(k|k-1) \tag{6.4.62}$$

We can simplify the equation for the predicted error variance by using $\mathbf{V}_{\tilde{\mathbf{x}}}(k|k)$. Substitution of Eqs. (6.4.61) and (6.4.62) in Eq. (6.4.47) yields

$$\mathbf{V}_{\tilde{\mathbf{x}}}(k+1|k) = \boldsymbol{\phi}(k+1, k)\mathbf{V}_{\tilde{\mathbf{x}}}(k|k)\boldsymbol{\phi}^T(k+1, k) + \boldsymbol{\Gamma}(k)\mathbf{V}_{\mathbf{w}}(k)\boldsymbol{\Gamma}^T(k) \tag{6.4.63}$$

These algorithms are summarized in Table 6.4.1b.

Example 6.4.1 Consider the scalar message model given by

$$\begin{aligned} x(k+1) &= .8x(k) + w(k) \\ z(k) &= h(k)x(k) + v(k) \end{aligned}$$

where $w(k)$ and $v(k)$ are stationary, zero mean, white with unity covariance. $h(k)$ is a periodic sequence as follows

$$h(k) = \begin{cases} .1 & k \text{ odd} \\ 1 & k \text{ even} \end{cases}$$

The one-stage predictor equations as obtained from Table 6.4.1 are given by

$$\hat{x}(k+1|k) = .8\hat{x}(k|k-1) + K(k+1, k)[z(k) - h(k)\hat{x}(k|k-1)]$$

$$K(k+1, k) = .8V_{\tilde{x}}(k|k-1)h(k)[h^2(k)V_{\tilde{x}}(k|k-1) + 1]^{-1}$$

$$V_{\tilde{x}}(k+1|k) = .64V_{\tilde{x}}(k|k-1)[h^2(k)V_{\tilde{x}}(k|k-1) + 1]^{-1} + 1$$

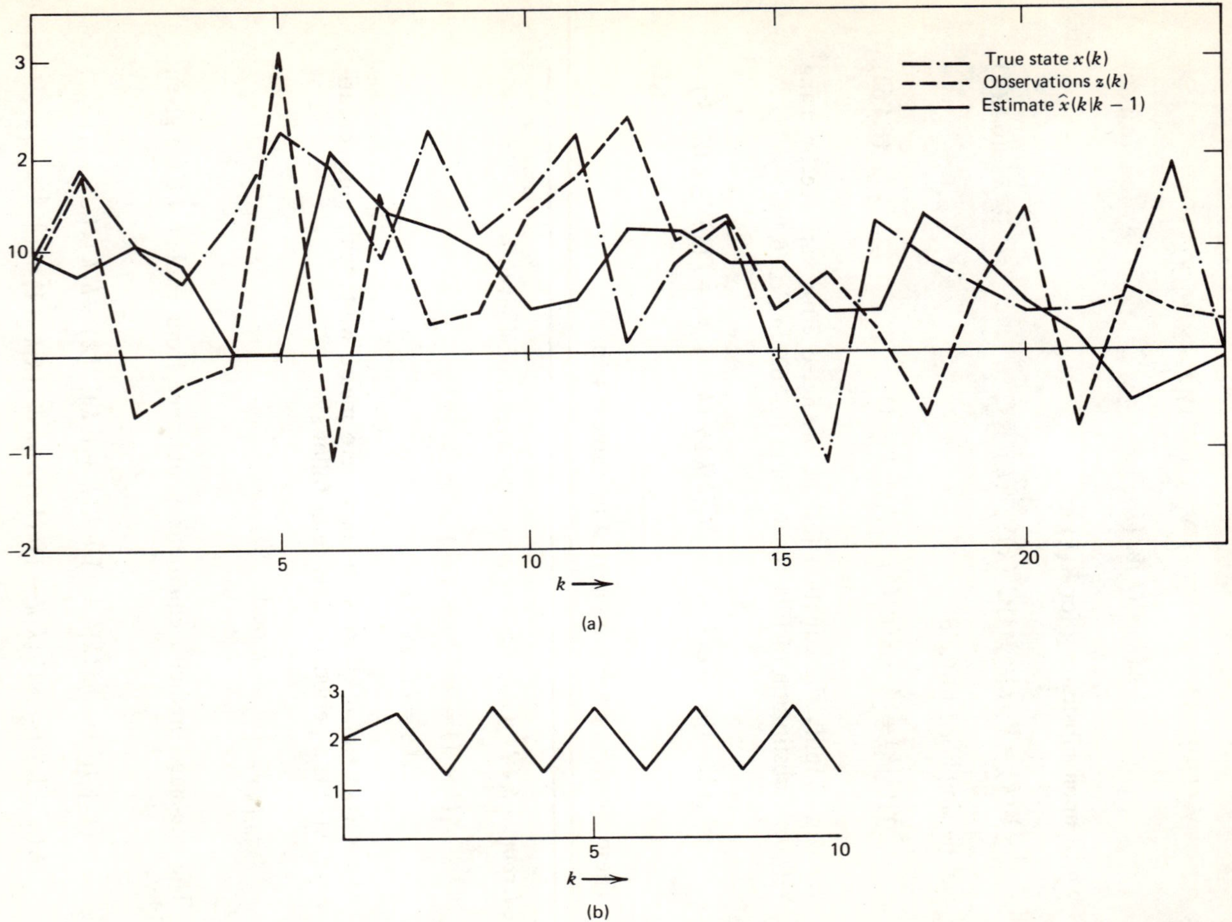

Fig. 6.4.2 Simulation results for discrete Kalman filter example. (a) True state and estimated values. (b) Filter gain.

Note that the equations for the error variance $V_{\tilde{x}}(k+1|k)$ and the filter gain $K(k+1, k)$ can be processed independently of the observations. Figure 6.4.2a shows the true state and the estimate for a single realization of the input and observation noise processes as obtained by simulation on a digital computer. Since the odd-numbered measurements have a relatively larger noise component, we should expect that these measurements do not play as significant a role in updating our estimate as the even-numbered measurements do: the value of the gain $K(k+1, k)$ should be smaller for odd values of k than for even k. In fact, as can be seen from Fig. 6.4.2b, the gain becomes periodic after a fairly short transient period.

We note that the estimate $\hat{x}(k)$ follows the true value $x(k)$ fairly reasonably for this particular simulation. We may not obtain similar results with other realizations of the input $w(k)$ or the observation noise $v(k)$ since $x(k)$ and $\hat{x}(k)$ are close only in the mean square sense. If, however, we average the results of a large number of simulations using a different realization of $w(k)$ and $v(k)$ in each simulation, we may expect the average values of true state and the estimate to be close to each other.

Continuous-Time Filter

We now derive the analogous estimator equations for the continuous time problem. If we define the set of observations over the interval $t_0 \leqslant \tau < t$ by

$$\mathbf{Z}(t) = \{\mathbf{z}(\tau) : t_0 \leqslant \tau < t\} \tag{6.4.64}$$

we see from Eq. (6.4.10) that the estimate $\hat{\mathbf{g}}(t|t_f)$ is given by

$$\hat{\mathbf{g}}(t|t_f) = \hat{\mathrm{E}}\{\mathbf{g}(t)|\mathbf{Z}(t_f)\} = \int_{t_0}^{t_f} \mathrm{E}\{\mathbf{g}(t)\boldsymbol{\nu}^T(\tau)\}\mathbf{V}_{\nu}^{-1}(\tau)\boldsymbol{\nu}(\tau)\, d\tau \tag{6.4.65}$$

Analogous to our derivation in the discrete-time case, we will show that the innovations $\boldsymbol{\nu}(t)$ for the observation model Eq. (6.4.1) is given by

$$\boldsymbol{\nu}(t) = \mathbf{z}(t) - \hat{\mathbf{y}}(t|t) \tag{6.4.66}$$

where

$$\hat{\mathbf{y}}(t|t) = \hat{\mathrm{E}}\{\mathbf{y}(t)|\mathbf{Z}(t)\} \tag{6.4.67}$$

We note that Eq. (6.4.66) can also be written as

$$\boldsymbol{\nu}(t) = \tilde{\mathbf{y}}(t|t) + \mathbf{v}(t) \tag{6.4.68}$$

To establish that $\boldsymbol{\nu}(t)$ is white, we note that for $t > s$

$$\begin{aligned} \mathrm{E}\{\boldsymbol{\nu}(t)\boldsymbol{\nu}^T(s)\} &= \mathrm{E}\{[\tilde{\mathbf{y}}(t|t) + \mathbf{v}(t)]\boldsymbol{\nu}^T(s)\} \\ &= \mathrm{E}\{\mathbf{v}(t)\boldsymbol{\nu}^T(s)\} \text{ by orthogonality} \\ &= \mathrm{E}\{\mathbf{v}(t)[\tilde{\mathbf{y}}(s|s) + \mathbf{v}(s)]^T\} \\ &= \mathrm{E}\{\mathbf{v}(t)\mathbf{v}^T(s)\} \text{ by causality} \\ &= 0 \end{aligned} \qquad \textbf{(6.4.69)}$$

It can be seen that this result holds for $s > t$ also. For $t = s$, we have

$$\begin{aligned} \mathrm{E}\{\boldsymbol{\nu}(t)\boldsymbol{\nu}^T(t)\} &= \mathrm{E}\{[\tilde{\mathbf{y}}(t|t) + \mathbf{v}(t)][\tilde{\mathbf{y}}(t|t) + \mathbf{v}(t)]^T\} \\ &= \mathrm{E}\{\tilde{\mathbf{y}}(t|t)\tilde{\mathbf{y}}^T(t|t)\} + \mathrm{E}\{\mathbf{v}(t)\mathbf{v}^T(t)\} \end{aligned}$$

We note that in the preceding expression, $\mathrm{E}\{\mathbf{v}(t)\mathbf{v}^T(t)\}$ is an impulse of strength $\mathbf{V}_\mathbf{v}(t)$. The variance of the error, $\mathrm{E}\{\tilde{\mathbf{y}}(t|t)\tilde{\mathbf{y}}^T(t|t)\}$ is a finite quantity (as otherwise there would be no solution to the minimization problem) and can be neglected in comparison with the other term. We conclude, therefore, that the variance of $\boldsymbol{\nu}(t)$ is also an impulse of strength $\mathbf{V}_\mathbf{v}(t)$. Thus, we have

$$\mathrm{E}\{\boldsymbol{\nu}(t)\boldsymbol{\nu}^T(s)\} = \mathbf{V}_\mathbf{v}(t)\delta(t-s) \qquad \textbf{(6.4.70)}$$

We point out that in the discrete-time problem, $\mathbf{V}_\mathbf{v}(k)$ is a finite quantity and hence $\mathbf{V}_{\tilde{\mathbf{y}}}(k|k-1)$ cannot be neglected.

In order to derive a differential equation for the estimate, we will need to assume a specific model for the signal process. We assume again that the signal $\mathbf{y}(t)$ is the output of a linear system driven by white noise:

$$\begin{aligned} \dot{\mathbf{x}}(t) &= \mathbf{A}(t)\mathbf{x}(t) + \mathbf{B}(t)\mathbf{w}(t) \\ \mathbf{y}(t) &= \mathbf{C}(t)\mathbf{x}(t) \end{aligned} \qquad \textbf{(6.4.71)}$$

where $\mathbf{w}(t)$ is zero mean, white with covariance

$$\mathrm{E}\{\mathbf{w}(t)\mathbf{w}^T(s)\} = \mathbf{V}_\mathbf{w}(t)\delta(t-s) \qquad \textbf{(6.4.72)}$$

and is uncorrelated with the observation noise $\mathbf{v}(t)$. The initial mean and variance of the state are assumed known

$$\mathrm{E}\{\mathbf{x}(t_0)\} = \boldsymbol{\mu}_\mathbf{x}(t_0) \qquad \text{and} \quad \operatorname{cov} \mathbf{x}(t_0) = \mathbf{V}_\mathbf{x}(t_0)$$

The function we are seeking to estimate is the state vector $\mathbf{x}(t)$ so that $\mathbf{g}(t) = \mathbf{x}(t)$. We will restrict our discussion to filtered estimates.

From Eq. (6.4.65), for $t_f = t$, we have

$$\hat{\mathbf{x}}(t|t) = \int_{t_0}^{t} \mathrm{E}\{\mathbf{x}(t)\boldsymbol{\nu}^T(\tau)\}\mathbf{V}_{\nu}^{-1}(\tau)\boldsymbol{\nu}(\tau)\, d\tau \tag{6.4.73}$$

Differentiating both sides with respect to t yields

$$\begin{aligned}\dot{\hat{\mathbf{x}}}(t|t) &= \frac{d}{dt}\int_{t_0}^{t} \mathrm{E}\{\mathbf{x}(t)\boldsymbol{\nu}^T(\tau)\}\mathbf{V}_{\nu}^{-1}(\tau)\boldsymbol{\nu}(\tau)\, d\tau \\ &= \int_{t_0}^{t} \mathrm{E}\{\dot{\mathbf{x}}(t)\boldsymbol{\nu}^T(\tau)\}\mathbf{V}_{\nu}^{-1}(\tau)\boldsymbol{\nu}(\tau)\, d\tau + \mathrm{E}\{\mathbf{x}(t)\boldsymbol{\nu}^T(t)\}\mathbf{V}_{\nu}^{-1}(t)\boldsymbol{\nu}(t)\end{aligned} \tag{6.4.74}$$

Substituting for $\dot{\mathbf{x}}(t)$ from Eq. (6.4.71) and noting that by causality $\mathrm{E}\{\mathbf{w}(t)\boldsymbol{\nu}^T(\tau)\} = 0$, we can write

$$\dot{\hat{\mathbf{x}}}(t|t) = \mathbf{A}(t)\int_{t_0}^{t} \mathrm{E}\{\mathbf{x}(t)\boldsymbol{\nu}^T(\tau)\}\mathbf{V}_{\nu}^{-1}(\tau)\, d\tau + \mathrm{E}\{\mathbf{x}(t)\boldsymbol{\nu}^T(t)\}\mathbf{V}_{\nu}^{-1}(t)\boldsymbol{\nu}(t) \tag{6.4.75}$$

Using Eq. (6.4.73) then yields

$$\dot{\hat{\mathbf{x}}}(t|t) = \mathbf{A}(t)\hat{\mathbf{x}}(t|t) + \mathbf{K}(t)\boldsymbol{\nu}(t) \tag{6.4.76}$$

where we have defined

$$\mathbf{K}(t) = \mathrm{E}\{\mathbf{x}(t)\boldsymbol{\nu}^T(t)\}\mathbf{V}_{\nu}^{-1}(t) \tag{6.4.77}$$

Now from Eq. (6.4.68) and the assumptions on $\mathbf{v}(t)$, we can write

$$\begin{aligned}\mathrm{E}\{\mathbf{x}(t)\boldsymbol{\nu}^T(t)\} &= \mathrm{E}\{\mathbf{x}(t)[\mathbf{C}(t)\tilde{\mathbf{x}}(t|t) + \mathbf{v}(t)]^T\} \\ &= \mathrm{E}\{\mathbf{x}(t)\tilde{\mathbf{x}}^T(t|t)\}\mathbf{C}^T(t) \\ &= \mathrm{E}\{[\hat{\mathbf{x}}(t|t) + \tilde{\mathbf{x}}(t|t)]\tilde{\mathbf{x}}^T(t|t)\}\mathbf{C}^T(t) \\ &= \mathbf{V}_{\tilde{\mathbf{x}}}(t|t)\mathbf{C}^T(t)\end{aligned} \tag{6.4.78}$$

The last equality follows since $\tilde{\mathbf{x}}(t|t)$ is orthogonal to $\hat{\mathbf{x}}(t|t)$. Substitution of Eq. (6.4.78) in Eq. (6.4.77) then yields

$$\mathbf{K}(t) = \mathbf{V}_{\tilde{\mathbf{x}}}(t|t)\mathbf{C}^T(t)\mathbf{V}_{\nu}^{-1}(t) = \mathbf{V}_{\tilde{\mathbf{x}}}(t|t)\mathbf{C}^T(t)\mathbf{V}_{\mathbf{v}}^{-1}(t) \tag{6.4.79}$$

where the last equality follows from Eq. (6.4.70). $\mathbf{K}(t)$ is referred to as the Kalman gain for the continuous-filter case. All that remains is to determine $\mathbf{V}_{\tilde{\mathbf{x}}}(t|t)$. To this end, we subtract Eq. (6.4.76) from Eq. (6.4.71) to get

$$\begin{aligned}\dot{\tilde{\mathbf{x}}}(t|t) &= \dot{\mathbf{x}}(t) - \dot{\hat{\mathbf{x}}}(t|t)\\ &= \mathbf{A}(t)\tilde{\mathbf{x}}(t|t) + \mathbf{B}(t)\mathbf{w}(t) - \mathbf{K}(t)\boldsymbol{\nu}(t)\\ &= [\mathbf{A}(t) - \mathbf{K}(t)\mathbf{C}(t)]\tilde{\mathbf{x}}(t|t) + \mathbf{B}(t)\mathbf{w}(t) - \mathbf{K}(t)\mathbf{v}(t)\end{aligned} \tag{6.4.80}$$

We thus have, in view of the fact that $\mathbf{w}(\cdot)$ and $\mathbf{v}(\cdot)$ are white and uncorrelated, from Eq. (2.10.48)

$$\begin{aligned}\dot{\mathbf{V}}_{\tilde{\mathbf{x}}}(t|t) = [\mathbf{A}(t) - \mathbf{K}(t)\mathbf{C}(t)]\mathbf{V}_{\tilde{\mathbf{x}}}(t|t)\\ + \mathbf{V}_{\tilde{\mathbf{x}}}(t|t)[\mathbf{A}(t) - \mathbf{K}(t)\mathbf{C}(t)]^T\\ + \mathbf{B}(t)\mathbf{V}_{\mathbf{w}}(t)\mathbf{B}^T(t) + \mathbf{K}(t)\mathbf{V}_{\mathbf{v}}(t)\mathbf{K}^T(t)\end{aligned} \tag{6.4.81}$$

Substitution of Eq. (6.4.79) into Eq. (6.4.81) and simplification yields

$$\begin{aligned}\dot{\mathbf{V}}_{\tilde{\mathbf{x}}}(t|t) = \mathbf{A}(t)\mathbf{V}_{\tilde{\mathbf{x}}}(t|t) + \mathbf{V}_{\tilde{\mathbf{x}}}(t|t)\mathbf{A}^T(t) + \mathbf{B}(t)\mathbf{V}_{\mathbf{w}}(t)\mathbf{B}^T(t)\\ - \mathbf{V}_{\tilde{\mathbf{x}}}(t|t)\mathbf{C}^T(t)\mathbf{V}_{\mathbf{v}}^{-1}(t)\mathbf{C}(t)\mathbf{V}_{\tilde{\mathbf{x}}}(t|t)\end{aligned} \tag{6.4.82}$$

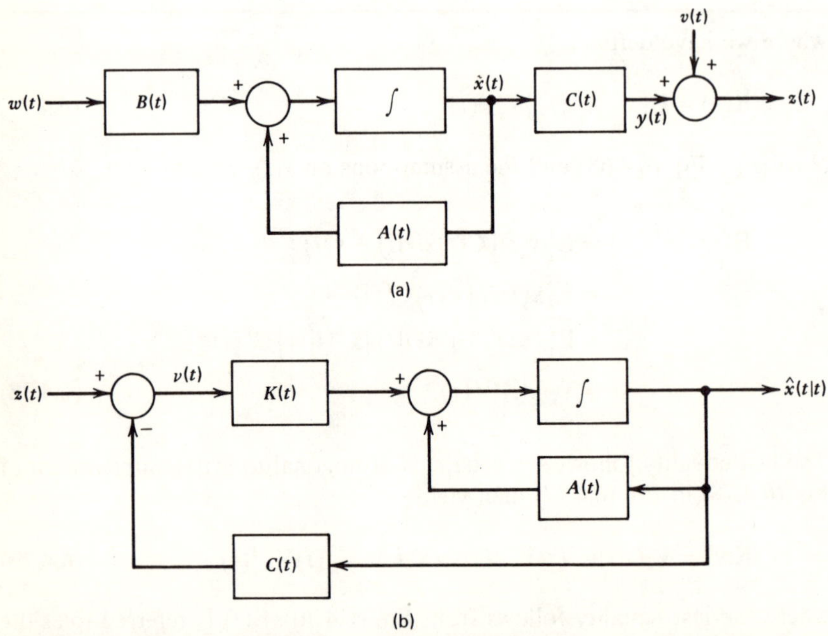

Fig. 6.4.3 Structure of continuous-time Kalman filter. (a) Message model. (b) Filter.

TABLE 6.4.2 CONTINUOUS-TIME KALMAN FILTER ALGORITHMS

Message Model:

$$\dot{\mathbf{x}}(t) = \mathbf{A}(t)\mathbf{x}(t) + \mathbf{B}(t)\mathbf{w}(t)$$
$$\mathbf{y}(t) = \mathbf{C}(t)\mathbf{x}(t) \tag{6.4.71}$$

Observation Model:

$$\mathbf{z}(t) = \mathbf{y}(t) + \mathbf{v}(t) \tag{6.4.1}$$

Prior Statistics:

$$E\{\mathbf{w}(t)\} = E\{\mathbf{v}(t)\} = 0, \; E\{\mathbf{x}(t_0)\} = \boldsymbol{\mu}_{\mathbf{x}}(t_0)$$
$$E\{\mathbf{w}(t)\mathbf{w}^T(s)\} = \mathbf{V}_{\mathbf{w}}(t)\delta(t-s), \; E\{\mathbf{v}(t)\mathbf{v}^T(s)\} = \mathbf{V}_{\mathbf{v}}(t)\delta(t-s)$$
$$E\{\mathbf{w}(t)\mathbf{v}^T(s)\} = E\{\mathbf{w}(t)\mathbf{x}^T(t_0)\} = E\{\mathbf{v}(t)\mathbf{x}^T(t_0)\} = 0$$
$$\operatorname{cov} \mathbf{x}(t_0) = \mathbf{V}_{\mathbf{x}}(t_0)$$

Filter Algorithm:

$$\dot{\hat{\mathbf{x}}}(t|t) = \mathbf{A}(t)\hat{\mathbf{x}}(t|t) + \mathbf{K}(t)\boldsymbol{\nu}(t) \tag{6.4.76}$$
$$\boldsymbol{\nu}(t) = \mathbf{z}(t) - \mathbf{C}(t)\hat{\mathbf{x}}(t|t)$$

Gain Algorithm:

$$\mathbf{K}(t) = \mathbf{V}_{\tilde{\mathbf{x}}}(t|t)\mathbf{C}^T(t)\mathbf{V}_{\mathbf{v}}^{-1}(t) \tag{6.4.79}$$

Variance Algorithm:

$$\dot{\mathbf{V}}_{\tilde{\mathbf{x}}}(t|t) = \mathbf{A}(t)\mathbf{V}_{\tilde{\mathbf{x}}}(t|t) + \mathbf{V}_{\tilde{\mathbf{x}}}(t|t)\mathbf{A}^T(t) + \mathbf{B}(t)\mathbf{V}_{\mathbf{w}}(t)\mathbf{B}^T(t) - \mathbf{V}_{\tilde{\mathbf{x}}}(t|t)\mathbf{C}^T(t)\mathbf{V}_{\mathbf{v}}^{-1}(t)\mathbf{C}(t)\mathbf{V}_{\tilde{\mathbf{x}}}(t|t) \tag{6.4.82}$$

Initial Conditions:

$$\hat{\mathbf{x}}(t_0) = \boldsymbol{\mu}_{\mathbf{x}}(t_0), \; \mathbf{V}_{\tilde{\mathbf{x}}}(t_0) = \mathbf{V}_{\mathbf{x}}(t_0) \tag{6.4.83}$$

Equation (6.4.82) is usually referred to as the matrix Riccati differential equation. It can easily be shown that for unbiased estimates, we must choose the initial conditions for Eqs. (6.4.75) and (6.4.81) as

$$\hat{\mathbf{x}}(t_0) = \boldsymbol{\mu}_{\mathbf{x}}(t_0) \quad \text{and} \quad \mathbf{V}_{\tilde{\mathbf{x}}}(t_0) = \mathbf{V}_{\mathbf{x}}(t_0) \tag{6.4.83}$$

The algorithms are summarized in Table 6.4.2. Figure 6.4.3 shows the structure of the Kalman filter.

Example 6.4.2 We consider the estimation of the state in a continuous-time model. The message and observation models are assumed to be as follows:

$$\dot{x}(t) = -x(t) + w(t) \qquad x(0) = 1$$

$$z(t) = \tfrac{1}{2}x(t) + v(t)$$

$$\mathrm{E}\{w(t)w(\tau)\} = 6\delta(t-\tau),\ \mathrm{E}\{v(t)v(\tau)\} = \tfrac{1}{2}\delta(t-\tau)$$

$$\mathrm{E}\{w(t)\} = E\{v(t)\} = \mathrm{E}\{w(t)v(\tau)\} = 0$$

The estimator equations for this problem are easily seen to be

$$\dot{\hat{x}}(t|t) = -\hat{x}(t|t) + K(t)\left[z(t) - \tfrac{1}{2}\hat{x}(t|t)\right]$$

with

$$K(t) = V_{\tilde{x}}(t|t)$$

The error variance equation is given by

$$\dot{V}_{\tilde{x}}(t|t) = -2V_{\tilde{x}}(t|t) - \tfrac{1}{2}V_{\tilde{x}}^2(t|t) + 6$$

While this equation can be solved using numerical techniques, it is instructive to obtain an analytical solution in this simple case. Since $x(0)$ is known (deterministic), we must choose $\hat{x}(0) = x(0)$ to obtain an unbiased estimate. It follows that $\tilde{x}(0)$ and $V_{\tilde{x}}(0)$ should be zero. Let us assume a solution of the form

$$V_{\tilde{x}}(t|t) = \alpha + \beta\tanh(\gamma t + \delta) = \alpha + \beta\tanh\theta$$

where α, β, γ, δ, and θ are unknown constants. Then

$$\dot{V}_{\tilde{x}}(t|t) = \beta\gamma(1 - \tanh^2\theta)$$

Substituting in the variance equation yields

$$\beta\gamma(1 - \tanh^2\theta) = -2(\alpha + \beta\tanh\theta) - \tfrac{1}{2}(\alpha^2 + \beta^2\tanh^2\theta + 2\alpha\beta\tanh\theta) + 6$$

Collecting like terms together and equating to zero, we have

$$\alpha = -2,\ \beta = 2\gamma,\ \gamma = \pm 2$$

Choosing $\gamma = +2$ (why?) yields

$$V_{\tilde{x}}(t|t) = -2 + 4\tanh(2t + \delta)$$
$$= -2 + 4\tanh\left(2t + \tanh^{-1}\tfrac{1}{2}\right)$$

where the last step is obtained by making use of the fact that $V_{\tilde{x}}(0) = 0$. The steady state value of the error variance is obtained by letting $t \to \infty$ and is given by

$$V_{\tilde{x}ss} = -2 + 4 = 2$$

It follows therefore that the filter gain $K(t)$ also approaches a constant value as $t \to \infty$, given by

$$K_{ss} = 2$$

The estimator equation then becomes

$$\dot{\hat{x}}(t|t) = -\hat{x}(t|t) + 2\left[z(t) - \tfrac{1}{2}\hat{x}(t|t)\right]$$
$$= -2\hat{x}(t|t) + 2z(t)$$

Laplace transformation of this equation yields

$$\hat{X}(s) = \frac{2}{s+2} Z(s)$$

so that the estimator is asymptotically equivalent to a time-invariant linear system with transfer function $2/(s+2)$.

The system was also simulated on a digital computer for different values of the initial condition $\hat{x}(0)$. The corresponding error variances are shown in Fig. 6.4.4a. As can be seen, the error variances approach the steady state value fairly rapidly. Since the filter gain tends to a constant value after an initial transient period, we might expect that the estimates for various initial values $x(0)$ tend to converge asymptotically. This is borne out by the plots of the estimate $\hat{x}(t)$ in Fig. 6.4.4b. We therefore expect that the estimate (at least after an initial transient period) is not significantly dependent on our choice of initial conditions. This is very desirable since the initial conditions in many cases represent no more than an educated guess on the part of the designer. Estimators whose performance does not depend on the choice of initial conditions (or more generally on the prior statistics) are called *robust estimators*.

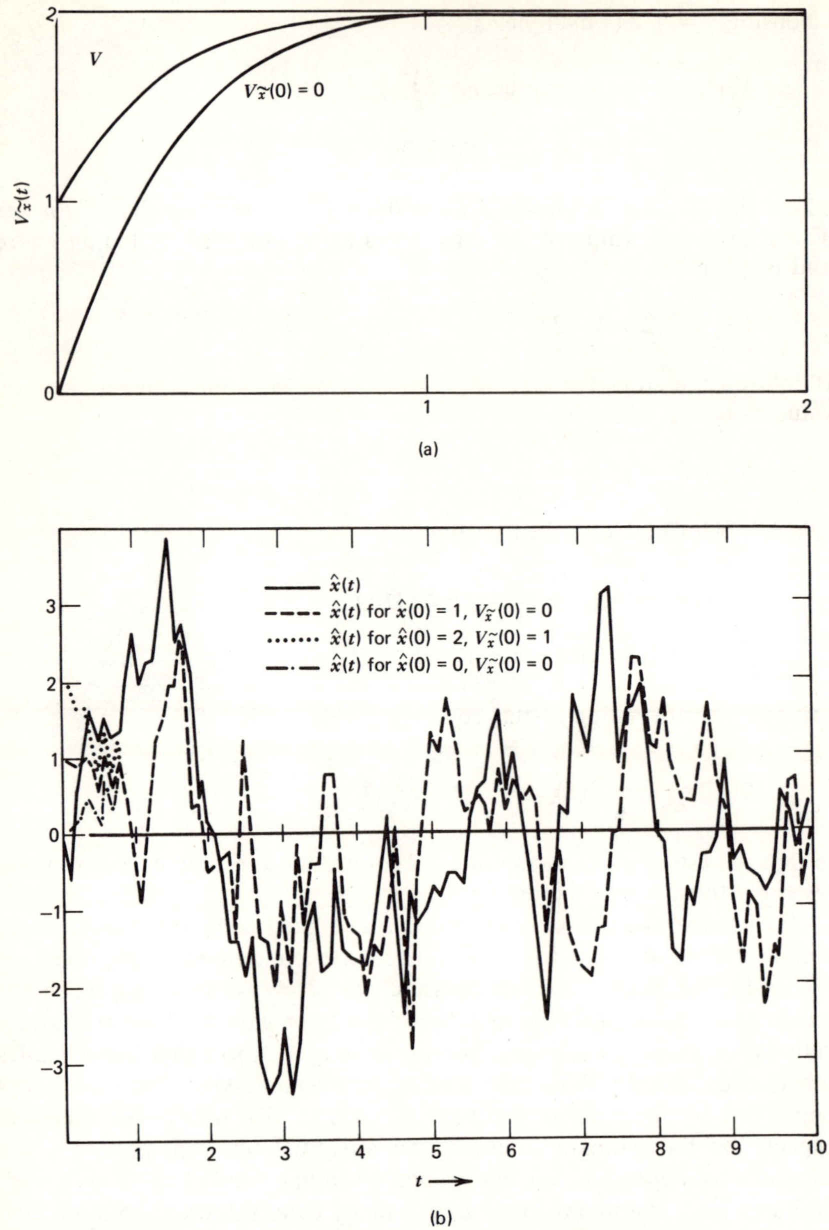

Fig. 6.4.4 Simulation results for Continuous-Time Kalman filter example. (a) Error variance. (b) True state and estimate for various initial conditions.

Continuous-Time Estimation with Discrete Observations

In many instances, even though the signal model is continuous, the observations are available only at discrete intervals of time. The relevant filtering algorithms follow easily from our previous discussions. We assume the message model of Eq. (6.4.71)

$$\begin{aligned}\dot{\mathbf{x}}(t) &= \mathbf{A}(t)\mathbf{x}(t) + \mathbf{B}(t)\mathbf{w}(t)\\ \mathbf{y}(t) &= \mathbf{C}(t)\mathbf{x}(t)\end{aligned} \tag{6.4.84}$$

for $t \geqslant t_0$.

The prior statistics of the noise input $\mathbf{w}(t)$ and the initial vector $\mathbf{x}(t_0)$ are as before. If we assume that noisy measurements of $\mathbf{y}(t)$ are available at discrete instants kT, the observation model becomes

$$\mathbf{z}(kT) = \mathbf{y}(kT) + \mathbf{v}(kT) \tag{6.4.85}$$

The solution to the differential equation of Eq. (6.4.84) over any interval $[\tau, t]$ is

$$\mathbf{x}(t) = \boldsymbol{\psi}(t, \tau)\mathbf{x}(\tau) + \int_\tau^t \boldsymbol{\psi}(t, \sigma)\mathbf{B}(\sigma)\mathbf{w}(\sigma)\, d\sigma \tag{6.4.86}$$

where $\boldsymbol{\psi}(\cdot, \cdot)$ is the state transition matrix corresponding to the matrix $\mathbf{A}(t)$. By setting $\tau = kT$ and $t = (k+1)T$ in Eq. (6.4.86), we obtain

$$\begin{aligned}\mathbf{x}[(k+1)T] &= \boldsymbol{\psi}[(k+1)T, kT]\mathbf{x}(kT)\\ &\quad + \int_{kT}^{(k+1)T} \boldsymbol{\psi}[(k+1)T, \sigma]\mathbf{B}(\sigma)\mathbf{w}(\sigma)\, d\sigma\end{aligned} \tag{6.4.87}$$

We now make the following definitions.

$$\mathbf{w}_1(kT) \triangleq \int_{kT}^{(k+1)T} \boldsymbol{\psi}[(k+1)T, \sigma]\mathbf{B}(\sigma)\mathbf{w}(\sigma)\, d\sigma \tag{6.4.88a}$$

$$\begin{aligned}\mathbf{V}_{\mathbf{w}_1}(kT) &\triangleq \int_{kT}^{(k+1)T} \boldsymbol{\psi}[(k+1)T, \sigma]\mathbf{B}(\sigma)\mathbf{V}_{\mathbf{w}}(\sigma)\mathbf{B}^T(\sigma)\\ &\qquad \times \boldsymbol{\psi}^T[(k+1)T, \sigma]\, d\sigma\end{aligned} \tag{6.4.88b}$$

Substitution in Eq. (6.4.87) then yields

$$\mathbf{x}[(k+1)T] = \boldsymbol{\psi}[(k+1)T, kT]\mathbf{x}(kT) + \mathbf{w}_1(kT) \tag{6.4.89a}$$

Also

$$\mathbf{y}(kT) = \mathbf{C}(kT)\mathbf{x}(kT) \tag{6.4.89b}$$

It can easily be verified that $\mathbf{w}_1(kT)$ is white, zero mean with

$$\operatorname{cov}[\mathbf{w}_1(kT), \mathbf{w}_1(jT)] = \mathbf{V}_{\mathbf{w}_1}(kT)\delta_{kj} \tag{6.4.90}$$

Comparison of Eqs. (6.4.89a, b) and (6.4.90) with Eqs. (6.4.19) and (6.4.21) shows that the algorithms of Table 6.4.1 can now be directly applied to this problem to obtain the estimates $\hat{\mathbf{x}}(kT|kT)$ and $\hat{\mathbf{y}}(kT|kT)$.

In order to obtain the estimate of the state in between observations, we note from Eq. (6.4.86) that in the interval $kT \leqslant t < (k+1)T$ we can write

$$\mathbf{x}(t) = \boldsymbol{\psi}(t, kT)\mathbf{x}(kT) + \int_{kT}^{t} \boldsymbol{\psi}(t, \sigma)\mathbf{B}(\sigma)\mathbf{w}(\sigma)\, d\sigma \tag{6.4.91}$$

Let us denote the observations up to time kT by $\mathbf{Z}(kT)$. Then it follows from Eq. (6.4.91) that

$$\begin{aligned}\hat{\mathrm{E}}[\mathbf{x}(t)|\mathbf{Z}(kT)] &= \boldsymbol{\psi}(t, kT)\hat{\mathrm{E}}[\mathbf{x}(kT)|\mathbf{Z}(kT)] \\ &\quad + \hat{\mathrm{E}}\left[\int_{kT}^{t} \boldsymbol{\psi}(t, \sigma)\mathbf{B}(\sigma)\mathbf{w}(\sigma)\, d\sigma \middle| \mathbf{Z}(kT)\right]\end{aligned} \tag{6.4.92}$$

where $\hat{\mathrm{E}}(\cdot)$ denotes the wide-sense conditional exception. The second term on the right-hand side of Eq. (6.4.93) is zero since $\mathbf{w}(\cdot)$ is a zero mean, white process and $\mathbf{w}(\sigma)$ is independent of $\mathbf{Z}(kT)$ for $\sigma > kT$. We therefore have

$$\begin{aligned}\hat{\mathbf{x}}(t|kT) &= \hat{\mathrm{E}}[\mathbf{x}(t)|\mathbf{Z}(kT)] \\ &= \boldsymbol{\psi}(t, kT)\hat{\mathbf{x}}(kT|kT)\end{aligned} \tag{6.4.93}$$

The estimate of the signal between observations follows as

$$\hat{\mathbf{y}}(t|kT) = \mathbf{C}(t)\hat{\mathbf{x}}(t|kT) \tag{6.4.94}$$

Equation (6.4.93) can be used to obtain the initial condition for processing the observation $\mathbf{z}[(k+1)T]$. We have

$$\hat{\mathbf{x}}(kT+T|kT) = \boldsymbol{\psi}(kT+T, kT)\hat{\mathbf{x}}(kT|kT) \tag{6.4.95}$$

The predicted error covariance can easily be shown to be

$$\begin{aligned}\mathbf{V}_{\tilde{\mathbf{x}}}(kT+T|kT) &= \mathbf{x}(kT+T, kT)\mathbf{V}_{\tilde{\mathbf{x}}}(kT|kT)\boldsymbol{\psi}^T(kT+T, kT) \\ &\quad + \mathbf{V}_{\mathbf{w}_1}(kT)\end{aligned} \tag{6.4.96}$$

where $\mathbf{V}_{\mathbf{w}_1}(kT)$ is defined in Eq. (6.4.89b).

Various extensions of the algorithms presented here are available in the literature. These extensions consider noise processes with nonzero means, cases where the input $\mathbf{w}(t)$ and the observation noise $\mathbf{v}(t)$ are correlated and the case where the input to the message model is a known signal plus a random process. The colored noise problem has also been considered in the literature. We do not consider these extensions here but invite the interested reader to look through the appropriate references listed at the end of this chapter [3–7].

We will now briefly consider the relation between the Wiener and Kalman filters.

6.5 RELATION BETWEEN THE KALMAN AND WIENER FILTERS

The Kalman filtering algorithms are rather general in nature. The Wiener filter, on the other hand, was derived for the estimation of stationary processes. It is of interest to explore the relation between the two filters when the underlying assumptions are the same. Specifically, we establish in this section that the Kalman filter reduces to the classical Wiener filter when the processes involved are stationary and the observation interval is semi-infinite. Incidentally, we are also able to solve the spectral factorization problem for vector processes that we encountered in Section 6.3. We restrict our discussion to the continuous-time filter. Since we have considered only white observation noise in the Kalman filter problem, we assume that the noise in the Wiener formulation is also white and uncorrelated with the signal. The estimate that is desired is that of the signal $\mathbf{y}(t)$.

Let us consider the solution using the Kalman filter for the problem under consideration. Since we are assuming stationary processes, it is clear that the matrices $\mathbf{A}$, $\mathbf{B}$, and $\mathbf{C}$ and the covariance kernels $\mathbf{V}_\mathbf{w}$ and $\mathbf{V}_\mathbf{v}$ in Eq. (6.4.70) and Eq. (6.4.71) are constant. Since the estimate $\hat{\mathbf{x}}(t|t)$ will now be asymptotically a stationary process, the error variance $\mathbf{V}_{\tilde{\mathbf{x}}}(t)$ must be asymptotically a constant. The error variance can thus be obtained as the steady state solution to Eq. (6.4.82) by setting the derivative equal to zero. Thus the error variance satisfies the algebraic equation

$$\mathbf{A}\mathbf{V}_{\tilde{\mathbf{x}}} + \mathbf{V}_{\tilde{\mathbf{x}}}\mathbf{A}^T + \mathbf{B}\mathbf{V}_\mathbf{w}\mathbf{B}^T - \mathbf{V}_{\tilde{\mathbf{x}}}\mathbf{C}^T\mathbf{V}_\mathbf{v}^{-1}\mathbf{C}\mathbf{V}_{\tilde{\mathbf{x}}} = 0 \tag{6.5.1}$$

This is a quadratic algebraic equation and is a reduced Riccati equation. Since the equation is nonlinear, it has many solutions. However since $\mathbf{V}_{\tilde{\mathbf{x}}}$ is a covariance matrix, it must be positive definite; thus we pick as our solution the (unique) positive definite solution to Eq. (6.5.1) [3].

The solution of the algebraic equation above is tedious and difficult. We may rather seek the solution to the algebraic equation as the steady state solution to the corresponding Riccati differential equation which is comparatively straightforward to solve on a computer.

We will now discuss the relationship of the stationary Kalman filter with the Wiener filter. The appropriate message and observation models are

$$\begin{aligned}\dot{\mathbf{x}}(t) &= \mathbf{A}\mathbf{x}(t) + \mathbf{B}\mathbf{w}(t)\\ \mathbf{y}(t) &= \mathbf{C}\mathbf{x}(t)\end{aligned} \tag{6.5.2}$$

and

$$\mathbf{z}(t) = \mathbf{y}(t) + \mathbf{v}(t) \tag{6.5.3}$$

The transfer matrix relating the output $\mathbf{z}(t)$ to the input $\mathbf{w}(t)$ is given by

$$\mathbf{G}(s) = \mathbf{C}[s\mathbf{I} - \mathbf{A}]^{-1}\mathbf{B} = \mathbf{C}\boldsymbol{\Omega}(s)\mathbf{B} \tag{6.5.4}$$

where

$$\boldsymbol{\Omega}(s) = [s\mathbf{I} - \mathbf{A}]^{-1} \tag{6.5.5}$$

The filter equations are

$$\begin{aligned}\dot{\hat{\mathbf{x}}}(t|t) &= \mathbf{A}\hat{\mathbf{x}}(t|t) + \mathbf{K}[\mathbf{z}(t) - \mathbf{C}\hat{\mathbf{x}}(t|t)]\\ &= [\mathbf{A} - \mathbf{K}\mathbf{C}]\hat{\mathbf{x}}(t|t) + \mathbf{K}\mathbf{z}(t)\\ \mathbf{y}(t) &= \mathbf{C}\hat{\mathbf{x}}(t|t)\end{aligned} \tag{6.5.6}$$

where

$$\mathbf{K} = \mathbf{V}_{\tilde{\mathbf{x}}}\mathbf{C}^T\mathbf{V}_{\mathbf{v}}^{-1} \tag{6.5.7}$$

with $\mathbf{V}_{\tilde{\mathbf{x}}}$ the positive definite solution to the algebraic equation

$$\mathbf{A}\mathbf{V}_{\tilde{\mathbf{x}}} + \mathbf{V}_{\tilde{\mathbf{x}}}A^T - \mathbf{V}_{\tilde{\mathbf{x}}}C^T\mathbf{V}_{\mathbf{v}}^{-1}\mathbf{C}\mathbf{V}_{\tilde{\mathbf{x}}} + \mathbf{B}\mathbf{V}_{\mathbf{w}}\mathbf{B}^T = 0 \tag{6.5.8}$$

The transfer function of the filter as obtained from Eq. (6.5.6) is

$$\mathbf{G}_{fK}(s) = \mathbf{C}[s\mathbf{I} - \mathbf{A} - \mathbf{K}\mathbf{C})]^{-1}\mathbf{K} \tag{6.5.9}$$

From our discussion in Section 6.3, (Eq. (6.3.52)) we see that the Wiener filter is determined in terms of the spectrum of the observed signal $\mathbf{z}(t)$. For the model of Eq. (6.5.2), since the input signal is white with variance

$\mathbf{V_w}$, the power spectrum of the signal output is given by

$$\begin{aligned}\mathbf{\Phi_y}(s) &= \mathbf{G}(s)\mathbf{V_w}\mathbf{G}^T(-s)\\ &= \mathbf{C}\mathbf{\Omega}(s)\mathbf{B}\mathbf{V_w}\mathbf{B}^T\mathbf{\Omega}^T(-s)\mathbf{C}^T\end{aligned} \tag{6.5.10}$$

The spectrum of the observed signal is

$$\begin{aligned}\mathbf{\Phi_z}(s) &= \mathbf{\Phi_y}(s) + \mathbf{\Phi_v}(s)\\ &= \mathbf{C}\mathbf{\Omega}(s)\mathbf{B}\mathbf{V_w}\mathbf{B}^T\mathbf{\Omega}^T(-s)\mathbf{C}^T + \mathbf{V_v}\end{aligned} \tag{6.5.11}$$

Since the desired estimate is $\hat{\mathbf{y}}(t)$ and since $\mathbf{y}(\cdot)$ and $\mathbf{v}(\cdot)$ are uncorrelated, from Eq. (6.3.52), the Wiener filter for this problem is specified by

$$\mathbf{G}_{fW}(s) = \left\{\mathbf{\Phi_y}(s)[\mathbf{\Phi}^-(s)]^{-1}\right\}_+[\mathbf{\Phi}^+(s)]^{-1} \tag{6.5.12}$$

where $\mathbf{\Phi}^+(s)$ is obtained from a spectral factorization of $\mathbf{\Phi}_z(s)$

$$\mathbf{\Phi_z}(s) = \mathbf{\Phi}^+(s)\mathbf{\Phi}^-(s) \tag{6.5.13}$$

Since $\mathbf{V_v}$ is a positive definite matrix, we can write

$$\mathbf{V_v} = (\mathbf{V_v})^{1/2}(\mathbf{V_v})^{1/2} \tag{6.5.14}$$

where $(\mathbf{V_v})^{1/2}$ is a positive definite symmetric matrix.We claim that $\mathbf{\Phi}^+(s)$ is given by

$$\mathbf{\Phi}^+(s) = \mathbf{V_v}^{1/2} + \mathbf{C}\mathbf{\Omega}(s)\mathbf{V_{\tilde{x}}}\mathbf{C}^T\mathbf{V_v}^{-1/2} \tag{6.5.15}$$

We verify this by direct computation. We have

$$\begin{aligned}\mathbf{\Phi}^+(s)\mathbf{\Phi}^-(s) &= \left[\mathbf{V_v}^{1/2} + \mathbf{C}\mathbf{\Omega}(s)\mathbf{V_{\tilde{x}}}\mathbf{C}^T\mathbf{V_v}^{-1/2}\right]\\ &\quad\times\left[\mathbf{V_v}^{1/2} + \mathbf{C}\mathbf{\Omega}(-s)\mathbf{V_{\tilde{x}}}\mathbf{C}^T\mathbf{V_v}^{-1/2}\right]^T\\ &= \mathbf{V_v} + \left\{\mathbf{C}\mathbf{\Omega}(s)\mathbf{V_{\tilde{x}}}\mathbf{C}^T + \mathbf{C}\mathbf{V_{\tilde{x}}}\mathbf{\Omega}^T(-s)\mathbf{C}^T\right.\\ &\quad\left.+\mathbf{C}\mathbf{\Omega}(s)\mathbf{V_{\tilde{x}}}\mathbf{C}^T\mathbf{V_v}^{-1}\mathbf{C}\mathbf{V_{\tilde{x}}}\mathbf{\Omega}^T(-s)\mathbf{C}^T\right\}\end{aligned} \tag{6.5.16}$$

The second term on the right side may be written as

$$\mathbf{C}\mathbf{\Omega}(s)\left\{\mathbf{V_{\tilde{x}}}[\mathbf{\Omega}^T(-s)]^{-1} + [\mathbf{\Omega}(s)]^{-1}\mathbf{V_{\tilde{x}}} + \mathbf{V_{\tilde{x}}}\mathbf{C}^T\mathbf{V_v}^{-1}\mathbf{C}\mathbf{V_{\tilde{x}}}\right\}\mathbf{\Omega}^T(-s)\mathbf{C}^T \tag{6.5.17}$$

By noting that $\boldsymbol{\Omega}^{-1}(s) = (s\mathbf{I} - \mathbf{A})$ and $\boldsymbol{\Omega}^{-T}(-s) = (-s\mathbf{I} - \mathbf{A})^T$ and canceling out the resulting $s\mathbf{V}_{\tilde{\mathbf{x}}}$ and $-s\mathbf{V}_{\tilde{\mathbf{x}}}$ terms, we can write the expression in Eq. (6.5.17) as

$$\mathbf{C}\boldsymbol{\Omega}(s)\left[-\mathbf{V}_{\tilde{\mathbf{x}}}\mathbf{A}^T - \mathbf{A}\mathbf{V}_{\tilde{\mathbf{x}}} + \mathbf{V}_{\tilde{\mathbf{x}}}\mathbf{C}^T\mathbf{V}_{\mathbf{v}}^{-1}\mathbf{C}\mathbf{V}_{\tilde{\mathbf{x}}}\right]\boldsymbol{\Omega}^T(-s)\mathbf{C}^T \tag{6.5.18}$$

Substitution of Eq. (6.5.8) in Eq. (6.5.18) then enables us to write this expression as

$$\mathbf{C}\boldsymbol{\Omega}(s)\mathbf{B}\mathbf{V}_{\mathbf{w}}\mathbf{B}^T\boldsymbol{\Omega}^T(-s)\mathbf{C}^T \tag{6.5.19}$$

We can thus write

$$\begin{aligned}\boldsymbol{\Phi}^+(s)\boldsymbol{\Phi}^-(s) &= \mathbf{V}_{\mathbf{v}} + \mathbf{C}\boldsymbol{\Omega}(s)\mathbf{B}\mathbf{V}_{\mathbf{w}}\mathbf{B}^T\boldsymbol{\Omega}^T(-s)\mathbf{C}^T \\ &= \boldsymbol{\Phi}_{\mathbf{z}}(s)\end{aligned} \tag{6.5.20}$$

It remains to show that $\boldsymbol{\Phi}^+(s)$ is analytic in the right half-plane. This will be true if det $\boldsymbol{\Phi}^+(s)$ has no poles in the right half-plane. We see from Eq. (6.5.15) that

$$\begin{aligned}\det \boldsymbol{\Phi}^+(s) &= \det\left\{\left[\mathbf{I} + \mathbf{C}\boldsymbol{\Omega}(s)\mathbf{V}_{\tilde{\mathbf{x}}}\mathbf{C}^T\mathbf{V}_{\mathbf{v}}^{-1}\right]\mathbf{V}_{\mathbf{v}}^{1/2}\right\} \\ &= \det\left\{\left[\mathbf{I} + \mathbf{C}\boldsymbol{\Omega}(s)\mathbf{K}\right]\right\} \det\ \left(V_{\mathbf{v}}^{1/2}\right)\end{aligned} \tag{6.5.21}$$

where we have used Eq. (6.5.7). By use of a matrix identity (see Appendix C) Eq. (6.5.21) can be written as

$$\begin{aligned}\det \boldsymbol{\Phi}^+(s) &= \det \boldsymbol{\Omega}(s) \det \left[(s\mathbf{I} - \mathbf{A}) + \mathbf{K}\mathbf{C}\right] \det \mathbf{V}_{\mathbf{v}}^{1/2} \\ &= \frac{\det\left[s\mathbf{I} - \mathbf{A} + \mathbf{K}\mathbf{C}\right]}{\det\ (s\mathbf{I} - \mathbf{A})} \det \mathbf{V}_{\mathbf{v}}^{1/2}\end{aligned} \tag{6.5.22}$$

Since we have assumed that our signal process has finite variance (implying that $\mathbf{A}$ is a stable matrix) and since the error variance is bounded, it can be shown that the estimator must also be a stable structure [13,17]. Thus the poles of $(s\mathbf{I} - \mathbf{A} + \mathbf{K}\mathbf{C})$ will all lie in the left half-plane implying that roots of det $\boldsymbol{\Phi}^+(s)$ will also be in the left half-plane, thus establishing our result.

We note that the solution to the spectral factorization problem as given in Eq. (6.5.15) involves the solution of the nonlinear algebraic equation Eq. (6.5.8).

To obtain the Wiener filter, we now substitute Eqs. (6.5.10) and (6.5.13) in Eq. (6.5.12) to get

$$\mathbf{G}_{fW}(s) = \left\{\mathbf{C}\boldsymbol{\Omega}(s)\mathbf{B}\mathbf{V}_{\mathbf{w}}\mathbf{B}^T\boldsymbol{\Omega}^T(-s)\mathbf{C}^T\left[\boldsymbol{\Phi}^-(s)\right]^{-1}\right\}_+\left[\boldsymbol{\Phi}^+(s)\right]^{-1} \tag{6.5.23}$$

From Eqs. (6.5.19) and (6.5.16), we can write

$$\mathbf{C}\boldsymbol{\Omega}(s)\mathbf{B}\mathbf{V}_{\mathbf{w}}\mathbf{B}^T\boldsymbol{\Omega}^T(-s)\mathbf{C}^T = \boldsymbol{\Phi}^+(s)\boldsymbol{\Phi}^-(s) - \mathbf{V}_{\mathbf{v}} \tag{6.5.24}$$

Thus we can write from Eq. (6.5.23)

$$\mathbf{G}_{fW}(s) = \left\{\boldsymbol{\Phi}^+(s) - \mathbf{V}_{\mathbf{v}}\left[\boldsymbol{\Phi}^-(s)\right]^{-1}\right\}_+\left[\boldsymbol{\Phi}^+(s)\right]^{-1} \tag{6.5.25}$$

By use of a matrix lemma (Appendix C) we can write

$$\begin{aligned}
\boldsymbol{\Phi}^-(s) &= \left[\mathbf{V}_{\mathbf{v}}^{1/2} + \mathbf{V}_{\mathbf{v}}^{-1/2}\mathbf{C}\mathbf{V}_{\tilde{\mathbf{x}}}\boldsymbol{\Omega}^T(-s)\mathbf{C}^T\right]^{-1} \\
&= \mathbf{V}_{\mathbf{v}}^{-1/2} - \mathbf{V}_{\mathbf{v}}^{-1}\mathbf{C}\mathbf{V}_{\tilde{\mathbf{x}}}\left(\mathbf{I} + \boldsymbol{\Omega}^T(-s)\mathbf{C}^T\mathbf{V}_{\mathbf{v}}^{-1}\mathbf{C}\mathbf{V}_{\tilde{\mathbf{x}}}\right)^{-1}\boldsymbol{\Omega}^T(-s)\mathbf{C}^T\mathbf{V}_{\mathbf{v}}^{-1/2} \\
&= \mathbf{V}_{\mathbf{v}}^{-1/2} - \mathbf{K}^T\left(-s\mathbf{I} - \mathbf{A} + \mathbf{C}^T\mathbf{K}^T\right)^{-1}\mathbf{C}^T\mathbf{V}_{\mathbf{v}}^{-1/2}
\end{aligned} \tag{6.5.26}$$

where we have used Eq. (6.5.7).

Substituting in Eq. (6.5.25) then yields

$$\begin{aligned}
\mathbf{G}_{fW}(s) = &\left[\boldsymbol{\Phi}^+(s) - \mathbf{V}_{\mathbf{v}}^{1/2} - \mathbf{V}_{\mathbf{v}}\mathbf{K}^T\left(-s\mathbf{I} - \mathbf{A} + \mathbf{C}^T\mathbf{K}^T\right)^{-1}\mathbf{C}^T\mathbf{V}_{\mathbf{v}}^{-1/2}\right]_+ \\
&\times\left[\boldsymbol{\Phi}^+(s)\right]^{-1}
\end{aligned} \tag{6.5.27}$$

From Eq. (6.5.15) we see that the first two terms in the braces on the right-hand side of Eq. (6.5.27) are together equal to $\mathbf{C}\boldsymbol{\Omega}(s)\mathbf{K}\mathbf{V}_{\mathbf{v}}^{1/2}$. Since the second term in the braces corresponds to a noncausal function (it has poles in the right half-plane), Eq. (6.5.27) reduces to

$$\mathbf{G}_{fW}(s) = \mathbf{C}\boldsymbol{\Omega}(s)\mathbf{K}\mathbf{V}_{\mathbf{v}}^{1/2}\left[\mathbf{V}_{\mathbf{v}}^{1/2} + \mathbf{C}\boldsymbol{\Omega}(s)\mathbf{K}\mathbf{V}_{\mathbf{v}}^{1/2}\right]^{-1} \tag{6.5.28}$$

Using Eq. (6.5.24) then yields

$$\begin{aligned}
\mathbf{G}_{fW}(s) &= \mathbf{C}\boldsymbol{\Omega}(s)\mathbf{K}\mathbf{V}_{\mathbf{v}}^{1/2}\left[\mathbf{V}_{\mathbf{v}}^{-1/2} - \mathbf{V}_{\mathbf{v}}^{-1/2}\mathbf{C}(s\mathbf{I} - \mathbf{A} + \mathbf{K}\mathbf{C})^{-1}\mathbf{K}\right] \\
&= \mathbf{C}\boldsymbol{\Omega}(s)\left[\mathbf{I} - \mathbf{K}\mathbf{C}(s\mathbf{I} - \mathbf{A} + \mathbf{K}\mathbf{C})^{-1}\right]\mathbf{K} \\
&= \mathbf{C}\left[s\mathbf{I} - \mathbf{A} + \mathbf{K}\mathbf{C}\right]^{-1}\mathbf{K}
\end{aligned} \tag{6.5.29}$$

Comparison with Eq. (6.5.9) shows that the Wiener filter is identical with the Kalman filter.

6.6 NONLINEAR ESTIMATION

We now present a brief discussion of estimation in certain nonlinear signal models. We consider two problems. First, we present algorithms for the approximate estimation of the signals in a nonlinear state-space model using extensions of the Kalman filter algorithms. In the second type of problem, the transmitted signal is a function of the waveform to be estimated. We discuss the use of the MAP estimator when the transmitted signal is observed in Gaussian noise.

Nonlinear State Models

We consider the following message and observation models:

Message model: $\dot{\mathbf{x}}(t) = \mathbf{f}[\mathbf{x}(t), t] + \mathbf{g}[\mathbf{x}(t), t]\mathbf{w}(t)$

$$\mathbf{y}(t) = \mathbf{h}[\mathbf{x}(t), t] \tag{6.6.1}$$

Observation model: $\mathbf{z}(t) = \mathbf{y}(t) + \mathbf{v}(t)$ for $t \geqslant t_0$ **(6.6.2)**

The statistics of the processes $\mathbf{w}(t)$ and $\mathbf{v}(t)$ and the initial state $\mathbf{x}(t_0)$ are assumed to be

$$E\{\mathbf{x}(t_0)\} = \boldsymbol{\mu}_{\mathbf{x}}(t_0) \qquad E\{\mathbf{w}(t)\} = E\{\mathbf{v}(t)\} = \mathbf{0}$$

$$\text{cov}\{\mathbf{w}(t), \mathbf{w}(\tau)\} = \mathbf{V}_{\mathbf{w}}(t)\delta(t-\tau), \ \text{cov}\{\mathbf{v}(t), \mathbf{v}(\tau)\} = \mathbf{V}_{\mathbf{v}}(t)\delta(t-\tau)$$

$$\text{cov}\{\mathbf{w}(t), \mathbf{v}(\tau)\} = \text{cov}\{\mathbf{w}(t), \mathbf{x}(t_0)\} = \text{cov}\{\mathbf{v}(t), \mathbf{x}(t_0)\} = \mathbf{0} \tag{6.6.3}$$

For mean square estimation, the estimate based on the set of observations over the interval $[t_0, t_f]$ can be obtained as the mean of the conditional density functions of $\mathbf{x}(t)$ or $\mathbf{y}(t)$. Unfortunately, in most cases, the determination of these density functions is a formidable task. The conditional density functions are usually obtained as the solutions to certain partial differential equations. These equations have been derived by Stratanovich [18] and by Kushner [19,20] for the case when all the processes involved are Gaussian and by Snyder [21] for Poisson processes. The solution of these equations, however, is not straightforward. While a representation of the estimator using the innovations approach has been obtained in [22], the determination of the actual filter is still difficult in most cases. In most applications, we are therefore constrained to seek an approximate solution to our estimation problem. In this section, we discuss two approximate schemes which have a simple structure and have been widely applied.

Linearized Kalman Filter

A common technique for solving nonlinear problems is to linearize around a known (nominal) trajectory. The perturbation equations so obtained will be linear and hence an approximate solution to the nonlinear problem can be easily obtained as the nominal value plus the perturbation term.

Let us denote by $\mathbf{x}_n(t)$ the solution to Eq. (6.6.1) with $\mathbf{w}(t)=0$. That is

$$\dot{\mathbf{x}}_n(t)=\mathbf{f}(\mathbf{x}_n(t), t), \quad \mathbf{x}_n(t_0)=\boldsymbol{\mu}_{\mathbf{x}}(t_0) \tag{6.6.4}$$

Let

$$\mathbf{y}_n(t)=\mathbf{h}[\mathbf{x}_n(t), t] \tag{6.6.5}$$

and

$$\delta\mathbf{z}(t)=\mathbf{z}(t)-\mathbf{y}_n(t) \tag{6.6.6}$$

If we write $\mathbf{x}(t)$ as $\mathbf{x}_n(t)+\delta\mathbf{x}(t)$, it can easily be verified that $\delta\mathbf{x}(t)$ satisfies the equation

$$\begin{aligned}\delta\dot{\mathbf{x}}(t)=&\,\mathbf{f}(\mathbf{x}_n(t)+\delta\mathbf{x}(t), t)-\mathbf{f}(\mathbf{x}_n(t), t)\\ &+\mathbf{g}[\mathbf{x}_n(t)+\delta\mathbf{x}(t), t]\mathbf{w}(t)\end{aligned}$$

Assuming that $\mathbf{f}(\cdot)$ and $\mathbf{g}(\cdot)$ are differentiable in their arguments, we can expand $\mathbf{f}(\cdot,\cdot)$ and $\mathbf{g}(\cdot,\cdot)$ in Taylor's series about $\mathbf{x}_n(t)$. If we assume that $\delta\mathbf{x}(t)$ is small enough, we can neglect higher-order terms and write

$$\delta\dot{\mathbf{x}}(t)\approx\frac{\partial\mathbf{f}[\mathbf{x}_n(t), t]}{\partial\mathbf{x}_n(t)}\delta\mathbf{x}(t)+\mathbf{g}[\mathbf{x}_n(t), t]\mathbf{w}(t) \tag{6.6.7}$$

Similarly substituting Eqs. (6.6.2) and (6.6.5) in Eq. (6.6.6), expanding $\mathbf{h}(\mathbf{x}(t), t)$ in a Taylor's series about $\mathbf{x}_n(t)$ and neglecting higher-order terms yields

$$\delta\mathbf{z}(t)\simeq\frac{\partial\mathbf{h}[\mathbf{x}_n(t), t]}{\partial\mathbf{x}_n(t)}\delta\mathbf{x}(t)+\mathbf{v}(t) \tag{6.6.8}$$

Equations (6.6.7) and (6.6.8) are now in a form where the linear Kalman filter algorithms can be used to obtain an estimate of $\delta\mathbf{x}(t)$. From Table 6.4.2, we can write

$$\delta\dot{\hat{\mathbf{x}}}(t|t)=\frac{\partial\mathbf{f}[\mathbf{x}_n(t), t]}{\partial\mathbf{x}_n(t)}\delta\hat{\mathbf{x}}(t|t)+\mathbf{K}(t)\left[\delta\mathbf{z}(t)-\frac{\partial\mathbf{h}[\mathbf{x}_n(t), t]}{\partial\mathbf{x}_n(t)}\delta\hat{\mathbf{x}}(t|t)\right] \tag{6.6.9}$$

where

$$\mathbf{K}(t) = \mathbf{V}_{\delta\tilde{\mathbf{x}}}(t|t)\frac{\partial \mathbf{h}^T[\mathbf{x}_n(t), t]}{\partial \mathbf{x}_n(t)}\mathbf{V}_{\mathbf{v}}^{-1}(t) \tag{6.6.10}$$

and

$$\begin{aligned}\frac{d}{dt}\mathbf{V}_{\delta\tilde{\mathbf{x}}}(t|t) = {}& \frac{\partial \mathbf{f}[\mathbf{x}_n(t), t]}{\partial \mathbf{x}_n(t)}\mathbf{V}_{\delta\tilde{\mathbf{x}}}(t|t) + \mathbf{V}_{\delta\tilde{\mathbf{x}}}(t|t)\frac{\partial \mathbf{f}^T[\mathbf{x}_n(t), t]}{\partial \mathbf{x}_n(t)} \\ &- \mathbf{V}_{\delta\tilde{\mathbf{x}}}(t|t)\frac{\partial \mathbf{h}^T[\mathbf{x}_n(t), t]}{\partial \mathbf{x}_n(t)}\mathbf{V}_{\mathbf{v}}^{-1}\frac{\partial \mathbf{h}[\mathbf{x}_n(t), t]}{\partial \mathbf{x}_n(t)}\mathbf{V}_{\delta\tilde{\mathbf{x}}}(t|t) \\ &+ \mathbf{g}[\mathbf{x}_n(t), t]\mathbf{V}_{\mathbf{w}}(t)\mathbf{g}^T[\mathbf{x}_n(t), t]\end{aligned} \tag{6.6.11}$$

where

$$\delta\tilde{\mathbf{x}}(t|t) = \delta\mathbf{x}(t) - \delta\hat{\mathbf{x}}(t|t)$$

and

$$\mathbf{V}_{\delta\tilde{\mathbf{x}}}(t|t) = \operatorname{cov}[\delta\tilde{\mathbf{x}}(t|t), \delta\tilde{\mathbf{x}}(t|t)]$$

These equations are known as the linearized Kalman algorithms. The estimate $\hat{\mathbf{x}}(t|t)$ can be determined as

$$\hat{\mathbf{x}}(t|t) = \mathbf{x}_n(t) + \delta\hat{\mathbf{x}}(t|t)$$

Extended Kalman Filter Algorithm

We can now write an expression for $\hat{x}(t|t)$ by adding Eqs. (6.6.4) and (6.6.9). We then have

$$\begin{aligned}\dot{\hat{\mathbf{x}}}(t|t) &= \dot{\mathbf{x}}_n(t) + \delta\dot{\hat{\mathbf{x}}}(t|t) \\ &= \mathbf{f}(\mathbf{x}_n(t), t) + \frac{\partial \mathbf{f}[\mathbf{x}_n(t), t]}{\partial \mathbf{x}_n(t)}\delta\hat{\mathbf{x}}(t|t) \\ &\quad + \mathbf{K}(t)\left[\mathbf{z}(t) - \mathbf{y}_n(t) - \frac{\partial \mathbf{h}[\mathbf{x}_n(t), t]}{\partial \mathbf{x}_n(t)}\delta\hat{\mathbf{x}}(t|t)\right]\end{aligned} \tag{6.6.12}$$

where we have used Eq. (6.6.6).

By noting that

$$\mathbf{f}(\hat{\mathbf{x}}(t|t), t) \simeq \mathbf{f}(\mathbf{x}_n(t), t) + \frac{\partial \mathbf{f}[\mathbf{x}_n(t), t]}{\partial \mathbf{x}_n(t)}\delta\hat{\mathbf{x}}(t|t)$$

and

$$\mathbf{h}(\hat{\mathbf{x}}(t|t), t)) \simeq \mathbf{h}(\mathbf{x}_n(t), t) + \frac{\partial \mathbf{h}[\mathbf{x}_n(t), t]}{\partial \mathbf{x}_n(t)} \delta\hat{\mathbf{x}}(t|t)$$

we can write Eq. (6.6.12) as

$$\dot{\hat{\mathbf{x}}}(t|t) = \mathbf{f}(\hat{\mathbf{x}}(t|t), t) + \mathbf{K}(t)[\mathbf{z}(t) - \mathbf{h}(\hat{\mathbf{x}}(t|t)] \tag{6.6.13}$$

Since

$$\begin{aligned}\tilde{\mathbf{x}}(t|t) = \mathbf{x}(t) - \hat{\mathbf{x}}(t|t) &= [\mathbf{x}_n(t) + \delta\mathbf{x}(t)] - [\mathbf{x}_n(t) + \delta\hat{\mathbf{x}}(t|t)] \\ &= \delta\tilde{\mathbf{x}}(t|t)\end{aligned} \tag{6.6.14}$$

the gain and variance equations (6.6.10) and (6.6.11) can also be written in terms of $\tilde{\mathbf{x}}(t|t)$ as follows:

$$\mathbf{K}(t) = \mathbf{V}_{\tilde{\mathbf{x}}}(t|t) \frac{\partial \mathbf{h}^T[\hat{\mathbf{x}}(t|t), t]}{\partial \hat{\mathbf{x}}(t|t)} \mathbf{V}_{\mathbf{v}}^{-1}(t) \tag{6.6.15}$$

and

$$\begin{aligned}\dot{\mathbf{V}}_{\tilde{\mathbf{x}}}(t|t) = &\frac{\partial \mathbf{f}[\hat{\mathbf{x}}(t|t), t]}{\partial \hat{\mathbf{x}}(t|t)} \mathbf{V}_{\tilde{\mathbf{x}}}(t|t) + \mathbf{V}_{\tilde{\mathbf{x}}}(t|t) \frac{\partial \mathbf{f}^T[\hat{x}(t|t), t]}{\partial \hat{x}(t|t)} \\ &- \mathbf{V}_{\tilde{\mathbf{x}}}(t|t) \frac{\partial \mathbf{h}^T[\hat{\mathbf{x}}(t|t), t]}{\partial \hat{\mathbf{x}}(t|t)} \mathbf{V}_{\mathbf{v}}^{-1}(t) \frac{\partial \mathbf{h}[\hat{\mathbf{x}}(t|t), t]}{\partial \hat{\mathbf{x}}(t|t)} \mathbf{V}_{\tilde{\mathbf{x}}}(t|t) \\ &+ \mathbf{g}[\hat{\mathbf{x}}(t|t), t] \mathbf{V}_{\mathbf{w}}(t) \mathbf{g}^T[\hat{\mathbf{x}}(t|t), t]\end{aligned} \tag{6.6.16}$$

where we have used the fact that to first order

$$\frac{\partial \mathbf{f}[\mathbf{x}_n(t), t]}{\partial \mathbf{x}_n(t)} \simeq \frac{\partial \mathbf{f}[\hat{\mathbf{x}}(t|t), t]}{\partial \hat{\mathbf{x}}(t|t)}$$

and similarly for

$$\frac{\partial \mathbf{h}(\hat{\mathbf{x}}(t|t), t)}{\partial \hat{\mathbf{x}}(t|t)}$$

Equations (6.6.13), (6.6.15), and (6.6.16) constitute a complete set of approximate algorithms for evaluating the estimate $\hat{\mathbf{x}}(t|t)$. These algorithms are known as the extended Kalman algorithms and are summarized in

TABLE 6.6.1 EXTENDED KALMAN FILTER ALGORITHM: CONTINUOUS TIME

Message Model:

$$\dot{\mathbf{x}}(t) = \mathbf{f}[\mathbf{x}(t), t] + \mathbf{g}[\mathbf{x}(t), t]\mathbf{w}(t)$$

$$\mathbf{y}(t) = \mathbf{h}[\mathbf{x}(t), t] \tag{6.6.1}$$

Observations:

$$\mathbf{z}(t) = \mathbf{y}(t) + \mathbf{v}(t) \tag{6.6.2}$$

Prior Statistics:

$$\mathrm{E}\{\mathbf{w}(t)\} = \mathrm{E}\{\mathbf{v}(t)\} = \mathbf{0}$$

$$\operatorname{cov}\{\mathbf{w}(t), \mathbf{w}(\tau)\} = \mathbf{V}_{\mathbf{w}}(t)\delta(t-\tau)$$

$$\operatorname{cov}\{\mathbf{v}(t), \mathbf{v}(\tau)\} = \mathbf{V}_{\mathbf{v}}(t)\delta(t-\tau)$$

$$\operatorname{cov}\{\mathbf{w}(t), \mathbf{v}(\tau)\} = \operatorname{cov}\{\mathbf{w}(t), \mathbf{x}(t_0)\} = \operatorname{cov}\{\mathbf{v}(t), \mathbf{x}(t_0)\} = \mathbf{0} \tag{6.6.3}$$

Filter Algorithm:

$$\dot{\hat{\mathbf{x}}}(t|t) = \mathbf{f}(\hat{\mathbf{x}}(t|t), t) + \mathbf{K}(t)[\mathbf{z}(t) - \mathbf{h}(\hat{\mathbf{x}}(t|t), t) \tag{6.6.13}$$

Filter Gain:

$$\mathbf{K}(t) = \mathbf{V}_{\tilde{\mathbf{x}}}(t|t)\frac{\partial \mathbf{h}^T[\hat{\mathbf{x}}(t|t), t]}{\partial \hat{\mathbf{x}}(t|t)}\mathbf{V}_{\mathbf{v}}^{-1}(t) \tag{6.6.15}$$

Error Variance:

$$\begin{aligned}\dot{\mathbf{V}}_{\tilde{\mathbf{x}}}(t|t) = {}& \frac{\partial \mathbf{f}[\hat{\mathbf{x}}(t|t), t]}{\partial \hat{\mathbf{x}}(t|t)}\mathbf{V}_{\tilde{\mathbf{x}}}(t|t) + \mathbf{V}_{\tilde{\mathbf{x}}}(t|t)\frac{\partial \mathbf{f}^T[x(t|t), t]}{\partial \hat{x}(t|t)} \\ & - \mathbf{V}_{\tilde{\mathbf{x}}}(t|t)\frac{\partial \mathbf{h}^T[\hat{\mathbf{x}}(t|t), t]}{\partial \hat{\mathbf{x}}(t|t)}\mathbf{V}_{\mathbf{v}}^{-1}(t)\frac{\partial \mathbf{h}[\hat{\mathbf{x}}(t|t), t]}{\partial \hat{\mathbf{x}}(t|t)}\mathbf{V}_{\tilde{\mathbf{x}}}(t|t) \\ & + \mathbf{g}[\hat{\mathbf{x}}(t|t), t]\mathbf{V}_{\mathbf{w}}(t)\mathbf{g}^T[\hat{\mathbf{x}}(t|t), t]\end{aligned} \tag{6.6.16}$$

Initial Conditions:

$$\hat{\mathbf{x}}(t_0) = \mathrm{E}\{\mathbf{x}(t_0)\} = \boldsymbol{\mu}_{\mathbf{x}}(t_0) \text{ (known)}$$

$$\mathbf{V}_{\tilde{\mathbf{x}}}(t_0) = \mathbf{V}_{\tilde{\mathbf{x}}_0} \text{ (known)}$$

TABLE 6.6.2 EXTENDED KALMAN FILTER ALGORITHMS: DISCRETE TIME

Message Model:

$$\mathbf{x}(k+1) = \boldsymbol{\phi}[x(k), k] + \boldsymbol{\Gamma}[\mathbf{x}(k), k]\mathbf{w}(k)$$
$$\mathbf{y}(k) = \mathbf{h}[\mathbf{x}(k), k] \qquad \textbf{(6.6.17)}$$

Observations:

$$\mathbf{z}(k) = \mathbf{h}[\mathbf{x}(k), k] + \mathbf{v}(k) \qquad \textbf{(6.6.18)}$$

Prior Statistics:

$$\mathrm{E}\{\mathbf{w}(k)\} = \mathrm{E}\{\mathbf{v}(k)\} = 0$$
$$\mathrm{cov}\,\{\mathbf{w}(k), \mathbf{w}(j)\} = \mathbf{V}_{\mathbf{w}}(k)\delta_{kj}$$
$$\mathrm{cov}\,\{\mathbf{v}(k), \mathbf{v}(j)\} = \mathbf{V}_{\mathbf{v}}(k)\delta_{kj}$$
$$\mathrm{cov}\,\{\mathbf{w}(k), \mathbf{v}(j)\} = \mathrm{cov}\,\{\mathbf{w}(k), \mathbf{x}(k_0)\} = \mathrm{cov}\,\{\mathbf{v}(k), \mathbf{x}(k_0)\} = 0 \qquad \textbf{(6.6.19)}$$

Filter Algorithm:

$$\hat{x}(k+1|k+1) = \hat{\mathbf{x}}(k+1|k) + \mathbf{K}(k+1)\{\mathbf{z}(k+1) - \mathbf{h}[\hat{\mathbf{x}}(k+1|k), k+1]\} \qquad \textbf{(6.6.20)}$$

One-Stage Prediction:

$$\hat{\mathbf{x}}(k+1|k) = \boldsymbol{\phi}[\hat{\mathbf{x}}(k|k), k] \qquad \textbf{(6.6.21)}$$

Filter Gain:

$$\mathbf{K}(k+1) = \mathbf{V}_{\tilde{\mathbf{x}}}(k+1|k+1)\frac{\partial \mathbf{h}^T[\hat{\mathbf{x}}(k+1|k), k+1)}{\partial \hat{\mathbf{x}}(k+1|k)}\mathbf{V}_{\mathbf{v}}^{-1}(k+1) \qquad \textbf{(6.6.22)}$$

Error Variance:

$$\mathbf{V}_{\tilde{\mathbf{x}}}(k+1|k) = \frac{\partial \boldsymbol{\phi}[\hat{\mathbf{x}}(k|k), k]}{\partial \hat{\mathbf{x}}(k|k)}\mathbf{V}_{\tilde{\mathbf{x}}}(k|k)\frac{\partial \boldsymbol{\phi}^T[\hat{\mathbf{x}}(k|k), k]}{\partial \hat{\mathbf{x}}(k|k)} + \boldsymbol{\Gamma}[\hat{\mathbf{x}}(k|k)k]\mathbf{V}_{\mathbf{w}}(k)\boldsymbol{\Gamma}^T[\hat{\mathbf{x}}(k|k), k] \qquad \textbf{(6.6.23)}$$

$$\mathbf{V}_{\tilde{\mathbf{x}}}(k+1|k+1) = \mathbf{V}_{\tilde{\mathbf{x}}}(k+1|k) - \mathbf{V}_{\tilde{\mathbf{x}}}(k+1|k)\frac{\partial \mathbf{h}^T[\hat{\mathbf{x}}(k+1|k), k+1]}{\partial \hat{\mathbf{x}}(k+1|k)}$$
$$\cdot \left\{\frac{\partial \mathbf{h}[\hat{\mathbf{x}}(k+1|k), k+1]}{\partial \hat{\mathbf{x}}(k+1|k)}\mathbf{V}_{\tilde{\mathbf{x}}}(k+1|k)\frac{\partial \mathbf{h}^T[\hat{\mathbf{x}}(k+1|k), k+1]}{\partial \hat{\mathbf{x}}(k+1|k)} + \mathbf{V}_{\mathbf{v}}(k+1)\right\}^{-1}$$
$$\cdot \frac{\partial \mathbf{h}[\hat{\mathbf{x}}(k+1|k), k+1]}{\partial \hat{\mathbf{x}}(k+1|k)}\mathbf{V}_{\tilde{\mathbf{x}}}(k+1|k) \qquad \textbf{(6.6.24)}$$

Initial Conditions:

$$\hat{\mathbf{x}}(k_0) = \mathrm{E}\{\mathbf{x}(k_0)\} = \boldsymbol{\mu}_{\mathbf{x}}(k_0) \text{ (known)}$$
$$\mathbf{V}_{\tilde{\mathbf{x}}}(k_0) = \mathbf{V}_{\mathbf{x}_0} \text{ (known)} \qquad \textbf{(6.6.25)}$$

Table 6.6.1. The algorithms can be thought of as being obtained from the linearized algorithms by choosing $\hat{\mathbf{x}}(t|t)$ as the nominal trajectory. They also represent a first-order approximate solution to the partial differential equations for the conditional density corresponding to this estimation problem [3,4,6].

In our derivation of the linearized algorithms, we assumed that the perturbation $\delta\mathbf{x}(t)$ around the nominal trajectory $\mathbf{x}_n(t)$ was small. In certain applications, this is not a valid assumption. The linearized algorithms perform poorly under these conditions especially when the variance of the input process $\mathbf{V}_\mathbf{w}(t)$ is high. The extended Kalman algorithms yield fairly good estimates when the signal-to-noise ratio is large, and are easy to implement. Higher-order approximate algorithms have been obtained in the literature which yield better estimates, but at the expense of added complexity in the estimator structure. The interested reader is referred to the various excellent texts on estimation theory [3,4,6] for these algorithms (see also [23]).

The extended Kalman filter algorithms for discrete-time models can be derived along the same lines. The algorithms are summarized in Table 6.6.2.

Example 6.6.1 We consider the estimation of a modulated signal observed in noise. The message and observation models are assumed to be

$$x(k+1) = .8x(k) + w(k)$$
$$y(k) = x(k)$$
$$z(k) = A\cos(\omega_0 k + 0.5y(k)) + v(k)$$

The problem, therefore, is one of phase modulation. The processes $w(k)$ and $v(k)$ are zero mean, stationary with unity variance. The estimator equations as obtained from Table 6.6.2 are given by

$$\hat{x}(k+1|k+1) = \hat{x}(k+1|k) + K(k+1)\{z(k+1) - A\cos[\omega_0 k + 0.5\hat{x}(k+1|k)]\}$$
$$\hat{x}(k+1|k) = 0.8\hat{x}(k|k)$$
$$K(k+1) = .5AV_{\tilde{x}}(k+1|k+1)\sin[\omega_0(k+1) + 0.5\hat{x}(k+1|k)]$$
$$V_{\tilde{x}}(k+1|k) = .64V_{\tilde{x}}(k|k) + 1$$
$$V_{\mathbf{x}}(k+1|k+1) = V_{\tilde{x}}(k+1|k) - K^2(k+1)\{.25A^2V_{\tilde{x}}(k+1|k) \times \sin^2[\omega_0(k+1) + .5\hat{x}(k+1|k)] + 1\}^{-1}$$

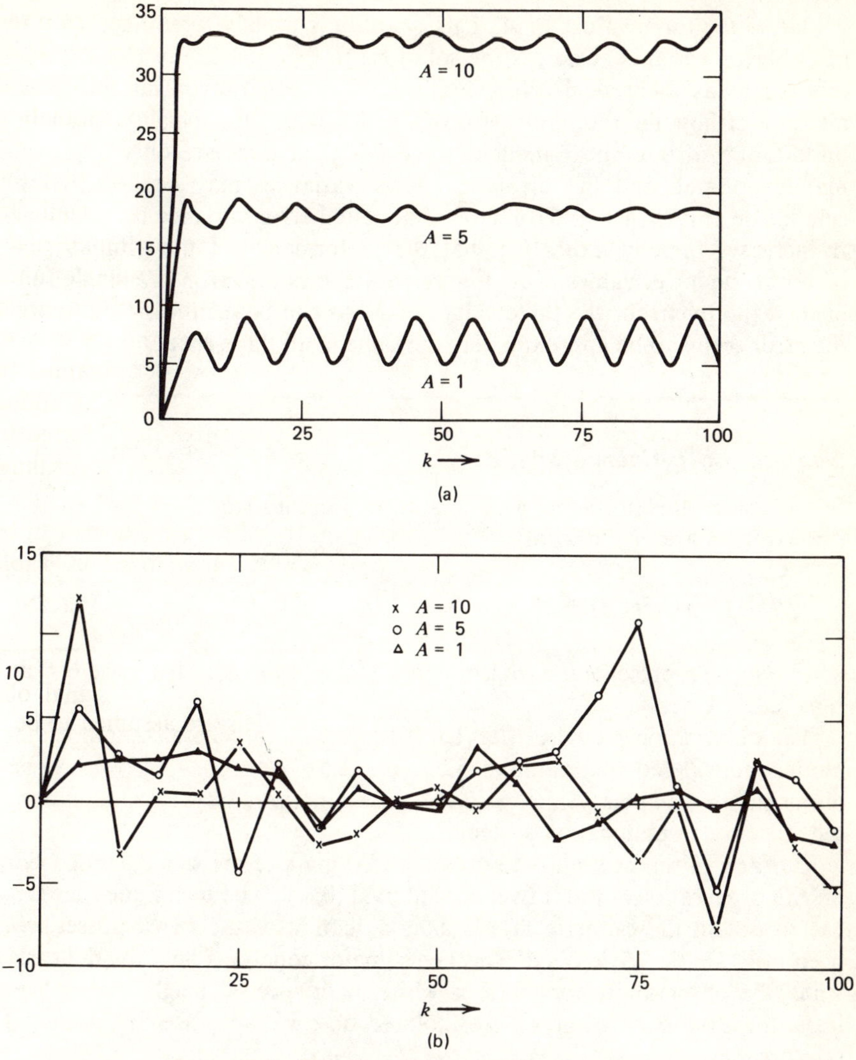

Fig. 6.6.1 Extended Kalman filter example. (a) Actual error variance for three values of A. (b) Error corresponding to a single run.

The system was simulated on a digital computer and the mean square error obtained by averaging the results of 5000 different runs of the example. The results are shown in Figure 6.6.1, where the quantity

$$\frac{1}{5000}\sum_{i=1}^{5000}\left[\tilde{x}^2(k|k)\right]_i$$

is plotted for three values of A. This quantity is usually referred to as the actual error variance, whereas the solutions to Eqs. (6.6.23) or (6.6.24) are referred to as theoretical error variances. The actual error variance is a measure of how the estimator performs with actual data. This is especially important with nonlinear models, since the algorithms are only approximately optimal, and the theoretical error variances may be deceptively small. The curves in Fig. 6.6.1 show that actual error variance increases as A increases thereby indicating that the performance of the estimator is better at the lower values of A. Figure 6.6.1 shows the error in a single run of the experiment for the three values of A. As can be seen from the figure the error seems to be approximately the same in all three cases.

Nonlinear MAP Estimation [2,24]

We consider the problem of estimating a random signal $x(t)$ when the observations are of the form

$$z(t) = y(x(t), t) + v(t) \qquad 0 \leqslant t \leqslant T \tag{6.6.26}$$

where $y(\cdot, \cdot)$ represents the transmitted signal and the additive noise $v(t)$ is zero mean, Gaussian.

The observation model of Eq. (6.6.16) represents a generalization of the models considered in Examples 5.5.2 and 5.5.3 (see Eq. (5.5.18)). As is discussed in Chapter 7, Eq. (6.6.25) arises in the study of modulation schemes in communication systems.

Our aim in this section is to obtain an estimate of the signal $x(t)$ based on the observations of $z(t)$ over the interval $[0, T]$. The technique that we use to obtain the estimate $\hat{x}(t)$ is a modification of the MAP procedure used in Example 5.5.3. We derive the estimator equations only for the case when the observation noise $v(t)$ is white, with spectral height $N_0/2$. We leave the extension to the case of colored observation noise to the interested reader.

As in Example (5.5.2), the first step in deriving the estimator equations is to sample the observation $z(t)$ so as to obtain N samples in the interval $[0, T]$. The sampled observations at instant t_k are given by

$$z_k = y_k + v_k \tag{6.6.27}$$

where

$$z_k = z(t_k), y_k = y(x_k, t_k), v_k = v(t_k)$$

and

$$x(t_k) = x_k \tag{6.6.28}$$

Let

$$\mathbf{z}_N = [z_1 z_2 \ldots z_N]^T, \mathbf{y}_N = [y_1 y_2 \ldots y_N]^T$$

and

$$\mathbf{x}_N = [x_1 x_2 \ldots x_N]^T$$

We will initially assume that $v(t)$ is lowpass with band width 2Ω. Following an argument similar to the one used in Example 5.5.2, the conditional density $p(\mathbf{z}_N|\mathbf{x}_N)$ follows from Eq. (5.5.21) as

$$p(\mathbf{z}_N|\mathbf{x}_N) = \left(\frac{1}{2\pi\sigma^2}\right)^{N/2} \exp\left\{-\frac{1}{2\sigma^2}\sum_{k=1}^{N}(z_k - y_k)^2\right\} \tag{6.6.29}$$

where

$$\sigma^2 = \frac{N_0\Omega}{2\pi}$$

In order to obtain the joint density function $p(\mathbf{z}_N, \mathbf{x}_N)$ required for obtaining the MAP estimate of $\mathbf{x}_N$, we assume that the samples $\{x_i\}$ are zero-mean and Gaussian distributed so that

$$p(\mathbf{x}_N) = \frac{1}{(2\pi)^{N/2}\det \mathbf{V}_{\mathbf{x}_N}} \exp\left\{-\frac{1}{2}\mathbf{x}_N^T \mathbf{V}_{\mathbf{x}_N}^{-1}\mathbf{x}_N\right\} \tag{6.6.30}$$

where

$$\mathbf{V}_{\mathbf{x}_N} = \mathrm{E}\{\mathbf{x}_N \mathbf{x}_N^T\} \tag{6.6.31}$$

We note that the (i, j)th element of $\mathbf{V}_{\mathbf{x}_N}$ is given by

$$(\mathbf{V}_{\mathbf{x}_N})_{ij} = \mathrm{E}\{x_i x_j\} = \mathrm{E}\{x(t_i)x(t_j)\} = V_x(t_i, t_j) \tag{6.6.32}$$

Combining Eqs. (6.6.29) and (6.6.30) yields for the joint density function

$$\begin{aligned} p(\mathbf{z}_N, \mathbf{x}_N) &= p(\mathbf{z}_N|\mathbf{x}_N)p(\mathbf{x}_N) \\ &= \frac{1}{(2\pi\sigma)^N \det V_{\mathbf{x}_N}} \exp\left\{-\frac{1}{2\sigma^2}[\mathbf{z}_N - \mathbf{y}_N]^T[\mathbf{z}_N - \mathbf{y}_N]\right. \\ &\quad \left. - \frac{1}{2}\mathbf{x}_N^T \mathbf{V}_{\mathbf{x}_N}^{-1}\mathbf{x}_N\right\} \end{aligned} \tag{6.6.33}$$

The MAP equation is obtained by partial differentiation of Eq. (6.6.23) with respect to $\mathbf{x}_N$ and by setting the result equal to zero. Noting that $\mathbf{y}_N$ is explicitly a function of $\mathbf{x}_N$, we obtain

$$\frac{1}{\sigma^2}\frac{\partial^T \mathbf{y}_N}{\partial \mathbf{x}_N}[\mathbf{z}_N - \mathbf{y}_N] - \mathbf{V}_{\mathbf{x}_N}^{-1}\mathbf{x}_N\Bigg|_{\mathbf{x}_N=(\hat{\mathbf{x}}_N)_{\mathrm{MAP}}} = 0 \tag{6.6.34}$$

Solving for $\mathbf{x}_N$ yields

$$(\hat{\mathbf{x}}_N)_{\mathrm{MAP}} = \frac{1}{\sigma^2}\mathbf{V}_{\mathbf{x}_N}\frac{\partial^T \mathbf{y}_N}{\partial \mathbf{x}_N}[\mathbf{z}_N - \mathbf{y}_N]\Bigg|_{\mathbf{x}_N=(\hat{\mathbf{x}}_N)_{\mathrm{MAP}}} \tag{6.6.35}$$

The Jacobian matrix $\partial \mathbf{y}_N / \partial \mathbf{x}_N$ can easily be shown to be a diagonal matrix

$$\frac{\partial \mathbf{y}_N}{\partial \mathbf{z}_N} = \begin{bmatrix} \frac{\partial y_1}{\partial x_1} & & & 0 \\ & \frac{\partial y_2}{\partial x_2} & & \\ & & \ddots & \\ 0 & & & \frac{\partial y_N}{\partial x_N} \end{bmatrix} \tag{6.6.36}$$

so that Eq. (6.6.35) can be written in component form as

$$(\hat{x}_k)_{\mathrm{MAP}} = \frac{1}{\sigma^2}\sum_{j=1}^{N} V_x(t_k, t_j)\frac{\partial y_j}{\partial x_j}(z_j - y_j)\Bigg|_{x_k=(\hat{x}_k)_{\mathrm{MAP}}} \tag{6.6.37}$$

We now let the number of samples, N, become infinity. In the limit the summation becomes an integral and the variance σ^2 becomes an impulse of strength $N_0/2$ (see Example 5.5.2). We therefore get the MAP equation as

$$\hat{x}_{\mathrm{MAP}}(t) = \frac{2}{N_0}\int_0^T V_x(t,\tau)\frac{\partial y(x(\tau),\tau)}{\partial x(\tau)}[z(\tau) - y(\tau)]\, d\tau\Bigg|_{\hat{x}_{\mathrm{MAP}}(\cdot)} \tag{6.6.38}$$

We discuss the solution of Eq. (6.6.38) for various special cases in Chapter 7. We point out that the estimate $\hat{x}_{\mathrm{MAP}}(\cdot)$ in Eq. (6.6.38) refers to the smoothed estimate based on the record of observations over the entire interval $[0, T]$.

6.7 SUMMARY

In this chapter, we have considered the estimation of stochastic processes observed in additive noise. We have for the most part restricted ourselves to linear mean square estimates of these processes. If the processes involved are all Gaussian, the linear estimate is the optimum estimate.

In Section 6.2, the basic concepts of linear mean square estimation of signal waveforms were presented. For continuous observations over an interval $[t_0, t_f]$, the optimum estimator is a linear time-varying system. The orthogonality principle was used in Section 6.3 to determine the optimum system. It was shown that the optimum system satisfies the Wiener–Hopf integral equation. As in Chapter 4, the solution to this equation is easily obtained if the observation noise is white. If the observation noise is colored, we can use a whitening filter to first convert the observation noise into white noise. The optimum filter operates on the whitened observations to yield the desired estimate. For stationary processes with rational spectra, the whitening filter can be determined by spectral factorization of the spectrum of the observations. These concepts extend to the case of discrete-time processes and multi-input/multi-output processes. However, in the latter case, the spectral factorization is not trivial.

Section 6.4 considered the important problem of estimation in linear state models using the Kalman filter. This algorithmic approach to the estimation problem was derived using the concept of innovations. Both discrete-time and continuous-time models were considered. The relation between the Kalman filter and the Wiener filter was explored in Section 6.5.

Nonlinear state models were considered in Section 6.6. In this case, the signal process will not be Gaussian and the optimum estimator is no longer linear. Since determination of the optimum estimator is not easy, two approximate methods for estimation were discussed. These algorithms were essentially obtained by linearizing about certain nominal trajectories and can be considered as first order approximations to the optimal estimator.

The chapter concluded with a derivation of the MAP estimator for certain signal models which are of interest in the study of modulation schemes in communication systems.

EXERCISES

6.1 Find the linear mean square estimate of $y(t+\lambda)$ in terms of $y(t)$ and its first two derivatives. That is,

$$\hat{y}(t+\lambda) = ay(t) + b\dot{y}(t) + c\ddot{y}(t)$$

6.2 Let $y(t)$ be a band-limited process such that

$$\Phi_y(\omega) = 0 \qquad |\omega| > \omega_c,\ \omega_c = \frac{\pi}{T}$$

We wish to obtain the mean square estimate of $x(t)$ in terms of its sample values at instants nT as

$$\hat{y}(nT) = \sum_{n=-\infty}^{\infty} a_n(t) y(nT)$$

Show that the mean square error is minimized if

$$a_n(t) = \frac{\sin \omega_c(t - nT)}{\omega_c(t - nT)}$$

(This is the sampling theorem for random processes.)

6.3 Find the optimum filter to estimate $y(t + \alpha)$ in the presence of noise when the signal and noise spectrums are given by

$$\Phi_y(s) = \frac{-s^2}{-s^2 + 2a^2},\ \Phi_v(s) = \frac{a^2}{-s^2 + a^2}$$

Assume that $y(\cdot)$ and $v(\cdot)$ are uncorrelated and that $z(u) = y(u) + v(u)$, $-\infty < u \leqslant t$. Sketch the minimum mean square error as a function of α and determine the irreducible error.

6.4 Determine the optimum filter to estimate $y(t)$ given that the signal and noise are uncorrelated and have spectral densities

$$\Phi_y(s) = \frac{-s^2}{(1 - s^2)^2} \qquad \Phi_v(s) = \frac{1}{2}$$

and

$$z(u) = y(u) + v(u) \qquad -\infty < u \leqslant t$$

6.5 Consider the control problem of Fig. E.6.1. in which the fixed plant has the transfer function

$$G_p(s) = \frac{1}{s(s+1)}$$

The signal input $y(t)$ has spectrum

$$\Phi_y(s) = \frac{A^2}{(-s^2 + a^2)(-s^2 + b^2)}$$

and is uncorrelated with the load disturbance $v(t)$ which has a spectrum

$$\Phi_v(s) = \frac{1}{-s^2 + d^2}$$

The gain K_T is assumed to be fixed.

(a) Find the optimum compensator $G_c(s)$ so that the output $c(t)$ follows the signal input in a mean square sense.

(b) Find $G_c(s)$ if the signal input to the plant $y_p(t)$ is constrained such that

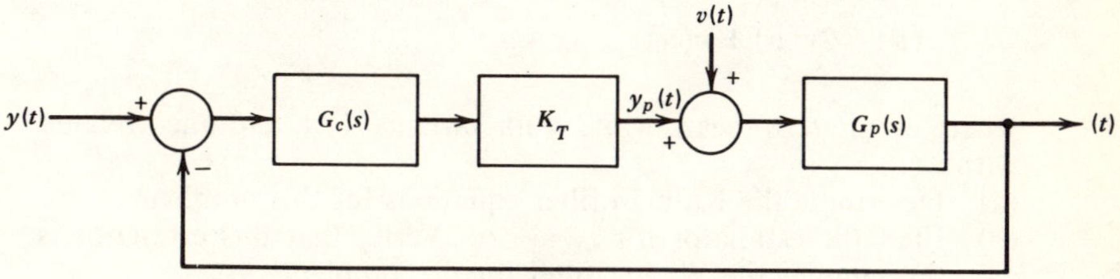

Fig. E.6.1

$$E\{y_p^2(t)\} < B \text{ for } B \text{ some positive constant.}$$

6.6 Find the optimum predictor for estimating $y(t + \alpha)$ when the signal spectrum is

$$\text{(a)} \quad \Phi_y(s) = \frac{2a}{-s^2 + a^2} \qquad \text{(b)} \quad \Phi_y(s) = \frac{-s^2 + 1}{s^4 + 2}$$

if $\Phi_v(s) = 1$ and $z(u) = y(u) + v(u)$ for $-\infty < u \leqslant t$.

6.7 The observed signal is

$$z(u) = y(u) + v(u) \qquad -\infty < u \leqslant t$$

(a) Find the optimum filter to estimate $\frac{d}{dt}y(t)$ when the signal and noise spectra are given by

$$\Phi_y(s) = \frac{1}{1 - s^2}, \quad \Phi_v(s) = \frac{1}{2}$$

Assume that the signal and noise are uncorrelated.

(b) Is $\widehat{\frac{d}{dt}y(t)} = \frac{d}{dt}\hat{y}(t)$?

6.8 Consider the following message and observation models:

$$\dot{x}_1(t) = x_2(t)$$

$$\dot{x}_2(t) = -2x_1(t) - 3x_2(t) + w(t)$$

where $w(t)$ is zero-mean, white, with unit variance. Let

$$y(t) = x_1(t)$$

and

$$z(u) = 2y(u) + v(u) \qquad t_0 \leqslant u \leqslant t$$

where $v(t)$ is zero-mean, white, with variance 1/4, and uncorrelated with $w(t)$.

(a) Determine the Kalman filter equations for this problem.

(b) Find the estimator for $t_0 = -\infty$. Verify that this estimator is the same as the Wiener filter for this problem.

6.9 In our derivation of the Kalman filter equations, we assumed that the input $\mathbf{w}(t)$ and the observation noise $\mathbf{v}(t)$ were uncorrelated. Derive the corresponding equations for the continuous-time problem when

$$E\{\mathbf{w}(t)\mathbf{v}^T(\tau)\} = \mathbf{V}_{\mathbf{wv}}(t)\delta(t-\tau)$$

(Note that in this case $\mathbf{x}(t)$ and $\mathbf{v}(t)$ will not be uncorrelated.)

6.10 Consider the following signal and observation models:

$$\dot{x}_1(t) = x_2(t)$$

$$\dot{x}_2(t) = -x_1(t) - 3x_2(t) + w(t)$$

where $w(t)$ is zero mean, white, with unity variance and

$$z(t) = x_1(t) + v(t)$$

where $v(t)$ is nonwhite (colored) noise. Suppose $v(t)$ can be modeled as the output of a linear system driven by white noise:

$$\dot{v}(t) = -v(t) + \xi(t)$$

where $\xi(t)$ is zero mean, white with unity covariance and uncorrelated with $w(t)$. We can then form an augmented message model by letting $v(t) = x_3(t)$ so that we have

$$\dot{x}_1(t) = x_2(t)$$
$$\dot{x}_2(t) = -x_1(t) - 3x_2(t) + w(t)$$
$$\dot{x}_3(t) = -x_3(t) + \xi(t)$$

However the message model now becomes

$$z(t) = x_1(t) + x_3(t)$$

and is hence noisefree. Since the continuous-time Kalman filter equations assume that all observations are corrupted by white noise (why?) we have to modify our observations to introduce a white component. We therefore use the derived observation $\dot{z}(t)$ since

$$\begin{aligned}\dot{z}(t) &= \dot{x}_1(t) + \dot{x}_3(t)\\ &= x_2(t) - x_3(t) + \xi(t)\end{aligned}$$

and has a white component.

(a) Derive the filter equations for this problem using the augmented message model and the derived observations. Note that the input and observation noises are correlated in this formulation.

(b) Derive the equations for the general model of Eqs. (6.4.71) and (6.4.1). Assume $\mathbf{v}(t)$ is colored and can be modeled as

$$\dot{\mathbf{v}}(t) = \mathbf{D}\mathbf{v}(t) + \boldsymbol{\xi}(t)$$

where $\boldsymbol{\xi}(t)$ is zero-mean, white with

$$\mathrm{E}\{\boldsymbol{\xi}(t)\boldsymbol{\xi}^T(s)\} = \mathbf{V}_\xi(t)\delta(t-s), \ \mathrm{E}\{\boldsymbol{\xi}(t)\mathbf{w}^T(s)\} = 0$$

6.11 Derive the estimation algorithms for the colored noise problem for the discrete-time models. Is there a difference between the continuous-time and discrete-time problems?

6.12 Consider the message model of Eq. (6.4.71) and the observation model

$$\mathbf{z}(u) = m(u)\mathbf{y}(u) + \mathbf{v}(u) \qquad t_0 \leqslant u \leqslant t$$

where $m(t)$ is a scalar white process with nonzero mean $\mu(t)$ and covariance kernel $V_m(t)$. Determine the linear minimum mean square estimation algorithms for this problem. Show that they reduce to the Kalman filter algorithms when $\mu(t) = 1$ and $V_m(t) = 0$.

6.13 Consider the message and observation models for the system

$$\dot{x}(t) = -x(t) + w(t)$$
$$z(t) = A \cos(w_0 t + .5x(t)) + v(t)$$

The input $w(t)$ is zero mean, white with unity variance as is the observation noise $v(t)$.

(a) Obtain the linearized Kalman filter and extended Kalman filter algorithms for this problem.

(b) Simulate the system on a computer and compare the performance of the algorithms for values of $A = 0.01$, 1, and 100.

6.14 Consider the following message and observation models in which α is an unknown constant parameter.

$$\dot{x}_1(t) = x_2(t)$$
$$\dot{x}_2(t) = -x_1(t) - \alpha x_2(t) + w(t)$$
$$z(t) = x_1(t) + v(t)$$

where $w(t)$ and $v(t)$ are zero-mean, white, with unity variance, and uncorrelated with each other. We can estimate the states and simultaneously identify the parameter α by considering $\alpha(t) = x_3(t)$ as a state and noting that

$$\dot{x}_3(t) = 0$$

(a) Set up the extended Kalman filter algorithms for this problem.

(b) Simulate the system on a computer to determine the effectiveness of the algorithm.

6.15 Set up the MAP equation for the estimation problem of Exercise 6.13.

6.16 Derive the MAP equation analagous to Eq. (6.6.38) for colored observation noise.

6.17 The output in a discrete-time, single-input, single-output system is modeled by the *autoregressive* (AR) equation

$$y(k) = \sum_{i=1}^{N} a_i y(k-i) + bw(k)$$

where $\{a_i\}$ and b are known constants. Assume that $w(k)$ is zero mean, Gaussian with $E\{w(k)w(j)\} = \sigma_w^2 \delta_{kj}$. Let the observation model be

$$z(k) = y(k) + v(k)$$

where $v(\cdot)$ is zero mean, Gaussian with $\mathrm{E}\{v(k)v(j)\} = \sigma_v^2\delta_{kj}$, uncorrelated with $w(\cdot)$. Find the estimate $\hat{y}(k|k-1)$. [25]

6.18 (Uncertain observations.) In some applications, we are not sure whether the observation is signal plus noise or noise alone. We can formulate the problem in terms of two hypotheses as.

$$H_1 : z(k) = y(k) + v(k) \qquad k = 1,2,\ldots,N$$
$$H_0 : z(k) = v(k)$$

The posterior density function of $y(k)$ can then be written as

$$p(y(k)|\mathbf{z}) = p(y(k)|\mathbf{z}, H_1)P(H_1|\mathbf{z}) + p(y(k)|\mathbf{z}, H_0)P(H_0|\mathbf{z})$$

Find the conditional mean estimate of $y(k)$ if all the processes are Gaussian.

6.19 (Divergence.) The Kalman filter algorithms assume that the system model and priors are known exactly. If there is a mismatch between the assumed model and the actual system, the estimates may become biased and the actual mean square error may be very large even though the theoretical error variance obtained by solving the error variance equation may be small. We illustrate this in this problem.

Assume that the actual message model is

$$x_a(k+1) = x_a(k) + m$$

where m is a constant. The observations are given by

$$z(k) = x_a(k) + v(k)$$

We design the estimator, however, by assuming that the message and observation models are

$$x_d(k+1) = x_d(k)$$
$$z(k) = x_d(k) + v(k)$$

Let $\tilde{x}(k|k)$ denote the actual error in the estimate:

$$\tilde{x}(k|k) = x_a(k) - \hat{x}_d(k|k)$$

Determine expressions for the propagation of the variance of $\tilde{x}(k|k)$. How does the actual error variance $V_{\tilde{x}}(k|k)$ compare with the theoretical error variance $V_{\tilde{x}_d}(k|k)$ at the end of N stages?

6.20 Consider the nonlinear message model of Eq. (6.6.1)

$$\dot{x}(t) = \mathbf{f}(\mathbf{x}(t), t) + \mathbf{g}(\mathbf{x}(t), t)\mathbf{w}(t)$$

but suppose that the observations are discrete, of the form

$$\mathbf{z}(k) = \mathbf{h}[\mathbf{x}(k), k] + \mathbf{v}(k)$$

where $x(k) = x(kT)$ for some T. Derive approximate algorithms for obtaining the estimate for this model.

6.21 Use the results of the previous problem to develop approximate estimation algorithms for the following problem. A radar located at a height RA from ground level observes the position of a freely falling object with constant drag coefficient. The vertical line of fall of the body is displaced from the radar location by a distance RD. Assume that gravity is neglected, and an exponential atmosphere is assumed. (See Fig. E.6.2.)

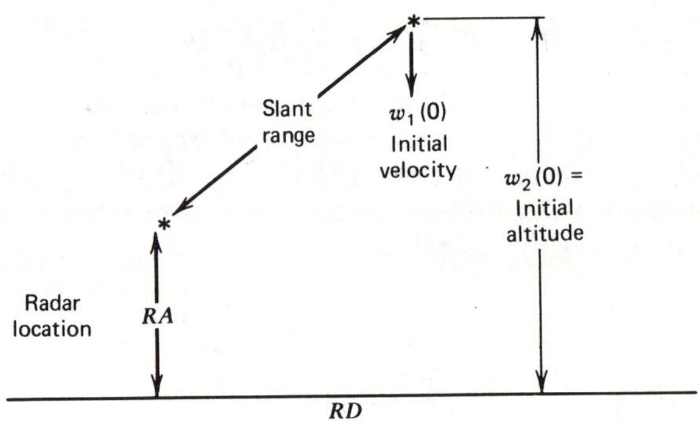

Fig. E.6.2

The state equations corresponding to the altitude, velocity, and drag coefficients are modeled as

$$\begin{aligned}
\dot{w}_1(t) &= -w_2(t) \\
\dot{w}_2(t) &= -k_1 \exp[-k_2 w_1(t) w_2^2(t) w_3(t)] \\
\dot{w}_3(t) &= 0
\end{aligned}$$

The observed quantity is the slant range so that the observations are

$$z(k) = \left(RD^2 + (w_1(k) - RA)^2\right)^{1/2} + v(k)$$

where $v(k)$ is an additive white noise sequence. Obtain the algorithm to estimate the states of the system. Simulate the system for

(i) $k_1 = 0$ (no drag, $RA = 250{,}000$ ft
$RD = 100{,}000$ ft (linear dynamics, nonlinear observations)

(ii) $k_1 = 0.5$ but $RA = RD = 0$ (nonlinear dynamics, linear observation). Assume the initial conditions:

$$w_1(0) = 300{,}000 \text{ ft}$$

$$w_2(0) = 20{,}000 \text{ ft/sec}$$

$$w_3(0) = 2 \times 10^{-2} \text{ ft/lb}$$

$$P(0) = \begin{bmatrix} 500 & 0 & 0 \\ 0 & 2 \times 10^4 & 0 \\ 0 & 0 & 6 \times 10^{-4} \end{bmatrix}$$

The noise variance is constant and equals 100 ft^2.

In your simulations, compute the sampled mean, and sampled mean square error using Monte Carlo methods. Compare your results with the actual trajectories obtained by numerically solving the system equations. (refer to [26] for a detailed exposition.)

6.22 Derive the extended discrete time Kalman filter algorithms of Table 6.6.2. [Hint: Expand the state equations in a Taylor series about $\hat{x}(k|k)$, and the observation about $\hat{x}(k|k-1)$.]

6.23 (Adaptive filter.) Consider the discrete-time model

$$\mathbf{x}(k+1) = \boldsymbol{\phi}\mathbf{x}(k) + \boldsymbol{\Gamma}\mathbf{w}(k)$$

with the observations

$$\mathbf{z}(k) = \mathbf{C}\mathbf{x}(k) + \mathbf{v}(k)$$

with usual assumptions on $\mathbf{w}$(k) and $\mathbf{v}$(k). However, we do not have exact knowledge of the matrices $\boldsymbol{\phi}$, $\boldsymbol{\Gamma}$, and $\mathbf{C}$. We assume instead, that these matrices may be denoted by a vector parameter $\boldsymbol{\theta}$ taking one of N values, $\boldsymbol{\theta}_1, \boldsymbol{\theta}_2, \ldots, \boldsymbol{\theta}_N$ with equal probabilities. In other words, we quantize the values the unknown parameters can take. Show that the conditional mean estimate of the state can be expressed as

$$\hat{\mathbf{x}}(k|k) = \sum_{i=1}^{N} \hat{\mathbf{x}}_i(k|k) P(\boldsymbol{\theta}_i|\mathbf{Z}(k))$$

where $\hat{\mathbf{x}}_i(k|k)$ is the conditional mean estimate of the system assuming that the model used corresponds to $\boldsymbol{\theta}_i$. $P(\boldsymbol{\theta}_i|\mathbf{Z}(k)$ is the a posteriori probability that $\boldsymbol{\theta}_i$ is the true vector parameter after observing $[\mathbf{z}(1), \mathbf{z}(2), \ldots, \mathbf{z}(k)] \triangleq \mathbf{Z}(k)$. Obtain the complete set of algorithms, and draw the block diagram of the estimator structure. What difficulties do you visualize in a practical implementation?

6.24 In this exercise, we consider the smoothing problem. We recall that the smoothing problem involves finding the estimate of a signal $\mathbf{y}(t)$ given observations over the interval $[t_0, t_f]$, where $t_0 \leqslant t \leqslant t_f$. We can distinguish three types of smoothing [5]. If t_0 and t_f are fixed, we have the fixed-interval smoothing problem. If t_0 and t are fixed, but t_f is increasing, we have the fixed-point smoothing problem. If $t = t_f - \Delta$ where $\Delta > 0$ is a fixed number and if t_f is increasing, we have fixed-lag smoothing.

We consider the message and observation models of Eq. (6.4.71) and Eq. (6.4.1). We recall from Eq. (6.4.65) that we can write

$$\hat{\mathbf{x}}(t|t_f) = \int_{t_0}^{t_f} \mathrm{E}\{\mathbf{x}(t)\boldsymbol{\nu}^T(\tau)\}\mathbf{V}_{\nu}^{-1}(\tau)\boldsymbol{\nu}(\tau)\, d\tau$$

It follows from Eq. (6.4.73) that for $t \leqslant t_f$, we can write the preceding as

$$\hat{\mathbf{x}}(t|t_f) = \hat{\mathbf{x}}(t|t) + \int_{t}^{t_f} \mathrm{E}\{\mathbf{x}(t)\boldsymbol{\nu}^T(\tau)\}\mathbf{V}_{\nu}^{-1}(\tau)\boldsymbol{\nu}(\tau)\, d\tau$$

Let $\psi_e(t, \tau)$ denote the state transition matrix corresponding to the matrix $[A(t) - K(t)C(t)]$ where $K(t)$ is the Kalman gain for the filtering problem for this model. Now define

$$\boldsymbol{\lambda}(t, t_f) = \int_{t}^{t_f} \psi_e^T(\tau, t)\mathbf{C}^T(\tau)\mathbf{V}_{\nu}^{-1}(\tau)\boldsymbol{\nu}(\tau)\, d\tau$$

(a) Show that

$$\hat{\mathbf{x}}(t|t_f) = \hat{\mathbf{x}}(t|t) + \mathbf{V}_{\tilde{\mathbf{x}}}(t|t)\boldsymbol{\lambda}(t, t_f)$$

(b) For the three cases of smoothing mentioned earlier, use the above equation to derive explicit algorithms for obtaining $\hat{\mathbf{x}}(t|t_f)$. (You may find [27] useful in this regard.)

REFERENCES

1. H. Cramer and M. R. Leadbetter, *Stationary and Related Stochastic Processes*. John Wiley and Sons, New York, 1967.
2. H. L. Van Trees, *Detection, Estimation and ModulationTheory, Part I*, John Wiley and Sons, New York, 1968.

3. A. P. Sage and J. L. Melsa, *Estimation Theory with Applications to Communications and Control*, McGraw-Hill Book Co., New York, 1971.
4. A. H. Jazwinski, *Stochastic Processes and Filtering Theory*, Academic Press, New York, 1970.
5. J. S. Meditch, *Stochastic Optimal Linear Estimation and Control*, McGraw-Hill Book Co., New York, 1969.
6. T. P. McGarty, *Stochastic System and State Estimation*, Wiley-Interscience, New York, 1974.
7. E. Wong, *Stochastic Processes in Information and Dynamical Systems*, McGraw-Hill Book Co., New York, 1971.
8. N. Wiener, *The Extrapolation, Interpolation and Smoothing of Stationary Time Series with Engineering Applications*, John Wiley and Sons, New York, 1962.
9. H. W. Bode and C. E. Shannon, A Simplified Derivation of Linear Least-Square Smoothing and Prediction Theory, *Proc. IRE*, **38**, April, 417–425 (1950).
10. S. S. L. Chang, *Synthesis of Optimum Control Systems*, McGraw-Hill Book Co., New York, 1961.
11. D. C. Youla, On the Factorization of Rational Matrices, *IEEE Trans. Infor. Theory*, **IT-7**, No. 3, July, 172–189 (1961).
12. B. D. O. Anderson, An Algebraic Solution to the Spectral Factorization Problem, *IEEE Trans. Automatic Control*, **AC-12**, No. 4, August, 410–414 (1967).
13. R. E Kalman, A New Approach to Linear Filtering and Prediction Problems, *Trans. ASME, J. Basic Eng.* **82D**, March, 34–45 (1960).
14. R. E Kalman and R. Bucy, New Results in Linear Filtering and Prediction Theory, *Trans. ASME, J. Basic Eng.*, **83D**, March, 95–108 (1961).
15. R. S. Bucy and P. D. Joseph, *Filtering For Stochastic Processes with Applications to Guidance*, Wiley-Interscience, New York, 1968.
16. T. Kailath, An Innovations Approach to Least Squares Estimation—Part I: Linear Filtering in Additive White Noise, *IEEE Trans. Automatic Control*, **AC-13**, No. 6, December, 646–655 (1968).
17. B. D. O. Anderson, Stability Properties of Kalman-Bucy Filters, *J. Franklin Inst.* **291**, 137–144.
18. R. L. Stratanovich, On the Theory of Optimal Nonlinear Filtering of Random Functions, *Theory Probabil. Appl.* (*USSR*), **4**, 223–225 (1959).
19. H. J. Kushner, On the Differential Equations Satisfied by Conditional Probability Densities of Markov Processes, with Applications, *SIAM J. Control, Ser. A*, **2**, No. 1, 106–119 (1962).
20. H. J. Kushner, Nonlinear Filtering: The Exact Dynamical Equations Satisfied by the Conditional Mode, *IEEE Trans. Automatic Control*, **AC-12**, No. 3, June, 262–267 (1967).
21. D. L. Snyder, Filtering and Detection for Doubly Stochastic Poisson Processes, *IEEE Trans. Info. Theory*, **IT-18**, January, 91–102 (1972).
22. P. A. Frost and T. Kailath, An Innovations Approach to Least Squares Estimation—Part III: Nonlinear Estimation in White Gaussian Noise, *IEEE Trans. Automatic Control*, **AC-16**, No. 3, June, 217–226 (1971).

23. R. S. Bucy, C. Hecht, and K. D. Senne, An Engineer's Guide to Building Nonlinear Filters, *F. J. Seiler Research Lab. Report* No. SRL-TR-72-004.
24. D. J. Sakrison, *Notes on Analog Comunication*, Van Nostrand Reinhold Co., New York, 1970.
25. R. L. Kashyap, A New Method of Recursive Estimation in Discrete Linear Systems, *IEEE Trans. Automatic Control*, **AC-15**, No. 11, February, 18–24 (1970).
26. R. P. Wishner, J. A. Tabaczynski, and M. Athans, A Comparison of Three Nonlinear Filters, *Automatica*, **5**, 487–496 (1969).
27. T. Kailath and P. Frost, An Innovations Approach to Least-Squares Estimation—Part II: Linear Smoothing in Additive White Noise, *IEEE Trans. Automatic Control*, **AC-13**, No. 6, December, 655–660 (1960).

CHAPTER 7

APPLICATIONS TO COMMUNICATION SYSTEMS

7.1 INTRODUCTION

A communication system can be loosely defined as one designed to transmit information in an efficient manner. The usual goal of a communication system is to eliminate or minimize the effects of unwanted disturbances while maximizing the amount of information being transmitted. Communication systems may be digital or analog. In analog systems, the information to be transmitted (message signal) is a continuous waveform. While the message signal can be transmitted directly, in most applications, particularly those involving transmission over long distances, the message signal modulates a carrier, a sinusoidal signal of known amplitude and frequency. The message may modulate either the amplitude or angle of the carrier waveform.

In a digital communication system, the information to be transmitted is discrete, as for example with the output of a digital computer. In many instances, analog data is sampled and quantized before transmission so that the information is in a discrete form. In both cases, the data is usually encoded into a sequence of symbols from a finite alphabet. With each symbol of the alphabet, we associate a pulse $y_i(t)$, $i = 1 \ldots M$ of duration T, where $\{y_i(t)\}$ are a set of known waveforms. The transmission of a symbol is then accomplished by the transmission of the associated waveform.

The transmitted signals in most cases undergo distortion over the transmitting medium. This distortion can usually be modeled as additive noise at the receiving end. In the case of an analog communication system, the receiver therefore must be designed to recover (estimate) the message signal. In a digital communication system, the receiver must decide as to which waveform was transmitted over each interval of duration T and thus decide which symbol was actually sent. The sequence of symbols at the reciever is then decoded to recover the message.

In this chapter, we will consider the application of the techniques of detection and estimation to the design of optimal communication systems. The problem of encoding and decoding discrete information is a significant step in digital communication systems. We will not, however, discuss coding in this chapter as it is outside the scope of the present book. The interested reader is referred to [5–7] for a treatment of this subject.

We start our discussions in this chapter with a consideration of various digital transmission schemes. We derive optimal receivers for these systems and evaluate their performance. We also discuss the problem of intersymbol interference which arises when the duration of a pulse waveform at the receiver exceeds the spacing between transmitted pulses so that two or more pulses interact to produce distortion at the receiver.

7.2 DIGITAL COMMUNICATION [1–4]

Figure 7.2.1 illustrates the basic components of a digital communication system.

In this section we discuss some basic digital schemes and derive the associated receiver structures. We restrict our discussions to binary channels in which the alphabet consists of the symbols 0 and 1. Thus the output of the encoder will be a sequence of 1's and 0's appearing at a rate of one every T seconds. Associated with a 0 is the signal $y_0(t)$ and with a 1, the signal $y_1(t)$.

We first consider baseband transmission schemes in which the waveforms $y_0(t)$ and $y_1(t)$ are transmitted directly. For communication over long distances, as indicated earlier, the waveforms modulate a high-

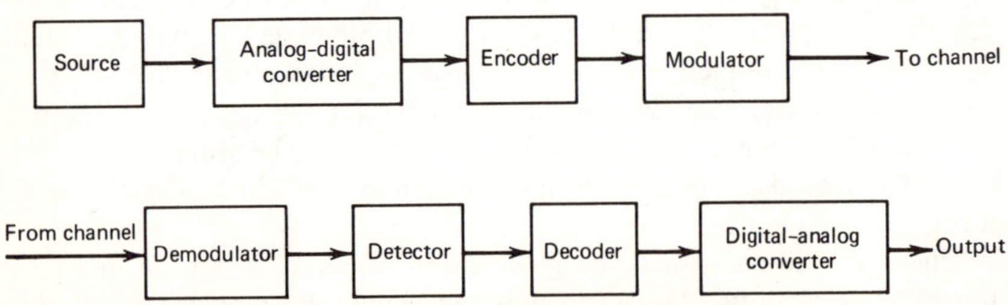

Fig. 7.2.1 Block diagram of digital communication system.

frequency carrier. In the latter case, if a local reference is available at the receiver for demodulation which is in phase with the transmitted carrier, the system is said to be *coherent*. Otherwise, it is said to be *noncoherent*. Similarly if a periodic signal (referred to as a clock) is available at the receiver, which is in synchronism with the transmitted sequence of pulses, the system is said to be *synchronous*. If such a clock is not available or necessary, the system is said to be *asynchronous*.

7.2.1 Baseband Digital Communication

Consider the binary digital communication system in which the signal pulse $y_1(t)$ represents a 1 and $y_0(t)$ represents a 0. At the receiver, the transmitted signal is assumed to be observed in additive white Gaussian noise $v(t)$ of spectral density $N_0/2$. Over each signaling interval of T seconds (bit interval), the receiver must make a decision as to whether $y_1(t)$ or $y_0(t)$ was transmitted.

If we cast the problem into the decision framework of Chapters 3 and 4 the observations can be modeled as

$$\begin{aligned} H_1 &: z(t) = y_1(t) + v(t) \qquad 0 \leqslant t \leqslant T \\ H_0 &: z(t) = y_0(t) + v(t) \qquad 0 \leqslant t \leqslant T \end{aligned} \tag{7.2.1}$$

In most communication problems an error in detecting either 1 or 0 is equally important. If we assume that the prior probabilities for the two are also equal, a natural choice is the minimum probability of error criterion discussed in Section 3.3. We recall that the threshold for the test under this criterion is unity. In terms of the log-likelihood ratio, the test is (see Eq. (4.2.28)),

$$\int_0^T z(t)\left[y_1(t) - y_0(t)\right] dt \underset{H_0}{\overset{H_1}{\gtrless}} \frac{1}{2}\int_{t_0}^{t_f}\left[y_1^2(t) - y_0^2(t)\right] dt \tag{7.2.2}$$

The test can thus be implemented by correlating the observation $z(t)$ with

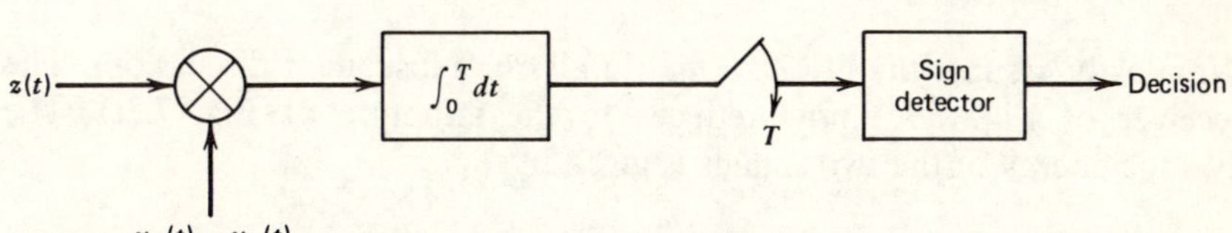

Fig. 7.2.2 Receiver structure for baseband binary digital communication system with equal energy signals.

the difference signal $[y_1(t) - y_0(t)]$ and comparing the correlator output at time T with a threshold. If the signals are of equal energy, the threshold is zero. In this case we only need to sense the sign of the correlator output at time T. The receiver structure is shown in Fig. 7.2.2.

Let us now consider the polar binary system in which the signals are

$$\begin{aligned} y_1(t) &= \sqrt{\frac{E}{T}} \qquad 0 \leqslant t \leqslant T \\ y_0(t) &= -\sqrt{\frac{E}{T}} \qquad 0 \leqslant t \leqslant T \end{aligned} \tag{7.2.3}$$

The waveforms of Eq. (7.2.3) are referred to as non-return-to-zero (NRZ) waveforms. Other formats include the return to zero (RZ) in which

$$y_1(t) = \begin{cases} \sqrt{\dfrac{E}{T}} & 0 \leqslant t \leqslant \tau \\ 0 & \tau < t \leqslant T \end{cases}$$

$$y_0(t) = \begin{cases} -\sqrt{\dfrac{E}{T}} & 0 \leqslant t \leqslant \tau \\ 0 & \tau < t \leqslant T \end{cases}$$

while in the split phase (Manchester) format

$$y_1(t) = \begin{cases} \sqrt{\dfrac{E}{T}} & 0 \leqslant t < T/2 \\ -\sqrt{\dfrac{E}{T}} & T/2 \leqslant t \leqslant T \end{cases}$$

$$y_0(t) = \begin{cases} -\sqrt{\dfrac{E}{T}} & 0 \leqslant t < T/2 \\ \sqrt{\dfrac{E}{T}} & T/2 \leqslant t \leqslant T \end{cases}$$

We mostly assume the NRZ format in all our subsequent discussions. The receiver of Fig. 7.2.2 now reduces to the structure of Fig. 7.2.3. The average energy of the two signals is given by

$$\epsilon = \frac{1}{2}\int_0^T y_1^2(t)\,dt + \frac{1}{2}\int_0^T y_0^2(t)\,dt = E \tag{7.2.4}$$

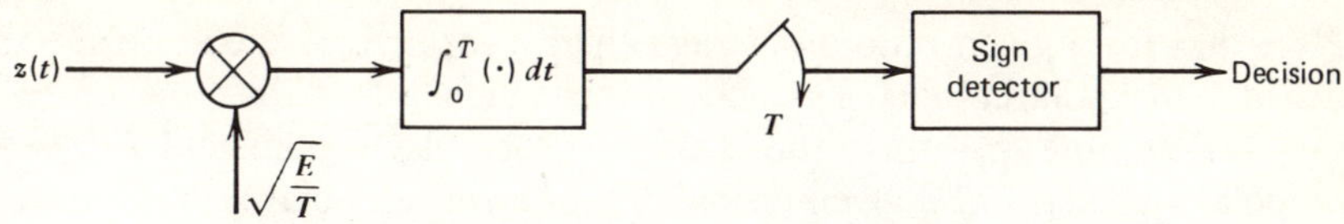

Fig. 7.2.3 Optimum receiver for polar binary signals.

while their correlation is

$$\bar{\rho} = \frac{1}{\epsilon}\int_0^T y_1(t)y_0(t)\,dt = -1 \tag{7.2.5}$$

so that the system is an ideal binary communication system. We recall that the probability of error for this problem is just P_F or P_M and is obtained from Eq. (4.2.51) as

$$P_e = \int_{\left[\frac{\epsilon(1-\bar{\rho})}{N_0}\right]^{1/2}}^{\infty} \frac{1}{\sqrt{2\pi}} \exp(-u^2/2)\,du \tag{7.2.6}$$

Substitution of Eqs. (7.2.4) and (7.2.5) thus yields the expression for the probability of error as

$$P_e = \int_{\left[\frac{2E}{N_0}\right]^{1/2}}^{\infty} \frac{1}{\sqrt{2\pi}} \exp(-u^2/2)\,du = \operatorname{erfc}_*\left[\left(\frac{2E}{N_0}\right)^{1/2}\right] \tag{7.2.7}$$

Thus for binary polar signals, the probability of error decreases as $2E/N_0$, the ratio of the average energy per digit to noise power spectral density, increases.

For the case of unipolar binary signals, 1 is represented by transmitting a rectangular pulse of height $\sqrt{E/T}$ and 0 by nontransmission. The decision rule is given by Eq. (7.2.2) where the threshold is now given by

$$\gamma = \frac{1}{2}\int_0^T [y_1^2(t) - y_0^2(t)]\,dt = \frac{E}{2} \tag{7.2.8}$$

The average signal energy ϵ is now equal to $E/2$ while the correlation coefficient ρ is zero. The probability of error follows from Eq. (7.2.6) as

$$P_e = \int_{\left(\frac{E}{2N_0}\right)^{1/2}}^{\infty} \frac{1}{\sqrt{2\pi}} e^{-u^2/2}\,du = \operatorname{erfc}_*\left[\left(\frac{E}{2N_0}\right)^{1/2}\right] \tag{7.2.9}$$

Note that for a given ratio of average energy per digit to the noise spectral density, the polar binary signal performs much better than the unipolar signal. (We note that this ratio is $2E/N_0$ for polar signals and E/N_0 for unipolar signals.) For a given probability of error, unipolar signal transmission requires 3 db greater average energy per bit-to-noise power density ratio than bipolar transmission. In fact we note that the binary polar set is an optimal signal set since the correlation between the two signals is -1 and therefore yields the lowest probability of error for a given E/N_0 ratio. The binary polar set belongs to a class of signals known as simplex signals. A set of M signals are said to be simplex if they are equally likely with correlation coefficient $-(1/(M-1))$. Even though such a signal set is optimum, orthogonal signal sets are more widely used in practice.

7.2.2 *R-F Digital Transmission*

Long-haul data transmission systems rarely employ baseband transmission. Usually the digital data modulate a sinusoidal carrier, and regenerative repeaters are used at intervals along the transmission path to maintain a certain signal-to-noise ratio. The basic digital modulation schemes are amplitude shift keying (ASK), frequency shift keying (FSK) and phase shift keying (PSK). In the binary case ASK takes the form of transmission for 1 and no transmission for 0, and is called on–off keying (OOK).

In this section, we derive the structure and performance of the optimum receivers for these various cases. We consider the operation of some of these receivers in fading media and indicate methods to improve their performance. The coherent detection systems to be considered first follow from our development of correlation receivers for binary communication in Section 4.2. Our discussion of noncoherent systems follows as an extension of the results of Section 4.6.

Coherent Systems

We recall that in a coherent system, a local reference is available at the receiver which is in phase with the transmitted carrier. Since the carrier phase is known, for purposes of analyzing coherent systems we can, without loss of generality, set the carrier phase equal to zero. We now proceed to analyze the various schemes.

The on–off keying system (OOK) employs $y_1(t)=\sqrt{2E/T}\,\sin\omega_c t$ for "mark" signal (digital 1) and $y_0(t)=0$ for "space" signal (digital 0) over each one digit interval of T sec. We again assume that the received signal $z(t)$ consists of the transmitted signal $y_i(t)$, $i=0, 1$ corrupted by additive white Gaussian noise of power spectral density $N_0/2$. The optimum

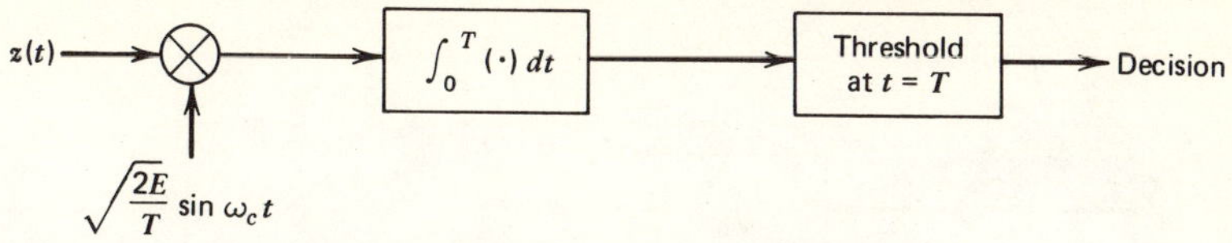

Fig. 7.2.4 Optimum receiver for **OOK.**

receiver is the correlation receiver and takes the form shown in Fig. 7.2.4. For the case of equal probability of occurrence of mark and space and minimum probability of error criterion, it can be seen from Eq. (4.2.28) that the threshold is $E/2$. This case is similar to unipolar baseband transmission. Again, the average energy per digit is $E/2$ and its ratio to the noise power spectral density is E/N_0. Since the correlation coefficient $\bar{\rho}$ is zero, the probability of error is obtained from Eq. (7.2.6) as

$$P_e = \int_{\left(\frac{E}{2N_0}\right)^{1/2}}^{\infty} \frac{1}{\sqrt{2\pi}} \exp\left[\frac{-u^2}{2}\right] du$$

$$= \text{erfc}_*\left[\left(\frac{E}{2N_0}\right)^{1/2}\right] \qquad \textbf{(7.2.10)}$$

The receiver in Fig. 7.2.4 can be practically implemented as a multiplier (mixer) followed by a low pass filter, or as a bandpass filter with center frequency ω_c.

The frequency shift keying system (FSK) employs

$$y_1(t) = \sqrt{2E/T}\ \sin \omega_1 t$$

for mark, and $y_0(t) = \sqrt{2E/T}\ \sin \omega_0 t$ for space signals, over each interval of T sec. Usually the carrier frequencies are chosen such that there are an integral number of cycles of each signal and the beat signal in a duration of T sec. In this case, the correlation coefficient $\bar{\rho}$ is given by

$$\bar{\rho} = \frac{1}{\epsilon} \int_0^T y_1(t) y_0(t)\, dt = \frac{1}{\epsilon} \frac{2E}{T} \int_0^T \sin \omega_1 t \sin \omega_0 t\, dt = 0 \qquad \textbf{(7.2.11)}$$

so that the two signals are orthogonal. This digital transmission system is known as orthogonal FSK.

The received signal $z(t)$ is either $y_1(t)$ or $y_0(t)$ additively corrupted by white noise of spectral density $N_0/2$. The optimum receiver consists of two correlators and the difference signal is compared to a threshold as shown in Fig. 7.2.5. Again for a practical implementation, the correlators can be

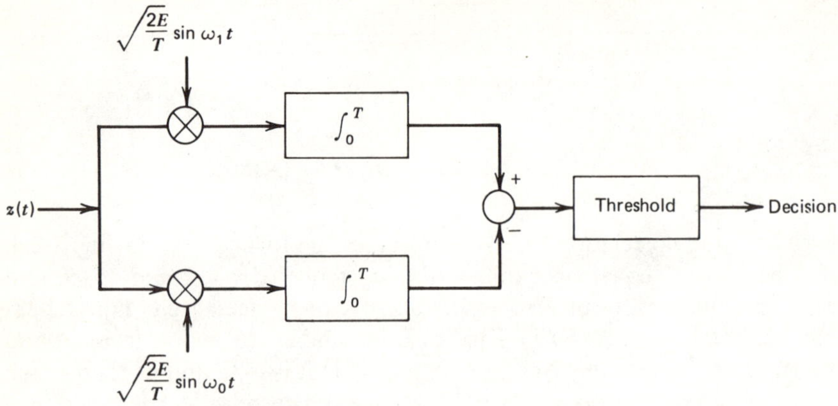

Fig. 7.2.5 Optimum coherent **FSK system.**

replaced by mixers followed by low pass filters or by two bandpass filters at appropriate center frequencies. The probability of error can be obtained from Eq. (7.2.6) by noting that the average energy is

$$\begin{aligned} \epsilon &= \frac{1}{2}\int_0^T \left[y_1^2(t) + y_0^2(t)\right] dt \\ &= \frac{E}{T}\int_0^T \left(\sin^2\omega_1 t + \sin^2\omega_0 t\right) dt \qquad \textbf{(7.2.12)} \\ &= E \end{aligned}$$

so that

$$P_e = \operatorname{erfc}_*\left[\left(\frac{E}{N_0}\right)^{1/2}\right] \qquad \textbf{(7.2.13)}$$

The average energy per digit to noise spectral density ratio for FSK is $2E/N_0$ as compared to E/N_0 for OOK. Comparison of Eq. (7.2.10) and (7.2.13), therefore, shows that for the same ratio of average energy per digit to noise spectral density, coherent OOK performs as well as orthogonal FSK. However, orthogonal FSK is preferred in practice because of its performance in fading channels. This is discussed in the latter part of this section.

In the case of phase shift keying, the signals for mark and space are $y_1(t) = \sqrt{2E/T}\,\sin\omega_c t$ and $y_0(t) = -\sqrt{2E/T}\,\sin\omega_c t$ respectively. This is also referred to as *phase reversal keying* (PRK). Since the correlation coefficient is -1, the system is an ideal system. The average signal energy is E. The optimum receiver in the presence of additive white noise of spectral density $N_0/2$ is shown in Fig. 7.2.6.

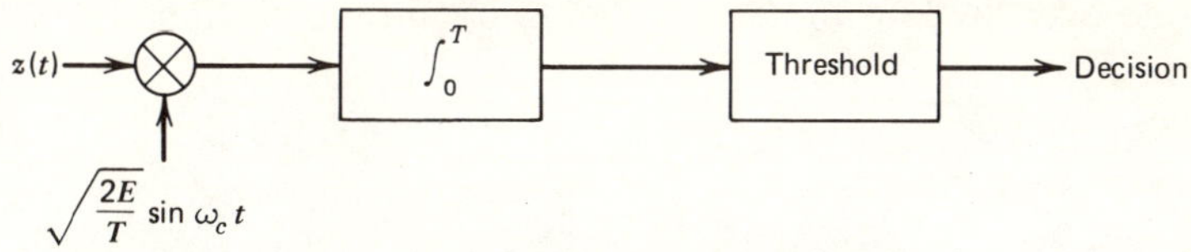

Fig. 7.2.6 Optimum receiver for **PSK.**

We point out that a PSK system, by its very nature, must be coherent, since it would be impossible to convey information in the phase of a carrier which is completely random. The probability of error easily follows from Eq. (7.2.6) as

$$P_e = \text{erfc}_*\left(\frac{2E}{N_0}\right)^{1/2} \tag{7.2.14}$$

For a given acceptable level of probability of error, the PRK system requires 3 db less average energy per digit than orthogonal FSK as can be seen by comparing Eqs. (7.2.11) and (7.2.13). If at all a coherent system is implementable, PRK is the obvious choice.

Noncoherent Systems

In noncoherent systems, the phase of the carrier is not available at the receiver. As such, we may expect the performance of a noncoherent system to be degraded in comparison to the corresponding coherent system. However, because of their simplicity, noncoherent systems are widely used in many applications.

In the case of noncoherent OOK, the received signals under the two hypotheses are

$$H_1 : z(t) = \sqrt{\frac{2E}{T}} \sin(\omega_c t + \theta) + v(t)$$

and

$$H_0 : z(t) = v(t) \tag{7.2.15}$$

over each time interval of T sec. In the case of noncoherent systems θ is an unknown parameter. The additive noise is again assumed to be zero-mean, Gaussian, white with spectral density $N_0/2$.

If θ is a random variable with known probability density function $p(\theta)$, we can first determine the conditional likelihood ratio $\Lambda(z|\theta)$ and then integrate over the known density $p(\theta)$ to obtain the likelihood $\Lambda(z)$ as

$$\Lambda(z(t)) = \int_{-\pi}^{\pi} \frac{1}{2\pi} \Lambda(z(t)|\theta) p(\theta)\, d\theta \tag{7.2.16}$$

The conditional likelihood ratio easily follows from Eq. (4.2.26) as

$$\Lambda(z(t)|\theta) = \exp\left[\frac{2}{N_0}\left\{\int_0^T z(t)\sqrt{\frac{2E}{T}} \sin(\omega_c t + \theta)\, dt\right\} - \frac{E}{N_0}\right] \tag{7.2.17}$$

Let us now define the following quantities:

$$L_s \triangleq \int_0^T \sqrt{\frac{2}{T}}\; z(t) \sin \omega_c t\, dt$$

and

$$L_c \triangleq \int_0^T \sqrt{\frac{2}{T}}\; z(t) \cos \omega_c t\, dt \tag{7.2.18}$$

It can easily be verified that L_s and L_c are Gaussian random variables under the two hypotheses with means and variances as follows:

$$\mathrm{E}\{L_s|H_0\} = \mathrm{E}\{L_c|H_0\} = 0, \;\; \mathrm{var}\{L_s|H_0\} = \mathrm{var}\{L_c|H_0\} = \frac{N_0}{2}$$

$$\mathrm{E}\{L_s|H_1, \;\theta\} = \sqrt{E}\; \cos\theta, \;\; \mathrm{E}\{L_c|H_1, \theta\} = \sqrt{E}\; \sin\theta$$

$$\mathrm{var}\,[L_s|H_1, \theta] = \mathrm{var}\,[L_c|H_1, \theta] = \frac{N_0}{2}$$

Let us now assume that θ is uniformly distributed in $[-\pi, \pi]$. It then follows that in terms of L_s and L_c, we can write the likelihood ratio as

$$\begin{aligned}\Lambda(z(t)) &= \frac{1}{2\pi}\int_{-\pi}^{\pi}\left[\exp\left\{\frac{2\sqrt{E}}{N_0}(L_c \sin\theta + L_s \cos\theta) - \frac{E}{N_0}\right\}\right] d\theta \\ &= I_0\left[\frac{2\sqrt{E}}{N_0}(L_c^2 + L_s^2)^{1/2}\right] \exp\left[-\frac{E}{N_0}\right]\end{aligned} \tag{7.2.19}$$

where $I_0(\cdot)$ is the modified Bessel function of the first kind. The decision rule, for equiprobable mark and space therefore is

$$\ln\left\{I_0\left[\frac{2\sqrt{E}}{N_0}\left(L_c^2+L_s^2\right)^{1/2}\right]\right\}\underset{H_0}{\overset{H_1}{\gtrless}}\gamma',\text{ some threshold} \tag{7.2.20}$$

For small values of x, $\ln I_0(x)\simeq x^2/4$, whereas for large values of x, $\ln I_0(x)\simeq x$. It therefore follows that for either low signal-to-noise ratios or high signal-to-noise ratios the decision rule is approximately given by

$$\left(L_c^2+L_s^2\right)^{1/2}\underset{H_0}{\overset{H_1}{\gtrless}}\gamma \tag{7.2.21}$$

and can be implemented as a bandpass filter in cascade with an envelope detector. The structure of the receiver is shown in Fig. 7.2.7. The new threshold γ is calculated according to the approximation used. We note that $l=(L_c^2+L_s^2)^{1/2}$ is a sufficient statistic.

In order to determine the probability of error, we need to know $p(l|H_0)$ and $p(l|H_1)$. Since L_s and L_c are Gaussian random variables, it easily follows that

$$p(l|H_0)=\begin{cases}\dfrac{2l}{N_0}e^{-l^2/N_0} & l\geqslant 0\\ 0 & l<0\end{cases} \tag{7.2.22a}$$

and

$$p(l|H_1)=\int_{-\pi}^{\pi}p(l|H_1,\theta)p(\theta)\,d\theta \tag{7.2.22b}$$

where

$$p(l|H_1,\theta)=\begin{cases}\dfrac{2l}{N_0}\exp\left(-\dfrac{l^2+E}{N_0}\right)I_0\left(\dfrac{2lE}{N_0}\right) & l\geqslant 0,\ -\pi\leqslant\theta\leqslant\pi\\ 0 & l<0\end{cases} \tag{7.2.23}$$

The probability of error P_{e_1} when a mark was transmitted is then given by

$$\begin{aligned}P_{e_1}&=\int_0^{\gamma}p(l|H_1)\,dl=1-\int_{\gamma}^{\infty}p(l|H_1)\,dl\\&=1-\int_{\gamma}^{\infty}\frac{2}{N_0}l\exp\left(-\frac{l^2+E}{N_0}\right)I_0\left(\frac{2lE}{N_0}\right)dl\\&=1-Q\left[\left(\frac{2E}{N_0}\right)^{1/2},\left(\frac{2}{N_0}\right)^{1/2}\gamma\right]\end{aligned} \tag{7.2.24}$$

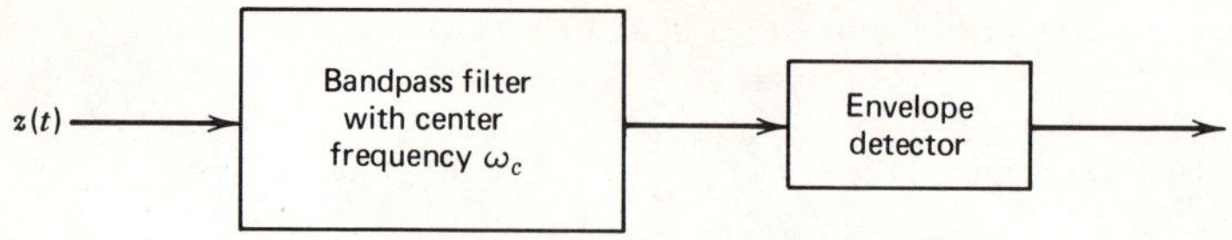

Fig. 7.2.7 Noncoherent **OOK** *receiver*

where $Q(a, b)$ are Marcum's Q-functions defined as

$$Q(a, b) = \int_b^\infty z \exp\left(-\frac{z^2 + a^2}{2}\right) I_0(az)\, dz \tag{7.2.25}$$

and are tabulated in [9]–[11]. Similarly the probability of error when space was transmitted can be determined as

$$P_{e_0} = \int_\gamma^\infty p(l|H_0)\, dl$$
$$= \exp\left[\frac{-\gamma^2}{N_0}\right] \tag{7.2.26}$$

The total probability of error can therefore be written as

$$P_e = \frac{1}{2}\left\{1 + \exp\left[\frac{-\gamma^2}{N_0}\right] - Q\left[\left(\frac{2E}{N_0}\right)^{1/2}, \left(\frac{2}{N_0}\right)^{1/2}\gamma\right]\right\} \tag{7.2.27}$$

In order for the probability of error to be minimum, the threshold γ has to be chosen depending on the factor $2E/N_0$. This optimum threshold can be obtained by setting the partial derivative of P_e in Eq. (7.2.27) with respect to γ equal to zero. We then obtain the transcendental equation

$$\exp\left(-\frac{E}{N_0}\right) \cdot I_0\left[\left(\frac{2E}{N_0}\right)^{1/2}\left(\frac{2}{N_0}\right)^{1/2}\gamma\right] = 1 \tag{7.2.28}$$

from which we can solve for $(2/N_0)^{1/2}\gamma$, the normalized threshold. An excellent analytic approximation to the solution of Eq. (7.2.28) is [2]

$$\left(\frac{2}{N_0}\right)^{1/2}\gamma = \left[2 + \frac{E}{2N_0}\right]^{1/2} \tag{7.2.29}$$

The expression for probability of error, Eq. (7.2.27) is rather complex. However, for high values of the signal-to-noise ratio E/N_0, we can obtain an approximate expression for P_e by noting that

$$Q(a, b) \simeq 1 - \mathrm{erfc}_*(a - b) \tag{7.2.30}$$

for large values of a and b [2, 3]. Substitution in Eq. (7.2.27) yields

$$P_e \simeq \frac{1}{2}\left\{ \exp\left[\frac{-\gamma^2}{N_0}\right] + \mathrm{erfc}_*\left[\left(\frac{E}{2N_0}\right)^{1/2}\right] \right. \tag{7.2.31}$$

For large values of signal-to-noise ratio, it follows from Eq. (7.2.29) that the optimum threshold satisfies

$$\sqrt{\frac{2}{N_0}}\ \gamma = \sqrt{\frac{E}{2N_0}} \tag{7.2.32}$$

Furthermore, we can approximate $\mathrm{erfc}_*(x)$ for large values of x as

$$\mathrm{erfc}_* x \simeq \frac{1}{\sqrt{2\pi}\, x} \exp\left(-\frac{x^2}{2}\right) \tag{7.2.33}$$

Substitution of these results in Eq. (7.2.31) yields the probability of error for noncoherent OOK for high signal-to-noise ratios as

$$P_e \simeq \frac{1}{2} \exp\left[-\left(\frac{E}{4N_0}\right)^{1/2}\right] \tag{7.2.34}$$

Thus the error rate decreases exponentially with SNR. Note that for a given probability of error and a given noise spectral density, the optimum noncoherent OOK requires about one dB more average energy than coherrent OOK.

For binary, noncoherent FSK, the received signals under the two hypotheses are

$$H_1 : z(t) = \sqrt{\frac{2E}{T}} \sin(\omega_1 t + \theta) + v(t) \qquad 0 \leqslant t \leqslant T$$

$$H_0 : z(t) = \sqrt{\frac{2E}{T}} \sin(\omega_0 t + \theta) + v(t) \qquad 0 \leqslant t \leqslant T$$

Usually the frequencies ω_1 and ω_0 are well separated. The optimum receiver for this case can be easily derived by introducing a "null hypothesis" (no signal) as a dummy hypothesis and following the same reasoning as in the case of noncoherent OOK.

Under the assumption that the phase is uniformly distributed in $(-\pi, \pi)$, it follows that the optimum receiver can be implemented by two bandpass filters with center frequencies corresponding to the mark and space frequencies, followed by envelope detectors. The envelopes are compared once every information pulse (T sec), and a decision is made in favor of the larger envelope. Figure 7.2.8 shows the noncoherent FSK receiver. We derive the performance of the receiver under the assumption that the mark and space signals are orthogonal with equal energy. For purposes of analysis, let us assume that a mark signal ($y_1(t)$) was transmitted. Then the output of the upper detector represents the envelope v_1 of the mark signal plus noise, while the output of the lower detector is the envelope v_0 of noise alone. An error is committed if v_0 exceeds v_1 when mark is transmitted. Let P_{e_1} denote this error probability. Similar to Eq. (7.2.21), we can obtain the following expressions for the probability density functions $P(v_1|H_1)$ and $P(v_0|H_1)$:

$$P(v_1|H_1) = \frac{2v_1}{N_0} \exp\left[\frac{-v_1^2 + E}{N_0}\right] I_0\left(\frac{2\sqrt{e}}{N_0} v_1\right) \qquad v_1 \geqslant 0 \qquad (7.2.35a)$$

and

$$P(v_0|H_1) = \frac{2v_0}{N_0} \exp\left[\frac{-v_0^2}{N_0}\right] \qquad v_0 \geqslant 0 \qquad (7.2.35b)$$

so that we have

$$\begin{aligned} P_{e_1} &= \text{Prob.}\left[v_0 > v_1 | H_1\right] \\ &= \int_0^\infty \left[\int_{v_1}^\infty p(v_0|H_1)\, dv_0\right] p(v_1|H_1)\, dv_1 \end{aligned} \qquad (7.2.36)$$

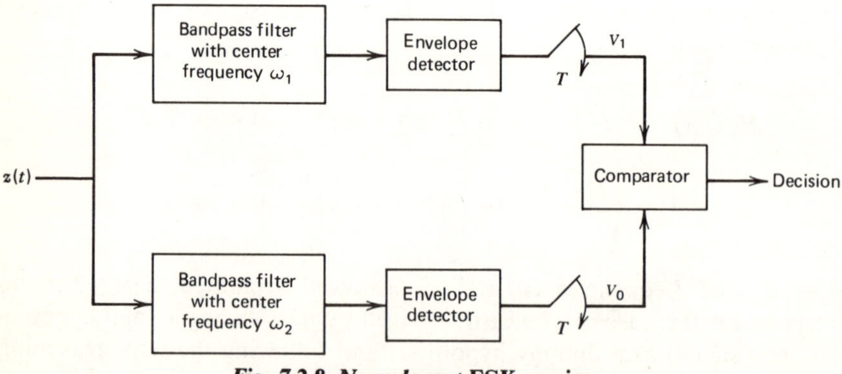

Fig. 7.2.8 Noncoherent **FSK** *receiver.*

Substituting for the density functions in Eq. (7.2.36) from Eq. (7.2.35) and using integral relations of Q functions [2], we obtain

$$P_{e_1} = \frac{1}{2} \exp\left[\frac{-E}{2N_0}\right]$$

Similarly, we can establish that

$$P_{e_0} = \frac{1}{2} \exp\left[\frac{-E}{2N_0}\right]$$

Therefore, the total probability of error P_e is:

$$P_e = \frac{1}{2} P_{e_1} + \frac{1}{2} P_{e_0} = \frac{1}{2} \exp\left[\frac{-E}{2N_0}\right] \tag{7.2.37}$$

We note that for a given E/N_0, the noncoherent FSK performs better than noncoherent OOK. But one has to be careful in stating this, because the FSK system uses double the energy. Therefore, for a given ratio of average energy per digit to noise power spectral density, noncoherent FSK and OOK perform almost identically, except at low values of this ratio.

We can obtain a comparison of the performance of a noncoherent FSK system with the coherent FSK system for large SNR by using the approximation of Eq. (7.2.33) in Eq. (7.2.13). This yields

$$P_e^{\text{FSK-COH}} \simeq \frac{1}{(2\pi E/N_0)^{1/2}} \exp\left[\frac{-E}{2N_0}\right] \tag{7.2.38}$$

Comparison of Eqs. (7.2.38) and (7.2.37) shows that for high SNR, coherent FSK performs better than noncoherent FSK. However for increasingly large SNR, since the behavior of P_e is dominated by the exponential term, the difference in performance between the two becomes negligibly small.

In practice noncoherent FSK is widely used because (*i*) it has a simple structure, (*ii*) its threshold does not depend on the E/N_0 ratio, (*iii*) its performance is comparable to coherent methods at high SNR, and as we shall see later (*iv*) its performance is better in fading channels than that of noncoherent OOK.

Differential Phase-Shift Keying

We have seen earlier that PRK offers the best performance among all binary digital transmission schemes. However PRK requires a phase reference at the receiver. If the phase of the carrier is essentially constant from interval to interval or bears a known relationship, we can obtain a reference for demodulating PRK by using the carrier phase of the previous signaling interval. A system which employs such a scheme is *differential phase-shift keying* (DPSK) [6,7]. Given a message sequence of binary digits, the encoded sequence is formed as follows. An arbitrary reference binary digit, say 1, is chosen as the first digit of the encoded sequence. Subsequent digits in the encoded sequence are formed by comparing a digit in the message sequence with the previous digit of the encoded sequence. For example if the first digit of the message sequence is a 1, this is compared with the reference digit of the encoded sequence. If the two are the same, the second digit of the encoded sequence is chosen as 1. If they are different, the second digit of the encoded sequence is chosen as 0. This now becomes the reference digit for the next digit of the message sequence. The encoded message sequence then phase-shift keys a carrier with phases 0 and π. Table 7.2.1 shows a typical message sequence and the corresponding encoded sequence and carrier phase. The receiver for DPSK is shown in Fig. 7.2.9. In order to illustrate the functioning of the receiver, let us consider the encoded message sequence of Table 7.2.1, assuming noisefree transmission. Since both the reference bit and the first encoded bit are 1, the signal input to the correlator as well as the reference input will both be $\sqrt{2E/T}\sin\omega_c t$. The output of the correlator is

$$l = \int_0^T \frac{2E}{T}\sin^2\omega_c t = E$$

and the receiver decides that a 1 was transmitted. During the next bit interval, since the transmitted bit is a 0, the signal input to the correlator is $-\sqrt{2E/T}\sin\omega_c t$ whereas the reference input is $\sqrt{2E/T}\sin\omega_c t$. The correlator output now is

$$l = -\int_0^T \frac{2E}{T}\sin^2\omega_c t\,dt = -E$$

and the receiver decides that a 0 was transmitted and so on for the rest of the encoded sequence.

TABLE 7.2.1 DIFFERENTIAL PHASE SHIFT KEYED SEQUENCE

Message Sequence:		1	0	1	0	1	1	1	0	0	1
Encoded Sequence:	1 Reference digit	1	0	0	1	1	1	1	0	1	1
Carrier Phase:	0	0	π	π	0	0	0	0	π	0	0

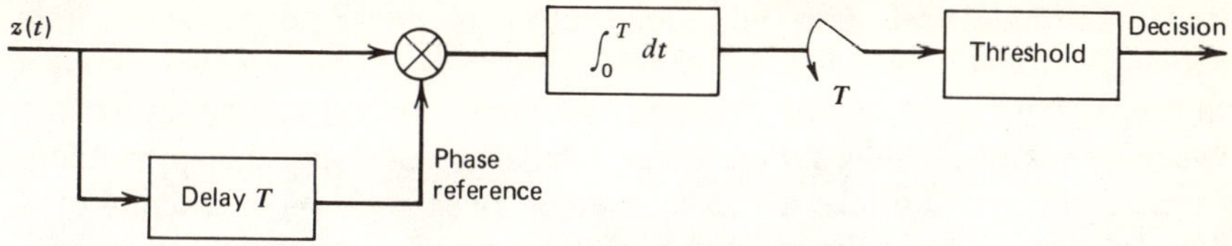

Fig. 7.2.9 Receiver for **DPSK**

Thus, in the absence of noise, use of DPSK yields same performance as PRK while obviating the need for a reference phase at the receiver. Noise in the received signal, however, affects both the signal input to the correlator and the reference, whereas in PRK only the signal input is corrupted by noise. We might therefore expect that the performance of DPSK in the presence of channel noise is not comparable to that of PRK. In fact the probability of error for DPSK is given by [2]

$$P_e = \frac{1}{2}\exp\left[-\frac{E}{N_0}\right] \tag{7.2.39}$$

Comparison with Eq. (7.2.37) for noncoherent FSK shows that DPSK requires 3 dB less SNR for the same error rate.

It is also useful at this point to consider the performance of PRK when the reference phase at the receiver is not correctly known. Let us assume that the transmitted signals are $\pm\sqrt{2E/T}\sin\omega_c t$, and that the reference signal is out of phase synchronism by an amount $\Delta\phi$. The reference signal is therefore $\sqrt{2E/T}\sin(\omega_c t + \Delta\phi)$. We can then show (see Exercise 7.8) that the probability of error is

$$P_e = \operatorname{erfc}_*\left[\left(\frac{2E\cos^2\Delta\phi}{N_0}\right)^{1/2}\right] \tag{7.2.40}$$

This expression is useful in designing synchronization circuits.

7.2.3 Fading Channel Performance

Long distance r-f telegraphy and many other systems utilizing ionospheric/tropospheric reflections are often subject to fading, a phenomenon caused by multipath propagation and addition of random phasors. Fortunately, it is possible to obtain a reasonable statistical model for this phenomenon, and to calculate the performance of digital systems on this basis.

The fading model often utilized is that of the nonselective, slow fading, Rayleigh model.† This assumes that the envelope of an otherwise known sinusoid is Rayleigh distributed and the phase uniformly distributed between $-\pi$ and π. Slow fading means that the envelope, though random from bit to bit, assumes a constant value over each bit interval. For a detailed discussion of the fading phenomenon and its modeling, the reader is referred to Schwartz, Bennet, and Stein [2]. In the case of OOK, the received waveforms can be modeled as

$$\begin{aligned} H_1 &: z(t) = \sqrt{\frac{2E}{T}} \sin(\omega_c t + \theta) + v(t) \\ H_0 &: z(t) = v(t) \end{aligned} \tag{7.2.41}$$

where now E and θ are random variables. The additive noise $v(t)$ is again assumed to be zero-mean, Gaussian, white with spectral density $N_0/2$.

As indicated earlier, we will assume that the envelope $\mathrm{v} = \sqrt{E}$ of the received signal is Rayleigh distributed and that the phase is uniformly distributed in $(-\pi, \pi)$. Let the mean square value of the received signal be $\bar{E}$, so that $E\{\mathrm{v}^2\} = \bar{E}$. It is obvious that this corresponds to the case of noncoherent OOK except that the received envelope is now random with probability density function

$$p(\mathrm{v}) = \frac{2\mathrm{v}}{\bar{E}} \exp\left(\frac{-\mathrm{v}^2}{\bar{E}}\right) \tag{7.2.42}$$

Therefore, we can expect the receiver structure to be the same as for noncoherent OOK, except that its performance will vary depending on v. An optimum receiver will use a *varying threshold* depending upon the received envelope (recall that the optimum threshold for noncoherent OOK depends on the ratio E/N_0), and will have to be adaptive. We can however, use the ratio of mean received energy per bit to noise spectral density, $\bar{E}/N_0$, to calculate a *fixed optimum threshold*.

The performance of the OOK receiver in fading media can be analyzed by determining the probability of error for a given received envelope and then averaging over the known density function $p(\mathrm{v})$. Let E/N_0 be denoted by R for convenience. Then

$$p(R) = \begin{cases} \dfrac{1}{R_0} \exp\left(\dfrac{-R}{R_0}\right) & 0 < R < \infty \\ 0 & \text{otherwise} \end{cases} \tag{7.2.43}$$

†This corresponds to a fluctuating point target model in the radar context as discussed in Section 8.2 where we use a different approach in determining the performance of the system using complex signal analysis.

where $R_0 = \bar{E}/N_0$. Then the probability of error expression becomes

$$P_e = \int_0^\infty P_e(R)p(R)\,dR \tag{7.2.44}$$

where $P_e(R)$ is the probability of error for any particular value of $R = E/N_0$. Using Eq. (7.2.27) in the preceding expression, and evaluating the integral we have:

$$P_e = \frac{1}{2}\left[1 - \exp\left(-\frac{\gamma^2}{N_0}\frac{1}{1+\dfrac{\bar{E}}{N_0}}\right)\right] + \frac{1}{2}\exp\left(\frac{-\gamma^2}{N_0}\right) \tag{7.2.45}$$

The optimum fixed threshold is obtained by differentiating Eq. (7.2.45) with respect to γ, and setting the result to zero. We then obtain

$$\gamma_{\text{opt. fixed}} = \left(\frac{N_0}{2}\right)^{1/2}\left[2\left(1+\frac{N_0}{\bar{E}}\right)\ln\left(1+\frac{\bar{E}}{N_0}\right)\right]^{1/2} \tag{7.2.46}$$

For large values of $\bar{E}/N_0$, this can be approximated as

$$\gamma_{\text{opt. fixed}} \simeq \left(\frac{N_0}{2}\right)^{1/2}\left[2\ln\left(\frac{\bar{E}}{N_0}\right)\right]^{1/2} \tag{7.2.47}$$

Using this value of the optimum fixed threshold in Eq. (7.2.45) yields the probability of error for high SNR as

$$P_e^{\text{opt. fixed}} = \frac{1}{2}\,\frac{\ln\left(\dfrac{\bar{E}}{N_0}\right)}{(\bar{E}/N_0)} \tag{7.2.48}$$

Note that the probability of error varies inversely with the ratio of average energy per digit to the noise spectral density for large values of this ratio. Unfortunately, this is true of all systems operating in fading media. Even a varying optimum threshold, for large E/N_0 ratio, does not help to improve this characteristic. The probability of error with a varying threshold can be approximated [2] as

$$P_e^{\text{varying opt.}} \simeq \left(\frac{1.63}{(\bar{E}/N_0)}\right) \tag{7.2.49}$$

Similarly, in the case of binary noncoherent orthogonal FSK, we can obtain the probability of error by averaging P_e in Eq. (7.2.37) over the density function in Eq. (7.2.44). We then obtain

$$\begin{aligned} P_e &= \int_0^\infty \frac{1}{R_0} \exp\left(\frac{-R}{R_0}\right) \frac{1}{2} \exp\left(\frac{-R}{2}\right) dR \\ &= \frac{1}{2+R_0} \\ &= \frac{1}{2+\bar{E}/N_0} \end{aligned} \tag{7.2.50}$$

Note that again the error rate varies inversely with $\bar{E}/N_0$ for large values of $\bar{E}/N_0$, as compared to the nonfading case where an exponential decrease in error rate is achieved.

If it is possible to operate coherently in a slowly fading medium, that is, the phase is estimated perfectly by means such as phase locking, it is possible to get approximately 3 dB improvement over noncoherent techniques. The corresponding results for the coherent schemes can be obtained by averaging the relevant probability of error expressions over the density function of Eq. (7.2.42). Thus the probability of error in the case of coherent FSK obtained by averaging P_e in Eq. (7.2.13) is

$$P_e = \frac{1}{2}\left[1 - \left(\frac{1}{1+2N_0/\bar{E}}\right)^{1/2}\right] \tag{7.2.51}$$

By averaging P_e in Eq. (7.2.14) we obtain the expression for PRK as

$$P_e = \frac{1}{2}\left[1 - \left(\frac{1}{1+N_0/\bar{E}}\right)^{1/2}\right] \tag{7.2.52}$$

We note that the performance of PRK is 3 dB better than that of FSK reemphasizing the fact that if coherent detection is at all possible, the optimum system is PRK.

Thus, at large values of the ratio of average energy per bit to noise spectral density, all these systems have an error rate inversely proportional to the ratio $\bar{E}/N_0$. In comparison, with steady signals, the error rates decrease exponentially with this ratio. We can qualitatively argue that the degradation due to fading is due to the finite probability of very low signal levels even when the mean energy is high, causing the probability of error to be close to 0.5. The obvious solution is to modify the statistics of the received signal envelope such that the probability density function is small for small values of the envelope. Diversity combining [12,16] has proved to be a very practical tool in this context.

Comparison of Digital Transmission Systems

We can use the error probability expressions derived in the preceding discussions to obtain a comparison of the different digital transmission schemes. In order to do so in a convenient fashion the probabilities of error for various systems are tabulated in Table 7.2.2 with δ, the ratio of average energy per bit to noise spectral density as a parameter.

Figure 7.2.10 shows the error rate curves for various δ for the systems shown in Table 7.2.2. It can be observed from Table 7.2.2 that OOK and FSK (orthogonal signalling) systems perform identically for almost all δ. The differences arise only in noncoherent and fading situations. We can explain this by noting that digital frequency modulation is not really a wide-band noise reduction system, but can be thought as two interleaved OOK signals of differing carrier frequencies . It is also seen that the noncoherent FSK performs only as well as noncoherent OOK. However, the former is preferred because its threshold is signal independent, a distinct advantage in fading conditions. The curves in Fig. 7.2.10 also emphasize that whenever possible, it is worthwhile measuring the phase in both fading and nonfading situations.

7.2.4 Synchronization

A coherent communication system requires various levels of synchronization for efficient operation. In baseband transmission with known signal shapes, the beginning and terminating times of the signals must be known at the receiver. For r-f systems, the carrier phase must also be known. In addition, if the bits are grouped into words, the beginning and termination times of the words must be known.

We first consider carrier synchronization. If the transmitted signal contains a component at carrier frequency, we can obtain carrier coherence very simply by locking onto this component. A commonly used device for achieving locking is the phase lock loop (PLL) shown in Fig. 7.2.11a. The system generally consists of a phase detector, loop filter, loop amplifier and a voltage controlled oscillator (VCO). In order to understand the operation of the PLL, let us assume that the input signal is given by

$$y_r(t) = E_r \sin\left[\omega_c t + \theta(t)\right] \tag{7.2.53}$$

TABLE 7.2.2

SYSTEM	MODE COHERENT/NONCOHERENT	CHANNEL FADING/NONFADING	PROBABILITY OF ERROR, P_e
(a) OOK	COHERENT	NONFADING	$\operatorname{erfc}_*\left[\left(\frac{\delta}{2}\right)^{1/2}\right]$
(b) OOK	NONCOHERENT[c]	NONFADING	$\frac{1}{2}\exp\left[\frac{-\delta}{4}\right]$
(c) OOK	NONCOHERENT[b]	FADING	$\frac{1}{2}\frac{\ln\delta}{\delta}$
(d) FSK[d]	COHERENT	NONFADING	$\operatorname{erfc}_*\left[\left(\frac{\delta}{2}\right)^{1/2}\right]$
(e) FSK[d]	COHERENT[a]	FADING	$\frac{1}{2}\left[1-\frac{1}{\sqrt{1+4/\delta}}\right]$
(f) FSK[d]	NONCOHERENT	NONFADING	$\frac{1}{2}\exp\left[\frac{-\delta}{4}\right]$
(g) FSK[d]	NONCOHERENT	FADING	$\frac{1}{2+\delta/2}$
(h) PSK	COHERENT	NONFADING	$\operatorname{erfc}_*[\delta^{1/2}]$
(i) PSK	COHERENT[+]	FADING	$\frac{1}{2}\left[1-\frac{1}{\sqrt{1+2/\delta}}\right]$

[a] Assumes perfect measurement of phase in fading media.
[b] Uses optimum fixed threshold, based on Rayleigh slow fading.
[c] Uses optimum fixed threshold, and P_e is the asymptotic value.
[d] Uses orthogonal signals.

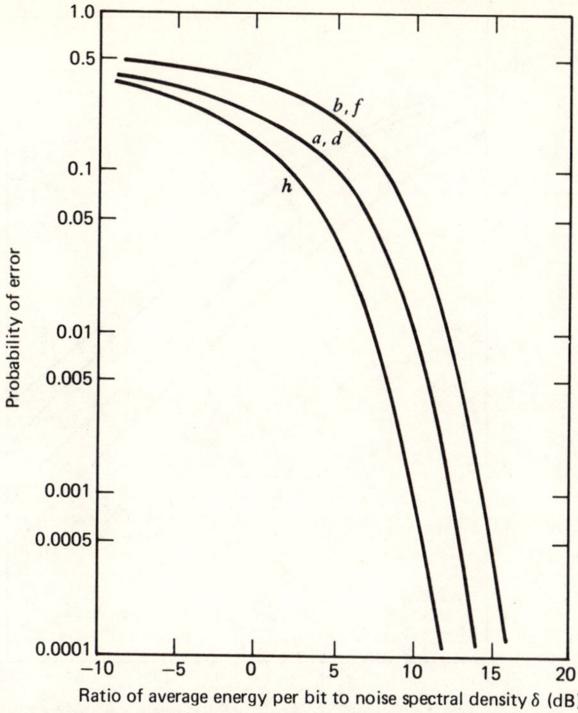

Fig. 7.2.10a Error rate curves for various r-f digital communication schemes in nonfading channels (see Table 7.2.1 for key to index).

and that the VCO output is

$$e_0(t) = E_c \sin\left[\omega_c t + \phi(t)\right] \tag{7.2.54}$$

While there are many types of phase detectors, we assume for our purposes that the phase detector is a multiplier followed by a lowpass filter which removes the second harmonic in the output of the multiplier. The phase detector output then becomes

$$e_d(t) = \tfrac{1}{2}E_c E_r K_d \sin\left[\theta(t) - \phi(t)\right] \tag{7.2.55}$$

where K_d is a constant associated with the multiplier. The VCO is essentially a frequency modulator, with the frequency deviation of the output $d\phi/dt$ being proportional to the VCO input $e_v(t)$. Thus

$$\frac{d\phi(t)}{dt} = K_v e_v(t) \tag{7.2.56}$$

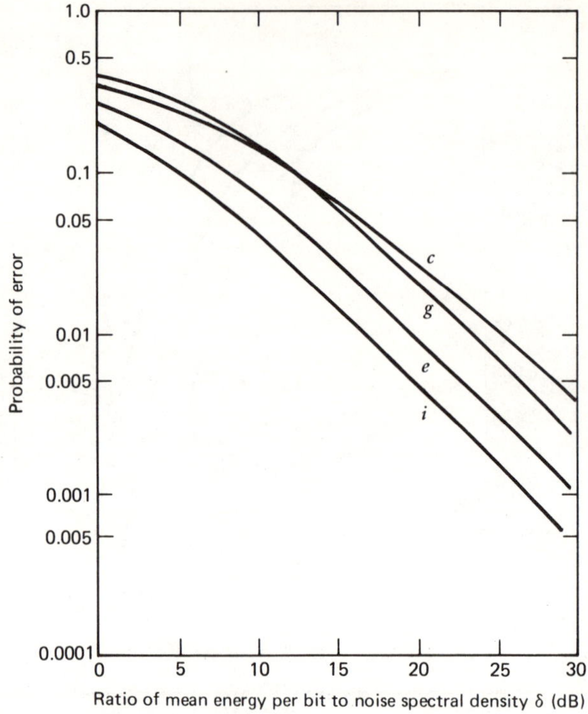

Fig. 7.2.10b Error rate curves for various r-f digital communication schemes in Rayleigh fading channel (see Table 7.2.1 for key to index).

where K_v is the VCO constant. If we assume that the loop filter and amplifier in the PLL can be replaced by a gain, we can combine Eqs. (7.2.55) and (7.2.56) to obtain

$$\frac{d\phi(t)}{dt} = K \sin\left[\theta(t) - \phi(t)\right] \tag{7.2.57}$$

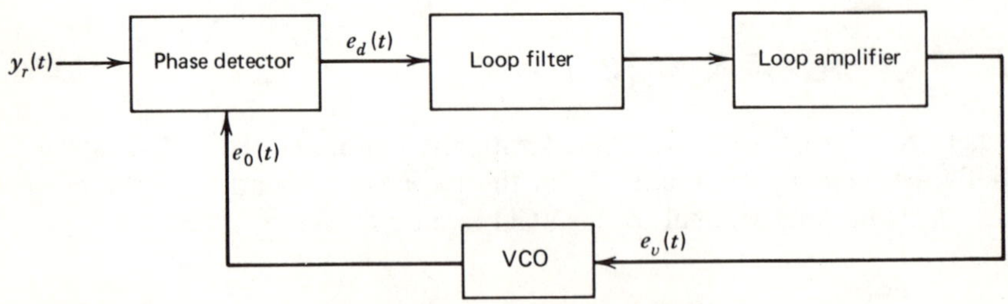

Fig. 7.2.11a The phase lock loop.

From Eq. (7.2.57), it is clear that if $\theta(t) > \phi(t)$, $d\phi(t)/dt$ will be positive so that $\phi(t)$ will increase. Similarly when $\theta(t) < \phi(t)$, $d\phi(t)/dt$ will be negative and $\phi(t)$ decreases. In either case, $\phi(t)$ is driven in the direction of $\theta(t)$ until $\phi(t) = \theta(t)$ and lock is achieved. The VCO phase $\phi(t)$ thus provides a good estimate of the carrier phase $\theta(t)$. It can also be seen from Eq. (7.2.57) that if $\theta(t)$ and $\phi(t)$ differ by more than $\pi/2$ radians, the VCO phase will not track the carrier phase. A detailed analysis of the PLL is beyond the scope of this book and the interested reader is referred to [17–19] for discussions of its performance and design.

In the absence of a carrier component in the transmitted signal, (as for example with PRK), we can use a Costas loop for synchronization. Figure 7.2.11b shows a block diagram of a Costas loop. For an assumed input of the type $y_r = E_r \cos \omega_c t$, the signals at the various points in the loop shown in Fig. 7.2.11b can easily be derived. The output from the lowpass filter preceding the VCO is essentially the dc value of the input. This signal drives the VCO such that θ is driven to zero and locking occurs. For a detailed discussion of these and other synchronization devices such as the decision-feedback loop and an analysis of their performance in the presence of noise, the reader is referred to [20,21].

For bit synchronization, three general methods are used. Synchronization can be derived, for example, by slaving the transmitter and receiver to the same clock, by using a separate synchronization signal (pilot clock) and by self-synchronization based on the modulation itself. Again, the reader is referred to [20,21] for detailed analysis of the timing misalignment of these systems. We can also use a MAP estimation technique for estimating the timing misalignment. We illustrate this for the case of a binary PRK

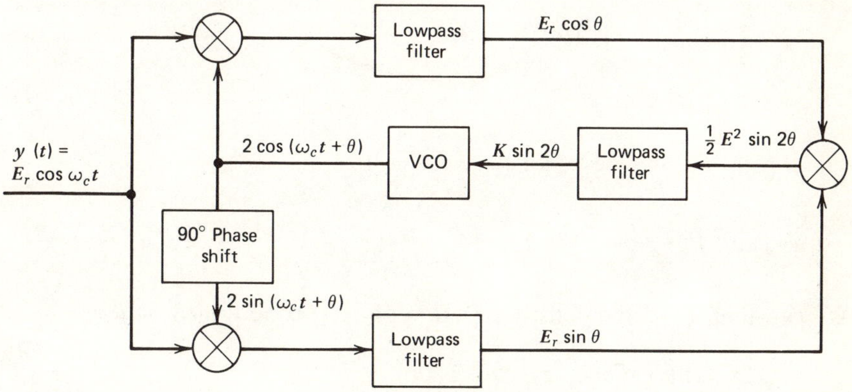

Fig. 7.2.11b Costas loop.

system [22]. Let us assume the transmitted signals

$$y_1(t) = x(t) \qquad \text{for Mark}$$
$$y_0(t) = -x(t) \qquad \text{for Space} \tag{7.2.58}$$

We assume that $y_1(t)$ and $y_0(t)$ are equiprobable. Let θ denote the timing misalignment. The received signal can then be modeled as

$$H_1 : z(t) = y_1(t, \theta) + v(t)$$
$$H_0 : z(t) = y_0(t, \theta) + v(t) \tag{7.2.59}$$

where $v(t)$ is stationary, white, Gaussian with mean zero and variance $N_0/2$.

We have considered a similar problem in our discussions in Chapter 5 (see Example 5.5.2). We can use a similar approach here to obtain an estimate $\hat{\theta}$ of the bit misalignment. To find $\hat{\theta}_{\mathrm{MAP}}$, we expand $z(t)$ over any bit interval in a set of orthonormal basis functions $\{g_k(t)\}$ and let $z_K(t)$ denote the K-term approximation to $z(t)$:

$$z_K(t) = \sum_{k=1}^{K} z_k g_k(t) \qquad \theta + jT \leqslant t < \theta + (j+1)T \tag{7.2.60}$$

where

$$z_k = \int_{jT}^{(j+1)T} z(t) g_k(t)\, dt$$

It easily follows that the coefficients $\{z_k\}$ are a set of independent, Gaussian random variables with the following moments:

$$\mathrm{E}\{z_k | H_i\} = y_{ki}(\theta) \qquad i = 0, 1$$
$$\mathrm{var}\,\{z_k | H_i\} = \frac{N_0}{2} \qquad i = 0, 1 \tag{7.2.61}$$

where from Eq. (7.2.58), we have

$$y_{k1}(\theta) = -y_{k0}(\theta) = \int_{\theta + jT}^{\theta + (j+1)T} x(t, \theta) g_k(t)\, dt \triangleq x_k(\theta) \tag{7.2.62}$$

We can then write the density function of $z_K(t)$ conditioned on θ as

$$p(z_K(t)|\theta) = p(z_1, z_2, \ldots, z_K | \theta)$$
$$= \sum_{i=0}^{1} p(z_1, z_2, \ldots, z_K | \theta, H_i) P(H_i) \tag{7.2.63}$$

Under the assumption that Mark and Space are equally likely, and since $\{z_K\}$ under each hypotheses are a set of Gaussian, independent variables with means and variances given by Eq. (7.2.61), we can write

$$
\begin{aligned}
p(z_K(t)|\theta) &= \frac{1}{2}\left\{ \prod_{i=1}^{K} \frac{1}{\sqrt{\pi N_0}} \exp\left[-\frac{(z_k - x_k(\theta))^2}{N_0} \right] \right.\\
&\quad \left. + \prod_{i=1}^{K} \frac{1}{\sqrt{\pi N_0}} \exp\left[-\frac{(z_k + x_k(\theta))^2}{N_0} \right] \right\}\\
&= \frac{1}{2(\pi N_0)^{K/2}} \left\{ \exp\left[-\frac{1}{N_0} \sum_{k=1}^{K} (z_k - x_k(\theta))^2 \right] \right.\\
&\quad \left. + \exp\left[-\frac{1}{N_0} \sum_{k=1}^{K} (z_k + x_k(\theta))^2 \right] \right\}\\
&= \frac{1}{2(\pi N_0)^{K/2}} \exp\left[-\frac{1}{N_0} \sum_{k=1}^{K} z^2(k) \right] \exp\left[-\frac{1}{N_0} \sum_{k=1}^{K} x_k^2(\theta) \right]\\
&\quad \cdot \cosh\left[\frac{1}{N_0} \sum_{k=1}^{K} z_k x_k(\theta) \right]
\end{aligned}
\tag{7.2.64}
$$

Now

$$
p(\theta|z_K(t)) = \frac{p(z_K(t)|\theta)p(\theta)}{p(z_K(t))} \tag{7.2.65}
$$

If we assume θ is uniformly distributed in $[0, T]$, since the denominator in Eq. (7.2.65) is just a normalizing factor, it follows that the MAP estimate of θ can be found by maximizing the expression on the right side of Eq. (7.2.64) as K tends to infinity. Now we have

$$
\begin{aligned}
\lim_{K\to\infty} \sum_{k=1}^{K} x_k^2(\theta) &= \lim_{K\to\infty} \int_{\theta+jT}^{\theta+(j+1)T} x_K^2(t)\, dt\\
&= E
\end{aligned}
\tag{7.2.66}
$$

where E represents the energy in the transmitted signal. Furthermore

$$
\lim_{K\to\infty} \sum_{k=1}^{K} z_k x_k(\theta) = \int_{\theta+jT}^{\theta+(j+1)T} z(t)x(t)\, dt \tag{7.2.67}
$$

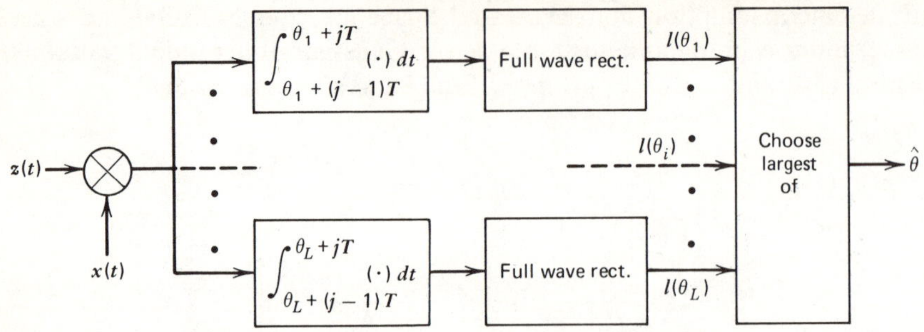

Fig. 7.2.12 Single-interval MAP estimator for bit misalignment.

Since cosh x is a nondecreasing function of its argument, it follows that

$$\hat{\theta}_{\text{MAP}} = \max_{\theta} \left| \frac{1}{N_0} \int_{\theta+jT}^{\theta+(j+1)T} z(t)x(t)\, dt \right| \tag{7.2.68}$$

A practical implementation of this scheme can be obtained by assuming that θ takes on a set of L discrete values in the interval $[0, T]$. The expression in Eq. (7.2.68) is then evaluated for each of these L values and $\hat{\theta}_{\text{MAP}}$ is chosen as that value which corresponds to the maximum. The structure of the estimator is shown in Fig. 7.2.12.

7.2.5 Intersymbol Interference in Digital Communication

In our discussions so far, we have considered two sources of error in digital communication systems, namely additive noise and fading over the transmission medium. A significant assumption underlying our discussions was that the channel did not possess any memory so that the received signal over any bit interval could be considered independently of the signals in other intervals. Such an assumption is not valid in the case of high-speed transmission of digital data, as for example, in communication between computers. The received signal over any bit interval will have contributions from transmissions over several intervals. This smearing of the transmitted signal affects the decision process rather detrimentally. This phenomenon is known as *intersymbol interference* (ISI).

In this section we formulate the model for a basic digital communication system with ISI and determine the increase in the probability of error in detecting a bit. In order to counter this effect, the observed signal may first be passed through a filter called the equalizer whose characteristics are the inverse of the channel characteristics. If the equalizer is exactly matched to the channel, the combination of channel and equalizer is just a gain so that there is no intersymbol interference present at the output of the equalizer. We confine our discussions to bit by bit detection even though many of our results can be extended to sequences of bits. An efficient algorithm due to Viterbi has been successfully used for this purpose [26]. Since this technique is outside the scope of the present text, the reader is invited to the cited reference.

To illustrate the various concepts involved, we only consider a baseband system. The extensions to modulated systems carry over in principle, and details can be found in the references [24]. Let $\{a_n\}, n = -\infty, \ldots, -2, -1, \ldots, \infty$ be a binary white sequence taking values 1 or -1 with equal probability. This sequence can be thought of as a representation of the input bits of a binary baseband digital communication system shown in Fig. 7.2.13. In Fig. 7.2.13, $h(t)$ represents the impulse response of the channel, $v(t)$ the additive white noise. For purposes of analysis we can consider the input to be an impulse train with weights a_n. Then the received signal can be synchronously sampled every T seconds to decide whether the bit transmitted was 1 or -1. It follows that the received signal is

$$\begin{aligned} z(t) &= y(t) + v(t) \\ &= \sum_{n=-\infty}^{\infty} a_n h(t - nT) + v(t) \end{aligned} \tag{7.2.69}$$

Consider an instant of sampling, say t_0, corresponding to the transmission of a_0. This sample is represented by

$$z_0 = a_0 h_0 + \sum_{n=-\infty}^{\infty}{}' \; a_n h_{-n} + v_n \tag{7.2.70}$$

where we represent $h(t_0)$ by h_0 and $h(t_0 - nT)$ by h_{-n}, and Σ' indicates the exclusion of the a_0 term. Equation (7.2.70) brings out the effects of channel memory and additive noise distinctly. The desired signal corresponds to the first term, the second term due to channel memory is the ISI effect, and the last term is the additive white noise. Even in the absence of additive noise (which is rarely the case), the second term can be significant

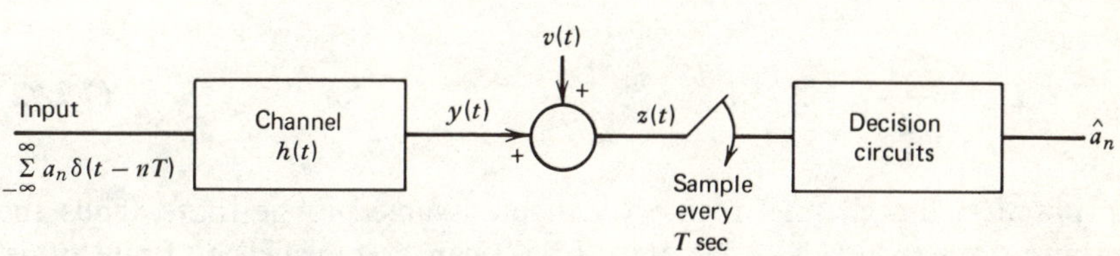

Fig. 7.2.13 Baseband digital communication system with intersymbol interference.

enough to cause an erroneous decision. Thus we see that ISI is a major source of error in high speed data transmission over memory channels, and its effect needs to be reduced in order to have a reliable system.

An interesting way of viewing Eq. (7.2.70) is to note that the ISI term

$$\sum_{n=-\infty}^{\infty}{}' a_n h_{-n}$$

is a correlated random sequence and the detection problem can be considered as one of detecting a known signal in colored noise discussed in Chapter 4. Therefore, we can expect the "optimum" receiver structure to be a whitening filter and a suitable matched filter. However, the design of such receivers in practice is subject to several constraints and certain approximations may have to be resorted to in order to obtain a design that minimizes the effect of ISI.

We recall from our discussions in Chapter 4 that the performance of a receiver in the presence of colored noise is dependent on the waveforms of the signals being transmitted. Thus by proper design of the transmitted waveforms, we can reduce the probability of error. Because we can consider ISI as colored noise, we can use the approach to advantage. If we evaluated the eigenfunction corresponding to the smallest eigenvalue of the colored noise process $\Sigma' a_n h(t-nT)$ and use it to transmit our binary data, detectability should indeed increase. In practice this evaluation is rather difficult, and what is more, we may end up with a waveform too complex to be implemented. Usually the waveforms are designed to have null response at sampling instants other than the one under consideration. An example is the raised cosine waveform [24].

We now turn to the problem of designing an equalizer for reducing the effect of ISI. A criterion for designing equalizers is to minimize the peak distortion D_p defined as

$$D_p = \frac{1}{h_0} \sum_{n=-\infty}^{\infty} |h_n| \tag{7.2.71}$$

Alternatively we can also seek to minimize the mean-square distortion D_{ms} defined as

$$D_{\text{ms}} = \frac{1}{|h_0|^2} \sum_{n=-\infty}^{\infty} |h_n|^2 \tag{7.2.72}$$

In practice, the channel memory can be assumed to be finite. Thus the infinite sums in Eqs. (7.2.71) and (7.2.72) can be replaced by finite sums. The equalizer is chosen to be a tapped delay line with say L taps, and the weights of the taps are adjusted such that either D_p or D_{ms} is minimum.

In the presence of additive noise at the receiver, we may choose the equalizer taps so as to minimize the mean square error between the input sequence and the equalizer output. The main components of this scheme are shown in Fig. 7.2.14. Thus, given the observations

$$z_k = \sum_{n=-N}^{N} h_n a_{k-n} + v_k \tag{7.2.73}$$

our problem is to design the equalizer gains c_l, $l = -L, -L+1, \ldots, 0, \ldots, L$ such that the equalizer output

$$\hat{a}_{n-D} = \sum_{l=-L}^{L} c_l z_{n-l} \tag{7.2.74}$$

approximates the transmitted symbol a_{n-D} in the mean square sense. In most cases, L may be chosen equal to N. If we denote the estimation error by $\tilde{a}_{n-D}$, we can use the orthogonal projection principle to obtain

$$\mathrm{E}\left[\tilde{a}_{n-D} z_{n-q}\right] = 0 \qquad q = -L, \ldots, 0, \ldots, L \tag{7.2.75}$$

where

$$\begin{aligned} \tilde{a}_{n-D} &= a_{n-D} - \hat{a}_{n-D} \\ &= a_{n-D} - \sum_{l=-L}^{L} c_l z_{n-l} \end{aligned} \tag{7.2.76}$$

Substitution of Eq. (7.2.76) in Eq. (7.2.75) yields the set of equations

$$\sum_{l=-L}^{L} c_l \mathrm{E}\{z_{n-l} z_{n-q}\} = \mathrm{E}\{a_{n-D} z_{n-q}\} \tag{7.2.77}$$

The right side of Eq. (7.2.77) is the cross-correlation between the delayed input and the unequalized channel output (observation) while the expectation on the left side corresponds to the autocorrelation function of the unequalized channel output. We can evaluate these expressions by using Eq.

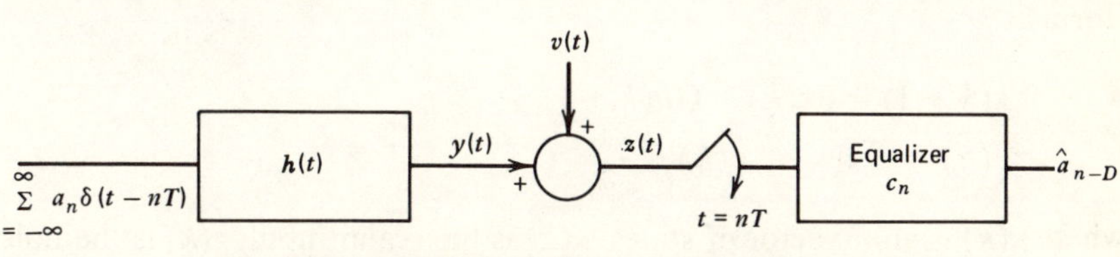

Fig. 7.2.14 Equalized digital communication system.

(7.2.73). We thus have

$$
\begin{aligned}
\mathrm{E}\{a_{n-D}z_{n-q}\} &= \mathrm{E}\left\{a_{n-D}\sum_{m=-N}^{N} h_m a_{n-m-q}\right\} \\
&= \sum_{m=-N}^{N} h_m \mathrm{E}\{a_{n-D}a_{n-m-q}\} \\
&= h_{D-q}
\end{aligned}
\tag{7.2.78}
$$

where the last step follows from the assumed independence of the transmitted signals. Similarly we can show that

$$
\mathrm{E}\{z_{n-l}z_{n-q}\} = \left[\sum_{m=-N}^{N} h_m h_{l+q-m}\right] + \sigma_v^2 \tag{7.2.79}
$$

Substitution of Eqs. (7.2.78) and (7.2.79) in Eq. (7.2.77) yields

$$
\sum_{l=-L}^{L} c_l \left[\sum_{m=-N}^{N} h_m h_{l+q-m} + \sigma_v^2\right] = h_{D-q} \quad \cdot\, q = -L \ldots L \tag{7.2.80}
$$

Equation (7.2.80) constitutes a set of $(2L+1)$ simultaneous algebraic equations which can be solved to obtain the optimal tap gains $\{c_l\}$.

It is left as an exercise to the reader to show that in the absence of additive white noise, the MMSE approach is the same as that of minimizing the mean square distortion.

We can also obtain a sequential solution to the problem of equalizer design by using the Kalman filter discussed in Chapter 6. Let us for the moment assume that the channel model is the same as in Eq. (7.2.73) but that the interference is one-sided, that is, only from past transmitted symbols. In this case $h_{-n} = 0$ for $n = 1, 2, \ldots, N$. The channel model is shown in Fig. 7.2.15. The input to the delay line is a string of independent random binary digits $u(k)$ where $u(k) = a_k$. The output is corrupted by stationary additive white noise $v(k)$, with variance V_v. We assume that $u(k)$ and $v(k)$ are independent for all time. The dynamic equations of the channel follow from the figure and can be represented in state variable form as

$$
\begin{aligned}
\mathbf{x}(k+1) &= \mathbf{F}\mathbf{x}(k) + \mathbf{G}u(k+1) \\
z(k) &= \mathbf{h}^T\mathbf{x}(k) + v(k)
\end{aligned}
\tag{7.2.81}
$$

where $\mathbf{x}(k)$ is an n-vector of states, $u(k)$ is the scalar input, $z(k)$ is the noise

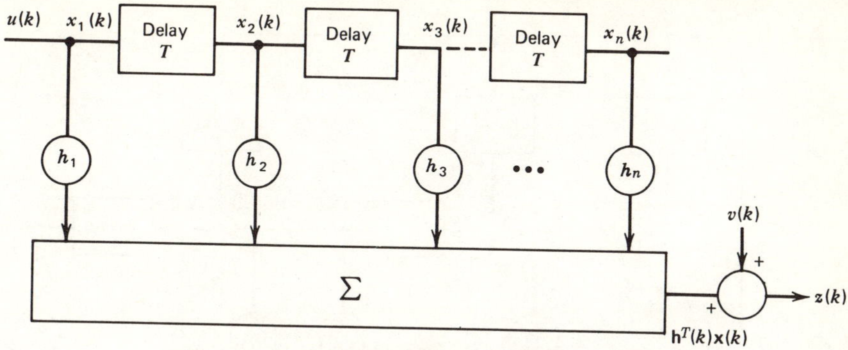

Fig 7.2.15 **ISI** *channel model for Kalman equalizer.*

corrupted output, and the matrix **F** and the vectors **G** and **h** are

$$\mathbf{F} = \begin{bmatrix} 0 & 0 & \cdots & 0 \\ 1 & 0 & \cdots & 0 \\ 0 & 1 & \cdots & 0 \\ \cdot & & & \\ \cdot & & & \\ \cdot & & & \\ 0 & 0 & \cdots 1 & 0 \end{bmatrix} \quad \mathbf{G} = \begin{bmatrix} 1 \\ 0 \\ 0 \\ \cdot \\ \cdot \\ \cdot \\ 0 \end{bmatrix} \quad \mathbf{h} = \begin{bmatrix} h_1 \\ h_2 \\ \cdot \\ \cdot \\ \cdot \\ \cdot \\ h_n \end{bmatrix}$$

The problem is to obtain an on-line equalization algorithm, given the input statistics and noise statistics, which yields an estimate of the message $u(k)$ at some delayed time $(k + D)$, $0 \leqslant D \leqslant N - 1$.

Since at any time k, we have the observation record $\mathbf{Z}(k) = \{z(j) : j < k\}$, the preceding problem statement can be rephrased as one of on-line estimation of the $(D + 1)$st component of $\mathbf{x}(k)$, that is, $x_{1+D}(k) = u(k - D)$, from these measurements. The Kalman filter algorithms discussed in Section 6.4 can be used to yield the minimum variance estimate of the state vector $\mathbf{x}(k)$, thereby giving a direct solution to our problem. The structure of the equalizer (Kalman filter) easily follows from our discussions in Chapter 6 and is shown in Fig. 7.2.16. The detailed derivation of the structure and the associated equations is left as an exercise for the reader. Note that an advantage of this equalizer is that the estimates $\hat{a}_{n-D}$ are obtained simultaneously for all D such that $0 \leqslant D \leqslant N - 1$.

In our discussions so far, we have assumed that the impulse response of the channel $h(t)$ or its sample values h_n are known. In practice this may be only approximately true because the channel characteristics may vary with time. In many cases, the samples h_n may not be known. In such cases, the

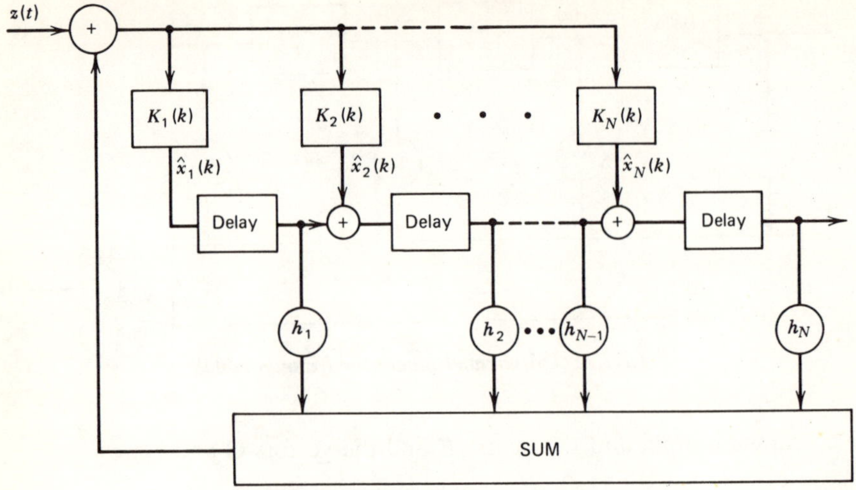

Fig. 7.2.16 Kalman equalizer for ISI *channels.*

design of the equalizer will have to be based on some adaptive technique that identifies the samples h_n directly or indirectly, while simultaneously providing equalization. We now discuss two methods of adaptive equalization. In the first method, a maximum likelihood estimate of the channel weighting sequence is used. The technique has been widely used in several applications. The other uses an extended Kalman filter formulation.

Adaptive Equalization: The Lucky Equalizer [23]

We recall from our earlier discussion, that, ideally, the equalizer must be chosen such that the equalized system (channel and equalizer in cascade) represents a pure gain (or a pure delay). If we denote by $\{h_k\}$ the weighting sequence of the equalized system, we must have

$$h_k = \begin{cases} h_0 & k = 0 \\ 0 & \text{otherwise} \end{cases} \tag{7.2.82}$$

The Lucky equalizer is a tapped delay line filter with variable taps in which the tap gains are adjusted so that the weighting sequence of the equalized system approximates h_k in Eq. (7.2.82). Since the channel characteristics are usually varying, the tap gains will have to be readjusted at frequent intervals of time. In order to obtain an algorithm for adjusting the tap gains periodically, let us assume that the channel characteristics are slowly varying so that they may be assumed to be constant over an interval of K observations. Figure 7.2.17 shows a block diagram of the scheme. If we denote the transmitted sequence by a_n, the observations at the receiver

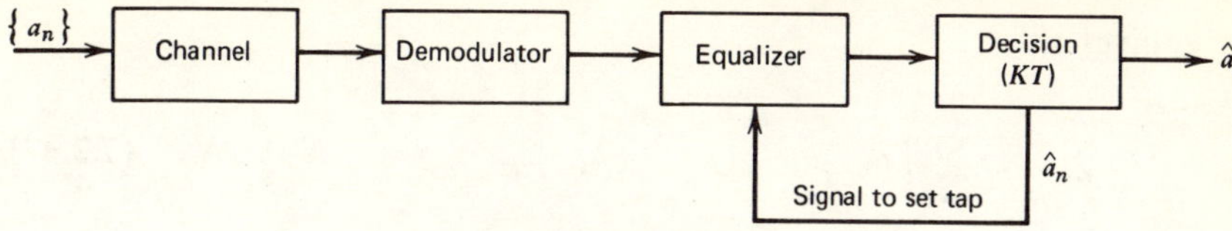

Fig. 7.2.17 Adaptive equalizer scheme.

(at the output of the equalizer) can be modeled as

$$z_k = \sum_{n=-N}^{N} h_n a_{k-n} + v_k \tag{7.2.83}$$

where the noise samples v_k are assumed to be independent, identically distributed, Gaussian random variables with mean zero and variance σ^2. Here, we are assuming that the impulse response of the equalized channel can be adequately represented by $(2N+1)$ samples. Over any K signaling intervals, we now seek the ML estimate of the weighting sequence $\{h_k\}$ for $k = -N, -N+1, \ldots, N-1, N$. In order to do so conveniently, let us define the $(2N+1)$-vector $\mathbf{h}$ and the K-vector $\mathbf{z}$ as

$$\mathbf{h} = [h_{-N}\; h_{-N+1} \ldots h_N]^T$$
$$\mathbf{z} = [z_1\; z_2 \ldots z_K]^T$$

The conditional density $p(\mathbf{z}|\mathbf{h})$ can easily be written from our assumptions as

$$p(\mathbf{z}|\mathbf{h}) = \prod_{k=1}^{K} \frac{1}{\sqrt{2\pi}\,\sigma} \exp\left[-\frac{\left(z_k - \sum_{n=-N}^{N} a_n h_{k-n}\right)^2}{2\sigma^2} \right] \tag{7.2.84}$$

The likelihood function $\ln p(\mathbf{z}|\mathbf{h})$ is

$$\ln p(\mathbf{z}|\mathbf{h}) = -\sum_{k=1}^{K} \frac{1}{2\sigma^2}\left(z_k - \sum_{n=-N}^{N} a_n h_{k-n}\right)^2 + \ln\left[\frac{1}{2\pi\sigma^2}\right]^{K/2} \tag{7.2.85}$$

The maximum likelihood estimates of the $(2N+1)$ response values $h_j, j = -N \ldots N$ can thus be determined by solving the set of $(2N+1)$ simultaneous equations obtained by setting $(\partial \ln p(\mathbf{z}|\mathbf{h})/dh_j)|_{h_j=\hat{h}_j} = 0$. We therefore have

$$\sum_{k=1}^{K} a_{k-j}\left[z_k - \sum_{n=-N}^{N} a_n \hat{h}_{k-n}\right] = 0 \quad \text{for } j = -N \ldots N \tag{7.2.86}$$

or equivalently

$$\sum_{n=-N}^{N} \hat{h}_n \sum_{k=1}^{K} [a_{k-j}a_{k-n}] = \sum_{k=1}^{K} a_{k-j}z_k \qquad j=-N\ldots N \tag{7.2.87}$$

Equation (7.2.87) is a set of simultaneous linear algebraic equations which can be solved to obtain the estimates $\hat{h}_n$. We can simplify the equation if we assume that the transmitted symbols are independent. In this case, for sufficiently large K, we can write

$$\frac{1}{K}\sum_{k=1}^{K} a_{k-j}a_{k-n} \simeq S\delta_{nj} \tag{7.2.88}$$

where S denotes the average power in the transmitted sequence. The estimates $\hat{h}_n$ are then obtained as

$$\hat{h}_n = \frac{1}{KS}\sum_{k=1}^{K} a_{k-j}z_k \qquad j=-N\ldots N \tag{7.2.89}$$

The algorithm for adjusting the tap gains c_n in the equalizer is now straightforward. For perfect equalization, we require that $h_0 = 1$ and $h_n = 0$ for $n \neq 0$. Therefore, for $n = 0$, if h_n is positive we decrease the corresponding tap gain by a fixed amount Δ whereas if h_n is negative, we increase the tap gain by Δ. Figure 7.2.18 shows the schematic of the adaptive equalization arrangement. The determination of $\hat{h}_n$ requires that we have the transmitted sequence available at the receiver. If the probability of error is small, the output of the detector $\hat{a}_n$ can be used in place of a_n in Eq. (7.2.89).

Ideally $h_0 = 1$ and $h_n = 0$, $n \neq 0$, so that the estimates $\hat{h}_n$ for $n \neq 0$ will be fairly small. Thus in order to obtain reasonably converging estimates, the number of observations K must be fairly large. We can overcome this problem by estimating $h_0' = h_0 - 1$ rather than estimating h_0 itself. If we now define $h_n' = h_n$ for $n \neq 0$, the estimates $\hat{h}_n'$ will be of the same order of magnitude for all n. The likelihood function of Eq. (7.2.85) can easily be written in terms of h_n' as

$$\ln p(\mathbf{z}|\mathbf{h}') = \sum_{k=1}^{K} \frac{1}{2\sigma^2}\left(z_k - a_k - \sum_{n=-N}^{N} h_n' a_{k-n}\right)^2 + \ln\left(\frac{1}{2\pi\sigma}\right)^{K/2} \tag{7.2.90}$$

so that the ML estimates are given by

$$\hat{h}_n' = \frac{1}{KS}\sum_{k=1}^{K} a_{k-n}e_k \qquad n=-N\ldots N \tag{7.2.91}$$

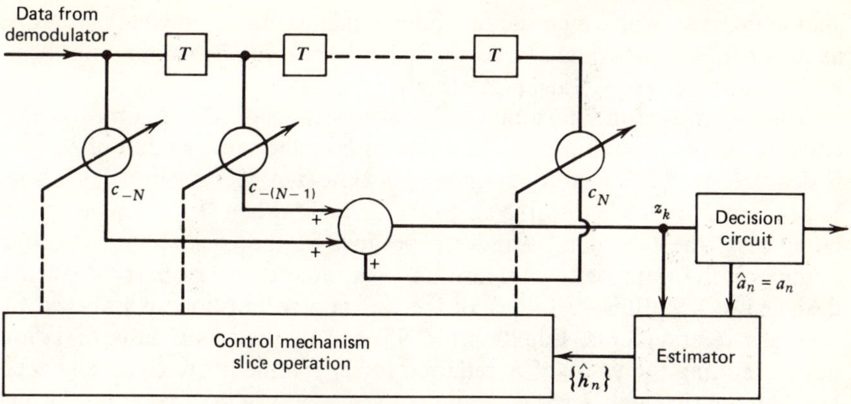

Fig. 7.2.18 Adaptive equalizer mechanism.

where

$$e_k = z_k - a_k \tag{7.2.92}$$

The moments of the estimates can be easily determined. We have from Eqs. (7.2.91) and (7.2.83)

$$\begin{aligned}
\mathrm{E}\{\hat{h}'_n\} &= \mathrm{E}\left\{\frac{1}{KS}\sum_{k=1}^{K} a_{k-n}e_k\right\} \\
&= \frac{1}{KS}\sum_{k=1}^{K}\mathrm{E}\left\{a_{k-n}\left[\sum_{m=-N}^{N} h_n a_{k-m} + v_k - a_k\right]\right\} \\
&= \begin{cases} h_n & n \neq 0 \\ h_n - 1 & n = 0 \end{cases}
\end{aligned} \tag{7.2.93}$$

where the last step follows from the assumed independence of the sequence $\{a_k\}$. It follows therefore that the estimates are unbiased.

The variance of the estimates is given by

$$\operatorname{var}\left[\hat{h}'_n\right] = \mathrm{E}\left\{\left(\hat{h}'_n\right)^2\right\} - \mathrm{E}^2\{h'_n\} \tag{7.2.94}$$

and can be reduced, after considerable algebraic manipulation to

$$\operatorname{var}\left[\hat{h}'_n\right] = \frac{1}{K}\sum_{\substack{n=-N \\ n\neq 0}}^{N} h_n^2 + \frac{2}{K^2}\sum_{j=1}^{K}(K-j)h_{n+j}h_{n-j} + \frac{\sigma^2}{KS} \tag{7.2.95}$$

As the interval over which the tap gains remain constant increases, that is, as K becomes increasingly large, the gains $\hat{h}_n'$ become Gaussian with mean and variance given by Eqs. (7.2.93) and (7.2.95).

The quantity fundamental to the accuracy and setting time of this equalizer is the probability that the sign of $\hat{h}_n$ is the same as that of h_n, and is denoted by P_c. For large K, since $\hat{h}_n$ is Gaussian with mean h_n, it is easy to see that $P_c \simeq 1$ when $|h_n|$ is large and $P_c \simeq 0.5$ when $|h_n|$ is small. In the latter case, the tap gain "wanders" because of the oscillatory corrective nature of the gain setting algorithm. The adaptive equalizer is usually designed to keep this "wander" of the tap gain within bounds imposed by accuracy requirements. Equation (7.2.95) can be used as a design relation in determining the value of K required for any value of N. Generally each tap gain will have an error of about 0.5Δ, where Δ is the magnitude of the adjustment every K samples. Thus each of the gains h_n will have a magnitude of about 0.5Δ. If $|h_n| = 0.5\Delta$, and the adjustment is made in the wrong direction, the corresponding error will be approximately 1.5Δ. We should design the equalizer so that the probability of each h_n being 1.5Δ is fairly small while the probability of h_n being close to 0.5Δ is large. It can be shown [23] from Eq. (7.2.95) that for $h_n = \pm 0.5\Delta$ with equal probability, for $P_c = 0.99$, K must be chosen to be approximately equal to $10.8N$ under no-noise conditions (see also Exercises 7.20–7.22 at the end of the chapter).

Adaptive Equalization Using Extended Kalman Filter

We earlier formulated the design of an equalizer using a Kalman filter where we assumed that the channel sample response $\{h_n\}$ was known. If this knowledge is not available, and if $\{h_n\}$ are varying with time, the equalizer algorithms must be adaptive.

By treating the coefficients (tap gains) of the channel as additional states, we can formulate a nonlinear estimation problem. Since the tap gains are assumed constant, they can be modeled by the difference equation

$$\mathbf{h}(k+1) = \mathbf{h}(k) \tag{7.2.96}$$

If we now augment the message model of Eq. (7.2.81) by this equation we obtain the set of equations

$$\begin{bmatrix} \mathbf{x}(k+1) \\ \hline \mathbf{h}(k+1) \end{bmatrix} = \left[\begin{array}{c|c} \mathbf{F} & \mathbf{0} \\ \hline \mathbf{0} & \mathbf{I} \end{array}\right] \begin{bmatrix} \mathbf{x}(k) \\ \hline \mathbf{h}(k) \end{bmatrix} + \begin{bmatrix} \mathbf{G} \\ \hline \mathbf{0} \end{bmatrix} u(k)$$

and

$$z(k) = \mathbf{h}^T(k)\mathbf{x}(k) + v(k)$$

We note that our message model is still linear in the state equations, but is nonlinear in the observations. We can use the extended Kalman filter algorithms of Section 6.5, to estimate the states $\mathbf{x}(k)$ and the coefficients $\mathbf{h}$. The resulting algorithms are as follows:

$$\begin{bmatrix} \hat{\mathbf{x}}(k|k) \\ \hat{\mathbf{h}}(k|k) \end{bmatrix} = \begin{bmatrix} \hat{\mathbf{x}}(k|k-1) \\ \hat{\mathbf{h}}(k|k-1) \end{bmatrix} + \mathbf{K}(k)\left[z(k) - \hat{\mathbf{h}}^T(k|k-1)\hat{\mathbf{x}}(k|k-1)\right] \tag{7.2.97}$$

$$\begin{aligned} \mathbf{K}(k) = \mathbf{V}_{\tilde{\mathbf{x}}_a}(k|k-1)\left[\hat{\mathbf{h}}(k|k-1) \vdots \hat{\mathbf{x}}(k|k-1)\right]^T \\ \times\Big\{\left[\hat{\mathbf{h}}(k|k-1) \vdots \hat{\mathbf{x}}(k|k-1)\right]\mathbf{V}_{\tilde{\mathbf{x}}_a}(k|k-1) \\ \left[\hat{\mathbf{h}}(k|k-1) \vdots \hat{\mathbf{x}}(k|k-1)\right]^T + V_v(k)\Big\}^{-1} \end{aligned} \tag{7.2.98}$$

$$\begin{aligned} \left[\mathbf{V}_{\tilde{\mathbf{x}}_a}(k|k-1) = \begin{bmatrix} \mathbf{F} & \mathbf{0} \\ \mathbf{0} & \mathbf{I} \end{bmatrix} \right. \\ \times \mathbf{V}_{\tilde{\mathbf{x}}_a}(k-1|k-1)\begin{bmatrix} \mathbf{F}^T & \mathbf{0} \\ \mathbf{0} & \mathbf{I} \end{bmatrix} + \begin{bmatrix} \mathbf{G} \\ \mathbf{0} \end{bmatrix}\left[\mathbf{G}^T \vdots \; \mathbf{0}\right] V_u(k) \end{aligned} \tag{7.2.99}$$

$$\left[\mathbf{V}_{\tilde{\mathbf{x}}_a}(k|k) = \left\{\mathbf{I} - \mathbf{K}(k)\left[\hat{\mathbf{h}}(k|k-1) \vdots \hat{\mathbf{x}}(k|k-1)\right]\right\}\mathbf{V}_{\tilde{\mathbf{x}}_a}(k|k-1)\right. \tag{7.2.100}$$

where $\mathbf{V}_{\tilde{\mathbf{x}}_a}(k|k)$ is the filtered error covariance of the estimate of the augmented state vector.

Equations (7.2.97)–(7.2.100) are the sequential adaptive equalizer algorithms. Reference [25] presents an example of equalization using extended Kalman filter demonstrating an error percentage almost equalling that of equalization with known channel. The algorithms are necessarily complex, but the advantage again is that a sequence of input symbols can be simultaneously estimated, but with different amount of delays. A complete performance analysis of this approach is not yet available, and only simulation results seem to indicate their validity and success. We point out, however, that the adaptive Kalman equalizer can become unstable.

7.3 ANALOG COMMUNICATION SYSTEMS

Wave form estimation concepts discussed in Chapter 6 play a significant role in obtaining the receiver structures of analog communication systems. In such systems, the message to be transmitted modulates the carrier linearly or nonlinearly. Modulation facilitates propagation in the atmospheric medium, and can, in certain cases, provide noise immunity. Amplitude modulations (AM) of different types constitute linear modulation. Frequency and phase modulation, (in general, termed angle modulation), are examples of nonlinear modulation.

We can approach the problem of estimation of signal waveforms in analog communications in two ways: (1) use the MAP approach of Section 6.6, and arrive at an integral equation, and (2) use the state variable techniques to arrive at Kalman filter type of algorithms. In the former case, the solution to the integral equation provides the estimate of the message. When used carefully, we obtain some practical receiver structures. In the latter case, we assume that the message is generated as the output of a linear system excited by white noise. Depending on the type of modulation, the observation model is either linear or nonlinear. Effects such as fading in the channel can be taken into account by suitably augmenting the message model. The model in such cases is usually nonlinear. In either case, we can apply the linear and nonlinear Kalman filter algorithms of Sections 6.4 and 6.6 to obtain appropriate receiver structures.

We present in this section both the approaches to certain specific situations. We use the MAP approach to obtain receiver structures for both amplitude and angle modulation. In the latter case, since the optimum filter turns out to be unrealizable, we obtain an approximate, physically implementable structure. In obtaining solutions using Kalman type algorithms, we discuss structures operating both in continuous time and in discrete time. The latter are implementable either as software on a computer or as special purpose hardware.

7.3.1 MAP Estimation of Modulated Signals [29,30]

We consider the estimation of a signal process $x(t)$ in an analog communication system. The signal $x(t)$ is assumed to be a sample function of a stochastic process and modulates either linearly or nonlinearly, a carrier. The transmitted signal $y(x(t), t)$ is observed in additive Gaussian noise $v(t)$ so that the observations are

$$z(t) = y(x(t), t) + v(t) \qquad t_0 \leqslant t \leqslant t_f \tag{7.3.1}$$

Depending on the form of $y(\cdot, \cdot)$, the observation model of Eq. (7.3.1) can be used to characterize several common analog modulation schemes. With

$$y(x(t), t) = \sqrt{2E}\, x(t) \sin \omega_c(t) \tag{7.3.2}$$

we have double sideband, suppressed carrier amplitude modulation (DSB-SC-AM). If the transmitted signal is

$$y(x(t), t) = \sqrt{2E}\, [1 + \alpha x(t)] \sin \omega_c t \tag{7.3.3}$$

(where α is a scalar referred to as the modulation index) we have conventional double sideband amplitude modulation (DSBAM) in which the carrier component is also transmitted. As noted earlier, in both these cases of amplitude modulation of the carrier, the transmitted signal is linear in the message process $x(t)$.

In angle modulation, either the phase or the frequency of the carrier is varied in accordance with the message signal $x(t)$. In phase modulation (PM), the transmitted signal is

$$y(x(t), t) = \sqrt{2E} \sin (\omega_c t + \beta x(t)) \tag{7.3.4}$$

where again β is the modulation index. The message $x(t)$ now appears nonlinearly in $y(x(t), t)$.

More generally, the message $x(t)$ may first be passed through a linear (usually time-invariant) filter (pre-emphasis network) which alters the spectrum of the message in some desirable fashion. The transmitted signal is

$$y(x(t), t) = \sqrt{2E} \sin \left(\omega_c t + \int_{t_0}^{t_f} h(t, \tau) x(\tau)\, d\tau\right) \tag{7.3.5}$$

where $h(t, \tau)$ is the impulse response of the filter.

Figure 7.3.1 shows the basic components of a general analog modulation system. If the preemphasis network is an integrator, Eq. (7.3.5) becomes

$$y(x(t), t) = \sqrt{2E} \sin \left(\omega_c t + \int_{t_0}^{t_f} x(\tau)\, d\tau\right) \tag{7.3.6}$$

We then have frequency modulation (FM).

Another example of a modulation scheme using preemphasis filtering is vestigial sideband amplitude modulation (VSBAM) which has been widely

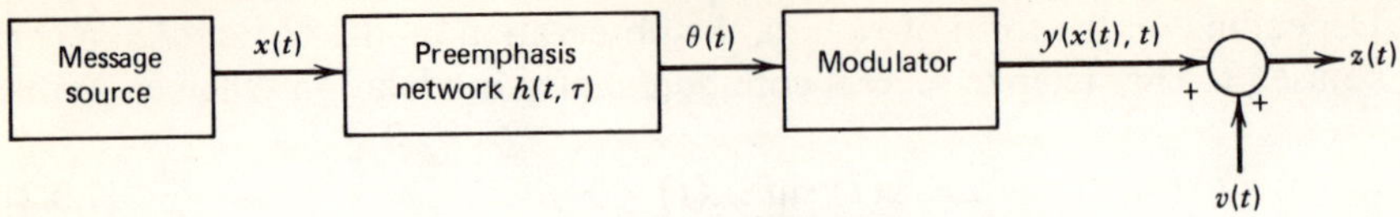

Fig. 7.3.1 Analog modulation system.

applied in the transmission of video signals. In VSBAM, the message $x(t)$ is passed through two linear time-invariant filters with impulse responses $h_c(t)$ and $h_s(t)$. The outputs of the two filters, $x_c(t)$ and $x_s(t)$, are used to amplitude modulate two carriers with the same frequency but in phase quadrature with each other (see Fig. 7.3.2). The transmitted signal is

$$y(x(t), t) = \sqrt{2E}\left[x_c(t)\cos\omega_c t + x_s(t)\sin\omega_c t\right] \tag{7.3.7}$$

In the special case when $x_c(t) = x(t)$ and $x_s(t) = \check{x}(t)$ where $\check{x}(t)$ denotes the Hilbert transform of $x(t)$ (see Section 2.5), the modulation scheme corresponds to single sideband amplitude modulation (SSBAM). In this case

$$H_c(j\omega) = 1$$

and

$$H_s(j\omega) = \begin{cases} -j & \omega > 0 \\ 0 & \omega = 0 \\ +j & \omega < 0 \end{cases}$$

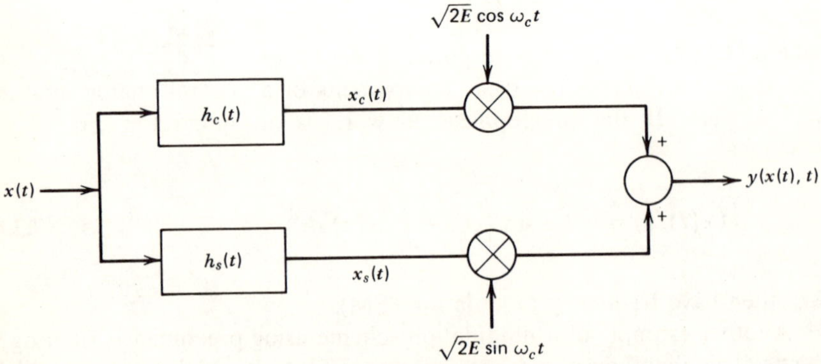

Fig. 7.3.2 **VSBAM** *system*

Under the assumption of a Gaussian message process, our starting point for the derivation of the estimator equations for the observation model of Eq. (7.3.1) is the MAP equation of Section 6.7. We recall that the equation is

$$\hat{x}(t) = \frac{2}{N_0}\int_{t_0}^{t_f} \phi_x(t, u)\frac{\partial y(\hat{x}(u), u)}{\partial \hat{x}(u)}\left[z(u) - y(\hat{x}(u), u)\right] du \qquad t_0 \leqslant t \leqslant t_f \tag{7.3.8}$$

A block diagram realization of the estimator of Eq. (7.3.8) is shown in Fig. 7.3.3. We note that the estimator is not physically realizable since the estimate in Eq. (7.3.8) is based on observations over the entire interval $[t_0, t_f]$. In terms of the notation used in Chapter 6, this implies that $\hat{x}(u)$ is actually the smoothed estimate $\hat{x}(u|t_f)$. Even if the observation interval is $[t_0, t]$, that is, $t_f = t$ the estimate $\hat{x}(u)$ depends on the observations $z(u)$, $t_0 \leqslant u \leqslant t$. We discuss this in some detail when we derive MAP receivers for angle modulated systems.

We now consider modulation systems with preemphasis filtering. Let $\theta(t)$ denote the output of the filter. We can show using similar procedures that the MAP equation yields

$$\hat{x}(t) = \frac{2}{N_0}\int_{t_0}^{t_f}\int_{t_0}^{t_f} \phi_x(t, \sigma)h(\sigma, u)\frac{\partial y(\hat{\theta}(u), u)}{\partial \hat{\theta}(u)}\left[z(u) - y(\hat{x}(u), u)\right] d\sigma\, du \tag{7.3.9}$$

where $h(\cdot, \cdot)$ denotes the impulse response of the preemphasis filter. Equation (7.3.9) can be derived from Eq. (7.3.8) by substituting for $y(\theta(t), t)$ appropriately and noting that the MAP estimate commutes over linear operations. That is,

$$\hat{\theta}(t) = \int_{t_0}^{t_f} h(t, u)\hat{x}(u)\, du$$

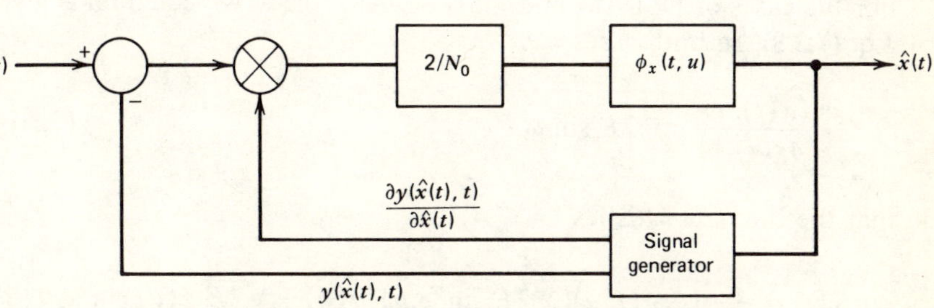

Fig. 7.3.3 **MAP** ***estimator for analog demodulation***

Equation (7.3.9) can be put in the same form as Eq. (7.3.8) by defining

$$\int_{t_0}^{t_f} \phi_x(t, \sigma) h(\sigma, u)\, d\sigma \triangleq h_x(t, u) \tag{7.3.10}$$

so that

$$\hat{x}(t) = \frac{2}{N_0} \int_{t_0}^{t_f} h_x(t, u) \frac{\partial y(\hat{x}(u), u)}{\partial \hat{x}(u)} [z(u) - y(\hat{x}(u), u)]\, du \tag{7.3.11}$$

We see that Eq. (7.3.11) is the same as Eq. (7.3.8) with $\phi_x(t, u)$ replaced by $h_x(t, u)$. Hence the structure of the estimator is as in Fig. 7.3.3 with $\phi_x(t, u)$ again replaced by $h_x(t, u)$.

We now discuss receiver receiver structures for the particular cases of amplitude and frequency modulation.

Amplitude Modulation

We consider the observation model of Eq. (7.3.1)

$$z(t) = y(x(t), t) + v(t) \qquad t_0 \leqslant t \leqslant t_f \tag{7.3.12}$$

where $v(t)$ is zero mean, Gaussian with spectral height $N_0/2$. The signal $y(x(t), t)$ for the various types of amplitude modulation is given by

DSBAM: $\quad y(x(t), t) = \sqrt{2E}\,(1 + x(t)) \sin \omega_c t \qquad$ **(7.3.13a)**

DSB-SC-AM: $\quad y(x(t), t) = \sqrt{2E}\, x(t) \sin \omega_c t \qquad$ **(7.3.13b)**

VSBAM: $\quad y(x(t), t) = \sqrt{E}\,[x_c(t) \cos \omega_c t + x_s(t) \sin \omega_c t)] \qquad$ **(7.3.13c)**

In all cases, we assume that the message bandwidth is much smaller than the carrier frequency.

For the cases of DSBAM and DSB-SC-AM, the MAP equation is given by Eq. (7.3.8). In both cases

$$\frac{\partial y(x(t), t)}{\partial x(t)} = \sqrt{2E} \sin \omega_c t \tag{7.3.14}$$

so that the estimate satisfies

$$\hat{x}(t) = \frac{2}{N_0} \int_{t_0}^{t_f} \phi_x(t, u) (\sqrt{2E} \sin \omega_c u) [z(u) - \sqrt{2E}\, \hat{x}(u) \sin \omega_c u]\, du \tag{7.3.15}$$

We can obtain a solution to this equation by assuming that $x(t)$ is stationary and that $t_0 = -\infty$ and $t_f = \infty$. Then Eq. (7.3.15) can be written as

$$\hat{x}(t) = \frac{2}{N_0}\int_{-\infty}^{\infty} \phi_x(t-u)\left(\sqrt{2E}\sin\omega_c u\right)\left[z(u) - \sqrt{2E}\,\hat{x}(u)\sin\omega_c u\right] du \tag{7.3.16}$$

The block diagram of the estimator follows from Fig. 7.3.3 and is shown in Fig. 7.3.4. In general, the loop filter will be lowpass with respect to the carrier frequency. As such the double frequency terms corresponding to $2E\hat{x}(t)\sin^2\omega_c t$ will be filtered out by the loop filter. By writing Eq. (7.3.16) as

$$\hat{x}(t) = \frac{2}{N_0}\int_{-\infty}^{\infty} \phi_x(t-u)\left[\sqrt{2E}\,z(u)\sin\omega_c u - E\hat{x}(u)\right] du + \frac{2}{N_0}\int_{-\infty}^{\infty} \phi_x(t-u)\hat{x}(u)E\cos 2\omega_c u\,du \tag{7.3.17}$$

and neglecting the double frequency term, we obtain the structure shown in Fig. 7.3.5a. Replacing the feedback loop by a single filter leads to the system of Fig. 7.3.5b. The estimate $\hat{x}(t)$ represents the estimate based on the entire set of observations over interval $(-\infty, \infty)$. Thus the optimum demodulator consists of a multiplier and an optimum unrealizable filter. We recall from our discussions in Chapter 6 that the estimate $\hat{x}(t)$ can be thought of as the output of an infinite-lag smoothing filter. Hence the noncausal filter can be replaced by a realizable filter and a (infinite) delay. A practical implementation of the structure can be obtained by using a delay of sufficient duration.

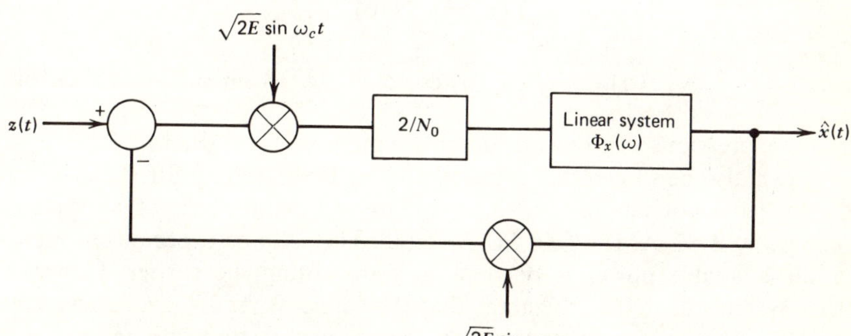

Fig. 7.3.4 **MAP** ***receiver structure for amplitude modulation.***

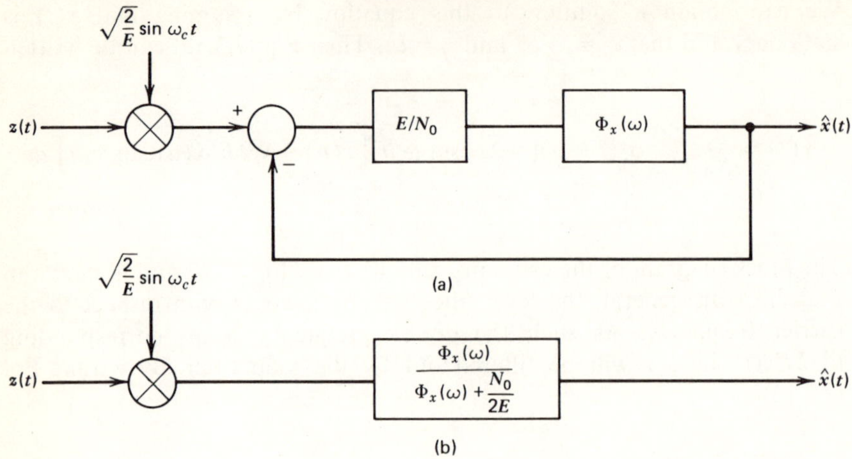

Fig. 7.3.5 Alternate structures for amplitude demodulation.

In the case of VSBAM, the relevant expression for the demodulator is given by Eq. (7.3.9). With a little algebraic manipulation it can be shown that the estimates $\hat{x}_c(t)$ and $\hat{x}_s(t)$ are given by

$$\hat{x}_c(t)=\frac{2}{N_0}\int_{t_0}^{t_f}\sqrt{E}\left\{\cos\omega_c u\left[\int_{t_0}^{t_f}h_c(t,v)\phi_x(v,u)\,dv\right]\right.$$
$$\left.+\sin\omega_c u\left[\int_{t_0}^{t_f}h_x(u,v)\phi_x(t,v)\,dv\right]\right\}$$
$$\cdot\left\{z(u)-\sqrt{E}\left[\hat{x}_c(u)\cos\omega_c u+\hat{x}_s(u)\sin\omega_c u\right]\right\}du \quad \textbf{(7.3.18a)}$$

$$\hat{x}_s(t)=\frac{2}{N_0}\int_{t_0}^{t_f}\sqrt{E}\left\{\cos\omega_c u\left[\int_{t_0}^{t_f}h_s(t,v)\phi_{\hat{x}}(v,u)\,dv\right]\right.$$
$$\left.+\sin\omega_c u\left[\int_{t_0}^{t_f}h_c(u,v)\phi_x(t,v)\,dv\right]\right\}$$
$$\cdot\left\{z(u)-\sqrt{E}\left[\hat{x}_c(u)\cos\omega_c u+\hat{x}_s(u)\sin\omega_c u\right]\right\}du \quad \textbf{(7.3.18b)}$$

These expressions can be considerably simplified for SSBAM.

It can also be shown that systems such as DSBAM or VSBAM in which the carrier is not suppressed lead to greater estimation errors than systems such as DSB-SC-AM or SSB-SC-AM [30]. This is to be expected since part of the available power is devoted to transmitting the carrier. However, these systems have the advantage that the carrier is readily available at the receiver. This is important since the demodulator structures assume that the carrier is known exactly at the receiver. For suppressed carrier systems the receiver must reconstruct the carrier from noisy observations of the transmitted signal. This can be done by including a pilot tone to facilitate reconstruction of the carrier. However, with noisy channels, there is the possibility of significant degradation of performance because of this.

Angle Modulation

We now consider the general angle modulation scheme depicted in Fig. 7.3.2. The message $x(t)$ is passed through a linear filter to obtain the phase function

$$\theta(t) = \int_{t_0}^{t_f} h(t, u)x(u)\, du \tag{7.3.19}$$

The transmitted signal is

$$y(x(t), t) = \sqrt{2E} \sin(\omega_c t + \theta(t)) \tag{7.3.20}$$

The observation model is assumed to be the same as in Eq. (7.3.12). Since

$$\frac{\partial y(x(t), t)}{\partial \theta(t)} = \sqrt{2E} \cos(\omega_c t + \theta(t)) \tag{7.3.21}$$

the MAP estimator follows from Eq. (7.3.8) as

$$\begin{aligned}\hat{x}(t) = \frac{2}{N_0} \int_{t_0}^{t_f} h_x(t, u)\left[\sqrt{2E} \cos\left(\omega_c u + \hat{\theta}(u)\right)\right] \\ \times \left[z(u) - \sqrt{2E} \sin\left(\omega_c u + \hat{\theta}(u)\right)\right] du\end{aligned} \tag{7.3.22}$$

where $h_x(t, u)$ was defined in Eq. (7.3.10). The structure of the estimator is shown in Fig. 7.3.6. Again, by noting that the loop filter usually acts as a lowpass filter in relation to the carrier frequency, we can neglect the double frequency terms in Eq. (7.3.22). We can then write the expression for the estimate as

$$\hat{x}(t) = \frac{2}{N_0} \int_{t_0}^{t_f} h_x(t, u)\left[\sqrt{2E} \cos\left(\omega_c u + \hat{\theta}(u)\right)\right] z(u)\, du \tag{7.3.23}$$

to obtain the simplified structure of Fig. 7.3.7. In the case of phase modulation, $h(t, u) = \delta(t - u)$ as there is no preemphasis filtering. It therefore follows from Eq. (7.3.10) that $h_x(t, u) = \phi_x(t, u)$ for this case. In the

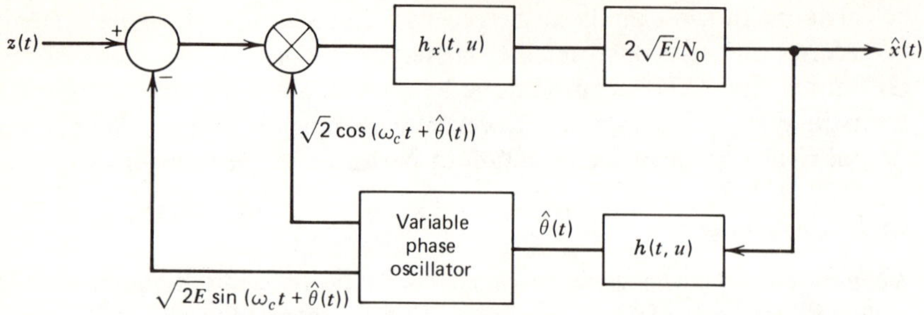

Fig. 7.3.6 **MAP** ***receiver structure for angle demodulation***

case of frequency modulation, the filter impulse response is unity corresponding to an integration in preemphasis filtering.

The filter $h_x(t, u)$ is again noncausal. We recall that for amplitude modulation, we were able to overcome this problem by introducing a delay. Unfortunately, we cannot do this here as the unrealizable filter is in the loop and hence the output of the filter must be fed back at the input in real time. In order to obtain a realizable estimator structure, we modify Eq. (7.3.23) to eliminate this problem. We restrict our discussion to the case of phase modulation and assume that the message process $x(t)$ is stationary. We further assume that the observation interval is $(-\infty, t)$. Under these conditions, the MAP estimate is given by

$$\hat{x}(t) = \frac{2}{N_0} \int_{-\infty}^{t} \phi_x(t-u)\left[\sqrt{2E}\cos\left(\omega_c u + \hat{x}(u)\right)\right] z(u)\, du \quad \textbf{(7.3.24)}$$

A practical implementation of this estimator can be obtained in terms of the PLL introduced earlier. In order to illustrate the application of the PLL in demodulation of angle modulated signals, we have redrawn Fig. 7.2.10 in the form of Fig. 7.3.8 in which the linear time-invariant system $f(t)$ represents the lowpass filter of the phase detector, the loop filter and

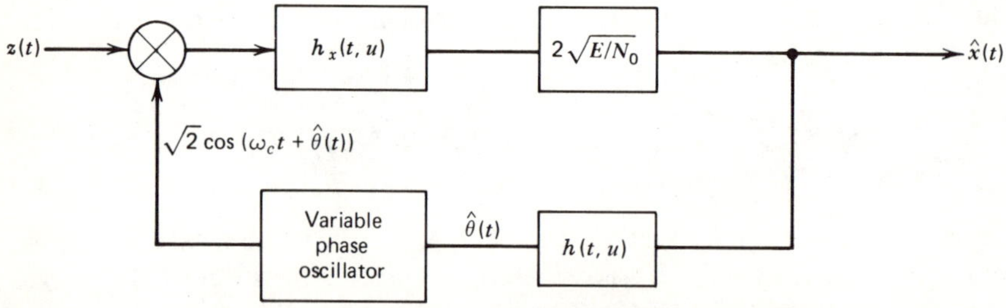

Fig. 7.3.7 ***Alternate structure for angle demodulation.***

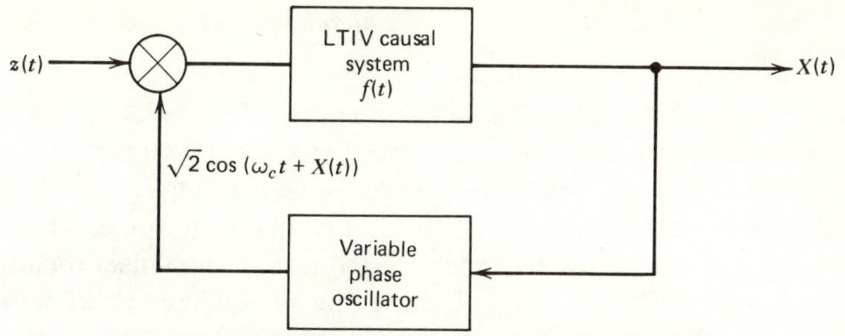

Fig. 7.3.8 The phase lock loop as demodulator.

the loop amplifier taken together. From the figure it easily follows that the output of the voltage controlled oscillator is

$$X(t)=\int_{-\infty}^{t} f(t-u)\left[\sqrt{2}\cos(\omega_c u+X(u)\right]z(u)\,du \tag{7.3.25}$$

If we set

$$f(t)=\frac{2\sqrt{E}}{N_0}\phi_x(t)$$

Eq. (7.3.25) becomes

$$X(t)=\frac{2}{N_0}\int_{-\infty}^{t}\phi_x(t-u)z(u)\left[\sqrt{2}\cos(\omega_c u+X(u))\right]du \tag{7.3.26}$$

While Eqs. (7.3.26) and (7.3.24) appear to be identical, there is a crucial difference between the two. As stated earlier, $\hat{x}(t)$ in Eq. (7.3.24) is the estimate based on the entire observation record. Thus $\hat{x}(u)$ on the right-hand side of Eq. (7.3.24) depends on the observation $z(s)$ for $-\infty<s\leqslant t$. However, $X(t)$ in Eq. (7.3.26) represents the output of a causal system and hence $X(u)$ in the right-hand side of Eq. (7.3.26) depends on $z(s)$ for $-\infty<s\leqslant u\leqslant t$. The PLL therefore, does not realize the MAP estimate. It is, however, important as a practical demodulator and has been widely used in several applications [21,29]. It can be shown that when the signal–noise ratio E/N_0 is large, the phase-lock loop behaves as an optimum demodulator for PM when the phase error is small. The phase-lock loop has superior performance as a demodulator even when the signal-noise ratio is small. If the device is to be used as an FM demodulator, the frequency estimate can be obtained by differentiating the phase estimate.

7.3.2 State-Variable Approach to Demodulation of Analog Modulated Signals [31]

We will now consider the state-variable approach to model some analog communication systems, and obtain receiver algorithms operating in real time. Specifically, we consider the message to be the output of a linear time invariant dynamical system driven by a white Gaussian noise. This output message is modulated by a carrier in a linear or nonlinear manner, as discussed earlier. The receiver observes the modulated signal corrupted with additive white noise. We can thus formulate the problem of demodulation of analog modulated signals in the framework of estimation of waveforms observed in noise discussed in Section 6.4. Unlike the MAP approach, the receivers obtained are physically realizable. In the case of nonlinear modulation, however, the receivers are only near optimal.

As indicated in Chapter 6, with nonlinear message and/or observation models, an explicit solution to the estimation problem is extremely difficult to find. In most cases, we have to resort to suboptimal solutions. The estimator structures increase in complexity as the order of approximation increases. In this section, we consider only a first-order approximate solution obtained by using the extended Kalman filter algorithms of Section 6.6 for continuous-time message models. We first present the extended Kalman filter algorithms for the case in which the message is generated by a linear dynamical system. Then, we apply the algorithms to the cases of DSB-SC-AM, PM and FM.

Extended Kalman Filter for Analog Demodulation

We assume that the message is the output of the linear dynamical system

$$\dot{\mathbf{x}}(t) = \mathbf{A}\mathbf{x}(t) + \mathbf{B}\mathbf{w}(t) \tag{7.3.27}$$

where $\mathbf{w}(t)$ is a Gaussian white noise with zero mean and covariance kernel $\mathbf{V}_{\mathbf{w}}(t)$. Here, A and B are constant matrices of appropriate order. The initial state $\mathbf{x}(0)$ is assumed Gaussian with known mean $E[\mathbf{x}(0)]$ and covariance $\mathbf{V}_{\mathbf{x}}(0)$.

The transmitted signal $y(\mathbf{x}(t), t)$ is a scalar, which can be either a linear or a nonlinear function of $\mathbf{x}(t)$. For example, in the case of double sideband suppressed carrier amplitude modulation (DSB-SC-AM), we have

$$y(\mathbf{x}(t), t) = \mathbf{c}^T\mathbf{x}(t) \cdot \sqrt{2}\, \sin \omega_c t \tag{7.3.28}$$

where $\mathbf{c}$ is a column vector, and ω_c is the carrier frequency in radians per second. In the case of phase modulation, the modulated signal is

$$y(\mathbf{x}(t), t) = \sqrt{2} \sin \left(\omega_c t + \mathbf{c}^T \mathbf{x}(t)\right) \tag{7.3.29}$$

In frequency modulation, the message signal $y_m(t)$ is integrated before modulation so that the transmitted signal is given by

$$y(\mathbf{x}(t), t) = \sqrt{2} \sin \left(\omega_c t + \beta \int_0^t y_m(\tau)\, d\tau\right) \tag{7.3.30}$$

We can put this in the same form as Eq. (7.3.29) by suitably augmenting the message model. Thus if our original message model is

$$\begin{aligned} \dot{\mathbf{x}}_m(t) &= \mathbf{A}_m \mathbf{x}_m(t) + \mathbf{B}_m \mathbf{w}(t) \\ y_m(t) &= \mathbf{c}_m^T \mathbf{x}_m(t) \end{aligned} \tag{7.3.31}$$

where $x_m(t)$ is an n-vector, we let

$$x_{m+1}(t) = y_m(t) \tag{7.3.32}$$

and define the augmented state vector $\mathbf{x}(t)$ and the matrices $\mathbf{A}$, $\mathbf{B}$, and $\mathbf{C}$ by

$$\mathbf{x}(t) = [\mathbf{x}_m(t)\ x_{m+1}(t)]^T$$

$$\mathbf{A} = \begin{bmatrix} \mathbf{A}_m & \mathbf{0} \\ \mathbf{c}_m^T & 0 \end{bmatrix} \quad \mathbf{B} = \begin{bmatrix} \mathbf{B}_m \\ 0 \end{bmatrix} \quad \text{and} \quad \mathbf{c}^T = [\mathbf{0} \quad \beta]$$

We then have

$$\dot{\mathbf{x}}(t) = \mathbf{A}\mathbf{x}(t) + \mathbf{B}\mathbf{w}(t)$$

and

$$y(\mathbf{x}(t), t) = \sqrt{2} \sin \left(\omega_c t + \mathbf{c}^T \mathbf{x}(t)\right)$$

which is similar to Eq. (7.3.29) for the case of phase modulation.

In all cases, the transmitted signal $y(\mathbf{x}(t), t)$ is observed at the receiver in the presence of additive Gaussian white noise $v(t)$. The additive noise $v(t)$ has zero mean and covariance kernel $V_v(t)$. Thus the observed signal becomes

$$z(t) = y(\mathbf{x}(t), t) + v(t) \tag{7.3.33}$$

Figure 7.3.9 shows the message and observation models corresponding to this formulation of the analog modulation system. The solution to the problem of analog demodulation thus reduces to determining the estimate of the message process $x(t)$. We see that the message and observation models of Eq. (7.3.27) and Eq. (7.3.33) are in a form suitable for applying the minimum mean square estimation algorithms of Section 6.4. Since the observation model is non-linear, we use the extended Kalman filter algorithms of Table 6.6.1. The estimation algorithms easily follow from the table as

$$\dot{\hat{\mathbf{x}}}(t|t) = \mathbf{A}\hat{\mathbf{x}}(t|t) + \mathbf{K}(t)[z(t) - y(\hat{\mathbf{x}}(t|t), t)] \quad (7.3.34)$$

The filter gain is

$$\mathbf{K}(t) = \mathbf{V}_{\mathbf{x}}(t|t)\left[\frac{\partial y(\hat{\mathbf{x}}(t|t), t)}{\partial \hat{\mathbf{x}}(t|t)}\right]^T V_v^{-1}(t) \quad (7.3.35)$$

The equation for the error variance follows as

$$\dot{\mathbf{V}}_{\tilde{\mathbf{x}}}(t|t) = \mathbf{A} V_{\tilde{\mathbf{x}}}(t|t) + V_{\tilde{x}}(t|t)\mathbf{A}^T + \mathbf{B}\mathbf{V}_{\mathbf{w}}(t)\mathbf{B}^T$$
$$- \mathbf{V}_{\tilde{\mathbf{x}}}(t|t)\left[\frac{\partial y(\hat{\mathbf{x}}(t|t), t)}{\partial \hat{\mathbf{x}}(t|t)}\right]^T V_v^{-1}(t)\left[\frac{\partial y(\hat{\mathbf{x}}(t|t), t)}{\partial \hat{\mathbf{x}}(t|t)}\right]\mathbf{V}_{\tilde{\mathbf{x}}}(t|t) \quad (7.3.36)$$

These results can easily be specialized to the cases of amplitude and angle modulation discussed previously. As noted earlier, the extended Kalman algorithms are only approximate.

In the case of amplitude modulation, the transmitted signal is given from Eq. (7.3.28) as

$$y(\mathbf{x}(t), t) = \mathbf{c}^T\mathbf{x}(t)\sqrt{2}\sin\omega_c t$$

and is linear in the observations. The algorithms of Eqs. (7.3.34)–(7.3.36)

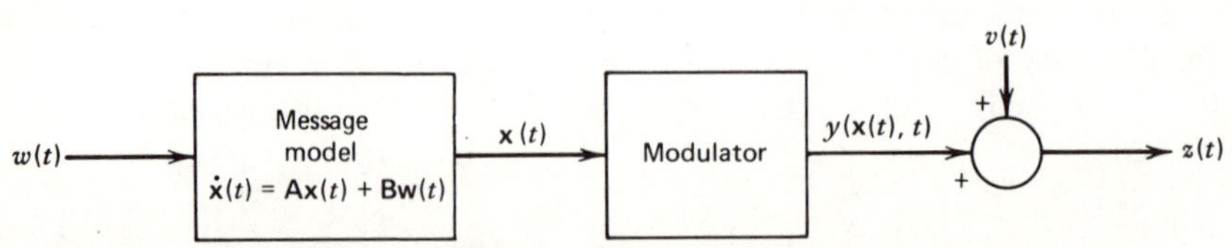

Fig. 7.3.9 Message and observation model for analog modulation systems.

are exact for this problem and reduce to

$$\dot{\hat{\mathbf{x}}}(t|t) = \mathbf{A}\hat{\mathbf{x}}(t|t) + \mathbf{K}(t)\left[z(t) - \mathbf{c}^T\hat{\mathbf{x}}(t|t)\sqrt{2}\sin\omega_c t\right] \tag{7.3.37a}$$

$$\mathbf{K}(t) = \mathbf{V}_{\tilde{\mathbf{x}}}(t|t)\mathbf{c}\left[\sqrt{2}\sin\omega_c t / V_v(t)\right] \tag{7.3.37b}$$

where

$$\begin{aligned}\dot{\mathbf{V}}_{\tilde{\mathbf{x}}}(t|t) &= \mathbf{A}\mathbf{V}_{\tilde{\mathbf{x}}}(t|t) + \mathbf{V}_{\tilde{\mathbf{x}}}(t|t)\mathbf{A}^T + \mathbf{B}\mathbf{V}_{\mathbf{w}}(t)\mathbf{B}^T \\ &\quad - \mathbf{V}_{\tilde{\mathbf{x}}}(t|t)\mathbf{c}\mathbf{c}^T\mathbf{V}_{\tilde{\mathbf{x}}}(t|t)\left[\sqrt{2}\sin^2\frac{\omega_c t}{V_v(t)}\right]\end{aligned} \tag{7.3.37c}$$

In the implementation of the estimator of Eq. (7.3.37), we note that the message of interest is usually baseband, disjoint with the carrier frequency ω_c. This allows us to neglect the double frequency terms in the expression for the error variance, Eq. (7.3.37c). Thus for amplitude modulation (DSB-SC-AM) an approximate expression for $\mathbf{V}_{\tilde{\mathbf{x}}}(t)$ is

$$\begin{aligned}\dot{\mathbf{V}}_{\tilde{\mathbf{x}}}(t|t) \simeq \mathbf{A}\mathbf{V}_{\tilde{\mathbf{x}}}(t|t) + \mathbf{V}_{\tilde{\mathbf{x}}}(t|t)\mathbf{A}^T + \mathbf{B}\mathbf{V}_{\tilde{\mathbf{w}}}(t)\mathbf{B}^T \\ - \frac{1}{V_v(t)}\mathbf{V}_{\tilde{\mathbf{x}}}(t|t)\mathbf{c}\,\mathbf{c}^T\mathbf{V}_{\tilde{\mathbf{x}}}(t|t)\end{aligned} \tag{7.3.38}$$

By defining the gain term

$$\mathbf{K}_1(t) \triangleq \frac{\mathbf{K}(t)}{\sqrt{2}\sin\omega_c t} = \frac{\mathbf{V}_{\tilde{\mathbf{x}}}(t|t)\mathbf{c}}{V_v(t)} \tag{7.3.39}$$

we can rewrite Eq. (7.3.37a) for $\hat{x}(t)$ as

$$\dot{\hat{\mathbf{x}}}(t|t) = \mathbf{A}\hat{\mathbf{x}}(t|t) + \mathbf{K}_1(t)\left[z(t)\sqrt{2}\sin\omega_c t - \mathbf{c}^T\hat{\mathbf{x}}(t|t)\right] \tag{7.3.40}$$

The estimator structure is shown in Fig. 7.3.10.

From the figure, it can be seen that the estimator (demodulator) can be interpreted as a mixer followed by a filter. The structure of the receiver is similar to the MAP demodulator of Fig. 7.3.5. The filter is, however, time-varying in the present case.

Angle Modulation

Angle modulation implies either phase or frequency modulation. As discussed earlier, the transmitted signal in both cases is described by essentially the same type of equation. Because of this, we present the algorithms for PM only, and comment on the receiver structure for FM.

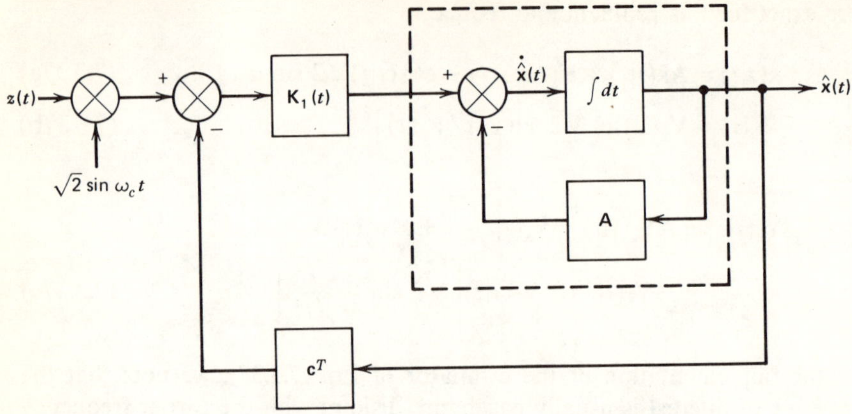

Fig. 7.3.10 **MMSE** *demodulator for amplitude modulation.*

In phase modulation, the transmitted signal is given by Eq. (7.3.29) so that the observations are of the form

$$z(t) = \sqrt{2}\sin\left(\omega_c t + \mathbf{c}^T\mathbf{x}(t)\right) + v(t) \tag{7.3.41}$$

The estimator equations can be written as

$$\dot{\hat{\mathbf{x}}}(t|t) = \mathbf{A}\hat{\mathbf{x}}(t|t) + \mathbf{K}(t)\left[z(t) - \sqrt{2}\sin\left(\omega_c t + \mathbf{c}^T\hat{\mathbf{x}}(t|t)\right)\right] \tag{7.3.42a}$$

where the gain is now given by

$$\mathbf{K}(t) = \mathbf{V}_{\tilde{\mathbf{x}}}(t|t)\mathbf{c}\left[\sqrt{2}\cos\frac{\left(\omega_c t + \mathbf{c}^T\hat{\mathbf{x}}(t|t)\right)}{V_v(t)}\right] \tag{7.3.42b}$$

with

$$\begin{aligned}\dot{\mathbf{V}}_{\tilde{\mathbf{x}}}(t|t) &= \mathbf{A}\mathbf{V}_{\tilde{\mathbf{x}}}(t|t) + \mathbf{V}_{\tilde{\mathbf{x}}}(t|t)\mathbf{A}^T + \mathbf{B}\mathbf{V}_{\mathbf{w}}(t)\mathbf{B}^T \\ &\quad - \mathbf{V}_{\tilde{\mathbf{x}}}(t|t)\mathbf{c}\mathbf{c}^T\mathbf{V}_{\tilde{\mathbf{x}}}(t|t)\left[2\cos^2\frac{\left(\omega_c t + \mathbf{c}^T\hat{\mathbf{x}}(t|t)\right)}{V_v(t)}\right]\end{aligned} \tag{7.3.42c}$$

If we assume that the message process $\mathbf{x}(t)$ has a baseband spectrum that is disjoint from the carrier frequency ω_c, some simplifications follow. In this case, we can again neglect the terms involving the second and higher harmonics of the carrier frequency in the demodulator equations. We can then write the error variance equation as

$$\dot{\mathbf{V}}_{\tilde{\mathbf{x}}}(t) = \mathbf{A}\mathbf{V}_{\tilde{\mathbf{x}}}(t|t) + \mathbf{V}_{\tilde{\mathbf{x}}}(t|t)\mathbf{A}^T + \mathbf{B}\mathbf{V}_{\mathbf{w}}(t)\mathbf{B}^T - \frac{1}{V_v(t)}\mathbf{V}_{\tilde{\mathbf{x}}}(t|t)\mathbf{c}\mathbf{c}^T\mathbf{V}_{\mathbf{x}}(t|t) \tag{7.3.43}$$

which is the same as Eq. (7.3.38) for AM.

Let us define the gain $\mathbf{K}_1(t)$ as

$$\mathbf{K}_1(t) \triangleq \frac{\mathbf{K}(t)}{\left(\sqrt{2}\cos\left(\omega_c t + \mathbf{c}^T\hat{\mathbf{x}}(t|t)\right)\right)} = \frac{\mathbf{V}_{\tilde{\mathbf{x}}}(t|t)\mathbf{c}}{V_v(t)} \tag{7.3.44}$$

Substituting Eq. (7.3.44) in Eq. (7.3.42a) and neglecting double frequency terms yields the estimator equation as

$$\mathring{\hat{\mathbf{x}}}(t|t) = \mathbf{A}\hat{\mathbf{x}}(t|t) + \mathbf{K}_1(t)z(t)\sqrt{2}\cos\left(\omega_c t + \mathbf{c}^T\hat{\mathbf{x}}(t|t)\right) \tag{7.3.45}$$

The structure of the estimator is shown in Fig. 7.3.11. It is clear from the figure that the demodulator for PM can be interpreted as a phase-lock loop and is similar to the MAP demodulator of Fig. 7.3.7 for angle modulated signals. The loop filter is again time-varying in this case.

Since the message model for FM is essentially the same as for PM, it is clear that the structure of Fig. 7.3.11 can be used for demodulation of FM signals. The estimate of the message can be recovered from $\hat{\mathbf{x}}(t|t)$.

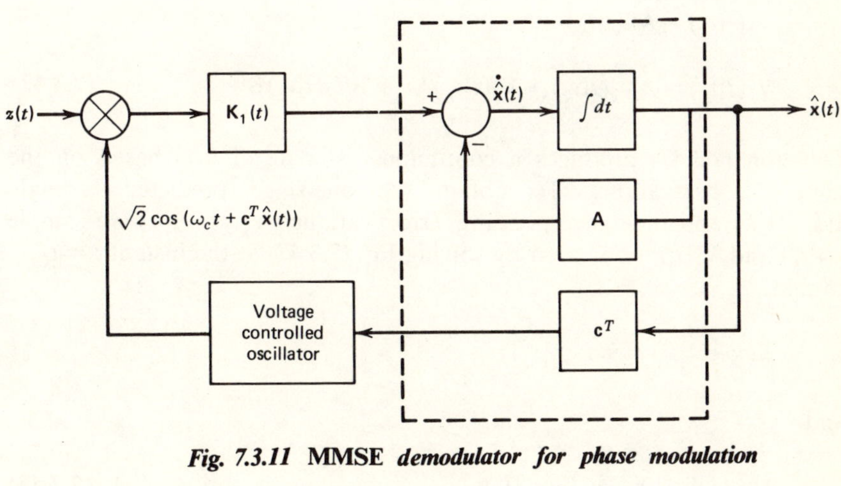

Fig. 7.3.11 MMSE *demodulator for phase modulation*

More complex structures which yield better performance can be obtained by using higher order terms in the approximate expressions for the estimates. A summary of results in this area can be found in [19].

With the increasing use of digital hardware in recent years, it becomes attractive to obtain a digital implementation of the receiver structures. The discrete-time equivalents of these structures can be obtained by sampling the observations $z(t)$ every T seconds to obtain the discrete samples

$$z(k) = y(\mathbf{x}(k), t_k) + v(k) \tag{7.3.46}$$

where we have used

$$z(t_k) = z(k),\ v(t_k) = v(k), \text{ etc.}$$

The variance of the observation noise sequence $v(k)$ is given by $(1/T)V_v(t_k)$. Let us denote this quantity by $V_v(k)$ for convenience. The estimator equations can be obtained by using the algorithms of Section 6.4 for continuous message models with discrete observations. We use the discrete extended Kalman filter algorithms of Section 6.6 to obtain the filtered estimator at sampling instants. In between observations we use the predicted estimate of the signal based on the last observation.

Let us assume that at any instant t_k, we have processed the observation $z(k)$ to obtain the filtered estimate $\hat{\mathbf{x}}(k|k)$ and the corresponding error covariance $\mathbf{V}_{\tilde{\mathbf{x}}}(k|k)$. We use these as initial conditions to solve the estimator equations during the intersampling interval $t_k \leqslant t \leqslant t_{k+1}$. These equations are

$$\dot{\hat{\mathbf{x}}}(t|t_k) = \mathbf{A}\hat{\mathbf{x}}(t|t_k)$$

$$\mathbf{V}_{\tilde{\mathbf{x}}}(t|t_k) = \mathbf{A}\mathbf{V}_{\tilde{\mathbf{x}}}(t|t_k) + \mathbf{V}_{\tilde{\mathbf{x}}}(t|t_k)\mathbf{A}^T + \mathbf{B}\mathbf{V}_{\tilde{\mathbf{x}}}(t|t_k)\mathbf{B}^T \tag{7.3.47}$$

Equation (7.3.47) produces a continuous estimate of $\mathbf{x}(t)$ based on the sampled observations. To obtain the one-stage predicted estimate $\hat{\mathbf{x}}(k+1|k)$ and the corresponding error variance $\mathbf{V}_{\tilde{\mathbf{x}}}(k+1|k)$ we sample $\hat{\mathbf{x}}(t|t_k)$ and $\mathbf{V}_{\tilde{\mathbf{x}}}(t|t_k)$ obtained by solving Eq. (7.3.47) at the instant $t = t_{k+1}$, so that

$$\hat{\mathbf{x}}(k+1|k) = \hat{\mathbf{x}}(t_{k+1}|t_k)$$

and

$$\mathbf{V}_{\tilde{\mathbf{x}}}(k+1|k) = \mathbf{V}_{\tilde{\mathbf{x}}}(t_{k+1}|t_k) \tag{7.3.48}$$

We use these as initial conditions in processing the new observation $z(k+1)$ to obtain $\hat{\mathbf{x}}(k+1|k+1)$. Use of the discrete extended Kalman filter algorithms yields the relevant equations as

$$\hat{\mathbf{x}}(k+1|k+1) = \hat{\mathbf{x}}(k+1|k) + \mathbf{K}(k+1)\big[z(k+1) - y(\hat{\mathbf{x}}(k+1|k), k+1)\big] \tag{7.3.49}$$

$$\mathbf{K}(k+1) = \mathbf{V}_{\tilde{\mathbf{x}}}(k+1|k+1)\frac{\partial y^T(\hat{\mathbf{x}}(k+1|k), k+1)}{\partial \hat{\mathbf{x}}(k+1|k)} V_v^{-1}(k+1) \tag{7.3.50}$$

$$\mathbf{V}_{\tilde{\mathbf{x}}}(k+1|k+1) = \mathbf{V}_{\tilde{\mathbf{x}}}(k+1|k) - \mathbf{V}_{\tilde{\mathbf{x}}}(k+1|k)\frac{\partial y^T(\hat{\mathbf{x}}(k+1|k), k+1)}{\partial \hat{\mathbf{x}}(k+1|k)} \times \big[V_{\tilde{y}}(k+1|k) + V_v(k+1)\big]^{-1} \times \frac{\partial y(\hat{\mathbf{x}}(k+1|k), k+1)}{\partial \hat{\mathbf{x}}(k+1|k)}\mathbf{V}_{\tilde{\mathbf{x}}}(k+1|k) \tag{7.3.51}$$

$$V_{\tilde{y}}(k+1|k) = \frac{\partial y(\hat{\mathbf{x}}(k+1|k), k+1)}{\partial \hat{\mathbf{x}}(k+1|k)} \times \mathbf{V}_{\tilde{\mathbf{x}}}(k+1|k)\frac{\partial y^T(\hat{\mathbf{x}}(k+1|k), k+1)}{\partial \hat{\mathbf{x}}(k+1|k)} \tag{7.3.52}$$

A block diagram schematic of the discrete time processor for continuous messages is given in Fig. 7.3.12.

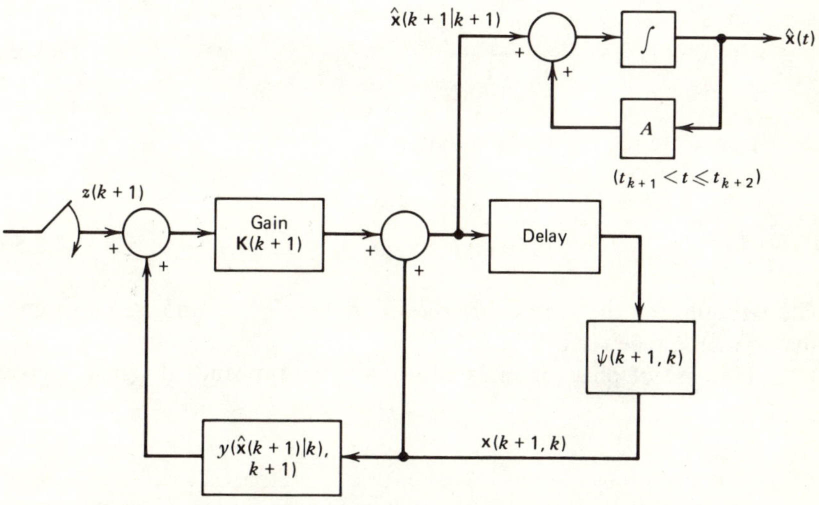

Fig. 7.3.12 Continuous–discrete demodulator for modulated signals.

We now specialize our results to the cases of amplitude and phase modulation. For amplitude modulation, the transmitted signal is as in Eq. (7.3.28), so that the sampled observations become

$$z(k) = \mathbf{c}^T\mathbf{x}(k)\sqrt{2}\sin\omega_c t_k + v(k) \tag{7.3.53}$$

The filtering algorithms follow in a straightforward manner from Eqs. (7.3.49)–(7.3.52) as†

$$\hat{\mathbf{x}}(k+1|k+1) = \hat{\mathbf{x}}(k+1|k) + \mathbf{K}(k+1)\big[z(k+1) - \mathbf{c}^T\hat{\mathbf{x}}(k+1|k)\sqrt{2}\sin\omega_c t_{k+1}\big] \tag{7.3.54a}$$

$$\mathbf{K}(k+1|k) = \mathbf{V}_{\tilde{\mathbf{x}}}(k+1|k)\mathbf{c}\big[V_{\tilde{y}}(k+1|k) + V_v(k+1)\big]^{-1} \times \sqrt{2}\sin\omega_c t_{k+1} \tag{7.3.54b}$$

$$\mathbf{V}_{\tilde{\mathbf{x}}}(k+1|k+1) = \mathbf{V}_{\tilde{\mathbf{x}}}(k+1|k) - \mathbf{V}_{\tilde{\mathbf{x}}}(k+1|k) \times\mathbf{c}\big[V_{\tilde{y}}(k+1|k) + V_v(k+1)\big]^{-1}\mathbf{c}^T\cdot 2\sin^2\omega_c t_{k+1} \tag{7.3.54c}$$

$$V_{\tilde{y}}(k+1|k) = \mathbf{c}^T V_{\tilde{\mathbf{x}}}(k+1|k)\mathbf{c}\cdot 2\sin^2\omega_c t_{k+1} \tag{7.3.54d}$$

As indicated earlier, the initial conditions for these algorithms are obtained by solving Eq. (7.3.47). We can again simplify the error variance algorithm by neglecting the double frequency terms. By defining the gain $\mathbf{K}_1(k+1)$ as

$$\mathbf{K}_1(k+1) = \frac{\mathbf{K}(k+1)}{\sqrt{2}\sin\omega_c t_{k+1}} \tag{7.3.55}$$

we can rewrite the estimator equation as

$$\hat{\mathbf{x}}(k+1|k+1) = \hat{\mathbf{x}}(k+1|k) + \mathbf{K}_1(k+1) \times\big[z(k)\sqrt{2}\sin\omega_c t_{k+1} - \mathbf{c}^T\hat{\mathbf{x}}(k+1|k)\big] \tag{7.3.56}$$

The structure of the receiver is shown in Fig. 7.3.13 and is the discrete analog of the receiver in Fig. 7.3.10.

In the case of phase modulated signals, the transmitted signal is given

†Equation (7.3.54b) is obtained from an alternative form of the gain expression using the one-step error covariance $\mathbf{V}_{\tilde{\mathbf{x}}}(k+1|k)$ derived from Eqs. (7.3.50) and (7.3.51) and a matrix inversion lemma.

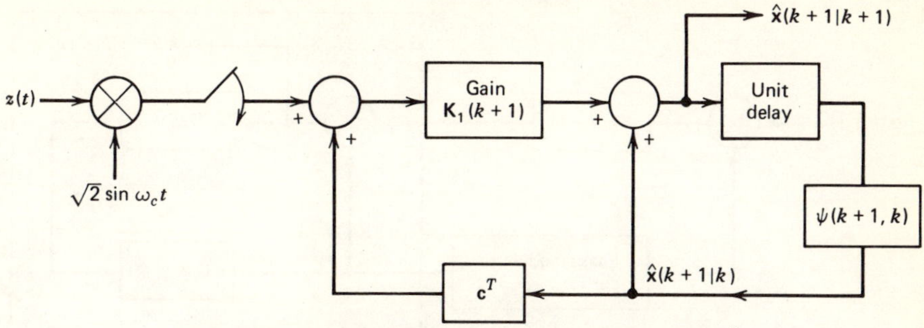

Fig. 7.3.13 Discrete-time receiver for DSB-SC-AM.

by Eq. (7.3.29). The sampled observations are

$$z(k) = \sqrt{2} \sin\left(\omega_c t_k + \mathbf{c}^T\mathbf{x}(k)\right) + v(k) \tag{7.3.57}$$

The estimator equations therefore are

$$\begin{aligned}\hat{\mathbf{x}}(k+1|k+1) = \hat{\mathbf{x}}(k+1|k) + \mathbf{K}(k+1)\\ \times\left[z(k) - \sqrt{2}\sin\left(\omega_c t_{k+1} + \mathbf{c}^T\hat{\mathbf{x}}(k+1|k)\right)\right]\end{aligned} \tag{7.3.58a}$$

$$\begin{aligned}\mathbf{K}(k+1) = \mathbf{V}_{\tilde{\mathbf{x}}}(k+1|k)\mathbf{c}\left[V_{\tilde{y}}(k+1|k) + V_v(k+1)\right]^{-1}\\ \times\sqrt{2}\cos\left(\omega_c t_{k+1} + \mathbf{c}^T\hat{\mathbf{x}}(k+1|k)\right)\end{aligned} \tag{7.3.58b}$$

$$\begin{aligned}\mathbf{V}_{\tilde{\mathbf{x}}}(k+1|k+1) = \mathbf{V}_{\tilde{\mathbf{x}}}(k+1|k) - \mathbf{V}_{\tilde{\mathbf{x}}}(k+1|k)\mathbf{c}\mathbf{c}^T\mathbf{V}_{\tilde{\mathbf{x}}}(k+1|k)\\ \times\left[V_{\tilde{y}}(k+1|k) + V_v(k+1)\right]^{-1}\\ \times 2\cos^2\left(\omega_c t_{k+1} + \mathbf{c}^T\hat{\mathbf{x}}(k+1|k)\right)\end{aligned} \tag{7.3.58c}$$

$$V_{\tilde{y}}(k+1|k) = \mathbf{c}^T\mathbf{V}_{\tilde{\mathbf{x}}}(k+1|k)\mathbf{c}\cdot 2\cos^2\left(\omega_c t_{k+1} + \mathbf{c}^T\hat{\mathbf{x}}(k+1|k)\right) \tag{7.3.58d}$$

By defining

$$\mathbf{K}_1(k+1) \triangleq \frac{\mathbf{K}(k+1)}{\sqrt{2}\cos\left(\omega_c t_{k+1} + \mathbf{c}^T\hat{\mathbf{x}}(k+1|k)\right)} \tag{7.3.59}$$

and again neglecting double frequency terms, the estimator equation can

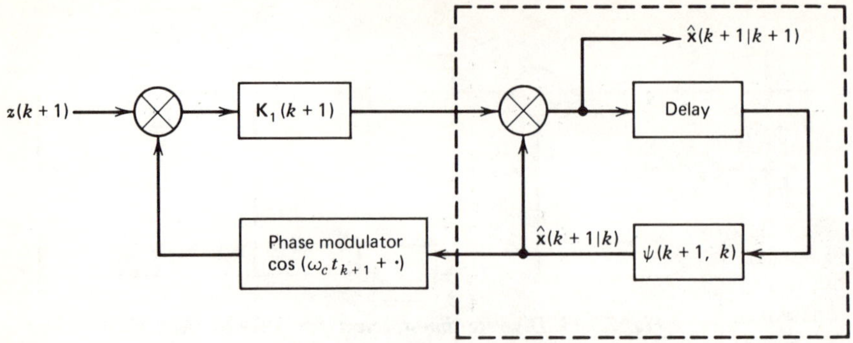

Fig. 7.3.14 Discrete-time angle demodulator.

be simplified as

$$\hat{\mathbf{x}}(k+1|k+1) = \hat{\mathbf{x}}(k+1|k) + \mathbf{K}_1(k+1)z(k) \cdot \sqrt{2}\cos\left(\omega_c t_{k+1} + \mathbf{c}^T\hat{\mathbf{x}}(k+1|k)\right) \quad \textbf{(7.3.60)}$$

We leave the details to the interested reader. Figure 7.3.14 shows the structure of the receiver and can be interpreted as a discrete version of the phase-lock loop.

7.4 SUMMARY

In this chapter, we have considered the application of detection and estimation techniques to the study of communication systems. We started with digital communication systems in Section 7.2. We derived optimal receiver structures and obtained expressions for the probability of error for both baseband and carrier systems. For coherent binary communication systems, the optimum system is PRK for a given average bit energy-to-noise ratio. The receiver consists of correlators followed by a threshold detector.

In the case of noncoherent systems, we essentially have a problem of detecting a signal with unknown parameters in the presence of noise. The unknown parameter in this case is the phase parameter θ. The optimum receiver is obtained by determining the conditional likelihood ratio $\Lambda(z|\theta)$ and averaging over the density function for θ. The receiver can generally be implemented by bandpass filters followed by envelope detectors.

As can be expected, the performance of noncoherent systems is not as good as that of the corresponding coherent systems. However, for very large SNR, the difference in performance between coherent FSK and noncoherent FSK becomes vanishingly small.

A system that does not require a separate phase reference is DPSK. While the performance of DPSK is not as good as PSK, it is better than that of noncoherent FSK. It hence represents a useful compromise between a fully coherent system and a noncoherent system.

In many communication channels, fading occurs due to random fluctuations of the transmission medium. This case is similar to noncoherent communication except that in addition to the phase the amplitude of the transmitted signals is also a random variable. A particularly useful channel model is the Rayleigh model. Expressions for P_e were derived for this model. The problem of synchronization was briefly discussed. The use of the phaselock loop for carrier synchronization was indicated and a MAP technique for estimating timing misalignment was discussed.

An important problem in digital communication is that of intersymbol interference which is caused by the fact that the channel acts as a system with memory. Thus the received pulse over any interval is distorted by contributions from previous pulses. An effective way of overcoming this problem is to pass the observations through an equalizer which is matched to the channel characteristics. The Lucky equalizer uses a tapped delay model of the channel while the Kalman filter equalizer uses a state variable description of the signal and channel characteristics. The design of adaptive equalizers was also considered.

Analog modulation was considered in Section 7.3. The MAP estimator discussed in Chapter 6 was used to obtain receiver structures for the demodulation of analog modulated signals. While the structure of the AM receiver is physically realizable, the receiver structure for angle modulation is not realizable. A good approximation to the MAP estimator in this case can be obtained by using a phase-lock loop.

An alternative approach to analog demodulation is to use a state variable description of the message process and use the minimum mean square estimation techniques of Chapter 6. Since the observation model is nonlinear, the extended Kalman filter algorithms were used to obtain appropriate receiver structures. The receivers for both AM and FM turn out to be physically realizable. The structures are very similar to the ones obtained using the MAP approach. The section concluded with a derivation of discrete structures for analog demodulation obtained by sampling the observations at equally spaced intervals of time. The receivers are the discrete analogs of the continuous-time structures.

EXERCISES

7.1 In a baseband binary polar system, E is 20 mV. Let $N_0 = 10^{-3}$ W/Hz. (a) What is P_e if the binary digits (bits) are transmitted at a rate of 5000 per second? (b) If the bit rate is increased to 10,000 bps, what should E be to give the same probability of error as in (a)? Assume NRZ waveforms.

7.2 In a baseband binary polar system, the amplitude of the transmitted signal over each T-second interval is either $+1$ or -1 with equal probability. If $N_0 = 10^{-3}$, for the minimum probability of error receiver (a) what is the minimum value of T such that $P_e = 10^{-5}$? (b) If we now want to halve the value of T, so that we double transmission rate, what should the amplitude of the signal be for the same P_e?

7.3 We want to build a digital data transmission system with a desired P_e of 10^{-3}. Compare the necessary signal-to-noise ratios required for PRK, OOK, FSK, noncoherent FSK, noncoherent OOK, and DPSK.

7.4 Derive the optimum receiver structure of a DPSK system under the usual assumptions of additive white noise. Prove that the bit error probability is given by $P_e = \frac{1}{2}\exp(-E/N_0)$. Note that 3 dB advantage of DPSK over noncoherent FSK arises because the total energy in the message bearing waveforms is twice that of FSK. (Why?)

7.5 In Chapter 4, we derived the minimum probability of error receiver for M-ary signal detection in white noise under the assumption that the signals were uncorrelated. In this problem we seek to extend our results to the case of correlated signals. The observations are;

$$H_i : z(t) = y_i(t) + v(t) \quad 0 \leqslant t \leqslant T \qquad i = 0, 1, 2, \ldots, M$$

where $v(t)$ is zero-mean, white, Gaussian with spectral density $N_0/2$. Assume that the signals $y_i(t)$ have equal energy with correlation coefficient

$$\rho_{ij} = \frac{1}{\epsilon}\int_0^T y_i(t) y_j(t)$$

where ϵ is the average energy of the signals. Find the optimum receiver and determine the probability of error assuming that the hypotheses are equally likely and the criterion is minimum probability of error.

7.6 In Exercise 7.5, assume that the signals $y_i(t)$ are

$$y_i(t) = \sqrt{E_i/T} \sin \omega_c t \quad 0 \leqslant t \leqslant T \qquad i = 1, 2, 3, 4$$

Find the minimum probability of error if there are an integral number of cycles of each waveform over the observation interval.

7.7 Specialize the results of Exercise 7.5 to M-ary ASK and M-ary PSK. In M-ary ASK, the transmitted signals are of the form

$$y_i(t) = \sqrt{E_i}\, s(t) \qquad i = 0, 1, 2, \ldots, M$$

where

$$\sqrt{E_i} = (i-1)A, \quad \int_0^T s^2(t)\, dt = 1$$

In M-ary PSK, the signals are

$$y_i(t) = \sqrt{2E/T} \cos\left(\omega_c t + \frac{2\pi i}{M}\right) \quad 0 \leqslant t \leqslant T$$
$$i = 0, 1, 2, \ldots, M-1$$

Assume that there are an integral number of cycles of the carrier in each signaling interval.

7.8 In a PRK system, the receiver has a phase error $\Delta\phi$ radians. Prove that the bit error probability assuming equal prior probabilities and minimum probability of error criterion is

$$P_e = \text{erfc}_*\left[\left(\frac{2E\cos^2\Delta\phi}{N_0}\right)^{1/2}\right]$$

7.9 Discuss the effect of an incorrect phase reference in M-ary PSK.

7.10 A binary communication system using orthonormal signals undergoing Rician fading can be modeled as follows:

$$\text{Mark: } z(t) = \sqrt{2}\,\alpha f_1(t)\cos[\omega_c t + \delta] + \sqrt{2}\, v f_1(t)\cos[\omega_c t + \theta] + v(t)$$

$$\text{Space: } z(t) = \sqrt{2}\,\alpha f_0(t)\cos[\omega_c t + \delta] + \sqrt{2}\, v f_0(t)\cos[\omega_c t + \theta] + v(t)$$

$$0 \leqslant t \leqslant T$$

α and δ are the amplitude and phase of the specular component of the signal and are assumed known. v is Rayleigh distributed with $\text{E}[v^2] = 2\sigma^2$, and θ is uniformly distributed. Obtain the optimum receiver structure, and prove that the bit error probability is

$$P_e = Q\left[\frac{\gamma}{(\beta+1/2)^{1/2}\beta^{1/2}}, \frac{\gamma(\beta+1)}{(\beta+2)^{1/2}\beta^{1/2}}\right]$$
$$-\left(\frac{\beta+1}{\beta+2}\right)\exp\left[-\frac{\gamma^2}{2}\left(\frac{\beta^2+2\beta+2}{\beta^2+2\beta}\right)\right]I_0\left[\gamma^2\frac{(\beta+1)}{\beta(\beta+2)}\right]$$

where $\beta \triangleq 2\sigma^2/N_0$, and $\gamma^2 = \alpha^2/\sigma^2$ (ratio of the energy in the specular component to the average energy in the random component). Note that when $\gamma = 0$, we have Rayleigh channel, and $\gamma = \infty$, we have completely known signal.

7.11 In transmitting phase information to the receiver for a coherent communication system, the information is usually corrupted by noise and must be estimated at the receiver. This uncertainty in phase may be modeled as a random variable. Consider partially coherent OOK in which the observations are

$$H_1 : z(t) = \sqrt{2E/T}\sin(\omega_c t + \theta) + v(t) \qquad 0 \leqslant t \leqslant T$$
$$H_0 : z(t) = v(t) \qquad 0 \leqslant t \leqslant T$$

where $v(t)$ is zero mean, white, Gaussian with spectral density $N_0/2$. Here θ is a random variable. It is clear that the decision rule can be obtained by finding the conditional likelihood ratio $\Lambda(z(t)|\theta)$ and averaging over $p(\theta)$. A useful model for θ is to assume that [32]

$$p(\theta : \Lambda_m) = \frac{\exp[\Lambda_m \cos\theta]}{2\pi I_0(\Lambda_m)} \qquad -\pi \leqslant \theta \leqslant \pi$$

where $I_0(\Lambda_m)$ is a modified Bessel function of the first kind. For $\Lambda_m = 0$, we see that $p(\theta) = 1/2\pi$ so that θ is uniformly distributed. For $\Lambda_m \to \infty$, $p(\theta)$ tends to an impulse so that we approach the known signal case. Find the optimum receiver for this problem assuming that the two hypotheses are equally likely. What is the minimum probability of error?

7.12 Extend the results of Exercise 7.11 to M-ary PSK in which the observations are

$$H_i\!: z(t) = \sqrt{2E/T}\sin\left(\omega_c t + \frac{2\pi i}{M} + \theta\right) + v(t) \qquad 0 \leqslant t \leqslant T$$
$$i = 1, 2, \ldots, M$$

7.13 Consider the dual diversity FSK system in Rayleigh medium. If mark was transmitted, the received signals are:

$$z_1(t) = \sqrt{\frac{2}{T} E_{11}} \sin(\omega_1 t + \theta_1) + v_1(t)$$

$$z_2(t) = \sqrt{\frac{2}{T} E_{21}} (\sin \omega_1 t + \theta_2) + v_2(t)$$

If space was transmitted, the received signals are

$$z_1(t) = \sqrt{\frac{2}{T} E_{10}} \sin(\omega_0 t + \phi_1) + v_1(t)$$

$$z_2(t) = \sqrt{\frac{2}{T} E_{20}} \sin(\omega_0 t + \phi_2) + v_2(t)$$

Over any observation interval $(0, T)$ the amplitude and phase are constant, but may vary from pulse to pulse (slow fading). The additive noises are Gaussian, independent, and zero mean with spectral density $N_0/2$. The amplitudes $\sqrt{E_{ji}}$ are identically and independently Rayleigh distributed with mean square value $\bar{E}$. The phase terms, θs and ϕs, are independent and identically uniformly distributed over $[0, 2\pi]$. Obtain the likelihood ratio test, and the optimum receiver structure. Show that the probability of error under the usual assumptions of equal prior probability and the criterion of minimum probability of error is

$$P_e = \frac{4 + 3\delta}{(2 + \delta)^3} \quad \text{where} \quad \delta = \frac{\bar{E}}{N_0}$$

Compare this result to the no diversity reception of FSK signaling in Rayleigh medium.

7.14 Extend the results of the previous problem for M-diversity operation.

7.15 Consider a baseband equiprobable binary (± 1) channel, whose impulse response samples are given by

$$h_{-3} = -0.077, \qquad h_{-2} = -0.355 \qquad h_{-1} = 0.059$$

$$h_0 = 1 \qquad h_1 = 0.059 \qquad h_2 = -0.273$$

and all other samples are zero. The additive noise samples are

Gaussian, white, with zero mean and variance σ_v^2. For each case, consider different signal to noise ratios. (The signal-to-noise ratio may be defined as $(1/\sigma_v^2)$ where unity in the numerator represents the variance of the binary input sequence, and therefore is the signal power.) Design a tap delay equalizer

(a) with 3 taps and delay $D = 2$
(b) with 5 taps and delay $D = 1$
(c) with 5 taps and delay $D = 2$
(d) with 5 taps and delay $D = 3$

Compute the mean square distortion and the mean square error of the equalized system. Present your results in a graphical form involving the various parameters. Discuss your results.

7.16 Apply the discrete Kalman filtering algorithms to design the equalizer for the channel described by Eq. (7.2.81), and derive the structure in Fig. 7.2.16.

7.17 Design a Kalman Equalizer for the channel defined in Exercise 7.15. Obtain the steady state mean square error using numerical computation. How do the results compare with the results of Exercise 7.15? [Hint: Redefine the channel sample response index for convenience. For example change h_{-3} to h_1, h_{-2} to h_2, etc.]

7.18 Show that the minimum mean square error design of a tapped delay line equalizer discussed in Section 7.2.5 reduces to the design for minimum mean square distortion under no noise conditions.

7.19 Assume that in a equiprobable binary (± 1) baseband digital communication system, the channel has a memory of $(2N + 1)$ symbols. The additive noise sequence is white, Gaussian, with zero mean and variance σ_v^2. Obtain a design procedure for a MMSE equalizer (no constraint on structure, such as nonrecursive or recursive) based on the discrete Weiner filtering technique presented in Chapter 6. Show that the equalizer has the structure shown in Fig. E.7.1 when D represents the delay. Identify the coefficients of the equalizer in the spectral factorization involved (Refer to [29] if you have difficulty.)

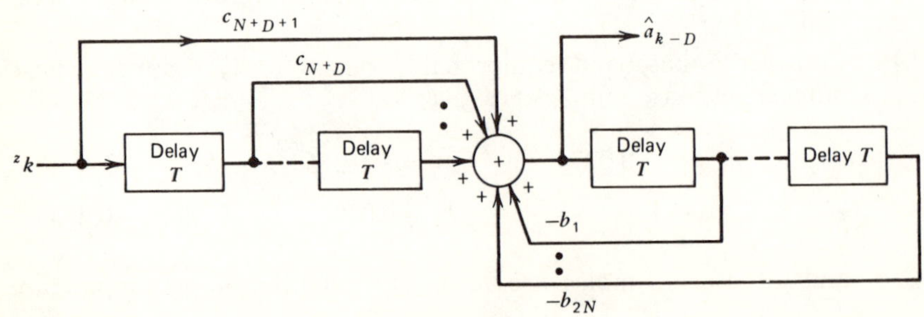

Fig. E.7.1

7.20 In the development of the maximum likelihood estimation approach to the adaptive equalizer, prove that

$$\text{var}\left[\hat{h}_j\right] = \frac{1}{K}\left[\frac{N^2}{2} + \frac{\sigma^2}{S}\right]$$

Consider that the variance is calculated with the probability of $h_j = 0.5\Delta$ or -0.5Δ being equal.

7.21 In the maximum likelihood estimation approach to adaptive equalization discussed in Section 7.2.5, assume that we require $P_e = 0.95$ when $h_j = \pm 0.5\Delta$ and $\Delta = 0.025$. Plot the number of samples required for averaging as a function of SNR (σ^2/S) in dB for different numbers of taps.

7.22 In Exercise 7.20, we require that $P_e = 0.98$ when $h_j = \pm 0.5\Delta$. Assume a SNR of 10 dB. Obtain a plot of K, the number of samples, versus Δ, step size for different N.

7.23 Obtain discrete demodulator algorithms for single sideband amplitude modulated signals (SSB-AM) observed in additive Gaussian, white noise.

7.24 The observations in an angle modulated system are

$$H_1 : z(t) = \sqrt{\frac{2E_1}{T}} \sin\left(\omega_1 t + \theta(t)\right) \qquad -\infty < t < \infty$$

$$H_2 : z(t) = \sqrt{\frac{2E_2}{T}} \sin\left(\omega_2 t + \theta(t)\right) \qquad -\infty < t < \infty$$

where $v_1(t)$ and $v_2(t)$ are zero mean, Gaussian, white with spectral densities $N_1/2$ and $N_2/2$, respectively. Assume $\theta(t)$ is a stationary, zero mean Gaussian process with spectrum $\Phi_\theta(\omega)$. Find the MAP estimate of $\theta(t)$ if the two hypotheses are equally likely.

7.25 In many analog communication channels, fading can be a serious problem. In digital communication, we assumed that channel fading is independent from one bit transmission to another. In the case of analog communication, such a slow fading model may be inappropriate. We can model the fading process as a mutliplicative process. For example, the received signal in an angle-modulated system with fading may be written as

$$z(t) = V_{ch}(t)\sqrt{2E} \sin\left(\omega_c t + \phi_{ch}(t)\right) + v(t)$$

where $V_{ch}(t)$ is a Rayleigh or Rician process and $\phi_{ch}(t)$ is uniform process, so that $V_{ch}(t)\cos\phi_{ch}(t)$ and $V_{ch}(t)\sin\phi_{ch}(t)$ are independent Gaussian variables. Assume that each has a covariance function $K_{ch}(t, u)$. Apply the MAP receiver theory developed in Section 7.3 to obtain an optimal receiver structure.

7.26 Assume in Exercise 7.25, that each of the quadrature processes with covariance $K_{ch}(t, u)$ can be represented in a state variable form. Obtain a receiver structure based on extended Kalman filter using discrete observations.

7.27 As a method of combating fading in analog communication, frequency diversity may be used. In this case, simultaneous transmission of message takes place over the channel at M different carrier frequencies. Hopefully, not all the channels will have fading. We can use all the M received signals to estimate the message. The received signal corresponding to the mth transmission can be modeled as

$$z^m(t) = V_{ch}^m(t)\sqrt{2E} \sin(\omega_c t + \theta(t) + \phi_{ch}^m(t)) + v^m(t) \qquad m = 1, 2, \ldots, M$$

where $V_{ch}^m(t)$ and $\phi_{ch}^m(t)$ correspond to the fading process and are independent and identically distributed for all m. The additive noise processes $v^m(t)$ are also independent and identically distributed over all m. Assuming that the fading process can be represented by a state model, obtain the approximately optimum receiver structure using discrete time observations.

7.28 Formulate the problem of fading reception for various amplitude modulation schemes such as DSBAM, DSB-SC-AM, VSBAM. Obtain the receiver structure using MAP estimation algorithms.

7.29 For Exercise 7.28, obtain the discrete time demodulator algorithm. [Hint: Refer to [33] for possible solution approaches.]

7.30 Consider the discrete time demodulation of DSBSC amplitude modulated signals with finite lag [34]. The signal process is

$$\dot{x}(t) = -\alpha x(t) + (2\alpha)^{1/2} u(t)$$

obtained from the message spectrum $\Phi_x(\omega) = 2\alpha/(\omega^2 + \alpha^2)$. With discretization, the message and transmitted process become

$$x(k+1) = x(k) \exp(-2T) + (2\alpha T)^{1/2} u(k)$$

and

$$y(k) = \sqrt{2}\, x(k) \sin \omega_c t_k + T^{-1/2} \eta(k)$$

Obtain the receiver structure that estimates $x(k - L)$, where L corresponds to a lag of LT seconds. Show that the mean square error is

$$P_e(\tau) = P_e(0) - \frac{(\Lambda/\pi)P_e(0)}{[1 + \Lambda/\pi P_e(0)]} \cdot \left\{1 - \exp\left[1 - 2\alpha\tau\left(1 + \frac{\Lambda}{\pi} P_e(0)\right)\right]\right\}$$

where $\tau = LT$ sec, $\Lambda = 2\pi/V_v\alpha$ (SNR), and $P_e(0)$ is the mean square error of the lag receiver and is given by

$$P_e(0) = \frac{\left(1 + \frac{2\Lambda}{\pi}\right)^{1/2} - 1}{(\Lambda/\pi)}$$

Plot $P_e(\tau)$ as a function of τ. What conclusions can you reach from these plots regarding proper choice of τ?

REFERENCES

1. J. M. Wozencraft and I. M. Jacobs, *Principles of Communication Engineering*, John Wiley and Sons, New York, 1965.
2. M. Schwartz, W. R. Bennett, and S. Stein, *Communication Systems and Techniques*, McGraw-Hill Book Co., New York, 1966.
3. H. L. Van Trees, *Detection, Estimation and Modulation Theory: Part I*, John Wiley and Sons, New York, 1968.
4. D. J. Sakrison, *Communication Theory: Transmission of Waveforms and Digital Information*, John Wiley and Sons, New York, 1968.
5. R. G. Gallager, *Information Theory and Reliable Communication*, John Wiley and Sons, New York, 1968.
6. E. R. Berlekamp, *Algebraic Coding Theory*, McGraw-Hill Book Co., New York, 1968.
7. S. Lin, *An Introduction to Error-Correcting Codes*, Prentice Hall, Englewood Cliffs, N. J., 1970.
8. M. L. Dolez, E. T. Heald, and D. L. Martin, Binary Data Transmission Techniques for Linear Systems, *Proc. IRE*, **45**, No. 5, May, 656–667 (1957).
9. J. I. Marcum, *Table of Q Functions*, Rand Corporation Rpt. RM-339, January 1, 1950.
10. A. R. Di Donato and M. I. Jarnagin, A Method for Computing the Circular Coverage Function, *Math. Comp. 16*, July, 347–355 (1955).
11. D. E. Johansen, New Techniques for Machine Computation of the Q-functions, Truncated Normal Deviates, and Matrix Eigenvalues, *Sylvania ARL, Waltham, Massachusetts, Sci. Rept.*, No. 2, July (1961).

12. D. G. Brennan, Linear Diversity Combining Techniques, *Proc. IRE*, **47**, No. 6, June, 1075–1102 (1959).
13. J. N. Pierce and S. Stein, Multiple Diversity with Non-Independent Fading, *Proc. IRE*, **48**, No. 1, January, 89–104 (1960).
14. G. L. Turin, On Optimal Diversity Reception I, *IRE Trans. Info. Theory*, **IT-7**, No. 3, July, 311–329 (1960).
15. G. L. Turin, On Optimal Diversity Reception II, *IRE Trans. Com. Syst.*, **CS-10**, No. 1, March, 22–31 (1962).
16. B. Barrow, Diversity Combination of Fading Signals with Unequal Mean Strengths, *IEEE Trans. Com. Syst.*, **CS-11**, No. 1, March, 73–78 (1963).
17. A. J. Viterbi, *Principles of Coherent Communication*, McGraw-Hill Book Co., New York, 1966.
18. J. Klapper and J. T. Frankle, *Phase-Locked and Frequency-Feedback Systems*, Academic Press, New York, 1972.
19. S. C. Gupta, Phase-Locked Loops, *Proc. IEEE*, **63**, No. 2, February, 291–306 (1975).
20. J. J. Stiffler, *Theory of Synchronous Communications*, Prentice-Hall, Englewood Cliffs, N. J., 1971.
21. W. C. Lindsey and M. K. Simon, *Telecommunication Systems Engineering*, Prentice-Hall, Englewood Cliffs, N. J., 1973.
22. A. L. McBride and A. P. Sage, Optimum Estimation of Bit Synchronization, *IEEE Trans. Aerospace and Electronic Syst.*, **AE-5**, No. 3, May, 525–536 (1969).
23. R. W. Lucky, Techniques for Adaptive Equalization of Digital Communication Systems, *Bell Systems Tech. J.*, February, 255–286 (1966).
24. R. W. Lucky, J. Salz, and E. J. Weldon, *Principles of Data Communication*, McGraw-Hill Book Co., New York, 1968.
25. R. E. Lawrence and H. Kaufman, The Kalman Filter for the Equalization of a Digital Communications Channel, *IEEE Trans. Com. Tech.*, **Com-19**, No. 6, Dec., 1137–1141 (1971).
26. G. D. Forney, Maximum Likelihood Sequence Estimation of Digital Sequences in the Presence of Intersymbol Interference, *IEEE Trans. Info. Theory*, **IT-18**, No. 3, May, 363–377 (1972).
27. S. M. Fitch and L. Kurz, Recursive Equlization in Data Transmission–A Design Procedure and Performance Evaluation, *IEEE Trans. Commun.*, **Com-23**, No. 5, May, 546–550 (1975).
28. A. Luvison and G. Pirani, Algorithmic and Computational Aspects in the Optimization of a Data Communication System, *IFAC Stochastic Control Symposium*, Budapest, Hungary, Sept. 1974.
29. H. L. Van Trees, *Detection, Estimation, and Modulation Theory, Pt. II Nonlinear Modulation Theory*, John Wiley and Sons, New York, 1971.
30. D. J. Sakrison, *Notes on Analog Communication*, Van Nostrand Reinhold Co., New York, 1970.
31. D. L. Snyder, The State Variable Approach to Analog Communication Theory, *IEEE Trans. Info. Theory*, **IT-14**, No. 1, January, 94–104 (1968).

32. A. J. Viterbi, Optimum Detection and Signal Selection for Partially Coherent Binary Communication, *IEEE Trans. Info. Theory*, **IT-11**, No. 2, April, 239–246 (1965).
33. P. K. Rajasekaran, N. Satyanarayana, and M. D. Srinath, Optimum Linear Estimation of Stochastic Signals in the Presence of Multiplicative Noise, *IEEE Trans. Aerospace Electronic Syst.*, **AES-7**, No. 3, May, 462–468 (1971).
34. S. Prasad and A. K. Mahlanabis, Finite Lag Receivers for Analog Communication, *IEEE Trans. Commun.*, **Com-23**, No. 2, February, 204–213 (1975).

CHAPTER 8

APPLICATIONS TO RADAR SYSTEMS

8.1 INTRODUCTION

Radar is an acronym for *radio detection and ranging*. It is an active device that operates by radiating electromagnetic energy, and detecting the presence and character of the echo returned from reflecting objects. The primary objectives of earlier radars were to detect the presence of a target at large distances and to locate the position of the target. These capabilities have been further extended to obtain information regarding target velocity, its shape, size, and to include mechanisms of automatic detection.

Radars can generally be classified broadly into two functional categories: search radars and tracking radars. As the name implies, a search radar is planned and designed with emphasis on detecting target presence. It may also provide some coarse information regarding the target. A tracking radar emphasizes information extraction, that is, parameter estimation, from the target echo. It tries to estimate with sufficient accuracy the significant target parameters such as range, range rate, angle, and angular velocity once the traget presence is established. However, in present-day radars with electronic scanning arrays performing both the functions adequately, such a binary classification of radars is not always possible.

In this chapter, we apply the results of our development of detection and estimation theory to the search and track problems of radar. We will first discuss the radar detection problem and subsequently consider the problem of estimating certain parameters of interest.

In order to detect the presence or absence of a target, the radar transmitter sends a pulse of electromagnetic energy to illuminate the target. The volume of space covered by the pulse is determined by the azimuth and elevation beamwidths of the transmitting antenna and is referred to as angular beamwidth. By scanning any volume in space in steps of angular beamwidths, the radar tries to search for the target. The total volume scanned is called search volume or surveillance region. Usually the search radars transmit a sequence of narrow r-f pulses every T_p seconds (referred to as interpulse interval), and receive the delayed echoes from the target. The target range is determined in terms of the delay τ as

$$R = \frac{c\tau}{2}$$

where c is the velocity of electromagnetic radiation in free space. For a pulse width of T sec, and a delay of T sec, the range $cT/2$ defines a cell called the range resolution cell.† A search radar effectively examines or interrogates every resolution cell in the surveillance region. In other words, every range element of $cT/2$ in each angular beamwidth is tested for target presence.

If a target is detected in a resolution cell, we know its location within the accuracies of a range element (or range cell $cT/2$) and the beamwidths in elevation and azimuth. This information is rather gross for many tactical applications. By further processing the echo return of a resolution cell, the various parameters can be estimated. The usual parameters of interest are range, range rate, angular position, and angular velocities, and can be estimated using maximum likelihood or maximum a posteriori probability estimation techniques. In recent times, sequential estimation algorithms have been used for dynamic target tracking. Such tracking can also take into account target maneuvers.

We start our discussions with radar target models in Section 8.2. We use the complex envelope representations of bandpass signals (see Chapter 2) in our model as they are convenient in representing the different physical phenomena. We show that depending on the target model, the return signal can be characterized as completely known, completely known except for some random parameters or as a realization of a stochastic process.

8.2 RADAR TARGET MODELS

We assume that a radar transmits a r-f signal which is modulated in both amplitude and phase. We can represent this signal as

$$y_T(t) = \sqrt{2E_T}\, f(t) \cos\left[\omega_c t + \theta_1(t)\right] \qquad 0 \leqslant t \leqslant T \tag{8.2.1}$$

†The range resolution cell is more precisely defined in terms of the effective bandwidth T_{eff} of the basic waveforms (see [15] for details).

Here $f(t)$ is the amplitude modulation which as discussed earlier, is usually a shaped pulse of finite duration T seconds. $\theta_1(t)$ is a phase modulation which is often used for considerations of increased resolution, ω_c is the carrier frequency, and E_T is the transmitted energy. We can represent this bandpass signal in terms of the complex envelope $\tilde{y}(t)$ as

$$y_T(t) = \sqrt{2}\ \mathrm{Re}\left\{\sqrt{E_T}\,\tilde{y}(t)\exp(j\omega_c t)\right\} \tag{8.2.2}$$

where

$$\tilde{y}(t) = f(t)\exp\left[j\theta_1(t)\right]$$

For convenience, we assume that the complex envelope is normalized such that

$$\int_{-\infty}^{\infty} |\tilde{y}(t)|^2\,dt = 1$$

We recall from our discussions in Chapter 2 that the complex envelope $\tilde{y}(t)$ is a baseband signal and corresponds to the in-phase and quadrature components of the modulation (information-bearing) portion of the transmitted signal. We can therefore use the complex envelope $\tilde{y}(t)$ in our discussions without actually using $y_T(t)$.

In what follows, we discuss basically three types of models for the radar target. In the first type, the return signal is essentially completely known, in the second it is known except for some random parameters, and in the third it is a realization of a stochastic process.

8.2.1 Steady Point Target

We assume that the reflecting characteristic of the radar target is linear. The reflected signal energy from a target depends on its electrical echoing area, called radar cross section†[2]. By a point target we mean that all this echoing area is concentrated at a single point in space with an appropriate reflection gain. Suppose the target is located at a range R, and is stationary. In this case the returned signal $y_r(t)$ will be a delayed version of the transmitted signal with a deterministic gain $\tilde{b}$, so that we can write

$$y_r(t) = \sqrt{2}\ \mathrm{Re}\left\{\sqrt{E_T}\,\tilde{b}\tilde{y}(t-\tau)\exp(j\omega_c t)\right\} \tag{8.2.3}$$

†The radar cross section (RCS) is defined as the absolute value of the ratio of the reflected electric filed to the incident electric field, and is generally different from the physical area.

where τ is the time delay and corresponds to a target range $R = c\tau/2$. The reflection gain is a complex quantity, so that we can write

$$\tilde{b} = |\tilde{b}| \exp(j\theta_2) \tag{8.2.4}$$

Here the phase θ_2 includes the term due to carrier delay, $\omega_c\tau$. It is obvious from Eq. (8.2.3) that the received signal energy is

$$E_r = E_T|\tilde{b}|^2 \tag{8.2.5}$$

For nonstationary targets, the effect of target velocity can be taken into account by introducing a Doppler shift of ω_d radians/sec in the carrier frequency ω_c. While this represents only a first-order approximation, it is adequate in many situations. The radar return signal can then be represented as

$$y_r(t) = \sqrt{2}\,\mathrm{Re}\left\{\sqrt{E_T}\,\tilde{b}\tilde{y}(t-\tau)\exp\left[j(\omega_c+\omega_d)t\right]\right\} \tag{8.2.6}$$

Again, we have absorbed the term $\exp(j\omega_d\tau)$ in the phase θ_2 of the reflection gain $\tilde{b}$. It follows that the complex envelope of the return signal from a steady point target at a range $c\tau/2$ and moving with a velocity corresponding to a Doppler shift of ω_d rad/sec can be written as

$$\tilde{y}_r(t) = \sqrt{E_T}\,\tilde{b}\tilde{y}(t-\tau)\exp(j\omega_d\tau) \tag{8.2.7}$$

Equation (8.2.7) can be used to model the return from a steady point target for the case where $\tilde{b}$ is known as well as the case where $|\tilde{b}|$ is known but θ_2 is unknown.

8.2.2 *Slowly Fluctuating Point Target* [3]

For the case in which the target has many different physical reflecting surfaces with arbitrary orientation, we can model it as a point target whose reflecting characteristic is random. That is, the reflection gain $\tilde{b}$ is a complex random variable. We further assume that $\tilde{b}$ takes a particular value for the time duration of interest. That is, if we are observing one pulse return, $\tilde{b}$ remains constant between τ and $\tau + T$ seconds. This is usually valid when the target orientation does not change significantly during the time duration under consideration. Such a target is called a slowly fluctuating point target. The return signal for such a target can therfore be written as in Eq. (8.2.6) but with the modification that $\tilde{b}$ is random. The complex envelope of the return signal is again given by Eq.

(8.2.7) with $\tilde{b}$ random. The average return signal energy is now given by

$$E_r = E_T \mathrm{E}\left[|\tilde{b}|^2\right] \tag{8.2.8}$$

where $\mathrm{E}[|\tilde{b}|^2]$ is the mean square value of $\tilde{b}$.

A number of statistical descriptions have been used to characterize $\tilde{b}$ in the radar literature to account for various physical situations. Almost all of them assume that θ_2, the angle of $\tilde{b}$, is uniformly distributed in $[0, 2\pi]$, and postulate different probability density functions to describe the amplitude $|\tilde{b}|$.

One of the most widely used models assumes that the target is composed of a collection of independent identically distributed random scatterers. In this case, the reflection gain $\tilde{b}$ will be Gaussian distributed or equivalently the amplitude $|\tilde{b}|$ will have a Rayleigh distribution and the phase θ_2 will be uniformly distributed in $[0, 2\pi]$. If we denote the mean square value of $\tilde{b}$ as $2\sigma_b^2$ (or equivalently, the real and imaginary parts of $\tilde{b}$ as having zero mean with variance σ_b^2 each), the probability density of the amplitude of the reflection gain can be explicitly written as

$$p(|\tilde{b}|) = \begin{cases} \dfrac{|\tilde{b}|}{\sigma_b^2} \exp\left(-\dfrac{|\tilde{b}|^2}{2\sigma_b^2}\right) & |\tilde{b}| \geqslant 0 \\ 0 & \text{otherwise} \end{cases} \tag{8.2.9}$$

This model is due to Swerling [4] who has also presented various other models for radar returns. For example, if we assume that the target is composed of one large specular reflector and a number of random scatterers, we obtain the so-called one dominant plus Rayleigh density function (see Exercise 8.6 at the end of the chapter).

As we will see later, for a given average received energy, the detection performance for the case of a fluctuating target is degraded compared to that of a steady target. The performance of the detector can be improved by transmitting a number of pulses, say N, and processing the resulting N returns. In this case, the return signal can be modeled in one of three different ways depending on the physical situation and the assumptions. First let us assume that the reflection gain remains constant, albeit random, during the entire pulse burst length of NT_p seconds where T_p is the interpulse interval. This means that the radar has a pulse to pulse phase coherence. We write the complex envelope of the return signal corresponding to the ith pulse as

$$\tilde{y}_r(t) = \sum_{i=1}^{N} \tilde{b}\sqrt{E_T}\, \tilde{y}(t - iT_p - \tau) \exp\left[j\omega_d(t - iT_p)\right] \tag{8.2.10}$$

with the average signal energy

$$E_r = N \cdot E_T \mathrm{E}\left[|\tilde{b}|^2\right]$$

It can thus be seen that by transmitting a coherent pulse train of length N, we can increase the received signal energy N-fold to achieve better performance.

There are situations in which pulse to pulse coherence does not exist. This leads to the model

$$\tilde{y}_r(t) = \sum_{i=1}^{N} |\tilde{b}| \sqrt{E_T}\, \tilde{y}(t - iT_p - \tau) \exp\left[j\omega_d(t - iT_p) + j\theta_i\right] \quad \textbf{(8.2.11)}$$

where $\{\theta_i\}$ are a set of independent random variables, uniformly distributed in $[0, 2\pi]$. Note, however, that the amplitude $|\tilde{b}|$ is constant through the N pulse period. This corresponds to a slowly (scan to scan) fluctuating target [4], where one scan corresponds to transmitting N pulses.

The third situation arises when the target reflecting characteristics change with each pulse, but remain constant within a pulse period. In this case we can write the complex envelope of the returned signal as

$$\tilde{y}_r(t) = \sum_{i=1}^{N} \tilde{b}_i \sqrt{E_T}\, \tilde{y}(t - iT_p - \tau) \exp\left[j\omega_d(t - iT_p)\right] \quad \textbf{(8.2.12)}$$

where the reflection gains, $\tilde{b}_i$, are independent but identically distributed. That is the phase θ_{2i} are independent and uniformly distributed in $[0, 2\pi]$, and the amplitudes $|\tilde{b}_i|$ are independent and distributed according to Eq. (8.2.9) or some other density function [4]. This is the pulse to pulse or rapidly fluctuating model of Swerling's classical work. We will later find that with proper modifications, Eq. (8.2.12) is a good approximation to range spread targets to be discussed in the next section.

8.2.3 Spread Targets [3]

In this section we concern ourselves with targets that cannot be modeled as slowly fluctuating. Typically these targets exhibit time and/or frequency selective reflecting properties. Consequently, the complex envelope of the returned signal is a stochastic process. Since we have not developed the techniques for the detection of a stochastic process in noise, we discuss these models only briefly and refer the reader to selected references available in the literature.

Let us first consider the case of a target with *time-selective reflecting characteristic*. This can happen when a target is composed of many different physical shapes and its orientation changes rapidly with time. At different times the target exhibits different random reflecting gains, with the result that the returned signal is no longer an attenuated and delayed replica of the transmitted signal. In this case, the reflection gain will be time varying and can be modeled as a realization $\tilde{b}(t)$ of a stochastic process so that the return signal can be written as

$$y_r(t) = \sqrt{2}\,\mathrm{Re}\left\{\sqrt{E_T}\,\tilde{b}\left(t-\frac{\tau}{2}\right)\tilde{y}(t-\tau)\exp(j\omega_c t)\right\} \tag{8.2.13}$$

where τ is the delay corresponding to the target range. The delay of $\tau/2$ is introduced in the reflection gain to explicitly show that the reflected signal traversed the target range only once [3]. Equation (8.2.13) tacitly assumes that the phase of the complex stochastic process $\tilde{b}(t)$ is uniform and includes the delay in the carrier term, $\omega_c\tau$. We can now represent the complex envelope of the return signal as

$$\tilde{y}_r(t) = \sqrt{E_T}\,\tilde{b}\left(t-\frac{\tau}{2}\right)\tilde{y}(t-\tau) \qquad \tau \leqslant t \leqslant \tau + T \tag{8.2.14}$$

As can be seen from Eq. (8.2.13) the effect of the time-varying reflection gain can be thought of as an amplitude modulation, which spreads or widens the energy spectrum of the reflected signal. To make this point clearer, consider the energy spectrum of the return from a slowly fluctuating point target. It is centered at a frequency offset from the carrier by the target Doppler frequency ω_d, and has the same shape as that of the transmitted energy spectrum. In the case of a return from a target with time-selective reflecting characteristic, the effect of the multiplicative term $\tilde{b}(t)$ is to attenuate and spread the transmitted energy spectrum. The center frequency is now given by $\omega_c + \bar{\omega}_d$ where $\bar{\omega}_d$ is the average Doppler frequency to be defined shortly. Because of this spreading effect of the energy spectrum, which is mainly due to rapid changes in target orientation and hence velocity, we call such a target, a *Doppler spread target*. That is, its Doppler frequency is spread about an average frequency $\bar{\omega}_d$.

If we assume that $\tilde{b}(t)$ is a complex stationary process with zero mean and covariance function $\phi_{\tilde{b}}(\tau)$, then the power spectrum $\Phi_{\tilde{b}}(f)$ is known as the *Doppler scattering function*. The name derives from the fact that the energy spectrum of the transmitted signal is convolved with $\Phi_{\tilde{b}}(f)$ to produce the spectral spreading described earlier. The average Doppler frequency is then given by $\bar{\omega}_d = 2\pi\bar{f}_d$ with

$$\bar{f}_d \triangleq \frac{1}{2\sigma_{\tilde{b}}^2}\int_{-\infty}^{\infty} f\,\Phi_{\tilde{b}}(f)\,df \tag{8.2.15}$$

The average received energy is given by

$$\bar{E}_r = \phi_{\tilde{b}}(0)E_T = 2\sigma_b^2 E_T \tag{8.2.16}$$

As noted earlier the return signal from a Doppler spread target is a sample function of a stochastic process. Detection of such signals requires techniques beyond the scope of the present text. However, if we approximate the reflecting gain $\tilde{b}(t)$ by a piecewise constant function, we can use the techniques discussed in earlier chapters to obtain a solution. This is necessarily suboptimum. We proceed as follows. Assume that $\tilde{b}(t)$ is band-limited to W Hz. We can then approximate $\tilde{b}(t)$ as

$$\tilde{b}(t) = \sum_{i=1}^{N} \tilde{b}_i U\left(t - \frac{i}{W}\right) \tag{8.2.17}$$

where $\tilde{b}_i$ are a set of identically distributed independent random variables and $U(\cdot)$ denotes the unit step function. With this approximation, the return signal from a Doppler spread target can be seen to be similar to the signal in Eq. (8.2.7) with obvious correspondences and modifications.

We now consider a target which is extended in range but fluctuates slowly. That is, we assume that the target has a physical length L, and each point on the length of the target acts as a slowly fluctuating, independent, point target. In other words, the reflection gain depends on which point of the total target is reflecting, and the return signal will be a superposition of all such reflections. If we parametrically represent the reflection gain as $\tilde{b}(\tau)$ to denote dependence on τ, the delay corresponding to range, we can write the return signal as

$$y_r(t) = \sqrt{2}\,\text{Re}\left\{\sqrt{E_T}\int_{\tau_1}^{\tau_2} \tilde{b}(\tau)\tilde{y}(t-\tau)\exp(j\omega_c t)\,dt\right\} \qquad \tau_1 \leqslant t \leqslant T+\tau_2 \tag{8.2.18}$$

where τ_1 and τ_2 are, respectively, the delays in the return signals from the target-ends nearest and farthest from the transmitter. Note that the return signal in Eq. (8.2.18) is spread out in time. This is due to the convolution of the range dependent reflection gain with the transmitted signal. Such targets are called range spread targets. It is left to the reader to show that range spread targets indeed cause frequency selective reflecting characteristics. More detailed discussions can be found in the references [3,23].

The most general target model is both range and Doppler spread. From our previous discussions, we know that the reflecting gain is time varying in addition to being dependent on the range. Therefore, we can write the complex envelope of the return signal as

$$\tilde{y}_r(t) = \sqrt{E_T} \int_{\tau_1}^{\tau_2} \tilde{b}\left(t - \frac{\tau}{2}, \tau\right)\tilde{y}(t - \tau)\, d\tau \tag{8.2.19}$$

which is a zero mean complex stochastic process. We note that the singly spread target (Doppler or range) is only a special case of the doubly spread target model given by Eq. (8.2.19).

We are now in a position to use these models in developing the receiver structures for the detection of targets. We present only certain basic results, and develop related special results in problems.

8.3 TARGET DETECTION [4,6–8]

Target detection is a major function of a radar. The reflected signal from a target is a delayed attenuated version of the transmitted signal, which may also be offset in frequency by a Doppler shift. The radar observation of the reflected signal is corrupted by additive noise. The noise presents uncertainty in establishing the presence or absence of the target. Thus we are led to a hypothesis-testing problem. We use the results developed in Chapters 3 and 4 to solve this problem. We formulate the hypothesis-testing problem in terms of complex envelopes of the bandpass signals involved. Further, we also assume that the target to be detected is at zero range with zero Doppler offset (i.e., $\tau = 0$ and $\omega_d = 0$) without any loss of generality. The extension to nonzero range and Doppler frequency is fairly straightforward and is not discussed further in this section.

Let us denote the observed signal by $z(t)$. Then, under hypothesis H_1 with target present, we have

$$z(t) = y_r(t) + v(t) \qquad 0 \leqslant t \leqslant T$$

where $y_r(t)$ is the reflected pulse modeled as in the previous section and $v(t)$ is the observation noise. We assume that $v(t)$ is bandpass, Gaussian white noise with mean zero and spectral density $N_0/2$ and centered at the carrier frequency ω_c.

We can equivalently represent the received signal in terms of its complex envelope as

$$\tilde{z}(t) = \tilde{y}_r(t) + \tilde{v}(t), \qquad 0 \leqslant t \leqslant T \tag{8.3.1}$$

where $\tilde{y}_r(t)$ is the complex envelope of the reflected signal,

$$z(t) = \sqrt{2}\ \mathrm{Re}\ \{\tilde{z}(t) \exp\,(j\omega_c t)\}$$

and

$$v(t) = \sqrt{2}\ \mathrm{Re}\ \{\tilde{v}(t) \exp(j\omega_c t)\}$$

In the alternate hypothesis H_0, the observation consists of pure noise only and hence

$$\tilde{z}(t) = \tilde{v}(t) \qquad 0 \leqslant t \leqslant T \tag{8.3.2}$$

In the sequel, we are concerned with the hypothesis-testing problem when $\tilde{y}_r(t)$ is the complex envelope corresponding to the different target models discussed earlier. Specifically, we consider targets that can be modeled as (i) steady point targets (known and unknown random phase), and (ii) slowly fluctuating point targets. We also consider multiple pulse detection of such targets for achieving improved performance. We conclude the section with a brief discussion on the detection of spread targets.

In deriving the decision rules for the radar problem we use exclusively the Neyman–Pearson criterion. This is mainly because of the difficulty or lack of significance in assigning prior probabilities to target presence at a point in space. We are now ready to discuss the simplest radar detection problem, namely the completely known steady point target.

8.3.1 Steady Point Target: Known Phase

In this case, we model the complex envelope $\tilde{y}_r(t)$ of Eq. (8.3.1) as in Eq. (8.2.7) where the reflecting gain $\tilde{b}$ is completely known including its phase θ_2. The problem then corresponds to the problem of detecting a completely known signal discussed in Section 4.2. As discussed therein, the solution can be obtained by choosing a set of orthonormal functions $\{\tilde{g}_i(t)\}$ as basis functions and expanding $\tilde{z}(t)$ in terms of these functions. If we choose our first basis function to be

$$\tilde{g}_1(t) = \tilde{y}(t)$$

it follows from our discussions in Section 4.2 that the decision can be based entirely on the coefficient corresponding to $\tilde{g}_1(t)$. That is

$$\tilde{y}_1 = \int_0^T \tilde{z}(t)\tilde{y}^*(t)\, dt \tag{8.3.3}$$

is a sufficient statistic for this problem.

The correlation operation of Eq. (8.3.3) can also be implemented by using a matched filter with impulse response $y^*(T-t)$.

We can now write the likelihood ratio as

$$\Lambda(\tilde{y}_1) = \frac{p(\tilde{y}_1|H_1)}{p(\tilde{y}_1|H_0)} \tag{8.3.4}$$

and the test as

$$\Lambda(\tilde{y}_1) \underset{H_0}{\overset{H_1}{\gtrless}} \gamma \tag{8.3.5}$$

where γ is a threshold adjusted according to the level of P_F required.

The sufficient statistic is a linear functional of a complex Gaussian process $\tilde{z}(t)$ and is, therefore, a complex Gaussian random variable. The reader can easily establish that $\tilde{y}_1$ under both hypotheses is distributed normally with

$$p(\tilde{y}_1|H_1) = \frac{1}{\sqrt{\pi N_0}} \exp\left\{ -\frac{|\tilde{y}_1 - \sqrt{E_r} \exp(j\theta_2)|^2}{N_0} \right\} \tag{8.3.6}$$

and

$$p(\tilde{y}_1|H_0) = \frac{1}{\sqrt{\pi N_0}} \exp\left\{ -\frac{|\tilde{y}_1|^2}{N_0} \right\} \tag{8.3.7}$$

where E_r was defined in Eq. (8.2.5). Substitution of these probability expressions in Eq. (8.3.4), enables us to write the likelihood ratio test as

$$\frac{2\sqrt{E_r}\, \text{Re}\,[\tilde{y}_1 \exp(-j\theta_2)]}{N_0} \underset{H_0}{\overset{H_1}{\gtrless}} \ln \gamma + \frac{E_r}{N_0} \tag{8.3.8}$$

Denoting the leftside of the above equation by y_s, it follows from Eqs. (8.3.6) and (8.3.7) that

$$\text{E}[y_s|H_1] = \frac{2E_r}{N_0}, \qquad \text{var}\,[y_s|H_1] = \frac{2E_r}{N_0}$$

$$\text{E}[y_s|H_0] = 0 \quad \text{and} \quad \text{var}\,[y_s|H_0] = \frac{2E_r}{N_0} \tag{8.3.9}$$

The probability of false alarm is the probability that with the target not present, the test statistic y_s will exceed the threshold

$$\gamma_1 = \ln \gamma + \frac{E_r}{N_0}$$

Thus, we have

$$P_F = \int_{\gamma_1}^{\infty} p(y_s)\, dy_s = \int_{\gamma_1}^{\infty} \frac{1}{\sqrt{2\pi}\,\eta} \exp\left[\frac{-y_s^2}{2\eta^2}\right] dy_s$$

where

$$\eta^2 \triangleq \frac{2E_r}{N_0}$$

This can be evaluated by a simple transformation as in Example 3.3.2 to yield

$$P_F = \operatorname{erfc}_*\left(\frac{\gamma_1}{\eta}\right) = \operatorname{erfc}_*\left(\frac{\eta}{2} + \frac{\ln \gamma}{\eta}\right) \tag{8.3.10}$$

The probability of detection is the probability that y_s exceeds the threshold γ_1 when the target is present, and is given by

$$P_D = \int_{\gamma_1}^{\infty} \frac{1}{\sqrt{2\pi}\,\eta} \exp\left[-\frac{(y_s - \eta^2)^2}{2\eta^2}\right] dy_s$$

$$= \operatorname{erfc}_*\left[\frac{\ln \gamma}{\eta} - \frac{\eta}{2}\right] \tag{8.3.11}$$

The radar system design usually utilizes the expression for P_D and P_F to calculate the threshold γ and the required predetection peak signal to noise ratio η^2. For example, let us assume that we require that $P_F \leqslant 10^{-6}$ and $P_D = 0.6$ (or 60% as usually expressed in radar terminology). We can then solve Eqs. (8.3.10) and (8.3.11) simultaneously to arrive at values of γ and η. In this case the predetection peak signal to noise ratio required is 14.5 dB. In general, it is convenient to plot P_D in percentage against η^2 in dB with P_F as a parameter on the curve. Such plots are presented in Fig. 8.3.1. The predetection peak signal-to-noise ratio required for a given P_D and P_F specification in turn determines the transmitted energy requirements through a fundamental equation called the *radar equation* [2].

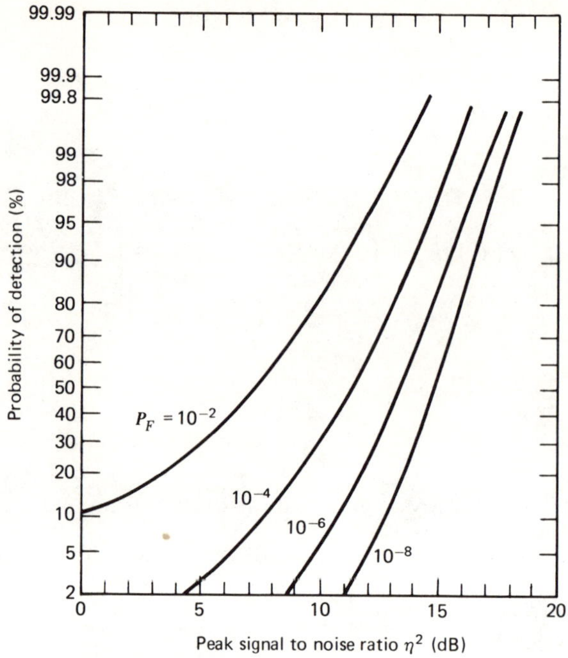

Fig. 8.3.1 Performance curves for single-pulse coherent detection.

8.3.2 Steady Point Target: Unknown Random Phase

In this case, we again use the complex envelope of the received signal, $\tilde{y}_r(t)$ as in Eq. (8.2.7) but with the assumption that θ_2, the phase of the reflection gain is random and is uniformly distributed in $[0, 2\pi]$.

The detection problem is now one of detecting a signal with unknown random parameter discussed in Section 4.6. We first calculate the likelihood ratio assuming θ_2 is known, and then average it over the density function of θ_2 to obtain the generalized likelihood ratio as

$$\Lambda(\tilde{y}_1) = \int_{\chi_{\theta_2}} \Lambda(\tilde{y}_1|\theta_2)p(\theta_2)\, d\theta_2 \tag{8.3.12}$$

where χ_{θ_2} is the space of the parameter θ_2.

From our discussions in the previous subsections, we can easily obtain from Eqs. (8.3.6) and (8.3.7) that

$$\Lambda(\tilde{y}_1|\theta_2) = \exp\left(-\frac{E_r}{N_0}\right) \exp\left\{\frac{2\sqrt{E_r}}{N_0}(\operatorname{Re}\tilde{y}_1 \cos\theta_2 + \operatorname{Im}\tilde{y}_1 \sin\theta_2)\right\} \tag{8.3.13}$$

Denoting $\operatorname{Re}\tilde{y}_1$ as L_c and $\operatorname{Im}\tilde{y}_1$ as L_s in the preceding expression, and using Eq. (8.3.12), we can write the generalized likelihood ratio as

$$\Lambda(\tilde{y}_1) = \exp\left(-\frac{E_r}{N_0}\right)\int_0^{2\pi}\frac{1}{2\pi}\exp\left[\frac{2\sqrt{E_r}}{N_0}(L_c\cos\theta_2 + L_s\sin\theta_2)\right]d\theta_2$$

$$= \exp\left(-\frac{E_r}{N_0}\right)I_0\left[\frac{2\sqrt{E_r}}{N_0}(L_c^2 + L_s^2)^{1/2}\right] \tag{8.3.14}$$

In Eq. (8.3.14), $I_0(\cdot)$ is a modified Bessel function of the first kind and order zero. The decision rule is to compare the likelihood ratio with a threshold. Alternately, $\ln\Lambda(\tilde{y}_1)$ can be compared with a threshold γ, so that the decision rule becomes

$$\ln I_0\left[\frac{2\sqrt{E_r}}{N_0}(L_c^2 + L_s^2)^{1/2}\right]\underset{H_0}{\overset{H_1}{\gtrless}}\frac{E_r}{N_0} + \gamma \triangleq \gamma_1$$

or

$$\ln I_0\left[\frac{2\sqrt{E_r}}{N_0}|\tilde{y}_1|\right]\underset{H_0}{\overset{H_1}{\gtrless}}\gamma_1 \tag{8.3.15}$$

The optimum receiver structure is shown in Fig. 8.3.2.

As in the case of noncoherent digital communication, the decision rule of Eq. (8.3.15) can be simplified. The logarithm of the modified Bessel function $\ln I_0(x)$ can be approximated by $x^2/4$ for very small values of x. With this approximation the decision rule of Eq. (8.3.15) can be implemented using a square law detector as shown in Fig. 8.3.3. The threshold will, of course, have to be modified. For large values of x, $\ln I_0(x)$ can be approximated by x. Therefore, for large peak signal to noise ratio, the square law detector in Fig. 8.3.3 can be replaced by a linear detector, and the threshold adjusted accordingly. Fortunately, most practical envelope detectors behave as square law devices for small signals and linearly for

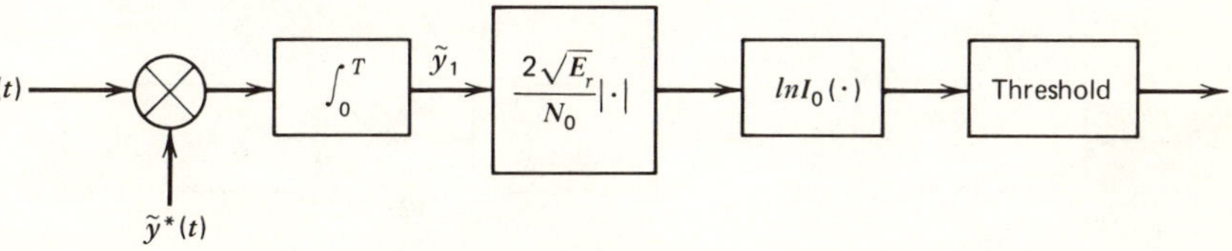

Fig. 8.3.2 Optimum receiver for detection of target return with unknown phase.

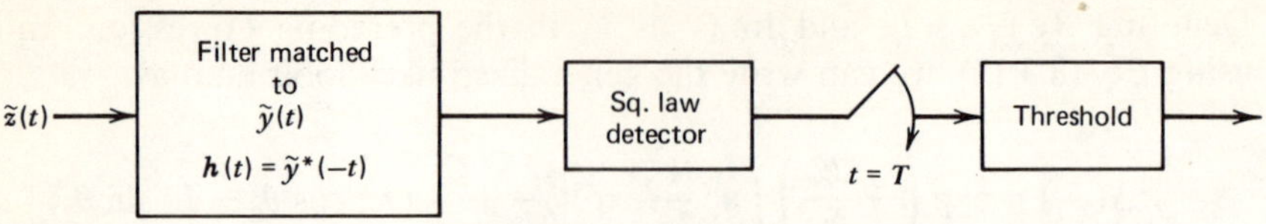

Fig. 8.3.3 A practical receiver for low SNR *for detection with unknown phase.*

large signals. Therefore we can replace the square law detector with a practical envelope detector to obtain an essentially optimum test statistic under small or large signal to noise ratios. This is shown in Fig. 8.3.4.

We now derive the performance of our radar receiver. In order to do this, we note that the decision law is

$$|\tilde{y}_1| \underset{H_0}{\overset{H_1}{\gtrless}} \gamma_2 \tag{8.3.16}$$

where γ_2 is a modified threshold. Let $l = |\tilde{y}_1|$. It is easy to show that the statistic l under the hypothesis H_0 is Rayleigh distributed as

$$p(l|H_0) = \begin{cases} \dfrac{2l}{N_0} \exp\left[-\dfrac{l^2}{N_0}\right], & l \geqslant 0 \\ 0 & \text{otherwise} \end{cases} \tag{8.3.17}$$

The probability of false alarm can then be obtained from Eqs. (8.3.16) and (8.3.17) as

$$\begin{aligned} P_F &= \text{Prob. } \{l > \gamma_2 | H_0\} \\ &= \int_{\gamma_2}^{\infty} p(l|H_0)\, dl = \int_{\gamma_2}^{\infty} \frac{2l}{N_0} \exp\left[-\frac{l^2}{N_0}\right] dl \\ &= \exp\left[-\frac{\gamma_2^2}{N_0}\right] \end{aligned} \tag{8.3.18}$$

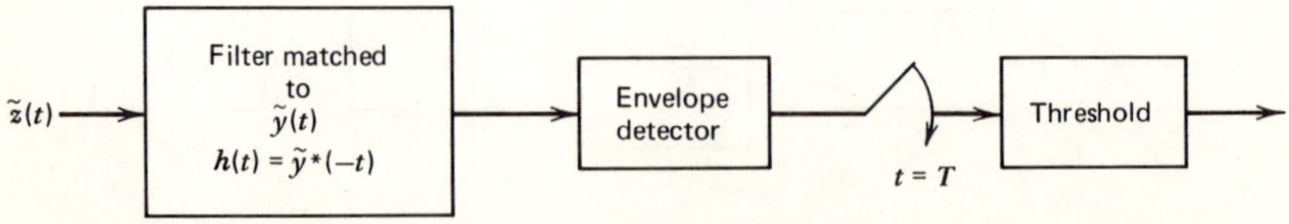

Fig. 8.3.4 A practical receiver for detection with unknown phase for all SNR.

To compute the probability of detection, we need to determine $p(l|H_1)$. In order to do this, we note that L_c and L_s, conditioned on θ_2, are Gaussian under hypothesis H_1 with the following moments:

$$\mathrm{E}[L_c|\theta_2, H_1] = \sqrt{E_r}\cos\theta_2,\ \mathrm{E}[L_s|\theta_2, H_1] = \sqrt{E_r}\sin\theta_2 \tag{8.3.19}$$

and

$$\mathrm{var}\,[L_c|\theta_2, H_1] = \mathrm{var}\,[L_s|\theta_2, H_1] = \frac{N_0}{2} \tag{8.3.20}$$

Therefore $p(l|H_1, \theta_2)$ is a Rician density given by

$$p(l|\theta_2, H_1) = \begin{cases} \dfrac{2}{N_0} l \exp\left[-\dfrac{(l^2+E_r)}{N_0}\right] I_0\left(\dfrac{2l\sqrt{E_r}}{N_0}\right) & l \geqslant 0 \\ 0 \quad \text{otherwise} \end{cases} \tag{8.3.21}$$

The density function $p(l|H_1)$ is obtained by averaging out $p(l|\theta_2, H_1)$ of Eq. (8.3.21) over the density function of θ_2. Because the expression in Eq. (8.3.21) is independent of θ_2, $p(l|H_1)$ is given by the same expression. The probability of detection is given by

$$\begin{aligned} P_D &= \text{Prob.}\ \{l \geqslant \gamma_2|H_1\} \\ &= \int_{\gamma_2}^{\infty} p(l|H_1)\,dl = \int_{\gamma_2}^{\infty} \frac{2}{N_0} l \exp\left[-\frac{(l^2+E_r)}{N_0}\right] I_0\left(\frac{2l\sqrt{E_r}}{N_0}\right) dl \\ &= Q\left[\left(\frac{2E_r}{N_0}\right)^{1/2}, \left(\frac{2}{N_0}\right)^{1/2}\gamma_2\right] \end{aligned} \tag{8.3.22}$$

where $Q[a, b]$ is Marcum's Q functions† and is available in a tabulated form [23], or can be computed using a recursive formula [25].

A relation between P_D and P_F can be obtained by expressing the threshold γ_2 in terms of P_F. We then have, from Eq. (8.3.18)

$$\gamma_2 = \sqrt{-N_0 \ln P_F}$$

†Marcum's Q-functions are defined as

$$Q\,[a, b] = \int_b^{\infty} z \exp\left(-\frac{z^2+a^2}{2}\right) I_0(az)\,dz$$

Substituting the preceding in the expression for P_D, we obtain

$$P_D = Q\left[\eta, (-2 \ln P_F)^{1/2}\right] \tag{8.3.23}$$

where η^2 is the peak signal-to-noise ratio $2E_r/N_0$. The curves of P_D versus η^2 with P_F as parameter is presented in Fig. 8.3.5. This curve is useful in determining the predetection peak signal-to-noise ratio required to attain a given performance level of P_D and P_F for noncoherent reception.

8.3.3 Slowly Fluctuating Point Target

For slowly fluctuating point targets, we model the reflection from the target according to Eq. (8.2.6) where $\tilde{b}$ is a complex random variable. Again the complex envelope of the target return $\tilde{y}_r(t)$ is given by Eq. (8.2.7).

We assume that the random variable $\tilde{b}$ is complex Gaussian with zero mean. In other words, we consider the target to be an assembly of random scatterers, and hence the absolute value of the gain $|\tilde{b}|$ is Rayleigh

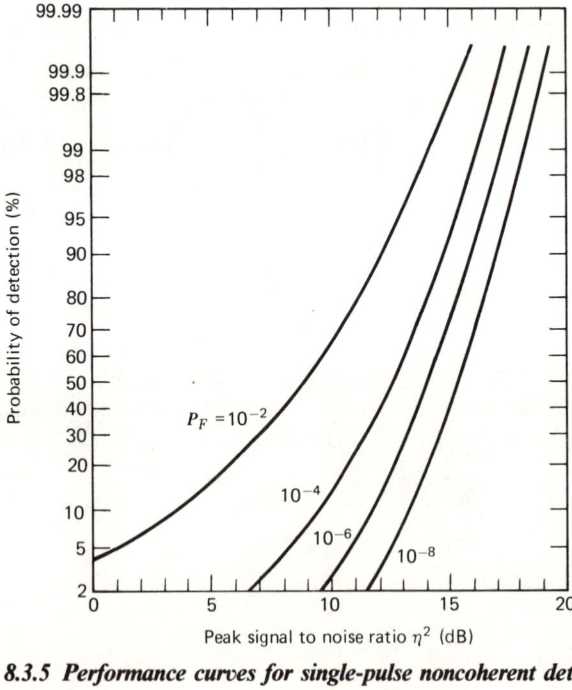

Fig. 8.3.5 Performance curves for single-pulse noncoherent detection.

distributed as in Eq. (8.2.9), and the phase θ_2 is uniformly distributed in $[0, 2\pi]$.

For convenience, we let

$$e = \sqrt{E_T}\,|\tilde{b}| = \sqrt{E_r}$$

Then e is also Rayleigh distributed as follows:

$$p(e) = \begin{cases} \dfrac{2e}{\bar{E}_r} \exp\left(-\dfrac{e^2}{\bar{E}_r}\right), & e \geqslant 0 \\ 0 & \text{otherwise} \end{cases} \tag{8.3.24}$$

where $\bar{E}_r$ is the average received energy and equals $2\sigma_b^2 E_T$.

The problem of detecting a target is now one of obtaining the generalized likelihood ratio and comparing it with a proper threshold.

Thus the optimum decision test can be implemented by first obtaining a likelihood ratio conditioned on the unknown random quantities e and θ_2, and then averaging it over e and θ_2 using appropriate density functions. In order to do this, we recognize that the likelihood ratio expression in Eq. (8.3.14) with $\sqrt{E_r}$ replaced by e gives us the likelihood ratio conditioned on e and averaged over θ_2. Therefore, we write:

$$\Lambda(\tilde{y}_1|e) = \exp\left[-\frac{e^2}{N_0}\right] I_0\left(\frac{2e}{N_0}|\tilde{y}_1|\right) \tag{8.3.25}$$

The likelihood ratio is obtained by averaging $\Lambda(\tilde{y}_1|e)$ over e as

$$\Lambda(\tilde{y}_1) = \int_0^\infty \exp\left[-\frac{e^2}{N_0}\right] I_0\left(\frac{2e}{N_0}l\right) p(e)\, de \tag{8.3.26}$$

where $l = |\tilde{y}_1|$. Let us denote the integral on the right side of Eq. (8.3.26) by $f(l)$. We note that $f(l)$ is monotonic in l. This is so irrespective of what the distribution of e is, because $I_0(\cdot)$ is an even function of its argument, and $p(e)$ is always positive as it is a density function. This simply shows that l, the envelope of the received waveform is indeed a sufficient statistic. Therefore, we can implement the optimum test by comparing $l = |\tilde{y}_1|$ with a threshold set according to P_F requirements. In other words, the test of Eq. (8.3.16) is indeed optimum even if the amplitude of the echo is a random variable. This shows that the receiver structure of Fig. 8.3.4 is optimum for fluctuating targets as well.

Because of the invariance of the receiver structure, we can obtain the probability of detection for fluctuating targets in a manner similar to our derivation of the likelihood ratio. We average the expression for P_D in Eq. (8.3.22) over the density function for the random variable e to obtain

$$P_D = \int_0^\infty P_D(e)p(e)\,de \tag{8.3.27}$$

where $P_D(e)$ is obtained from Eq. (8.3.22) as

$$P_D(e) = \int_{\gamma_2}^\infty \frac{2l}{N_0} \exp\left[-\frac{(l^2+e^2)}{N_0}\right] I_0\left(\frac{2el}{N_0}\right) dl \tag{8.3.28}$$

Here γ_2 is the threshold of the test. The evaluation of P_D is most conveniently carried out by defining the normalized variables

$$a = \left(\frac{2}{N_0}\right)^{1/2} l,\ \eta^2 = \left(\frac{2e^2}{N_0}\right),\ \text{and}\ \gamma_3 = (2/N_0)^{1/2}\gamma_2$$

We then have, for the probability of detection

$$P_D = \int_0^\infty P_D(\eta)p(\eta)\,d\eta \tag{8.3.29}$$

where

$$P_D(\eta) = \int_{\gamma_3}^\infty a \exp\left[-\frac{(a^2+\eta^2)}{2}\right] I_0[\eta a]\,da \tag{8.3.30}$$

For Rayleigh fluctuating targets (Eq. 8.3.24), it easily follows that the density function of the variable η is

$$p(\eta) = \begin{cases} \dfrac{2\eta}{\bar{\eta}^2} \exp\left(-\dfrac{\eta^2}{\bar{\eta}^2}\right), & \eta \geqslant 0 \\ 0 & \text{otherwise} \end{cases} \tag{8.3.31}$$

where $\bar{\eta}^2 = [2\bar{E}_r/N_0]$, the average peak signal to noise ratio. On substituting Eqs. (8.3.30) and (8.3.31) in Eq. (8.3.29) and interchanging the orders of integration, we obtain the expression for P_D as

$$P_D = \int_{\gamma_3}^\infty \frac{2a}{\bar{\eta}^2} \exp\left[-\frac{a^2}{2}\right] \int_0^\infty \eta \exp\left[-\frac{(2+\bar{\eta}^2)}{2\bar{\eta}^2}\eta^2\right] I_0[\eta a]\,d\eta\,da \tag{8.3.32}$$

The inner integral can be evaluated using standard formulas† to yield the expression for P_D as

$$P_D = \int_{\gamma_3}^{\infty} \frac{2a}{(2+\bar{\eta}^2)} \exp\left[-\frac{a^2}{(2+\bar{\eta}^2)}\right] da$$

$$= \exp\left[-\frac{\gamma_3^2}{(2+\bar{\eta}^2)}\right]$$

$$= \exp\left[-\frac{\gamma_2^2}{(N_0/2)(2+\bar{\eta}^2)}\right] \qquad \textbf{(8.3.33)}$$

The probability of false alarm is the same as for noncoherent detection, and is given by Eq. (8.3.18). Repeating the expression for convenience, we have

$$P_F = \exp\left[-\frac{\gamma_2^2}{N_0}\right] \qquad \textbf{(8.3.34)}$$

so that P_D can be written as

$$P_D = \exp\left[\frac{2\ln P_F}{(2+\bar{\eta}^2)}\right] = (P_F)^{\frac{1}{(1+\bar{\eta}^2/2)}} \qquad \textbf{(8.3.35)}$$

A plot of the probability of detection, P_D versus $\bar{\eta}^2$, the average peak signal to noise ratio, with P_F, the probability of false alarm for the Rayleigh fluctuating target, as parameter is shown in Fig. 8.3.6.

†The integral of interest here is the following integral [5, 8]:

$$\int_0^{\infty} t^{\mu-1} I_\nu(\alpha t) \exp\left[-p^2t^2\right] dt$$

$$= \frac{\Gamma\left(\frac{\mu+\nu}{2}\right)\left(\frac{\alpha}{2p}\right)^{\nu}}{2p^{\mu}\Gamma(\nu+1)} \exp\left(\frac{\alpha^2}{4p^2}\right) {}_1F_1\left(\frac{\nu-\mu}{2}+1, \nu+1; \frac{-\alpha^2}{4p^2}\right)$$

where $I_\nu(\cdot)$ is the modified Bessel function of the first kind and order ν, $\Gamma(\cdot)$ is the gamma function and ${}_1F_1(x_1, x_2, x_3)$ is the confluent hypergeometric function.

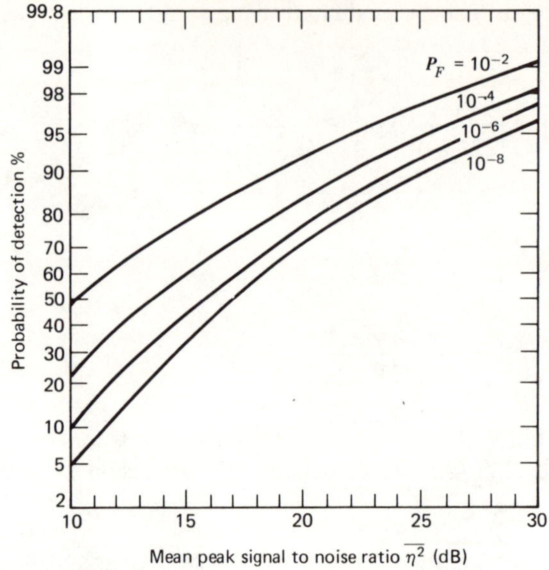

Fig. 8.3.6 Performance curves for single-pulse detection of a Rayleigh fluctuating (pulse to pulse) target.

8.3.4 Multiple Pulse Detection

We now consider radar systems that use several reflected echoes to make a decision regarding the presence of a target. We refer to this as multiple pulse detection. Because we are using a number of pulses, the average signal-to-noise ratio is significantly greater than that for a single pulse,† resulting in better detectability. Further, in the case of fluctuating targets, multiple pulse detection provides better performance in the useful ranges of P_D and P_F. This can be explained by observing that multiple pulse detection corresponds to time diversity in the case of digital communication.

In the echo pulse train, each of the pulses may be coherently related. Such a coherent pulse train corresponds to a completely known signal. In this case, the optimum filter is the same as for coherent detection based on single pulse, except that the matched filter is matched to the pulse train. Such a processing technique is known as coherent processing or predetection integration. If the average peak signal to noise ratio per pulse is $\bar{\eta}^2$, the average peak signal to noise ratio for a train of N coherent pulses is $N\bar{\eta}^2$.

†The signal-to-noise ratio will be increased only if the pulses are on the target. Usually, most targets stay within the radar beamwidth for the entire scan.

Therefore, for performance calculations, we can use the results of coherent single-pulse detection with $\bar{\eta}^2$ replaced by $N\bar{\eta}^2$. We can, therefore, calculate the peak SNR required for a given P_D and P_F from Fig. 8.3.1 and divide by N to obtain the peak SNR required per pulse for the same performance.

In many cases, we will have a pulse train in which the pulses are coherently related, but the initial phase of the pulse train is unknown and random. This implies that the different pulses of the pulse train maintain a known phase relationship. Such a pulse train can be thought of as a completely known signal with random phase. An optimum receiver for this case will then correspond to noncoherent detection based on single pulse return with the matched filter matched to the pulse train. That is, the receiver consists of a bandpass filter (or equivalently two quadrature correlators) matched to the pulse train and followed by an envelope detector, and threshold comparator. Again the performance can be calculated using Fig. 8.3.4 in a manner similar to that used for a coherent pulse train.

An important case of multiple pulse detection arises when the target reflecting characteristics are varying according to Eq. (8.2.12). Following arguments similar to the case of a single pulse and using the independence of the reflection gains $\tilde{b}_i$, we can write the likelihood ratio for this case as

$$\begin{aligned}\Lambda(\tilde{\mathbf{y}}_1) &= \prod_{i=1}^{N} \int_0^\infty \Lambda_i(y_{1i}|e_i)p(e_i)\, de_i \\ &= \prod_{i=1}^{N} \Lambda_i(\tilde{y}_{1i}) \end{aligned} \tag{8.3.36}$$

where e_i is the magnitude of $\tilde{b}_i$, $\tilde{y}_{1i}$ is the sufficient statistic (correlator output) for the ith pulse and $\Lambda_i(\tilde{y}_{1i}|e_i)$ is the likelihood ratio for the ith pulse conditioned on e_i but averaged over the phase θ_{2i}. Assuming that e_is are independent and identically Rayleigh distributed† according to Eq. (8.3.24), we can write with $l_i = |\tilde{y}_{1i}|$

$$\Lambda_i(\tilde{y}_{1i}) = \int_0^\infty \frac{2e_i}{\bar{E}_r} \exp\left[-\frac{e_i^2}{\bar{E}_r}\right] I_0\left[\frac{2e_i}{N_0} l_i\right] \exp\left[-\frac{e_i^2}{N_0}\right] de_i \tag{8.3.37}$$

The use of normalized variables as in Eq. (8.3.28) yields a simplified form for the per pulse likelihood ratio $\Lambda_i(\tilde{y}_{1i})$. Thus, in terms of the variable $a_i = (2/N_0)l_i$, we obtain

$$\Lambda_i(\tilde{y}_{1i}) = \frac{1}{2p^2\bar{\eta}^2} \exp\left[\frac{a_i^2}{4p^2}\right] \tag{8.3.38}$$

†This corresponds to Swerling II model in radar literature [4,8].

where

$$p^2 = \left(\frac{1}{2} + \frac{1}{\bar{\eta}^2}\right) \text{ and } \bar{\eta}^2 = \frac{2E_r}{N_0}$$

Thus $\bar{\eta}^2$ represents the mean peak SNR.

Substitution in Eq. (8.3.37) yields the generalized likelihood ratio for pulse to pulse Rayleigh fluctuating targets as

$$\Lambda(z(t)) = \left(\frac{1}{2p^2\bar{\eta}^2}\right)^N \exp\left[\frac{1}{4p^2}\sum_{i=1}^{N} a_i^2\right] \tag{8.3.39}$$

It follows that the decision rule can be implemented as

$$\sum_{i=1}^{N} a_i^2 \underset{H_0}{\overset{H_1}{\gtrless}} \gamma_1^2$$

or equivalently

$$\sum_{i=1}^{N} |\tilde{y}_{1i}|^2 \underset{H_0}{\overset{H_1}{\gtrless}} \gamma_2^2 \tag{8.3.40}$$

where γ_2^2 is the threshold. We conclude that the square law detector, followed by pulse train sampler and an integrator is indeed the optimum receiver structure for all SNR in the case of pulse to pulse fluctuating targets.

We can determine the performance of this detector by applying the results of Example 3.4.2. We note that $\tilde{y}_{1i}$ in Eq. (8.3.40) is a *complex* Gaussian random variable. To determine its moments under the two hypotheses, we note that $p(\tilde{y}_{1i}|H_i)$ can be obtained by averaging $p(\tilde{y}_{1i}|H_1, e_i, \theta_{2i})$ over the probability spaces of e_i and θ_{2i}. This yields

$$\mathrm{E}\{\tilde{y}_{1i}|H_1\} = 0 \qquad \text{and} \quad \mathrm{var}\,\{\tilde{y}_{1i}|H_1\} = \bar{E}_r + N_0$$

Under hypothesis H_0, $\tilde{y}_{1i}$ is purely noise, so that

$$\mathrm{E}\{\tilde{y}_{1i}|H_0\} = 0 \qquad \text{and} \quad \mathrm{var}\,\{\tilde{y}_{1i}|H_0\} = N_0$$

The left side of Eq. (8.3.40) represents the sum of the magnitude squares of N independent complex Gaussian random variables (or equivalently $2N$ real random variables) with the same mean but different variances under the two hypotheses. We can, therefore, use the results of Example (3.4.2) to write the probabilities of detection and of false alarm as

$$P_D = 1 - I\left[\frac{\gamma_2^2/N_0}{\sqrt{N}\,2\left(1+\frac{\bar{\eta}^2}{2}\right)}, N-1\right] \tag{8.3.41}$$

and

$$P_F = 1 - I\left[\frac{\gamma_2^2/N_0}{\sqrt{N}}, N-1\right] \tag{8.3.42}$$

where $I(\cdot, \cdot)$ is the incomplete gamma function defined in Chapter 3 as

$$I(m, n) = \int_0^{m\sqrt{1+n}} \frac{\exp(-u)u^n}{n!}\, du \tag{8.3.43}$$

and is tabulated [9].

Curves of P_D versus $\bar{\eta}^2$ for $N = 20$ and various values of P_F are shown in Fig. 8.3.7. The threshold levels required to plot these curves were obtained from the tables of Pachares [24].

8.3.5 Detection of Spread Targets: A Brief Discussion

From our earlier discussion on target models spread in range and/or Doppler, we find that the return signal is a stochastic process. This return signal is corrupted by additive Gaussian noise. Thus, we are led to the problem of detecting stochastic signals in noise, which we have not considered in our development so far. The solution usually calls for an estimate of the stochastic process to be correlated with the received signal to generate a sufficient statistic [3,10,11].

However, if we make a piecewise constant (albeit random) approximation to the reflecting characteristic $\tilde{b}(t)$ or the stochastic process $\tilde{y}_r(t)$, we can use the techniques developed to obtain a solution which is suboptimum. For example, assume that the target is Doppler spread, and the reflecting gain is approximated by Eq. (8.2.17). Then we can consider the complex envelope of the return signal as made up of N components:

$$\begin{aligned}\tilde{y}_r(t) &= \sqrt{E_T}\,\tilde{y}(t-\tau)\sum_{i=1}^{N}\tilde{b}_i u\left(t-\frac{i}{W}\right)\\ &= \sum_{i=1}^{N}\frac{\sqrt{E_T}}{N}\tilde{b}_i\tilde{y}_i(t-\tau)\end{aligned} \tag{8.3.44}$$

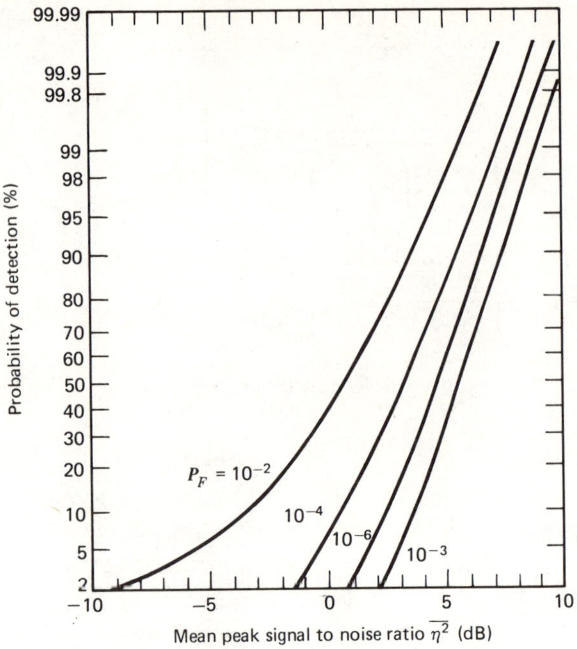

Fig. 8.3.7 Performance curves for multiple pulse detection of Rayleigh fluctuating (pulse to pulse) target.

where

$$\tilde{y}_i(t) = \tilde{y}(t), \ \frac{(i-1)}{W} \leqslant t \leqslant \frac{i}{W}, \qquad i = 1, 2, \ldots, N$$

If we now consider $\tilde{y}_i(t)$ to be the ith pulse in a sequence of N subpulses forming the pulse $\tilde{y}(t)$, we can use techniques similar to that used for multiple pulse detection of slowly fluctuating Rayleigh point targets discussed in Section 8.3.3. An important difference, however, is that the test statistic will now be a weighted sum of the statistics generated by correlating the received signal with the subpulse $\tilde{y}_i(t)$. The calculation of these weights and obtaining the receiver structure for this case as well as for range spread targets is left as an important exercise to the reader.

This concludes our discussion of the detection of radar targets. We have been selective in our material so as to illustrate the application of techniques to practical cases. A number of problems at the end deal with the application of detection theory to develop additional results on radar detection.

8.4 PARAMETER ESTIMATION IN RADAR SYSTEMS

In the previous section where we considered detection of targets, we assumed that the target range and velocity (or Doppler frequency shift) were known. Any such knowledge of these parameters is usually coarse, and proper techniques have to be used to obtain fine estimates. In addition to these parameters one may be interested in obtaining estimates of the angular coordinates of the target position.

In this section we consider maximum likelihood estimation of the delay and Doppler frequency (equivalently, range and velocity) of a steady point target with unknown random phase. We also obtain the Cramer–Rao bounds on the estimates. Exercises at the end of the chapter extend and develop additional results for the cases of slowly fluctuating point target and multiple pulse reception.

In the process of our development, we obtain an important quantity, called the *radar ambiguity function*. We briefly discuss this function and indicate the significant role it plays in waveform design.

8.4.1 Likelihood Function

The return signal from a steady point target is usually corrupted by additive Gaussian bandpass noise of spectral height $N_0/2$. We can represent the complex envelope of the noisy observation as

$$\tilde{z}(t) = \tilde{y}_r(t) + \tilde{v}(t) \qquad -\infty < t < \infty \tag{8.4.1}$$

where $\tilde{y}_r(t)$ was defined in Eq. (8.2.7). Further we assume that θ_2, the phase of the reflecting gain is a random variable uniformly distributed over $[0,2\pi]$.

We can expand $z(t)$ in an orthonormal series expansion as in the previous section. In this case, however, we choose the first orthonormal function as

$$\tilde{g}_1(t) = \tilde{y}(t-\tau)\exp(j\omega_d t) \tag{8.4.2}$$

In order to obtain the likelihood function, we determine the probability density function of the coefficient of $\tilde{g}_1(t)$ assuming that τ and ω_d are known. Once we obtain this function, we vary τ and ω_d to seek a maximum of the function. The values of τ and ω_d for which the maximum is obtained are the maximum likelihood estimates $\hat{\tau}$ and $\hat{\omega}_d$. As in Chapter 4 and 5, we can use the log-likelihood function in determining the estimates. We can readily write the log-likelihood function from Eq. (8.3.15) as

$$L_1(\tau, \omega_d) = K + \ln I_0\left[\frac{2\sqrt{E_r}}{N_0}|\tilde{y}_1|\right] \tag{8.4.3}$$

where $I_0(\cdot)$ is the modified Bessel function of the first kind and order zero, K is a constant not depending on either τ or ω_d, $E_r = E_T|\tilde{b}|^2$ and

$$\tilde{y}_1 = \int_{-\infty}^{\infty} \tilde{z}(t)\tilde{y}^*(t-\tau)\exp(-j\omega_d t)\,dt \tag{8.4.4}$$

The maximum likelihood estimates of τ and ω_d are then obtained by maximizing $L_1(\tau, \omega_d)$. Equivalently, we can maximize the function

$$L(\tau, \omega_d) = \ln I_0\left[\frac{2\sqrt{E_r}}{N_0}|\tilde{y}_1|\right] \tag{8.4.5}$$

to obtain our estimates.

Let us now consider a practical implementation of the maximum likelihood estimator. Because $\ln I_0(\cdot)$ is a monotonic function of its argument, we can seek the vector parameter $\boldsymbol{\alpha} = [\tau \vdots \omega_d]^T$ such that $|\tilde{y}_1|$ is maximum. That is, we implement the correlator of Eq. (8.4.4), and obtain the envelope of the output for each value of $\boldsymbol{\alpha}$ in its range. The estimate $\hat{\boldsymbol{\alpha}}$ will then correspond to the value of $\boldsymbol{\alpha}$ for which this is a maximum. This is clearly impractical except in special cases. To obtain a practical structure, let us first assume that ω_d is known. Then we can implement the correlator of Eq. (8.4.4) and monitor the envelope of its output, which is a function of τ. The time at which it peaks gives $\hat{\tau}$. However, if τ is known, but ω_d is unknown, we have to build a continuum of correlators to correspond to the values of ω_d in its expected range. We can overcome this problem by quantizing the range of ω_d into a set of M intervals centered around the frequencies $\omega_{d_1}, \omega_{d_2}, \ldots, \omega_{d_M}$. We perform M simultaneous correlation operations on the observation and observe the envelopes of the outputs at time τ. The estimate $\hat{\omega}_d$ will correspond to the maximum of these values. The practical implementation of the receiver when both τ and ω_d have to be estimated follows easily. We quantize the range of ω_d into M intervals to implement M correlators followed by envelope detectors. The outputs of the envelope detectors are monitored and a two-dimensional search for the maximum is conducted. The values of Doppler frequency and time for which this maximum occurs yield maximum likelihood estimates.

An important question that arises is how fine the quantization of the range of ω_d should be. We can obtain an answer to this as well as the theoretical maximum accuracy possible in terms of the Cramér–Rao bounds. We will find from our subsequent discussions that the quantization interval should not exceed one standard deviation of the estimate $\hat{\omega}_d$.

8.4.2 Cramér–Rao Bounds for Estimates of Delay and Doppler Frequency

We now derive the Cramér–Rao bounds on the estimates of τ and ω_d. We assume a high signal-to-noise ratio in order to simplify the derivation. This is not unreasonable, since, as we have seen earlier we require a fairly high signal-to-noise ratio to obtain reliable detection.

In order to determine the bounds, we proceed as in the previous section. That is, we first obtain the bound assuming that either τ or ω_d is known but not both and then extend our results to the case where both parameters are unknown.

Suppose that the target has an actual delay of τ_a and Doppler frequency ω_a, and we correlate the received signal envelope $z(t)$ as in Eq. (8.4.4). Then we can write

$$|\tilde{y}_1| = \left| \int_{-\infty}^{\infty} \left\{ \sqrt{E_r}\, \tilde{y}(t-\tau_a) \exp(j\omega_a t) \exp(j\theta_2) + \tilde{v}(t) \right\} \cdot \left\{ \tilde{y}^*(t-\tau) \exp(-j\omega_d t) \right\} dt \right| \tag{8.4.6}$$

Under the assumption of high signal-to-noise ratio we can write the likelihood function of Eq. (8.4.5) as

$$\begin{aligned} L(\tau, \omega_d) &= \ln I_0\left[\frac{2\sqrt{E_r}}{N_0} |\tilde{y}_1| \right] \\ &\simeq \frac{2\sqrt{E_r}}{N_0} |\tilde{y}_1| \end{aligned} \tag{8.4.7}$$

In general we can also assume that the signal and noise are uncorrelated so that

$$\frac{2\sqrt{E_r}}{N_0} |\tilde{y}_1| \simeq \frac{2E_r}{N_0} \left| \int_{-\infty}^{\infty} \tilde{y}(t-\tau_a)\tilde{y}^*(t-\tau) \exp\left[j(\omega_d - \omega_a)t \right] dt \right| \tag{8.4.8}$$

Let $\tau' = \tau - \tau_a$ denote the error in the estimate of delay and $\omega_d' = \omega_d - \omega_a$ denote the error in the estimate of the Doppler shift.

Let $r = t - \tau + \tau'/2$. We can then rewrite Eq. (8.4.7) in the form

$$L(\tau, \omega_d) = \frac{2\sqrt{E_r}}{N_0} |\tilde{y}_1| \simeq \frac{2E_r}{N_0} |\chi(\tau', \omega_d')| \tag{8.4.9a}$$

where

$$\chi(\tau', \omega_d') = \int_{-\infty}^{\infty} \tilde{y}(r + \tau'/2)\tilde{y}^*(r - \tau'/2) \exp(j\omega_d' r)\, dr \tag{8.4.9b}$$

The function $\chi(\tau', \omega_d')$ is sometimes called the time frequency autocorrelation function of $y(t)$ and sometimes called the ambiguity function† [1,3,15]. The function $|\chi(\tau', \omega_d')|^2$ is the response due to the signal mismatched in delay by τ' and in Doppler by ω_d' of a matched filter (matched to the signal $y(t - \tau) \exp(j\omega_d t)$). As such, the ambiguity function $\chi(\tau', \omega_d')$ can be expected to play an important role in distinguishing between a number of close objects. To illustrate this, consider two targets, one at the matched delay and Doppler with $|\tilde{b}| = 1$, and another mismatched by τ' and ω_d' with $|\tilde{b}| = p$. Then the "mismatched" target contributes an interfering energy of $p^2|\chi(\tau', \omega_d')|^2$, which may be larger or comparable to that of the true signal. This makes the problem of resolution difficult. In many modern radar systems, the transmitted signal $\tilde{y}(t)$ is carefully designed such that the ambiguity function exhibits desired nulls or minima at the locations corresponding to values of delay and Doppler frequency where interfering targets are to be expected. Thus, the ambiguity function is a practical tool to study the design of waveforms for target resolution. As we shall see later, waveform design can also improve the accuracy bounds of our estimates. In fact, we relate to these design parameters of waveforms through the ambiguity function $\chi(\tau', \omega_d')$.

We now return to the problem of deriving the accuracy bounds for delay and Doppler estimates.

If our interest is to estimate the delay when the Doppler frequency shift ω_d is known, we can set $\omega_d = \omega_a$ or $\omega_d' = 0$ in Eqs. (8.4.9a) and (8.4.9b) without loss of generality. Similarly if ω_d is to be estimated, we can set $\tau = \tau_a$ or $\tau' = 0$ in the foregoing equations.

We will first derive the bound on the estimate of the delay. Note that this is the same as the variance of τ'. In Chapter 5, we derived the Cramér–Rao bound for an unbiased estimate. Applying this bound for

†We purposely digress here from our main course to briefly discuss the significance of $\chi(\tau', \omega_d')$. A number of its properties are developed in exercises at the end.

obtaining the variance of τ, we have

$$\operatorname{var}\tau = \operatorname{var}\tau' \geqslant \frac{1}{\mathrm{E}\left\{\dfrac{\partial^2 L(\tau', 0)}{\partial \tau'^2}\right\}} \tag{8.4.10}$$

Assuming high SNR, we can approximate the above bound by using the log-likelihood function of Eq. (8.4.8). This yields

$$\operatorname{var}\tau' \geqslant \frac{-1}{\left(\dfrac{2E_r}{N_0}\right)\dfrac{\partial^2|\chi(\tau', 0)|}{\partial \tau'^2}} \tag{8.4.11}$$

In Eq. (8.4.11) the derivative must be evaluated at the true value of τ', that is, $\tau' = 0$.

In order to evaluate the derivative, we note that

$$|\chi(\tau', 0)| = [\chi(\tau', 0)\chi^*(\tau', 0)]^{1/2}$$

so that

$$\frac{\partial|\chi(\tau', 0)|}{\partial \tau'} = \frac{1}{2|\chi(\tau', 0)|}\left[\frac{\partial}{\partial \tau'}\chi(\tau', 0)\cdot\chi^*(\tau', 0) + \frac{\partial}{\partial \tau'}\chi^*(\tau', 0)\chi(\tau', 0)\right] \tag{8.4.12}$$

and

$$\frac{\partial^2|\chi(\tau', 0)|}{\partial \tau'^2} = \frac{1}{|\chi(\tau', 0)|}\operatorname{Re}\left[\chi(\tau', 0)\frac{\partial^2}{\partial \tau'^2}\chi^*(\tau', 0) + \frac{\partial}{\partial \tau'}\chi(\tau', 0)\frac{\partial}{\partial \tau'}\chi^*(\tau', 0)\right] - \frac{1}{|\chi(\tau', 0)|^3}\operatorname{Re}\left[\chi^*(\tau', 0)\frac{\partial}{\partial \tau'}\chi(\tau', 0)\right]^2 \tag{8.4.13}$$

Evaluating the derivative in Eq. (8.4.13) at $\tau' = 0$ we obtain

$$\left.\frac{\partial^2|\chi(\tau', 0)|}{\partial \tau'^2}\right|_{\tau'=0} = \left[\operatorname{Re}\left\{\frac{\partial^2}{\partial \tau'^2}\chi^*(\tau', 0)\right\} + \left|\frac{\partial}{\partial \tau'}\chi(\tau', 0)\right|^2 - \left[\operatorname{Re}\frac{\partial}{\partial \tau'}\chi(\tau', 0)\right]^2\right]_{\tau'=0} \tag{8.4.14}$$

In order to compute the right-hand side of Eq. (8.4.14), we note that

$$\begin{aligned}\frac{\partial}{\partial \tau'}\chi(\tau', 0)\bigg|_{\tau'=0} &= \frac{\partial}{\partial \tau'}\left\{\int_{-\infty}^{\infty} \tilde{y}\left(r - \frac{\tau'}{2}\right)\tilde{y}^*\left(r + \frac{\tau'}{2}\right) dr\right\}\bigg|_{\tau'=0} \\ &= \frac{\partial}{\partial \tau'}\left\{\int_{-\infty}^{\infty} \tilde{y}(t - \tau + \tau')\tilde{y}^*(t - \tau)\, dt\right\}\bigg|_{\tau'=0} \\ &= \int_{-\infty}^{\infty} \frac{\partial}{\partial \tau'}\{[\tilde{y}(t - \tau + \tau')]\tilde{y}^*(t - \tau)\, dt\}\bigg|_{\tau'=0} \\ &= -\int_{-\infty}^{\infty}\left[\frac{d}{dt}\tilde{y}(t)\right]\tilde{y}^*(t)\, dt \end{aligned} \quad \mathbf{(8.4.15)}$$

Similarly, we can show that

$$\frac{\partial^2}{\partial \tau'^2}\{\chi(\tau', 0)\}\bigg|_{\tau'=0} = \int_{-\infty}^{\infty}\left[\frac{d^2}{dt^2}\tilde{y}(t)\right]\tilde{y}^*(t)\, dt \quad \mathbf{(8.4.16)}$$

Let $G(j\omega)$ be the Fourier transform of $\tilde{y}(t)$. It then follows that

$$\frac{\partial}{\partial \tau'}\chi(\tau', 0)\bigg|_{\tau'=0} = j\int_{-\infty}^{\infty} \omega|G(j\omega)|^2 \frac{d\omega}{2\pi} \quad \mathbf{(8.4.17)}$$

and

$$\frac{\partial^2}{\partial \tau'^2}\chi(\tau', 0)\bigg|_{\tau'=0} = -\int_{-\infty}^{\infty} \omega^2|G(j\omega)|^2 \frac{d\omega}{2\pi} \quad \mathbf{(8.4.18)}$$

Using Eqs. (8.4.17) and (8.4.18) in Eq. (8.4.14), we get

$$\frac{\partial^2}{\partial \tau'^2}|\chi(\tau', 0)|\bigg|_{\tau'=0} = -\left\{\int_{-\infty}^{\infty} \omega^2|G(j\omega)|^2 \frac{d\omega}{2\pi} - \left|\int_{-\infty}^{\infty} \omega|G(j\omega)|^2 \frac{d\omega}{2\pi}\right|^2\right. \quad \mathbf{(8.4.19)}$$

By assumption

$$\int_{-\infty}^{\infty} |\tilde{y}(t)|^2\, dt = \int_{-\infty}^{\infty} |G(j\omega)^2| \frac{d\omega}{2\pi} = 1 \quad \mathbf{(8.4.20)}$$

so that we can write Eq. (8.4.19) as

$$\frac{\partial^2}{\partial \tau'^2}|\chi(\tau', 0)|\bigg|_{\tau'=0} = -\left\{\frac{\int_{-\infty}^{\infty} \omega^2|G(j\omega)|^2 \frac{d\omega}{2\pi} - \int_{-\infty}^{\infty} \omega|G(j\omega)|^2 \frac{d\omega}{2\pi}}{\int_{-\infty}^{\infty} |G(j\omega)|^2 \frac{d\omega}{2\pi}}\right\} \quad \mathbf{(8.4.21)}$$

The quantity inside the brackets on the right-hand side of Eq. (8.4.21) is a measure of the bandwidth of the signal $\tilde{y}(t)$, and is denoted by β^2. Using Eq. (8.4.21) in Eq. (8.4.11), we finally obtain

$$\operatorname{var}(\tau) = \operatorname{var}(\tau') > \left[\frac{2E_r}{N_0}\beta^2\right]^{-1} \tag{8.4.22}$$

We see from Eq. (8.4.22) that we can increase the accuracy of our estimate by increasing the energy E_r or making β^2 as large as possible. That is, signals with larger bandwidth can provide more accurate estimates of time delay.

For estimating the Doppler frequency, we assume $\tau' = 0$, that is, the delay is known exactly, and use the Cramér–Rao bound to obtain

$$\operatorname{var}\omega_d = \operatorname{var}\omega_d' \geqslant \frac{-1}{\left(\dfrac{2E_r}{N_0}\right)\dfrac{\partial^2}{\partial \omega_d'^2}|\chi(0, \omega_d')|} \tag{8.4.23}$$

Again the second derivate is evaluated at $\omega_d' = 0$. From Eq. (8.4.14) we can write

$$\begin{aligned}\left.\frac{\partial^2}{\partial \omega_d'^2}|\chi(0, \omega_d')|\right|_{\omega_d'=0} &= \left[\operatorname{Re}\left\{\frac{\partial^2}{\partial \omega_d'^2}\chi(0, \omega_d')\right\} + \left|\frac{\partial}{\partial \omega_d'}\chi(0, \omega_d')\right|^2\right. \\ &\qquad \left.\left. - \operatorname{Re}\left[\frac{\partial}{\partial \omega_d'}\chi(0, \omega_d')\right]^2\right]\right|_{\omega_d'=0}\end{aligned} \tag{8.4.24}$$

Since

$$\begin{aligned}\frac{\partial}{\partial \omega_d'}\chi(0, \omega_d') &= \frac{\partial}{\partial \omega_d'}\left[\int_{-\infty}^{\infty} |\tilde{y}(t-\tau)|^2 \exp(j\omega_d' t)\,dt\right] \\ &= j\int_{-\infty}^{\infty} t|\tilde{y}(t-\tau)|^2 \exp[j\omega_d' t]\,dt\end{aligned} \tag{8.4.25}$$

and

$$\frac{\partial^2}{\partial \omega_d'^2}\chi(0, \omega_d') = -\int_{-\infty}^{\infty} t^2|\tilde{y}(t-\tau)|^2 \exp[j\omega_d' t]\,dt \tag{8.4.26}$$

it follows that

$$\left.\frac{\partial}{\partial \omega_d'}\chi(0, \omega_d')\right|_{\omega_d'=0} = j\int_{-\infty}^{\infty} t|\tilde{y}(t)|^2\,dt \tag{8.4.27}$$

and

$$\left.\frac{\partial^2}{\partial \omega_d'^2}\chi(0, \omega_d')\right|_{\omega_d'=0} = -\int_{\infty}^{\infty} t^2|\tilde{y}(t)|^2\, dt \tag{8.4.28}$$

Similar to our definition of β^2, we can define the parameter t_d^2 as a measure of the duration of the transmitted waveform $\tilde{y}(t)$ by

$$t_d^2 = \frac{\displaystyle\int_{\infty}^{\infty} t^2|\tilde{y}(t)|^2\, dt - \left[\int_0^T t|\tilde{y}(t)|^2\, dt\right]^2}{\displaystyle\int_0^T |\tilde{y}(t)|^2\, dt} \tag{8.4.29}$$

Use of Eqs. (8.4.27)–(8.4.29) in Eq. (8.4.24) yields

$$\left.\frac{\partial^2}{\partial \omega_d'^2}|\chi(0, \omega_d')|\right|_{\omega_d'-0} = -t_d^2 \tag{8.4.30}$$

The bound on the estimate of Doppler frequency obtained by substituting Eq. (8.4.30) in Eq. (8.4.23) is

$$\text{var}\ \omega_d = \text{var}\ \omega_d' \geqslant \left[\frac{2E_r}{N_0}t_d^2\right]^{-1} \tag{8.4.31}$$

From Eq. (8.4.31) we see that in order to increase the accuracy of our estimate we must increase either the energy E_r or the parameter t_d^2 or both. The transmitter power available places a limit on the maximum value of E_r. However, the parameter t_d^2 can be increased by proper design of the waveform $\tilde{y}(t)$.

In the case of joint estimation of τ and ω_d, we need to obtain the Fisher information matrix discussed in Section 5.5 in order to obtain the variance bounds on the estimates. It is obvious from the previous development that

$$J_{11} = -\mathrm{E}\left\{\frac{\partial^2 L(\tau, \omega_d)}{\partial \tau'^2}\right\} = \frac{2E_r}{N_0}\beta^2 \tag{8.4.32}$$

and

$$J_{22} = -\mathrm{E}\left\{\frac{\partial^2}{\partial \omega_d'^2}L(\tau, \omega_d)\right\} = \frac{2E_r}{N_0}t_d^2 \tag{8.4.33}$$

It only remains to determine

$$J_{12} = \mathrm{E}\left\{ \frac{\partial^2}{\partial \tau' \partial \omega_d'} L(\tau, \omega_d) \right\} \tag{8.4.34}$$

in order to be able to calculate the variance bounds on the estimates using Eq. (5.5.53). Using the approximation in Eq. (8.4.8) J_{12} can be expressed as

$$J_{12} \simeq -\frac{2E_r}{N_0} \frac{\partial^2}{\partial \tau' \partial \omega_d'} |\chi(\tau', \omega_d')| \tag{8.4.35}$$

Using a derivation similar to the ones previously developed, we can show after some algebraic manipulations that J_{12} can be evaluated as

$$J_{12} = \frac{2E_r}{N_0}\left[\overline{\omega t} - \bar{\omega}\bar{t} \right] \tag{8.4.36}$$

where the quantities $\overline{\omega t}$, $\bar{\omega}$, and $\bar{t}$ are defined as

$$\overline{\omega t} = \mathrm{Re}\left\{ \frac{\partial^2}{\partial \tau' \partial \omega_d'} \chi(\tau', \omega_d') \Big|_{\substack{\tau'=0\\ \omega_d'=0}} \right\} = \int_0^T t \frac{d\theta_1(t)}{dt} |\tilde{y}(t)|^2 \, dt \tag{8.4.37}$$

$$\bar{\omega} = -j \frac{\partial}{\partial \tau'} \chi(\tau', \omega_d') \Big|_{\substack{\tau'=0\\ \omega_d'=0}} = \int_{-\infty}^{\infty} \omega |G(j\omega)|^2 \frac{d\omega}{2\pi} \tag{8.4.38}$$

and

$$\bar{t} = -j \frac{\partial}{\partial \omega_d'} \chi(\tau', \omega_d') \Big|_{\substack{\tau'=0\\ \omega_d'=0}} = \int_0^T t |\tilde{y}(t)|^2 \, dt \tag{8.4.39}$$

The inverse of the information matrix, $J^{-1} = \psi$ is then obtained as

$$\psi = \left[\frac{2E_r}{N_0} \right]^{-1} \begin{bmatrix} t_d^2 & \bar{\omega}\bar{t} - \overline{\omega t} \\ \bar{\omega}\bar{t} - \overline{\omega t} & \beta^2 \end{bmatrix} \left[\beta^2 t_d^2 - \left(\overline{\omega t} - \bar{\omega}\bar{t} \right)^2 \right]^{-1} \tag{8.4.40}$$

so that the bounds for the variance of the estimates follow as

$$\operatorname{var}(\tau) \geqslant \left[\frac{2E_r}{N_0} \right]^{-1} \frac{t_d^2}{\left[\beta^2 t_d^2 - \left(\overline{\omega t} - \bar{\omega}\bar{t} \right)^2 \right]} \tag{8.4.41}$$

and

$$\operatorname{var}(\omega_d) \geqslant \left[\frac{2E_r}{N_0}\right]^{-1} \frac{\beta^2}{\left[\beta^2 t_d^2 - \left(\overline{\omega t} - \bar{\omega}\bar{t}\right)^2\right]} \tag{8.4.42}$$

From Eqs. (8.4.41) and (8.4.42), we see that, for a given level of transmitted signal energy E_r, better estimation accuracy is obtained if the "time-bandwidth" product βt_d is large and the difference $(\overline{\omega t} - \bar{\omega}\bar{t})$ is small.

8.5 DYNAMIC TARGET TRACKING

The application of discrete-time recursive filtering and prediction techniques for tracking purposes in a radar system has been of significant interest in recent times. Such techniques are well suited for tracking both maneuvering and constant velocity targets. The radar provides noisy measurements of the target coordinates once every T seconds, where the scan rate is $1/T$ revolutions per second. These measurements are processed by the tracking filter to yield estimates of the target coordinates and their derivatives. Typical applications include airport surveillance radars (ASR) and air route surveillance radars (ARSR) and reentry vehicle tracking [16–19]. These techniques can also be applied in a multitarget environment and with multisensor returns, but decision techniques will have to be used to correlate the correct return to the correct target. This is a rather involved situation and is not discussed here.

In the first part of this section, we model the target by a linear system obtained by linearizing the actual target dynamics about the nominal range and angle coordinates. We then use the discrete Kalman filter algorithms of Chapter 6 to track the target. We consider both constant velocity and linear acceleration models.

Maneuver initiation in a target that has been flying a constant velocity trajectory causes loss of track since the track filter will no longer be matched to the actual target dynamics. In the second part of this section, we formulate the problem of maneuver detection as a composite hypothesis testing problem. Once maneuvering has been established, the parameters of the filter can be suitably updated to maintain track.

8.5.1 Tracking of Maneuvering and Constant Velocity Targets [16]

We recall that a radar system measures the range and angle every T seconds. Let us assume that at time kT, the target is at a range $R_0 + r(k)$, and angle $\theta_0 + \theta(k)$. One scan time later at time $(k+1)T$, the range and angle will be $R_0 + r(k+1)$ and $\theta_0 + \theta(k+1)$. Therefore, $r(k)$ and $\theta(k)$ are the deviations from a nominal value R_0 and θ_0. From the range and angle measurements, which are generally noisy, we would like to estimate the present range, range rate, bearing and bearing rate. In many applications we would also like to predict these quantities for tactical control purposes. Let us denote by $\mathbf{x}$ the six-component vector whose elements are the target range, velocity, acceleration, bearing, bearing rate, and bearing acceleration. To derive a set of state equations for this problem, we approximate the velocity and acceleration at time kT as

$$\dot{x}_1(k) = \frac{x_1(k+1) - x_1(k)}{T} = x_2(k)$$

and

$$\dot{x}_2(k) = \frac{x_2(k+1) - x_2(k)}{T} = x_3(k)$$

so that

$$x_1(k+1) = x_1(k) + Tx_2(k)$$

and

$$x_2(k+1) = x_2(k) + Tx_3(k)$$

We similarly model the acceleration at time $(k+1)T$ as

$$x_3(k+1) = \rho x_3(k) + w_1(k)$$

where $w_1(k)$ is a zero mean, white sequence with variance $\sigma_{w_1}^2$. This represents a maneuvering target with linear acceleration.

With similar approximations for bearing, bearing rate and bearing acceleration, the model for maneuvering targets can be written as

$$\mathbf{x}(k+1) = \mathbf{A}\mathbf{x}(k) + \mathbf{w}(k) \tag{8.5.1}$$

where

$$\mathbf{A} = \begin{bmatrix} 1 & T & 0 & 0 & 0 & 0 \\ 0 & 1 & T & 0 & 0 & 0 \\ 0 & 0 & \rho & 0 & 0 & 0 \\ 0 & 0 & 0 & 1 & T & 0 \\ 0 & 0 & 0 & 0 & 1 & T \\ 0 & 0 & 0 & 0 & 0 & \rho \end{bmatrix} \tag{8.5.2}$$

and

$$\mathbf{w}(k) = \begin{bmatrix} 0 \\ 0 \\ w_1(k) \\ 0 \\ 0 \\ w_2(k) \end{bmatrix} \tag{8.5.3}$$

where $w_2(k)$ is also a white sequence with variance $\sigma_{w_2}^2$ and uncorrelated with $w_1(k)$. These inputs lead to correlated acceleration terms $x_3(k)$ and $x_6(k)$ with variances $\sigma_{w_1}^2/(1-\rho^2)$ and $\sigma_{w_2}^2/(1-\rho^2)$.

The model of Eq. (8.5.1) is quite general in the sense that we can derive constant velocity, and uncorrelated acceleration models from it. If the target is on a constant velocity path, the acceleration is zero, so that $\rho = 0$ and $\sigma_{w_1}^2 = \sigma_{w_2}^2 = 0$. Wind gusts, and so on, usually affect the constant velocity path of targets, and can be modeled as white noise acceleration inputs. In this case, $\rho = 0$, making $x_3(k) = w_1(k)$ and $x_6(k) = w_2(k)$. In both the constant velocity and white noise acceleration cases, the system order reduces to four. Further, note that the equations for range variables and bearing variables are decoupled. We can use this decoupling to advantage to reduce the numerical computations involved.

The radar measures noisy versions of the range and bearing. The measurement model is thus given by

$$\mathbf{z}(k) = \mathbf{H}\mathbf{x}(k) + \mathbf{v}(k) \tag{8.5.4}$$

where

$$\mathbf{H} = \begin{bmatrix} 1 & 0 & 0 & 0 & 0 & 0 \\ 0 & 0 & 0 & 1 & 0 & 0 \end{bmatrix}$$

and

$$\mathbf{v}(k) = \begin{bmatrix} v_1(k) \\ v_2(k) \end{bmatrix}$$

In Eq. (8.5.4), $v_1(k)$ and $v_2(k)$ are assumed to be independent noise sequences uncorrelated with each other. Further, they are uncorrelated with $w_1(k)$ and $w_2(k)$.

The tracking algorithms are obtained in a straightforward manner by applying the discrete time Kalman filter algorithms of Chapter 6, with message and observation models given by Eqs. (8.5.1) and (8.5.4). In order to do this, we need to initialize the Kalman filter algorithms. In other words, we need the values of $\mathbf{V}_{\tilde{\mathbf{x}}}(k|k)$ and $\hat{\mathbf{x}}(k|k)$ for some k to start with. These are generally not available. We can, however, generate a suboptimal estimate and compute the associated error variance. These values can then be used to initialize the optimal algorithms.

To illustrate the initiation procedure, consider that the target is essentially on a constant velocity path, except for disturbance due to wind gust acceleration, that is, $\rho = 0$. In this case, the message model has a dimension of only 4. We wait for the first two measurements $\mathbf{z}(1)$ and $\mathbf{z}(2)$ to obtain the "estimate" $\hat{\mathbf{x}}(2|2)$ (which is ad hoc or suboptimal) as

$$\begin{aligned}
\hat{x}_1(2|2) &= z_1(2)\\
\hat{x}_2(2|2) &= \frac{1}{T}(z_1(2) - z_1(1))\\
\hat{x}_4(2|2) &= z_2(2)\\
\hat{x}_5(2|2) &= \frac{1}{T}(z_2(2) - z_2(1))
\end{aligned} \tag{8.5.5}$$

That is, the estimates are obtained by considering the measurements to be noise-free, and the velocities to be constant. The error is then given by

$$\tilde{\mathbf{x}}(2|2) = \mathbf{x}(2) - \hat{\mathbf{x}}(2|2)$$

so that the error variance $\mathbf{V}_{\tilde{\mathbf{x}}}(2|2)$ as obtained from Eqs. (8.5.1), (8.5.4), and (8.5.5) is

$$\mathbf{V}_{\tilde{\mathbf{x}}}(2|2) = \begin{bmatrix} \sigma_{v_1}^2 & \sigma_{v_1}^2/T & 0 & 0 \\ \sigma_{v_1}^2/T & \sigma_{w_1}^2 + \left(2\sigma_{v_1}^2/T^2\right) & 0 & 0 \\ 0 & 0 & \sigma_{v_2}^2 & \left(\sigma_{v_2}^2/T\right) \\ 0 & 0 & \sigma_{v_2}^2/T & \sigma_{w_2}^2 + \left(2\sigma_{v_2}^2/T^2\right) \end{bmatrix} \tag{8.5.6}$$

Equations (8.5.5) and (8.5.6) can be used to initialize the discrete Kalman filter algorithms to process the measurements from $k = 3$ onwards. A similar initialization procedure can be adopted for maneuvering or constant velocity targets.

Because of the difficulty in implementing the time-varying gain needed for the Kalman filter, often in practice a constant gain is used. Furthermore, a constant velocity model rather than a linear acceleration model is usually used in simple tracking systems. Such filters are called $\alpha - \beta$ filters, where α is the gain for range update and β the gain for velocity update. Optimum values of α and β have also been determined [20]. However, for maneuvering targets the $\alpha - \beta$ tracking filters may perform poorly.

8.5.2 *Detection of Target Maneuver* [19]

A target can be tracked accurately if we know the correct state representation of the target. If we use a constant velocity model to track a maneuvering target, we are likely to obtain poor tracking, particularly for tactical control purposes. It therefore becomes necessary to detect the time at which the target started maneuvering. Once this is known, we can switch the tracking filter to match the maneuvering state model to obtain proper track.

For simplicity of illustration, we will consider only the range coordinate of the model given by Eq. (8.5.1). We assume constant velocity path initially, and therefore let $\rho = 0$ and $\sigma_{w_1}^2 = 0$. A Kalman filter can be matched to this model to yield us the one-stage prediction estimates $\hat{x}_1(k|k-1)$ and $\hat{x}_2(k|k-1)$. From our discussions in Chapter 6, we know that the innovation sequence

$$\nu(k) = z(k) - \hat{x}_1(k|k-1) \tag{8.5.7}$$

is white with zero mean and covariance

$$V_\nu(k) = \sigma_{v_1}^2 + V_{\tilde{x}_1}(k|k-1) \tag{8.5.8}$$

where $V_{\tilde{x}_1}(k|k-1)$ is the error covariance of one-stage prediction estimate of range. In steady state, that is, after the tracking transients have died out, the tracking can be assumed exact, and consequently $\mathbf{V}_{\tilde{\mathbf{x}}}(k|k-1) \simeq \mathbf{0}$. Then, we can write

$$V_\nu(k) \simeq \sigma_{\nu_1}^2 \tag{8.5.9}$$

Now, suppose the target undergoes a maneuver starting at time $k_0 T$. This shows up as an addition of the acceleration term to the dynamical equations, and consequently manifests itself as an addition to range measurement. That is, for $k \geqslant k_0$, we can write the measurement equation as

$$z(k) = z_0(k) + \xi(k - k_0) \tag{8.5.10}$$

where $z_0(k)$ is the measurement that would have been obtained if there had been no maneuver and $\xi(k-k_0)$ is an unknown time function due to maneuver which is zero for $k \leqslant k_0$. Because the Kalman filter is linear, its range prediction estimate can be written as

$$\hat{x}_1(k|k-1) = \hat{x}_{10}(k|k-1) + \hat{x}_{1\xi}(k|k-1) \tag{8.5.11}$$

where $\hat{x}_{10}(k|k-1)$ is the estimate based on the measurement $z_0(k-1)$ and $\hat{x}_{1\xi}(k|k-1)$ is the estimate based on $\xi[(k-1)-k_0]$. The residual or innovation sequence, in turn, can be written as

$$\nu(k) = \nu_0(k) + \zeta(k - k_0) \tag{8.5.12}$$

where $\nu_0(k)$ is the innovation sequence that would have been obtained had there been no maneuver and

$$\zeta(k - k_0) = \xi(k - k_0) - \hat{x}_{1\zeta}(k|k-1) \tag{8.5.13}$$

Equation (8.5.12) indicates that whenever there is maneuver, it adds a bias term to the zero mean innovation sequence $\nu_0(k)$. Thus, to detect the occurrence of the maneuver, it is only necessary to monitor the bias. As long as no bias develops we continue to use the constant velocity tracker; when a bias is detected, we would switch to the tracker matched to the appropriate dynamics.

The detection of bias in the residual sequence can be posed as a statistical hypothesis testing problem as follows. Let the hypothesis H_1 indicate that a maneuver occurs and H_0 the hypothesis that no maneuver occurs. The innovations sequence under the two hypotheses can then be written as

$$\begin{aligned} H_1 &: \nu(k) = \nu_0(k) + \zeta(k - k_0) \\ H_0 &: \nu(k) = \nu_0(k) \end{aligned} \tag{8.5.14}$$

where $\nu_0(k)$ is zero mean and $\nu(k)$ has the same covariance under both hypotheses equal to $\sigma_{v_1}^2$. However, the mean $\zeta(k - k_0)$ is not known. It is reasonable to assume that this quantity varies as a square of the time elapsed since $k_0 T$. Then we can model the mean $\zeta(k - k_0)$ as

$$\zeta(k - k_0) = \mu T^2 (k - k_0)^2 \tag{8.5.15}$$

where μ is some unknown constant depending on the maneuver acceleration. Thus, detecting a maneuver is equivalent to detecting the presence of a deterministic signal of unknown amplitude and time of arrival in the background of zero mean white noise. We can further assume that the noise is Gaussian. The optimum test is to compute the generalized likelihood ratio and compare it with a threshold to make the decision. The log likelihood ratio, based on k observations, is

$$L(k) = \max_{\mu,\, k_0} \left[\sum_{n=0}^{k} \nu^2(n) - \sum_{n=0}^{k} \left[\nu(n-k) - \mu s(n - k_0) \right]^2 \right] \tag{8.5.16}$$

where

$$s(i) = \begin{cases} (iT)^2 & \text{for } i \geqslant 0 \\ 0 & \text{for } i < 0 \end{cases}$$

We now define the functions

$$\epsilon(k, k_0) \triangleq \sum_{n=0}^{k} s^2(n - k_0) \tag{8.5.17}$$

and

$$\rho(k, k_0) \triangleq \sum_{n=0}^{k} \nu(n) s(n - k_0) \tag{8.5.18}$$

In terms of these functions, the log-likelihood ratio of Eq. (8.5.16) can be rewritten as

$$\begin{aligned} L(k) &= \max_{\mu,\, k_0} \{2\mu\rho(k, k_0) - \mu^2\epsilon(k, k_0)\} \\ &= \max_{\mu,\, k_0} \left\{ \frac{\rho^2(k, k_0)}{\epsilon(k, k_0)} - \mu\epsilon^{1/2}(k, k_0) - \frac{\rho(k, k_0)}{\epsilon^{1/2}(k, k_0)} \right\}^2 \end{aligned}$$

The maximum likelihood estimate of $\mu(k)$ denoted by $\hat{\mu}(k)$, is then obtained from Eq. (8.5.19) as

$$\hat{\mu}(k) = \frac{\rho^2(k, k_0)}{\epsilon(k, k_0)} \tag{8.5.20}$$

This can be substituted in the log-likelihood ratio of Eq. (8.5.19) to obtain

$$L(k) = \max_{k_0} \frac{\rho^2(k, k_0)}{\epsilon(k, k_0)} \tag{8.5.21}$$

If we are at time kT and are testing for bias, we are looking for maneuver initiation times $(k-1)T$, $(k-2)T$, and so on; that is, for all past times. This is generally infeasible. Clearly, there is only a finite value of the past that we need to test, say $(k - M)T$. Thus the test in Eq. (8.5.21) is equivalent to testing for

$$k_0 = k - j, \qquad j = 1, 2, \ldots, M$$

and becomes

$$L(k) = \max_{j=1,2,\ldots,M} \frac{\rho^2(k, k-j)}{\epsilon(k, k-j)} \tag{8.5.22}$$

The log-likelihood ratio of Eq. (8.5.22) can be implemented in the form of a bank of M matched filters, followed by square law detectors. The largest of the outputs can then be compared with a threshold to make the decision. This is shown in Fig. 8.5.1. The log-likelihood ratio now becomes

$$L(k) = \max_{j=1,2,\ldots,M} \sum_{n=0}^{k} \nu(n) h_j(k-n) \tag{8.5.23}$$

where

$$h_j(l) = \begin{cases} \dfrac{s(j-l)}{\sqrt{\epsilon(j)}}, & l \geqslant 0 \\ 0 & l < 0 \end{cases} \qquad j = 1, 2, \ldots, M \tag{8.5.24}$$

and

$$\epsilon(j) = \epsilon(k, k-j)$$

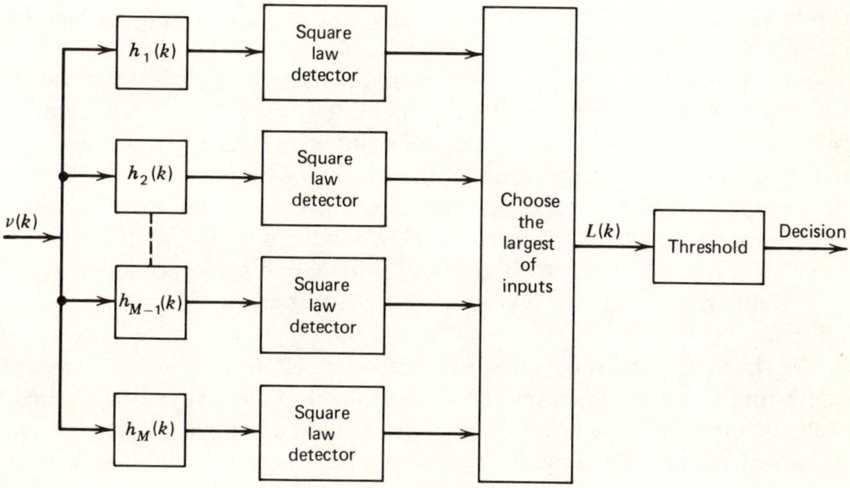

Fig. 8.5.1 Maneuver detector.

A suboptimum practical maneuver detector can be obtained by assuming $s(k) = \frac{1}{2}\mu k^2$, $k > 0$. The details can be found in [19].

8.6 SUMMARY

In this chapter we presented certain applications of detection theory, maximum likelihood estimation, and Kalman filtering to the radar detection and tracking problems. In order to do so, we developed some target models in Section 8.2 which used the complex representation of signals. The specific models considered were those for a steady point target, a slowly fluctuating point target, and spread targets.

In Section 8.3, we discussed the detection of radar targets in the presence of additive white Gaussian noise. The target models of Section 8.2 were used to obtain receiver structures and determine their performance. Because detection of stochastic signals in noise is outside the scope of the present text, we discussed the detection of spread targets only briefly. The results were again derived in terms of complex envelopes of bandpass signals.

In Section 8.4, we introduced the problem of parameter estimation in radar. Typical parameters of interest are target range and velocity, or equivalently the time delay and Doppler frequency shift. We used the maximum likelihood approach and obtained the likelihood function in terms of complex envelopes. We derived accuracy bounds on our estimates and related them to certain design parameters of the transmitted waveform. The concept of ambiguity function was introduced and its significance was discussed briefly.

Section 8.5 was concerned with continuous tracking of target range velocity and angular position. We illustrated the application of the Kalman filter to this problem. When a target begins maneuvering, track maintenance is difficult mainly because the tracker dynamics are no longer matched to that of the target. In such a case maneuver detection becomes important so that the dynamics may be modified properly to keep the target in track. We described a technique that combines hypothesis testing and Kalman filtering to evolve an adaptive tracking for maneuvering targets.

We have, for reasons of space, not covered in this chapter many interesting topics to which the theory developed in the text can be applied. Some of these have been left as exercises at the end of the chapter for the interested reader. The first is the detection of targets in the presence of colored noise, which arises naturally in the form of clutter in radar (Exercise 8.11). The techniques of Section 8.2 can be used with certain approximations to the case of spread targets (Exercise 8.10). Parameter estimation in the case of slowly fluctuating point targets and spread targets was not discussed. An important topic that has also not been covered is that of array processing in which several spatially distributed observations are available. The interested reader is again referred to the literature.

EXERCISES

8.1 Show that the range spread target modeled in Eq. (8.2.8) exhibits frequency selective reflecting characteristics. You may assume that $\tilde{b}(\tau)$ is a complex stochastic process in τ with zero mean, and is an uncorrelated process, that is, $E[\tilde{b}(\tau_1)\tilde{b}^*(\tau_2)] = \phi_{\tilde{b}}(\tau_1, \tau_2)\delta(\tau_1 - \tau_2)$. The corresponding power spectrum is known as *range scattering function*.

8.2 In Section 8.2.3 we obtained piecewise constant approximation to the Doppler spread target. Following an analogous development, obtain a piecewise constant approximation model of the range spread target. You may assume that $\tilde{b}(\tau)$ is a complex, zero mean, uncorrelated process as in the previous problem.

8.3 In a radar receiver employing single pulse detection, we require $P_F = 10^{-3}$ and $P_D = 0.8$. Calculate the average peak signal to noise ratio $\bar{\eta}^2$ to satisfy the requirements, and also the threshold for (i) a completely known signal, (ii) signal with unknown random phase, (iii) signal from a Rayleigh fluctuating target. What is the increase in $\bar{\eta}^2$ required if a $P_D = 0.95$ is specified?

8.4 The radar equation [2,8] relates the range of the target, transmitted energy (power), system parameters like antenna-gains, target cross section, and system losses to the received peak signal to noise ratio, $\bar{\eta}^2$, and is given by

$$\bar{\eta}^2 = 2\frac{E_T G_T G_R \lambda^2 \bar{\sigma}_c}{(4\pi)^3 R^4 N_0 L}$$

where

$\bar{\eta}^2$ = average peak signal to noise ratio
E_T = transmitted pulse energy
G_T = transmitter antenna gain
G_R = receiving antenna gain
λ = wavelength of transmission, corresponds to f_c
$\bar{\sigma}_c$ = average target cross section
R = target range
$N_0 = k_B T_{eq}$
k_B = Boltzman constant, 1.38×10^{-23} joule/°K
T_{eq} = equivalent input noise temperature
L = system loss factor

Design a radar system, assuming single-pulse detection and exactly known signal, to meet $P_F = 10^{-3}$ and $P_D = 0.8$ at a range of 200 nautical miles. The data available is:

$$
\begin{aligned}
G_T = G_R &= 30 \text{ dB} \\
f_c &= 1000 \text{ MHz} \\
T_{\text{eq}} &= 1000°\text{F} \\
L &= 3 \text{ dB} \\
\bar{\sigma}_c &= 1 \text{ m}^2
\end{aligned}
$$

Repeat the problem if the signal is of unknown random phase, and for the case of Rayleigh target fluctuation.

8.5 Design a radar system to meet $P_F = 10^{-2}$ and $P_D = 0.9$ at a range of 300 nautical miles. The system losses are 3 dB and the effective input noise temperature is 2000°F. Assume an average cross section of 1 m^2 for the target, and the transmitted rectangular pulse has at most 40% duty cycle. Consider the cases of fluctuating and nonfluctuating targets.

8.6 Consider a slowly fluctuating point target in which the reflecting gain $\tilde{b}$ is modeled as due to one specular or dominant scatterer plus a number of independent small scatterers. This gives rise to the one dominant plus Rayleigh model [4] for $c \triangleq |\tilde{b}|^2$ and a uniform density function for the phase θ_2. The density function of c is given by

$$
p(c) = \frac{c}{\sigma_b^2} \exp(-c/\sigma_b), \qquad c \geqslant 0
$$

Obtain the receiver structure for the detection of such a target from bandlimited white-noise-corrupted observations. Set up the performance equations and try to solve them. Consult Swerling [4] or Difranco and Rubin [8] if you have difficulty. Obtain the graphs relating P_D and $\bar{\eta}^2$ with P_F as parameter.

8.7 Consider a slowly fluctuating point target in which we model $\tilde{b}$ as a complex Gaussian random variable, but with nonzero mean and variance σ_b^2. This leads to the Rician density model for $|\tilde{b}|$ and a uniform density function for θ_2. Obtain the optimum Neyman–Pearson receiver under the usual assumptions. Set up and solve the performance equations. Draw plots of P_D versus $\bar{\eta}^2$ with P_F as parameter. Also study the effect of varying the mean value of $\tilde{b}$ and compare with the results obtained for Rayleigh fluctuating target (mean is zero for this case), exactly known steady point target, and steady point target with unknown random phase.

8.8 In Section 8.3 we studied detection of point targets that had unknown random phase, and targets with random amplitude and random phase (Rayleigh fluctuating target, for example). Very often it is desirable to measure the phase as accurately as possible and use it to improve performance. Consider the point target models in which $|\tilde{b}|$ is modeled as a Rayleigh or Rician density function and the phase θ_2 is exactly known. Obtain the optimum Neyman–Pearson receiver for detection under usual assumptions. Evaluate the expressions for P_D and P_F. Compare them with the results for Rayleigh fluctuating target, and target with unknown random phase. What do you conclude?

8.9 We considered essentially the optimum schemes for radar detection. A suboptimum radar detection technique is shown in Fig. E.8.1. It is called the *binary integrator* [21] or *double-threshold detection* [22], and consists of a single-pulse matched filter, followed by an envelope detector, an amplitude quantizer (first threshold), and a range gate. (This corresponds to sampling the output of the first threshold, so that the output is due to the return from a particular range resolution cell). Typically, the first threshold output is 1 or 0. The binary counter counts the number of 1's in a total of N pulses of the train. If the counter records more than a fixed number (say M, $M \leqslant N$), a target is declared present in the particular range cell. Show that this system has the following performance:

$$P_D = \sum_{k=M}^{N} \binom{N}{k} P_S^k (1 - (P_S)^{N-k})$$

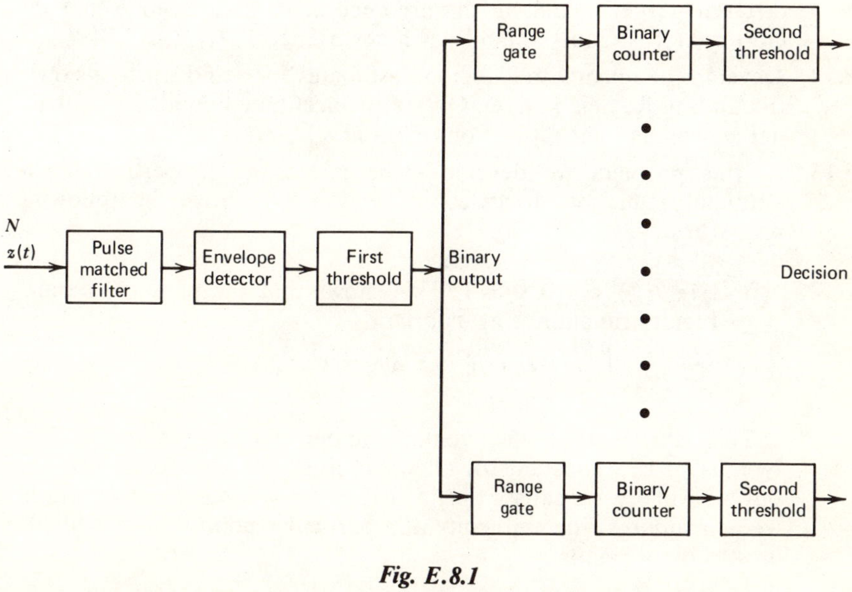

Fig. E.8.1

and

$$P_F = \sum_{k=M}^{N} \binom{N}{k} P_N^k (1 - (P_N)^{N-k})$$

where P_S is the probability that the output of the envelope detector exceeds the first threshold when a target is present (single-pulse basis); and P_N is the probability that the output of the envelope detector exceeds the first threshold when noise alone is present. Also obtain expressions for P_S and P_N. Consider the cases of nonfluctuating and Rayleigh fluctuating targets. Plot P_D versus $\bar{\eta}^2$ with $P_N = 10^{-2}$, and $N = 5$, 10, 100, and for different values of M.

8.10 Equation (8.2.17) represents a piecewise constant approximation to a Doppler spread target. We had indicated that an approach similar to multiple pulse detection of Rayleigh fluctuating targets discussed in Section 8.3.4 can be used to obtain an optimum receiver. Obtain this receiver under usual assumptions. Bring out the difference between this receiver and multiple pulse detection of Rayleigh fluctuating point target. Assume that $\tilde{b}_i$s in Eq. (8.2.17) are identically distributed zero mean complex Gaussian random variables with variance σ_b^2.

8.11 In many radar detection problems, the target return signal is masked by colored noise $\tilde{v}_c(t)$ in addition to the white noise $\tilde{v}(t)$. The colored noise is usually due to the returns from unwanted or interfering targets and is called clutter. Develop, using results from Chapter 4, the optimum Neyman–Pearson receivers for the detection of the different target models in the presence of clutter. Assume that the clutter $\tilde{v}_c(t)$ is Gaussian and has a covariance $V_{\tilde{v}_c}(t_1, t_2)$.

8.12 Develop the maximum likelihood estimates for τ and ω_d for a slowly fluctuating Rayleigh target. Obtain the accuracy bounds for individual as well as joint estimation of τ and ω_d.

8.13 In this problem we develop some interesting properties of the ambiguity function discussed in Section 8.4. Prove the following properties:

(i) $|\chi(\tau', \omega_d')| \leq \chi(0, 0) = 1$. This shows that the peak of the ambiguity function is at its origin.

(ii) $$\frac{1}{2\pi}\int_{-\infty}^{\infty}\int_{-\infty}^{\infty} |\chi(\tau', \omega_d')|^2 \, d\tau' \, d\omega_d' = 1$$

This shows that the volume under the ambiguity surface $|\chi(\tau', \omega_d')|^2$ is a constant for all waveforms. We can shape $\chi(\tau', \omega_d')$ but we will always satisfy this invariance of volume. In other words, we can suppress the ambiguity at a particular point, but it redistributes to other points.

(iii) If $\tilde{y}(t)$ has an ambiguity function $\chi(\tau', \omega_d')$, then $\tilde{y}(kt)$ has an ambiguity function $\dfrac{1}{\sqrt{k}}\chi\left(k\tau', \dfrac{\omega_d'}{k}\right)$

(iv) Let $\tilde{y}(t)$ and $\tilde{y}_1(t)$ be two functions such that $\tilde{y}_1(f)$ is the Fourier transform of $\tilde{y}(t)$. If $\chi(\tau', \omega_d')$ and $\chi_1(\tau', \omega_d')$ represent the ambiguity functions of $\tilde{y}(t)$ and $\tilde{y}_1(t)$ respectively, then

$$\chi_1(\omega_d', -\tau) = \chi(\tau, \omega_d)$$

(v) If $\chi_1(\tau', \omega_d')$ and $\chi_2(\tau', \omega_d')$ represent the ambiguity functions of $\tilde{y}_1(t)$ and $\tilde{y}_2(t)$, respectively, then the ambiguity function $\chi(\tau', \omega_d')$ of the function $\tilde{y}(t) = \tilde{y}_1(t)\tilde{y}_2(t)$ is given by

$$\chi(\tau', \omega_d') = \int_{-\infty}^{\infty} \chi(\tau', z)\chi_2(\tau', \omega_d - z)\, dz$$

8.14 In radar parameter estimation and resolution problems, the ambiguity function of Section 8.4 plays a significant role. Derive and plot the ambiguity functions of the following signals (you may use the properties derived in the previous problem):

(i) Rectangular pulse:

$$\tilde{y}_{11}(t) = \begin{cases} \dfrac{1}{T} & -T/2 < t < T/2 \\ 0 & \text{otherwise} \end{cases}$$

(ii) Gaussian pulse:

$$\tilde{y}(t) = \left(\frac{1}{\pi T^2}\right)^{1/4} \exp\left(-\frac{t^2}{2T^2}\right) \quad -\infty < t < \infty$$

(iii) Gaussian pulse with linear frequency modulation:

$$\tilde{y}(t) = \left(\frac{1}{\pi T^2}\right)^{1/4} \exp\left(-\frac{t^2}{2T^2}\right) \exp(-jkt^2)$$

where k is a constant.

(iv) Rectangular pulse with linear frequency modulation:

$$\tilde{y}(t) = \begin{cases} \dfrac{1}{T}\exp(jkt^2) & -T/2 < t < T/2 \\ 0 & \text{otherwise} \end{cases}$$

where k is a constant.

REFERENCES

1. P. M. Woodward, *Probability and Information Theory, with Applications to Radar*, Pergamon Press, Long Island City, New York, 1953.
2. M. I. Skolnik, Introduction to Radar, in *Radar Handbook*, McGraw-Hill, New York, 1970.
3. H. L. Van Trees, *Detection, Estimation and Modulation Theory Pt. III*, John Wiley & Sons, New York, 1971.
4. P. Swerling, Probability of Detection for Fluctuating Targets, *Rand Research Report*, **RM-1217**, March (1954). Also appeared in a special issue, *I.R.E. Trans. Info. Theory*, **IT-6**, April, 269–308 (1960).
5. G. N. Watson, *Theory of Bessel Functions*, Cambridge University Press, London/New York, 1944.
6. J. I. Marcum, A Statistical Theory of Target Detection by Pulsed Radar, *Rand Research Report*, **RM-754**, Dec. (1947). Also appeared in a special issue, *I.R.E. Trans. Info. Theory*, **IT-6**, April, 59–144, (1960).
7. J. I. Marcum, A Statistical Theory of Target Detection by Pulsed Radar, *Mathematical Appendix, Rand Research Report*, **RM-753**, July (1948). Also appeared in a special issue, *I.R.E. Trans. Info. Theory*, **IT-6**, April, 145–267, (1960).
8. J. V. DiFranco and W. L. Rubin, *Radar Detection*, Prentice Hall, Englewood Cliffs, N.J., 1968.
9. K. Pearson, *Tables of the Incomplete Gamma Function*, Cambridge University Press, London/New York, 1946.
10. F. C. Schweppe, Evaluation of Likelihood Functions for Gaussian Signals, *IEEE Trans. Info. Theory*, **IT-11**, No. 1, Jan., 61–70 (1965).
11. T. Kailath, A General Likelihood Ratio Formula for Random Signals in Gaussian Noise, *IEEE Trans. Info. Theory*, **IT-15**, No. 3, May, 350–361, (1969).
12. P. Swerling, Parameter Estimation Accuracy Formulas, *IEEE Trans. Info. Theory*, **IT-10**, No. 1, Oct., 302–313, (1964).
13. E. J. Kelly, I. S. Reed, and W. L. Root, Part II: The Accuracy of Radar Measurements, *J. Soc. Ind. Appl. Math.*, **8**, No. 3 (1960).
14. E. J. Kelly, The Radar Measurement of Range, Velocity, and Acceleration, *I.R.E Trans. Military Electron.*, April (1961).
15. A. W. Rihaczek, *Principles of High Resolution Radar*, McGraw-Hill Book Co., New York, 1969.
16. R. A. Singer and K. W. Behnke, Real Time Tracking Filter Evaluation and Selection for Tactical Applications, *IEEE Trans. Aerospace Electron. Syst.*, **AES-7**, Jan., 100–110 (1971).
17. R. K. Mehra, A Comparison of Several Nonlinear Filters for Re-entry Vehicle Tracking, *IEEE Trans. Automatic Control*, **AC-16**, No. 6, August, 307–319 (1971).
18. R. P. Wishner, R. E. Larson, and M. Athans, *Status of Radar Tracking Algorithms*, presented at the AFOSR and IEEE Symp. Nonlinear Estimation Theory and Its Applications, San Diego, Calif. Sept. 21–23, 1970.

19. R. J. McAulay and E. Denlinger, A Decision Directed Adaptive Tracker, *IEEE Trans. Aerospace Electron. Syst.*, **AES-9**, March, 229–236 (1973).
20. T. R. Benedict and G. W. Bordner, Synthesis of an Optimal Set of Radar Track-While-Scan Smoothing Equations, *IEEE Trans. Automatic Control*, **27**, July, 27–32 (1962).
21. J. V. Harrington: An Analysis of the Detection of Repeated Signals in Noise by Bianry Integration, *I.R.E. Trans. Info. Theory*, **IT-1**, No. 1, March, 1–9 (1955).
22. P. Swerling, The Double Threshold Method of Detection, *Rand Corp. Memo*, **RM-1008**, December (1952).
23. J. I. Marcum, Table of *Q*-Functions, *Rand Corp. Rep.*, **RM-339**, Jan. (1950).
24. J. Pachares, A Table of Bias Levels Useful in Radar Detection Problems, *I.R.E. Trans. Info. Theory*, **IT-14**, No. 1, March, 38–45 (1958).
25. L. E. Brennan and I. S. Reed, A Recursive Method of Computing the *Q* Function, *IEEE Trans. Info. Theory*, **IT-11**, No. 2, April, 312–313, (1965).

CHAPTER 9

MISCELLANEOUS APPLICATIONS

9.1 INTRODUCTION

The basic problems of detection and estimation arise in several areas besides the communication and radar applications discussed in the previous chapters. The solutions to the detection and estimation problems discussed earlier can be applied in areas such as control, pattern recognition, system identification, speech and picture processing. In this chapter we consider two specific applications, namely statistical pattern classification and system identification.

9.2 APPLICATION TO STATISTICAL PATTERN CLASSIFICATION

Automatic pattern recognition covers a wide variety of problems such as speech recognition, character recognition, finger-print classification, aerial and microphotograph processing, seismic signal analysis for both geophysical exploration and explosion monitoring, electrocardiogram (EKG), and electroencephalogram (EEG) recording analysis in medicine and physiology [1]. We can roughly describe the pattern recognition problem as follows. We are given several groups of objects such that the objects within each group have certain common attributes. We refer to the description of the object in terms of measurements of these attributes as a *pattern* and the measurements themselves as the *features* of the pattern. The several groups are called *pattern classes*. The basic problem of pattern recognition is to classify or label an object as belonging to one of the pattern classes by processing the features associated with the pattern.

The design of an automatic pattern recognition system involves several steps. The first is to represent the input data in a form suitable for machine classification of the object. We thus measure certain distinctive attributes of the object being classified and arrange the measurements in the form of a *pattern vector*. For example, in character recognition the area surrounding the character may be divided into a grid pattern and a value of 0 or 1 may be assigned to each cell depending on whether the cell contains a portion of the character. If the measurements are continuous waveforms as in the case of ECG analysis or speech recognition, the sample values of the waveform at a set of discrete instants of time can serve as the pattern vector. Alternatively, the waveforms can be expanded in a complete set of orthonormal functions and the coefficients of the expansion can be used as the pattern vector.

In most cases, not all the elements of a pattern vector have the same significance in the classification of the pattern. The measurements usually contain many irrelevant data as far as the classification problem is concerned. We may therefore seek to order the elements of the pattern vector in decreasing order of importance and retain only those elements which are most useful. This is usually referred to as *feature selection*.

The third step in pattern recognition systems is the determination of optimum procedures for classification of the patterns. If there are M pattern classes, we can consider the *pattern space* to be the sum of M regions each of which contains the pattern vectors of that class. The recognition problem thus consists of generating the decision boundaries which seperate the M pattern classes. The functions defining the decision boundaries are referred to as *discriminant functions*.

We can determine the discriminant functions in several ways. If we have complete prior information about the pattern classes, the determination of the discriminant functions is relatively straightforward. If there is little or no information regarding the classes, we can use a training or learning procedure to determine the characteristics of the patterns. There are several approaches to the classification of patterns and both deterministic and statistical algorithms are available in the literature. Here, we restrict ourselves to statistical methods. The reader is referred to the literature for other approaches [2–7]. In particular, we show how the techniques of detection and estimation discussed in the earlier chapters can be used. Even though we are concerned mainly with the classification problem, we briefly discuss the feature extraction problem in Section 9.2.5.

9.2.1 Basic Problem of Statistical Pattern Classification

In this section, we discuss the basic statistical pattern classification problem. As stated in the previous section, the problem is to classify patterns into one of several classes on the basis of certain measurements. Usually the number M of the pattern classes is known, but may be unknown in some cases [8]. The pattern is characterized by the feature vector which forms the basis for classification. In practice, within a particular class the features of a pattern show statistical variability, so that it is appropriate to treat the patterns (pattern vectors) as random vectors described by multivariate (vector) probability distribution or density functions. The simplest problem arises when these underlying distributions are exactly known and our machine is to classify a test pattern into one of the M pattern classes.

Often, the form of these distributions is known or can be assumed, but certain parameters associated with the distributions may be unknown. In this case, if certain patterns with known classifications are available, we can use these sample patterns to estimate or learn the unknown parameters of the distributions involved. These patterns with known classification which are used for learning purposes are referred to as training samples. As can be expected, a machine which needs to learn the associated distribution functions, will, in general, not perform as well as a machine classifying patterns with completely known distributions. The performance of the machine is usually evaluated by determining the probability of misclassifying the patterns. This mode of learning using training samples of known classification is termed *learning with a teacher* or *supervised learning*.

If training samples with known classification are not available, the samples with unknown classification can be used to learn or estimate the parameters of interest to design the classifier. The classifier is then said to work in a *nonsupervised mode* or *learn without a teacher*.

It is helpful to formulate mathematically the various aspects of the problem described previously. Let $\mathbf{x}$ denote the pattern vector

$$\mathbf{x}=[x_1, x_2, \ldots, x_n]^T$$

If $\mathbf{x}$ belongs to a pattern from the jth class H_j of M possible classes, we denote it as $\mathbf{x}^j$, and its components as x_i^j. The probability density function of $\mathbf{x}$ when it belongs to class H_j is represented as $p(\mathbf{x}|H_j)$. Patterns presented in sequence or as a batch are denoted $\mathbf{x}_1, \mathbf{x}_2, \mathbf{x}_3, \ldots$. If the probability density function is known except for certain parameters $\boldsymbol{\theta}_j$, it is represented as $p(\mathbf{x}|H_j, \boldsymbol{\theta}_j)$.

The simplest problem arises when $p(\mathbf{x}|H_j)$, $j=1, 2, \ldots, M$ are known, and a test pattern $\mathbf{x}$ has to be classified into one of the M categories. The decision rule (discriminant function) may be linear, quadratic, or otherwise depending on $p(\mathbf{x}|H_j)$.

The problem of supervised learning pattern classification may be stated as follows. For each class, we know either the density function $p(\mathbf{x}|H_j)$ or the conditional density function $p(\mathbf{x}|H_j, \boldsymbol{\theta}_j)$ where $\boldsymbol{\theta}_j$ represents a vector of unknown parameters. For each class in which the parameter $\boldsymbol{\theta}_j$ has to be learned, we are given a set of training patterns $\mathbf{Z}_{N_j} = [\mathbf{x}_1^j, \mathbf{x}_2^j, \ldots, \mathbf{x}_{N_j}^j]$ belonging to that class. We can assume the parameters $\boldsymbol{\theta}_j$ to be either deterministic or random variables. In the latter case, we assume a prior density function for the parameter. We then arrive at a decision rule to classify the test pattern $\mathbf{x}$. This is discussed in Section 9.2.3. The machine essentially operates in two modes. In the learning mode the unknown parameters are estimated using the samples with known classification. In the testing mode new samples are classified into the various classes on the basis of decision rules formulated at the end of the learning mode.

In nonsupervised learning, we are given a set of patterns with unknown classification, $\mathbf{Z}_N = [\mathbf{x}_1, \mathbf{x}_2, \ldots, \mathbf{x}_N]$. Learning the vector $\boldsymbol{\theta}_j$ for those classes with unknown parameters must proceed using these unclassified samples. Any new patterns which are presented to the machine may also be used to refine the estimates of the unknown parameters. In this case, there is no clear distinction between the learning and testing modes. Unfortunately, as we shall see later, the computational and storage requirements for unsupervised learning become exceedingly large as the number of sample patterns increases. We discuss this in Section 9.2.4 where we also present some techniques to overcome this difficulty.

We start our discussions with the problem of classification with known probability density functions. For clarity of presentation, we restrict most of our discussions to the case where we have only two pattern classes. The extension to M classes is fairly straightforward.

9.2.2 Pattern Classification with Known Densities

When the underlying density functions for the pattern classes are completely known, we can obtain the discriminant functions from our discussion of hypothesis testing problems in Section 3.3. In practice the discriminant functions are chosen so as to minimize the average probability of misclassification. This corresponds to the minimum probability of error criterion discussed in Section 3.3. If the prior probabilities of the pattern classes are not known, we can use either the Neyman–Pearson criterion or a minimax approach in designing our classifiers. In any case, the results can be obtained by a straightforward application of our discussions in Chapter 3.

Example 9.2.1 Let us consider the classification of a test pattern $\mathbf{x}$, into one of two classes. The densities $p(\mathbf{x}|H_j)$, $j = 1, 2$ are assumed to be Gaussian with means $\mathbf{m}_1$ and $\mathbf{m}_2$, and covariance matrices $\mathbf{V}_1$ and $\mathbf{V}_2$. We seek the decision rule so as to minimize the probability of misclassification. The rule follows easily from Eq. (3.3.21) as

$$\frac{p(\mathbf{x}|H_1)}{p(\mathbf{x}|H_2)} \underset{H_2}{\overset{H_1}{\gtrless}} \frac{P_2}{P_1} \tag{9.2.1}$$

where P_1 and P_2 are the prior probabilities of pattern classes 1 and 2, respectively. If we assume $P_1 = P_2 = 1/2$ then taking the logarithm of Eq. (9.2.1) and rearranging yields the decision rule as

$$\tfrac{1}{2}(\mathbf{x}-\mathbf{m}_2)^T\mathbf{V}_2^{-1}(\mathbf{x}-\mathbf{m}_2) - \tfrac{1}{2}(\mathbf{x}-\mathbf{m}_1)^T\mathbf{V}_1^{-1}(\mathbf{x}-\mathbf{m}_1)$$

$$+\tfrac{1}{2}\left[\ln|V_2| - \ln|V_1|\right] \underset{H_2}{\overset{H_1}{\gtrless}} 0 \tag{9.2.2}$$

We now consider some special cases. When the covariances are equal, that is, $\mathbf{V}_1 = \mathbf{V}_2 = \mathbf{V}$, Eq. (9.2.2) reduces to

$$(\mathbf{m}_1 - \mathbf{m}_2)^T\mathbf{V}^{-1}\mathbf{x} + \tfrac{1}{2}(\mathbf{m}_2^T\mathbf{V}^{-1}\mathbf{m}_2 - \mathbf{m}_1^T\mathbf{V}^{-1}\mathbf{m}_1) \underset{H_2}{\overset{H_1}{\gtrless}} 0 \tag{9.2.3}$$

We note that the left-hand side of Eq. (9.2.3) is linear in the features $\mathbf{x}$. The discriminant function is thus a weighted sum of the measured features $x_1, x_2, \ldots, x_n$ and a constant that depends on the known statistics of the two pattern classes. We can therefore write the discriminant function as

$$f(\mathbf{x}) = \mathbf{w}^T\mathbf{x} + w_0 \tag{9.2.4}$$

where $\mathbf{w}$ is an n-vector of weighting coefficients and w_0 is a constant. $\mathbf{w}$ and w_0 are obtained from Eq. (9.2.3) as

$$\mathbf{w}^T = (\mathbf{m}_1 - \mathbf{m}_2)^T\mathbf{V}^{-1} \quad \text{and} \quad w_0 = \tfrac{1}{2}(\mathbf{m}_2^T\mathbf{V}^{-1}\mathbf{m}_2 - \mathbf{m}_1^T\mathbf{V}^{-1}\mathbf{m}_1) \tag{9.2.5}$$

The decision rule in terms of $f(\mathbf{x})$ is

$$f(\mathbf{x}) \underset{H_2}{\overset{H_1}{\gtrless}} 0 \tag{9.2.6}$$

Because of their simplicity, linear discriminant functions as in Eq. (9.2.4) are widely used in many classifiers, even though such operation may be nonoptimal.

The probability of error (misclassification) easily follows (see Example 3.4.1). The sufficient statistic l equals $f(\mathbf{x})$ and is Gaussian under both hypotheses with moments

$$\begin{aligned} \mathrm{E}\{l|H_j\} &= \mathbf{w}^T\mathbf{m}_j + w_0 \triangleq \mu_j \qquad j=1,2 \\ \operatorname{var}\{l|H_j\} &= \mathbf{w}^T\mathbf{V}\mathbf{w} \triangleq \sigma_l^2 \end{aligned} \tag{9.2.7}$$

The probability of error easily follows as

$$P_e = \tfrac{1}{2}\int_{-\infty}^{0} p(l|H_1)\,dl + \tfrac{1}{2}\int_{0}^{\infty} p(l|H_2)\,dl \tag{9.2.8}$$

Let us now assume that $\mathbf{V} = \sigma^2\mathbf{I}$ where σ^2 is a scalar and $\mathbf{I}$ is an $n \times n$ identity matrix. The assumption is valid when the physical features are statistically independent with equal variances, σ^2. In such a case, the decision rule of Eq. (9.2.3) can be written as

$$\left(\mathbf{m}_1^T\mathbf{x} - \tfrac{1}{2}\mathbf{m}_1^T\mathbf{m}_1\right) - \left(\mathbf{m}_2^T\mathbf{x} - \tfrac{1}{2}\mathbf{m}_2^T\mathbf{m}_2\right) \underset{H_2}{\overset{H_1}{\gtrless}} 0 \tag{9.2.9}$$

and can be interpreted as a minimum distance classifier. That is, given a pattern vector $\mathbf{x}$, we measure the Euclidean distance $\|\mathbf{x} - \mathbf{m}_i\|$ from $\mathbf{x}$ to each of the mean vectors, and assign $\mathbf{x}$ to that class which corresponds to the least distance.

The probability of error easily follows from Eq. (9.2.8) by noting that for this case

$$\mu_1 = -\mu_2 = \frac{(\mathbf{m}_1 - \mathbf{m}_2)^T(\mathbf{m}_1 - \mathbf{m}_2)}{\sigma^2} \tag{9.2.10}$$

so that we have

$$P_e = \operatorname{erfc}_*\left(\frac{1}{\sigma^2}\sum_{i=1}^{n}(m_{1i} - m_{2i})\right) \tag{9.2.11}$$

The significance of each feature can be determined from an examination of the preceding expression. Those features for which the difference between the two means is largest are the most useful, while those features for which the mean is the same do not play a role in reducing P_e. It follows from Eq. (9.2.11) that we can decrease the probability of error by increasing n, that is, by introducing new independent features. In theory at least,

P_e can be reduced to an arbitrarily small quantity by increasing n without limit. In practice it has been found that P_e does decrease with increasing n up to a certain point. However, the inclusion of additional features beyond this point tends to degrade performance. This seems to be related to the fact that the number of samples in the design of any classifier is always finite. Since a detailed discussion of this is beyond the scope of this book the interested reader is referred to the literature for details [6].

For the case when the means of the two classes are equal but the variances are different, with $\mathbf{m}_1 = \mathbf{m}_2 = \mathbf{m}$, the decision rule becomes

$$\tfrac{1}{2}(\mathbf{x}-\mathbf{m})^T\mathbf{V}_2^{-1}(\mathbf{x}-\mathbf{m}) - \tfrac{1}{2}(\mathbf{x}-\mathbf{m})^T\mathbf{V}_1^{-1}(\mathbf{x}-\mathbf{m}) + \tfrac{1}{2}\left[\ln|V_2| - \ln|V_1|\right] \underset{H_1}{\overset{H_2}{\gtrless}} 0 \tag{9.2.12}$$

The discriminant function $f(\mathbf{x})$ is quadratic in $\mathbf{x}$ and can be written as

$$f(\mathbf{x}) = \mathbf{x}^T\mathbf{w}\mathbf{x} + w_0 \tag{9.2.13}$$

where $\mathbf{w}$ is an $n \times n$ matrix and w_0 is a scalar. The probability of misclassification can be computed as in Example 3.4.2.

9.2.3 *Supervised Learning Pattern Recognition*

In supervised learning pattern recognition, the class conditional densities are assumed to be known except for certain parameters. In order to facilitate learning of the unknown parameters, we have access to a certain number of patterns corresponding to each class for which the parameters have to be estimated. For simplicity, let us assume that we have two classes. Then the conditional densities of interest are $p(\mathbf{x}|H_j, \boldsymbol{\theta}_j)$, $j = 1, 2$. We also assume that we are given N_1 patterns from class 1 and N_2 patterns from class 2. The classification rule is obtained by computing the conditional likelihood ratio

$$\Lambda(\mathbf{x}|\mathbf{Z}_{N_1}, \mathbf{Z}_{N_2}) = \frac{p(\mathbf{x}|\mathbf{Z}_{N_1}, H_1)}{p(\mathbf{x}|\mathbf{Z}_{N_2}, H_2)} = \frac{p(\mathbf{x}|\mathbf{x}_1^1, \mathbf{x}_2^1, \ldots, \mathbf{x}_{N_1}^1, H_1)}{p(\mathbf{x}|\mathbf{x}_1^2, \mathbf{x}_2^2, \ldots, \mathbf{x}_{N_2}^2, H_2)} \tag{9.2.14}$$

We consider the case where $\boldsymbol{\theta}_j$ are random vectors with known a priori density $p(\boldsymbol{\theta}_j)$. In order to determine $p(\mathbf{x}|\mathbf{Z}_{N_j}, H_j)$, $j = 1, 2$, we use the conditional densities in each pattern class to write

$$p(\mathbf{x}|\mathbf{Z}_{N_j}, H_j) = \int_{\chi_{\boldsymbol{\theta}_j}} p(\mathbf{x}|\boldsymbol{\theta}_j, H_j)p(\boldsymbol{\theta}_j|\mathbf{Z}_{N_j})\, d\boldsymbol{\theta}_j, \qquad j = 1, 2 \tag{9.2.15}$$

where χ_{θ_j} is the space of the parameter vector

There still remains the problem of finding $p(\boldsymbol{\theta}_j|\mathbf{Z}_{N_j})$. We can determine $p(\boldsymbol{\theta}_j|\mathbf{Z}_{N_j})$ in a recursive manner from $p(\boldsymbol{\theta}_j|\mathbf{Z}_{N_j-1})$ by applying Bayes' rule:

$$p(\boldsymbol{\theta}_j|\mathbf{Z}_{N_j}) = p(\boldsymbol{\theta}_j|\mathbf{Z}_{N_j-1}, \mathbf{x}^j_{N_j})$$

$$= \frac{p(\mathbf{x}^j_{N_j}|\mathbf{Z}_{N_j-1}, \boldsymbol{\theta}_j)p(\boldsymbol{\theta}_j|\mathbf{Z}_{N_j-1})}{\int_{\chi_{\theta_j}} p(\mathbf{x}^j_{N_j}|\mathbf{Z}_{N_j-1}, \boldsymbol{\theta}_j)p(\boldsymbol{\theta}_j|\mathbf{Z}_{N_j-1})\, d\boldsymbol{\theta}_j} \tag{9.2.16}$$

If we assume that the patterns are conditionally independent, that is,

$$p(\mathbf{x}^j_{N_j}|\mathbf{Z}_{N_j-1}, \boldsymbol{\theta}_j) = p(\mathbf{x}^j_{N_j}|\boldsymbol{\theta}_j) \tag{9.2.17}$$

we can simplify the algorithm of Eq. (9.2.16) as

$$p(\boldsymbol{\theta}_j|\mathbf{Z}_{N_j}) = \frac{p(\mathbf{x}^j_{N_j}|\boldsymbol{\theta}_j)}{\int_{\chi_{\theta_j}} p(\mathbf{x}^j_{N_j}|\boldsymbol{\theta}_j)p(\boldsymbol{\theta}_j|\mathbf{Z}_{N_{j-1}})\, d\boldsymbol{\theta}_j} p(\boldsymbol{\theta}_j|\mathbf{Z}_{N_{j-1}}). \tag{9.2.18}$$

Equation (9.2.18) represents an algorithm for updating the density function for $\boldsymbol{\theta}_j$ as more patterns from class j are presented to the classifier in a recursive manner, starting from the known density $p(\boldsymbol{\theta}_j|H_j)$. As such it is usually referred to as a learning algorithm for $\boldsymbol{\theta}_j$. The implementation of the classification algorithm is now straightforward, at least conceptually. We determine the conditional density $p(\boldsymbol{\theta}_j|\mathbf{Z}_{N_j})$ recursively from Eq. (9.2.18) and use it in Eq. (9.2.15) to determine $p(\mathbf{x}|\mathbf{Z}_{N_j}, H_j), j = 1, 2$ for each pattern to be classified. We then compute the likelihood ratio of Eq. (9.2.14) and compare it with the threshold to determine the classification.

The computational feasibility of Eq. (9.2.18) depends critically on the existence of a finite dimensional sufficient statistic for the relevant prior and posterior density functions. Otherwise, one needs to quantize the parameter vector $\boldsymbol{\theta}_j$ to implement the algorithm practically (why?). Prior and posterior density functions that satisfy the requirement of a finite dimensional sufficient statistic are called conjugate or reproducing pairs, as stated in Chapter 5. In this case, assuming a reproducing form for $p(\boldsymbol{\theta})$ leads to considerable simplification of the computations. If an infinite number of training samples are available, the reproducing pair converges to the correct parameter value [9]. Again, in the evaluation of Eq. (9.2.15) we may need to quantize $\boldsymbol{\theta}_j$ over its range.

Example 9.2.2 We consider the case where the density function $p(\mathbf{x}|H_j, \boldsymbol{\theta}_j)$ is Gaussian with mean $\mathbf{m}$ and variance $\mathbf{V}$. For simplicity, we suppress the dependence of the various quantities on j, the category. We assume that this has to be carried out for each class, when relevant. The unknown parameter $\boldsymbol{\theta}$ can be either the mean $\mathbf{m}$ or the variance $\mathbf{V}$ or both and must be learned from the training pattern set $\mathbf{Z}_N = \{\mathbf{x}_1, \mathbf{x}_2, \ldots, \mathbf{x}_N\}$. In every case, we will assume an appropriate reproducing form for the prior density function for $\boldsymbol{\theta}$.

When $\boldsymbol{\theta}$ is the mean vector, we have

$$p(\mathbf{x}|\mathbf{m}) = \frac{1}{(2\pi)^{N/2}|\mathbf{V}|^{1/2}} \exp\left\{-\frac{1}{2}(\mathbf{x}-\mathbf{m})^T\mathbf{V}^{-1}(\mathbf{x}-\mathbf{m})\right\}$$

We choose for $\mathbf{m}$, a Gaussian prior probability function $p(\mathbf{m})$ of the form (see Section 5.7).

$$p(\mathbf{m}) = \frac{1}{(2\pi)^{N/2}|\phi_0|^{1/2}} \exp\left\{-\frac{1}{2}(\mathbf{m}-\mathbf{m}_0)^T\phi_0^{-1}(\mathbf{m}-\mathbf{m}_0)\right\}$$

where $\mathbf{m}_0$, ϕ_0 are the expected value and covariance matrix (usually an educated guess of the designer) of the parameter vector $\mathbf{m}$. The density function $p(\mathbf{m}|\mathbf{Z}_k)$ can then be determined by successive application of Bayes' rule as in Eq. (9.2.18). Because of the reproducing property of the Gaussian density, the density function $p(\mathbf{m}|\mathbf{Z}_k)$ is also Gaussian. The mean and variance of the density function $p(\mathbf{m}|\mathbf{Z}_k)$ can be computed using either the MAP approach or by application of the orthogonal projection principle and a modest amount of manipulations of matrix equations. These parameters can be sequentially updated using the following algorithms [10]: (see also Example 5.5.3)

$$\mathbf{m}_k = \mathbf{V}(\phi_{k-1}+\mathbf{V})^{-1}\mathbf{m}_{k-1} + \phi_{k-1}(\phi_{k-1}+\mathbf{V})^{-1}\mathbf{x}_k \tag{9.2.19a}$$

$$\phi_k = \mathbf{V}(\phi_{k-1}+\mathbf{V})^{-1}\phi_{k-1} \tag{9.2.19b}$$

Use of Eq. (9.2.15) then yields the density function of interest $p(\mathbf{x}|\mathbf{Z}_N, H_j)$ as a Gaussian density with mean $\mathbf{m}_N$ and covariance $(\mathbf{V}+\phi_N)$.

When $\boldsymbol{\theta}$ is the unknown covariance matrix $\mathbf{V}$, the appropriate prior density function to use is the Wishart density function with parameters $\mathbf{V}_0$ and $\mathbf{c}_0$. Let us without loss of generality, assume that $\mathbf{m}=0$. With $\mathbf{Q}=\mathbf{V}^{-1}$, we can write for the prior density

$$p(\mathbf{Q}) = \begin{cases} k_{n,c_0}\left|\dfrac{c_0}{2}\mathbf{V}_0\right|^{(c_0-1)/2} |\mathbf{Q}|^{(c_0-n-2)/2} \exp\left\{\dfrac{1}{2}\text{tr}\,[c_0\mathbf{V}_0\mathbf{Q}]\right\} & \text{on } \chi_{\mathbf{Q}} \\ 0, & \text{otherwise} \end{cases}$$

where χ_Q denotes the subset of the Euclidean space of dimension $\frac{1}{2}n(n+1)$. The matrix $\mathbf{Q}$ is positive definite, and k_{n, c_0} is the normalizing constant

$$k_{n, c_0} = \left[(\pi)^{n(n-1)/4} \prod_{\alpha=1}^{n} \Gamma\left(\frac{c_0 - \alpha}{2}\right) \right]^{-1}$$

Here $\mathbf{V}_0$ is a positive definite matrix reflecting the initial knowledge of $\mathbf{V}^{-1}$, c_0 is a scalar reflecting the confidence of the initial estimate $\mathbf{V}_0$, and $\Gamma(\cdot)$ is the gamma function. By successive applications of Bayes' formula, we can show that the a posteriori density function $p(\mathbf{Q}|\mathbf{Z}_N)$ is also Wishart with the parameters $\mathbf{V}_N$ and c_N replacing $\mathbf{V}_0$ and c_0, where

$$\mathbf{V}_N = \frac{c_0 \mathbf{V}_0 + \sum_{i=1}^{N} \mathbf{x}_i \mathbf{x}_i^T}{c_0 + N} \tag{9.2.20a}$$

$$c_N = c_0 + N \tag{9.2.20b}$$

Again the density function $p(\mathbf{x}|\mathbf{Z}_{N_j}, H_j)$ is calculated using Eq. (9.2.15).

When both the mean $\mathbf{m}$ and the covariance matrix $\mathbf{V}$ (or more appropriately, its inverse $\mathbf{V}^{-1}$) are unknown, we choose the prior density as Gauss–Wishart, that is, the mean $\mathbf{m}$ is Gaussian with prior mean $\mathbf{m}_0$ and covariance matrix $\phi_0 = d_0^{-1}\mathbf{V}$, and $\mathbf{V}^{-1} = \mathbf{Q}$ is distributed according to Wishart distribution with parameters c_0 and $\mathbf{V}_0$. Successive application of Bayes' formula yields the a posteriori density function to be Gauss–Wishart with parameters c_N, d_N, $\mathbf{m}_N$, and $\mathbf{V}_N$ replacing c_0, d_0, $\mathbf{m}_0$, and $\mathbf{V}_0$. The respective expressions are:

$$c_N = c_0 + N \tag{9.2.21a}$$

$$d_N = d_0 + N \tag{9.2.21b}$$

$$\mathbf{m}_N = \frac{d_0 \mathbf{m}_0 + \sum_{i=1}^{N} \mathbf{x}_i}{d_0 + N} \tag{9.2.21c}$$

$$\mathbf{V}_N = \frac{1}{c_0 + N} \Bigg\{ (c_0 \mathbf{V}_0 + d_0 \mathbf{m}_0 \mathbf{m}_0^T) + (N-1)\mathbf{S} + \frac{1}{N} \left[\sum_{i=1}^{N} \mathbf{x}_i \right] \left[\sum_{j=1}^{N} \mathbf{x}_j \right]^T - d_N \mathbf{m}_N \mathbf{m}_N^T \Bigg\} \tag{9.2.21d}$$

where

$$S = \frac{1}{N-1} \sum_{i=1}^{N} \left(\mathbf{x}_i - \frac{1}{N} \sum_{j=1}^{N} \mathbf{x}_j \right) \left(\mathbf{x}_i - \frac{1}{N} \sum_{j=1}^{N} \mathbf{x}_j \right)^T \qquad \textbf{(9.2.21e)}$$

Again the density function $p(\mathbf{x}|H_j, \mathbf{Z}_{N_j})$ can be calculated by using Eq. (9.2.15).

If the vectors $\boldsymbol{\theta}_j$ are nonrandom parameters, we can find a maximum-likelihood estimate of $\boldsymbol{\theta}_j$ for each j and use these as the true values of the parameters in computing the likelihood ratio. The decision rule then becomes

$$\Lambda(\mathbf{x}) = \frac{p(\mathbf{x}|H_1, \boldsymbol{\theta}_1)|_{\boldsymbol{\theta}_1 = \hat{\boldsymbol{\theta}}_{1ml}}}{p(\mathbf{x}|H_2, \boldsymbol{\theta}_2)|_{\boldsymbol{\theta}_2 = \hat{\boldsymbol{\theta}}_{2ml}}} \qquad \textbf{(9.2.22)}$$

The ML estimates based on the training samples are obtained in the usual manner (see Chapter 5) by maximizing the conditional density $p(\mathbf{Z}_{N_j}|\boldsymbol{\theta}_j)$, $j = 1, 2$. Since the samples are independent, we can write

$$p(\mathbf{Z}_{N_j}|\boldsymbol{\theta}_j) = \prod_{k=1}^{N_j} p(\mathbf{x}_k^j|\boldsymbol{\theta}_j) \qquad j = 1, 2 \qquad \textbf{(9.2.23)}$$

Taking logarithms and setting the gradient with respect to $\boldsymbol{\theta}_j$ equal to zero yields

$$\sum_{k=1}^{N_j} \frac{1}{p(\mathbf{x}_k^j|\boldsymbol{\theta}_j)} \nabla_{\boldsymbol{\theta}_j} p(\mathbf{x}_k^j|\boldsymbol{\theta}_j) \Bigg|_{\boldsymbol{\theta}_j = \boldsymbol{\theta}_{jml}} = 0 \qquad j = 1, 2 \qquad \textbf{(9.2.24)}$$

These usually constitute a set of nonlinear equations in the components of $\boldsymbol{\theta}_j$, which can be solved to obtain the ML estimates.

9.2.4 Nonsupervised Learning Pattern Recognition

In the previous section, we assumed that the classification of the training samples used to learn the parameters were known. In this section, we consider the problem of unsupervised learning in which the classifications of the training samples $\mathbf{Z}_N = \{\mathbf{x}_1, \mathbf{x}_2, \ldots, \mathbf{x}_N\}$ are not known. Again we assume that the form of the class conditional densities are known and only certain parameters associated with these densities are unknown, that is, that $p(\mathbf{x}|H_j, \boldsymbol{\theta}_j)$, $j = 1, \ldots, M$ are known. If we denote by $\boldsymbol{\theta}$, the vector of all unknown parameters,

$$\boldsymbol{\theta} = [\boldsymbol{\theta}_1, \boldsymbol{\theta}_2, \ldots, \boldsymbol{\theta}_M]^T$$

we can write the density function of a pattern $\mathbf{x}$ conditioned on the (unknown) parameter vector $\boldsymbol{\theta}$, by expanding in a *mixture density* as

$$p(\mathbf{x}|\boldsymbol{\theta}) = \sum_{j=1}^{M} p(\mathbf{x}|H_j, \boldsymbol{\theta}_j)P_j \tag{9.2.25}$$

where P_j denotes the prior probabilities of the various classes. The conditional densities $p(\mathbf{x}|H_j, \boldsymbol{\theta}_j)$ are referred to as the *component densities* and the prior probabilities P_j as the *mixing parameters*.

The problem of nonsupervised learning can now be stated as follows. Given samples corresponding to the mixture density of Eq. (9.2.25) we are required to estimate the unknown parameter vector, using these samples. Once $\boldsymbol{\theta}$ is known, the mixture can be decomposed into its components so that the problem reduces to pattern classification using known densities considered in Section 9.2.2.

A question that arises is whether we can estimate the vector $\boldsymbol{\theta}$ uniquely from the mixture. We say that the density $p(\mathbf{x}|\boldsymbol{\theta})$ is *identifiable* if for $\boldsymbol{\theta} \neq \boldsymbol{\theta}'$, there exists a pattern $\mathbf{x}$ for which $p(\mathbf{x}|\boldsymbol{\theta}) \neq p(\mathbf{x}|\boldsymbol{\theta}')$ [12–14]. In practice, most mixtures of commonly encountered distributions are identifiable. However, with discrete distributions such is not always the case. In the sequel, we assume only identifiable mixtures.

We now turn to the problem of designing the classifier. Again, we restrict ourselves to two pattern classes. We then write the decision rule for classifying the samples as

$$\Lambda(\mathbf{x}|\mathbf{Z}_N) \underset{H_2}{\overset{H_1}{\gtrless}} \gamma \tag{9.2.26}$$

where

$$\Lambda(\mathbf{x}|\mathbf{Z}_N) = \frac{p(\mathbf{x}|H_1, \mathbf{Z}_N)}{p(\mathbf{x}|H_2, \mathbf{Z}_N)} \tag{9.2.27}$$

and γ is the threshold.

In the case where $\boldsymbol{\theta}$ is a random vector with known a priori density $p(\boldsymbol{\theta})$, the density functions $p(\mathbf{x}|H_j, \mathbf{Z}_N)$ can be written as

$$p(\mathbf{x}|H_j, \mathbf{Z}_N) = \int_{X_\theta} p(\mathbf{x}|\boldsymbol{\theta}, H_j, \mathbf{Z}_N)p(\boldsymbol{\theta}|H_j, \mathbf{Z}_N)\, d\boldsymbol{\theta} \tag{9.2.28}$$

Since the selection of the test pattern $\mathbf{x}$ is independent of the set of training samples, $p(\mathbf{x}|\boldsymbol{\theta}, H_j, \mathbf{Z}_N) = p(\mathbf{x}|\boldsymbol{\theta}_j, H_j)$. We can also assume that $p(\boldsymbol{\theta}|H_j, \mathbf{Z}_N) = p(\boldsymbol{\theta}|\mathbf{Z}_N)$. We thus obtain

$$p(\mathbf{x}|H_j, \mathbf{Z}_N) = \int_{X_\theta} p(\mathbf{x}|\boldsymbol{\theta}_j, H_j)p(\boldsymbol{\theta}|\mathbf{Z}_N)\, d\boldsymbol{\theta} \tag{9.2.29}$$

It only remains to determine $p(\boldsymbol{\theta}|\mathbf{Z}_N)$ in order to complete the design of the classifier.

We can use Bayes' rule to write

$$p(\boldsymbol{\theta}|\mathbf{Z}_N) = \frac{p(\mathbf{Z}_N|\boldsymbol{\theta})p(\boldsymbol{\theta})}{\int_{X_\theta} p(\mathbf{Z}_N|\boldsymbol{\theta})p(\boldsymbol{\theta})\, d\boldsymbol{\theta}} \tag{9.2.30}$$

where

$$p(\mathbf{Z}_N|\boldsymbol{\theta}) = \prod_{i=1}^{N} p(\mathbf{x}_i|\boldsymbol{\theta}) \tag{9.2.31}$$

This can also be written in a recursive form analogous to Eq. (9.2.18) in the case of supervised learning as

$$p(\boldsymbol{\theta}|\mathbf{Z}_N) = \frac{p(\mathbf{x}_N|\boldsymbol{\theta})p(\boldsymbol{\theta}|\mathbf{Z}_{N-1})}{\int_{X_\theta} p(\mathbf{x}_N|\boldsymbol{\theta})p(\boldsymbol{\theta}|\mathbf{Z}_{N-1})\, d\boldsymbol{\theta}} \tag{9.2.32}$$

The computation of $p(\boldsymbol{\theta}|\mathbf{Z}_N)$ is again formidable. In the supervised learning problem, at least in the cases where a finite-dimensional sufficient statistic exists, we can find solutions which are computationally feasible. With unsupervised learning, since the samples are obtained from a mixture distribution, we cannot expect that a simple sufficient statistic exists. Hence, even if we assume a reproducing form for the prior density $p(\boldsymbol{\theta})$, because the multiplying factor $p(\mathbf{x}_N|\boldsymbol{\theta})$ is a mixture density, the reproducing form of $p(\boldsymbol{\theta}|\mathbf{Z}_N)$ is not usually preserved. This means that, in practice, we have to store the density functions $p(\boldsymbol{\theta}|\mathbf{Z}_k)$ for all values of $\boldsymbol{\theta}$, almost an impossible task [15]. We can overcome this by storing the density only at a discrete set of values for $\boldsymbol{\theta}$ and use the expression in Eq. (9.2.32) to obtain the updated probabilities for the various levels of $\boldsymbol{\theta}$. The computations can be quite cumbersome for even moderate sized problems.

We can also obtain an idea of the problems involved in the computation of $p(\boldsymbol{\theta}|\mathbf{Z}_N)$ by writing this density function in a slightly different manner [16]. We note that since there are two hypotheses and N patterns, there are 2^N possible ways of obtaining a sequence of N training samples. Let us order these sequences in any manner and denote the jth sequence by α_j. We can then write $p(\boldsymbol{\theta}|\mathbf{Z}_N)$ in a mixture distribution as

$$p(\boldsymbol{\theta}|\mathbf{Z}_N) = \sum_{j=1}^{2^N} p(\boldsymbol{\theta}|\mathbf{Z}_N, \alpha_j)p(\alpha_k|\mathbf{Z}_N) \tag{9.2.33}$$

As an example, let us assume that $N=2$. Then, the training set $\{\mathbf{x}_1, \mathbf{x}_2\}$ could be obtained in any of the following ways:

$\alpha_1 : (H_1, H_1)$ with probability $P(\alpha_1|\mathbf{Z}_2)$

$\alpha_2 : (H_1, H_2)$ with probability $P(\alpha_2|\mathbf{Z}_2)$

$\alpha_3 : (H_2, H_2)$ with probability $P(\alpha_3|\mathbf{Z}_2)$

$\alpha_4 : (H_2, H_1)$ with probability $P(\alpha_4|\mathbf{Z}_2)$

For any particular sequence α_j, since the classifications of the patterns are known, learning takes place in a supervised manner. Once the density functions $p(\boldsymbol{\theta}|\mathbf{Z}_N, \alpha_j)$ are determined for each j, Eq. (9.2.33) can be used to average them over the probability $P(\alpha_j|\mathbf{Z}_N)$ which is the probability that sequence α_j occured, given the data set $\mathbf{Z}_N$, to obtain the density $p(\boldsymbol{\theta}|\mathbf{Z}_N)$. This average density is used in Eq. (9.2.29) to obtain $p(\mathbf{x}|\mathbf{Z}_N, H_j)$, $j=1, 2$ for computing the likelihood ratio.

Even though this solution appears elegant, we encounter difficulties in implementing it in practice. As the number of samples in the training set increases, that is, N increases, the total number of sequences increases exponentially. Therefore the computational and storage requirements "explode" for large N. In the sequel, we discuss two suboptimal schemes which overcome this problem.

If $\boldsymbol{\theta}$ is a nonrandom parameter, as in the case of supervised learning, we can obtain a maximum likelihood estimate and use this as the true value in designing the classifier. The classification rule now becomes

$$\Lambda(\mathbf{x}) = \frac{p(\mathbf{x}|H_1, \boldsymbol{\theta}_1)|_{\boldsymbol{\theta}_1=\hat{\boldsymbol{\theta}}_{1ml}}}{p(\mathbf{x}|H_2, \boldsymbol{\theta}_2)|_{\boldsymbol{\theta}_2=\hat{\boldsymbol{\theta}}_{2ml}}} \tag{9.2.34}$$

To determine the ML estimates, we note from Eq. (9.2.25) that

$$p(\mathbf{x}|\boldsymbol{\theta}) = \sum_{j=1}^{2} p(\mathbf{x}|H_j, \boldsymbol{\theta}_j)P_j \tag{9.2.35}$$

Assuming independence of samples, we have

$$p(\mathbf{Z}_N|\boldsymbol{\theta}_j) = \prod_{k=1}^{N} p(\mathbf{x}_k|\boldsymbol{\theta})$$

Taking logarithms and setting the gradient with respect to $\boldsymbol{\theta}_j$ equal to zero yields

$$\sum_{k=1}^{N} \frac{1}{p(\mathbf{x}_k|\boldsymbol{\theta})} \nabla_{\boldsymbol{\theta}_j} \left[\sum_{j=1}^{2} p(\mathbf{x}_k|H_j, \boldsymbol{\theta}_j) P_j \right] \Bigg|_{\hat{\boldsymbol{\theta}}_{jml}} = 0 \tag{9.2.36}$$

where we have used Eq. (9.2.35)

By assuming that $\boldsymbol{\theta}_1$ and $\boldsymbol{\theta}_2$ are independent, and noting that $P(H_j) = P_j$ can be written as

$$P(H_j) = \frac{P(H_j|\mathbf{x}_k, \boldsymbol{\theta}_j) p(\mathbf{x}_k|\boldsymbol{\theta})}{p(\mathbf{x}_k|H_j, \boldsymbol{\theta}_j)} \qquad j = 1, 2 \tag{9.2.37}$$

we can write Eq. (9.2.36) as

$$\sum_{k=1}^{N} P(H_j|\mathbf{x}_k, \hat{\boldsymbol{\theta}}_{jml}) \nabla_{\boldsymbol{\theta}_j} \log p(\mathbf{x}_k|H_j, \hat{\boldsymbol{\theta}}_{jml}) = 0 \tag{9.2.38}$$

These again constitute a set of nonlinear equations for obtaining the ML estimates of $\boldsymbol{\theta}_1$ and $\boldsymbol{\theta}_2$.

Suboptimal Learning Procedures

It was pointed out earlier that the determination of $p(\boldsymbol{\theta}|\mathbf{Z}_N)$ involved an exploding memory problem. This is because the number of sequences in Eq. (9.2.34) over which one needs to average increases exponentially as the number of samples in the training set increases.

One way of reducing the computations is to limit the summation in Eq. (9.2.34) to a fixed number of sequences, preferably to a single sequence. As can be expected, such a technique is necessarily suboptimal in most cases. We will now discuss two such techniques namely the *decision-directed* technique [17] and the *probabilistic learning* technique [18].

In the decision directed method, when a pattern (sample) is presented to the classifier, a decision is made as to its classification. The classification can be made using any of the criteria discussed in Chapter 3. For example, we could use the maximum a posteriori criterion. This decision by the classifier is then accepted as the true classification for purposes of learning.

The learning samples are thereby partitioned into two sets $\mathbf{Z}_{N_1}$ and $\mathbf{Z}_{N_2}$ with known classification. Learning can then proceed in the normal supervised learning mode and in a sequential manner if so desired. Though the method is not too complicated to implement, it leads to asymptotically biased results.

The probabilistic teacher method is similar to the previous approach but differs in the manner in which the patterns are classified and learning proceeds. When each new sample is presented to the classifier, a random experiment is conducted with outcomes H_1 and H_2. The outcome of this experiment at stage k is the label l_k assigned to the pattern by the classifier. This label is then accepted as the true classification for purposes of learning. We will denote by α_{k-1} the sequence of past labels, and the probabilities associated with the random experiment at stage k by $P(H_j|\mathbf{x}_k, \mathbf{Z}_{k-1}, \alpha_{k-1}), j = 1, 2.$

We now briefly outline the manner in which the method is implemented. We assume $\boldsymbol{\theta}$ is a random vector with prior density $p(\boldsymbol{\theta})$. By using Bayes' rule, we can write the required probabilities $P(H_j|\mathbf{x}_k, \mathbf{Z}_{k-1}, \alpha_{k-1})$ as

$$P(H_j|\mathbf{x}_k, \mathbf{Z}_{k-1}, \alpha_{k-1}) = \frac{p(\mathbf{x}_k|H_j, \mathbf{Z}_{k-1}, \alpha_{k-1})P(H_j)}{\sum_{j=1}^{2} p(\mathbf{x}_k|H_j, \mathbf{Z}_{k-1}, \alpha_{k-1})P(H_j)} \qquad j = 1, 2. \tag{9.2.39}$$

The density $p(\mathbf{x}_k|H_j, \mathbf{Z}_{k-1}, \alpha_{k-1})$ can be obtained by integrating over the parameter space $\chi_{\boldsymbol{\theta}}$ as

$$p(\mathbf{x}_k|H_j, \mathbf{Z}_{k-1}, \alpha_{k-1}) = \int_{\chi_{\boldsymbol{\theta}}} p(\mathbf{x}_k|H_j, \boldsymbol{\theta}, \mathbf{Z}_{k-1}, \alpha_{k-1})p(\boldsymbol{\theta}|\mathbf{Z}_{k-1}, \alpha_{k-1})\, d\boldsymbol{\theta} \qquad j = 1, 2 \tag{9.2.40}$$

where $p(\boldsymbol{\theta}|\mathbf{Z}_{k-1}, \alpha_{k-1})$ is the updated density function of $\boldsymbol{\theta}$ based on the sequences $\mathbf{Z}_{k-1}$ and α_{k-1}. Assuming conditional independence of the samples, the dependence of the first term on $\mathbf{Z}_{k-1}$ and α_{k-1} in the expression on the right-hand side of Eq. (9.2.40) can be dropped. This is similar to our earlier assumptions. Once the density functions of Eq. (9.2.39) are computed, the random experiment can be performed. We then have the outcome l_k and hence the sequence α_k. Learning proceeds through the following recursive relation:

$$p(\boldsymbol{\theta}|\mathbf{Z}_k, \alpha_k) = \frac{p(\mathbf{x}_k|\boldsymbol{\theta}, l_k)}{\int_{\chi_{\boldsymbol{\theta}}} p(\mathbf{x}_k|\boldsymbol{\theta}, l_k)p(\boldsymbol{\theta}|\mathbf{Z}_{k-1}, \alpha_{k-1})} \cdot p(\boldsymbol{\theta}|\mathbf{Z}_{k-1}, \alpha_{k-1}) \tag{9.2.41}$$

Note that Eq. (9.2.41) is of the same form as Eq. (9.2.18) for supervised learning. We can use reproducing pairs for the prior density function of $\boldsymbol{\theta}$, so that we have to only update the finite dimensional sufficient statistic.

We conclude our consideration of pattern recognition with a brief discussion of the important problem of feature selection.

9.2.5 Feature Selection

In our discussion of the pattern classification problem, we have assumed that we can associate with each sample, a vector of observables (features). The classification is then determined by this feature vector. If the number of features is large, the computational requirements for correct classification become significant. It becomes necessary to select a fewer number of features from a given set in order to reduce the computational burden and still be able to classify the patterns satisfactorily. Alternatively we may seek to transform the feature vector to one of lower dimension. Once we find a reasonable solution to this problem, we need to consider only a feature vector of lower dimension. This problem is therefore referred to as dimensionality reduction. In this section, we will present an approach to feature selection which is based on a generalized form of the Karhunen–Loéve transformation.

Before considering the problem of dimensionality reduction for pattern classification, let us consider the problem of choosing a vector of reduced dimension, $\mathbf{x}(n)$, to adequately represent a vector $\mathbf{x}(r)$ of dimension $r > n$. If the elements of $\mathbf{x}(r)$ were uncorrelated, one way of choosing the reduced vector would be to retain those components of $\mathbf{x}(r)$ which have the largest variance and discard the others. In general, the elements of $\mathbf{x}(r)$ will be correlated. Wc may therefore seek to expand $\mathbf{x}(r)$ in a set of orthonormal basis vectors $\mathbf{g}_1, \mathbf{g}_2, \ldots, \mathbf{g}_r$ such that

$$\mathbf{x}(r) = \sum_{i=1}^{r} y_i \mathbf{g}_i \tag{9.2.42}$$

where

$$\mathbf{g}_i^T \mathbf{g}_j = \delta_{ij} \tag{9.2.43}$$

We wish to retain a subset $\{y_1, y_2, \ldots, y_n\}$ of the coefficients of the expansion and still be able to represent $\mathbf{x}(r)$. Let us define $\hat{\mathbf{x}}(r)$ as

$$\hat{\mathbf{x}}(r) = \sum_{i=1}^{n} y_i \mathbf{g}_i \tag{9.2.44}$$

We seek the set of orthonormal basis vectors $\{\mathbf{g}_i\}$ to minimize the mean square error in representing $\mathbf{x}(r)$, $\mathrm{E}\{\|\mathbf{x}(r)-\hat{\mathbf{x}}(r)\|^2\}$. It is clear from Eqs. (9.2.42) and (9.2.43) that

$$\begin{aligned}\mathrm{E}\{\|\mathbf{x}(r)-\hat{\mathbf{x}}(r)\|^2\} &= \mathrm{E}\left\{\left\|\sum_{i=n+1}^{r} y_i\mathbf{g}_i\right\|^2\right\}\\ &= \sum_{i=n+1}^{r} \mathrm{E}\{y_i^2\} \end{aligned} \tag{9.2.45}$$

Since

$$y_i = \mathbf{g}_i^T\mathbf{x}(r) \tag{9.2.46}$$

we can rewrite Eq. (9.2.45) as

$$\mathrm{E}\{\|\mathbf{x}(r)-\hat{\mathbf{x}}(r)\|^2\} = \sum_{i=n+1}^{r} \mathrm{E}\{\mathbf{g}_i^T V_{\mathbf{x}}\mathbf{g}_i\} \tag{9.2.47}$$

where $\mathbf{V}_{\mathbf{x}}$ is the covariance matrix of the vector $\mathbf{x}(r)$.

The problem therefore reduces to one of minimizing the right-hand side of Eq. (9.2.47) subject to the constraint

$$\mathbf{g}_i^T\mathbf{g}_i = 1 \tag{9.2.48}$$

We solve the problem by forming the augmented cost function

$$J = \sum_{i=n+1}^{r} \left[\mathrm{E}\{\mathbf{g}_i^T\mathbf{V}_{\mathbf{x}}\mathbf{g}_i\} - \lambda_i\mathbf{g}_i^T\mathbf{g}_i\right] \tag{9.2.49}$$

where the $\{\lambda_i\}$ are a set of Lagrange multipliers. The optimum $\mathbf{g}_i$ obtained by setting the gradient vector $\nabla_{\mathbf{g}_i} J$ equal to zero is

$$\mathbf{V}_{\mathbf{x}}\mathbf{g}_i = \lambda_i\mathbf{g}_i \tag{9.2.50}$$

Thus the constants λ_i are the eigenvalues of the covariance matrix of $\mathbf{x}(r)$ and the vectors $\mathbf{g}_i$ are the associated eigenvectors. The expansion in Eq. (9.2.42), therefore, is the discrete version of the Karhunen–Loéve expansion which we have encountered earlier.

The minimum mean square error is obtained from Eq. (9.2.47) as

$$\mathrm{E}\{\|\mathbf{x}(r)-\hat{\mathbf{x}}(r)\|^2\}_{\min} = \sum_{i=n+1}^{r} \lambda_i \tag{9.2.51}$$

It therefore follows that the optimum technique (in the mean square sense) to represent a vector $\mathbf{x}(r)$ of dimension r in terms of a fewer number n of components, is to order the eigenvalues of the covariance matrix in descending order of magnitude. We then expand $\mathbf{x}(r)$ in a Karhunen–Loéve expansion and retain only the first n coefficients. If we choose an $n \times r$ matrix $\mathbf{T}$ such that its rows are the first n eigenvectors of $\mathbf{V}_{\mathbf{x}}$, then the reduced vector $\mathbf{x}(n)$ is given by

$$\mathbf{x}(n) = \mathbf{T}\mathbf{x}(r)$$

Let us now extend the preceding discussion to the problem of dimensionality reduction in a pattern classification problem. We restrict our discussion to the binary classification problem in which we are required to choose between two pattern classes. Given the pattern vector $\mathbf{x}(r)$ of dimension r, we seek to obtain a vector $\mathbf{x}(n)$, of dimension n such that we are still able to discriminate between the two pattern classes.

Let $\mathbf{V}_1$ and $\mathbf{V}_2$ be the covariance matrices of the vectors belonging to the classes H_1 and H_2, respectively. Let P_1 and P_2 be the probabilities of occurrence of the two classes. We define a generalized covariance matrix $\mathbf{V}$ as

$$\mathbf{V} \triangleq P_1\mathbf{V}_1 + P_2\mathbf{V}_2 \tag{9.2.52}$$

Let $\mathbf{D}$ be a matrix of eigenvectors of $\mathbf{V}$ such that

$$\mathbf{D}^T\mathbf{V}\mathbf{D} = \mathbf{I} \tag{9.2.53}$$

where $\mathbf{I}$ is the identity matrix.

If we define the matrices $\mathbf{V}^i$, $i = 1, 2$ as

$$\mathbf{V}^i = \mathbf{D}^T(P_i\mathbf{V}_i)\mathbf{D} \tag{9.2.54}$$

we can write Eq. (9.2.54) as

$$\mathbf{V}^1 + \mathbf{V}^2 = \mathbf{I} \tag{9.2.55}$$

Let λ_j^i and $\mathbf{g}_j^i$ represent the eigenvalues and eigenvectors of $\mathbf{V}^i$ for $i = 1, 2$

$$\mathbf{V}^i\mathbf{g}_j^i = \lambda_j^i\mathbf{g}_j^i \qquad j = 1, 2, \ldots, r \tag{9.2.56}$$

It can easily be established that

$$0 \leqslant \lambda_j^i \leqslant 1 \qquad \text{for} \quad i = 1, 2; j = 1, 2, \ldots, r \tag{9.2.57}$$

From Eq. (9.2.55) we can therefore write

$$\mathbf{V}^2\mathbf{g}_j^1 = (\mathbf{I} - \mathbf{V}^1)\mathbf{g}_j^1 = (1 - \lambda_j^1)\mathbf{g}_j^1 \tag{9.2.58}$$

Equation (9.2.58) shows that if $\mathbf{g}_j^1$ is an eigenvector of $\mathbf{V}^1$ associated with the eigenvalue λ_j^1, it is also an eigenvector of $\mathbf{V}^2$ associated with the eigenvalue $(1 - \lambda_j^1)$. If we now arrange the eigenvalues of $\mathbf{V}^1$ in descending order, the eigenvalues of $\mathbf{V}^2$ will then be in increasing order. That is, the most important features of class H_1 are the least important ones of class H_2 and vice versa. We therefore choose a $n \times r$ matrix $\mathbf{T}_1$ such that its first n_1 rows are the first n_1 eigenvectors of $\mathbf{V}^1$ and the remaining $n_2 = (n - n_1)$ rows are the eigenvectors of $\mathbf{V}^2$. We can then choose the vector $\mathbf{x}(n)$ as

$$\mathbf{x}(n) = \mathbf{T}\mathbf{x}(r) \tag{9.2.59}$$

where

$$\mathbf{T} = \mathbf{T}_1\mathbf{D} \tag{9.2.60}$$

The procedure thus involves expanding the vector $\mathbf{x}(r)$ in the eigenvectors of the generalized covariance matrix $\mathbf{V}$ and retaining only the first n_1 terms (corresponding to the n_1 most significant features of the class H_1) and the last n_2 terms (corresponding to the n_2 most significant features of class H_2). One way of choosing n_1 and n_2 is to choose them as integers closest to P_1n and P_2n, respectively.

9.2.6 Section Summary

In this section, we have attempted to apply the techniques of estimation and detection to the problem of classifying patterns. In Section 9.2.1, we discussed the basic components of the pattern recognition problem. The several steps involved in the design of a pattern recognition system are feature selection, feature ordering and extraction, and pattern classification. Our attention was mostly concentrated on the classification problem. We assume that in every case the functional forms of the underlying density functions are known. The case when the densities are completely known was considered in Section 9.2.2. The techniques of hypothesis testing discussed in Chapter 3 are directly applicable to this problem and the results can be derived in a straightforward manner.

If the class-conditional densities are known only to within certain parameters, these parameters must be learned before we can classify the patterns. The problem is fairly straightforward if we are given sample (test) patterns whose classifications are known. This results in the supervised learning problem discussed in Section 9.2.3. The learning of the parameters is facilitated if, as discussed in Chapter 5, we assume a reproducing form for the prior densities.

When the classification of the test patterns is not known, the problem of learning becomes complicated. The unknown parameters of the class conditional densities must be estimated using these samples of unknown classification, that is, in a nonsupervised manner. This problem is considered in Section 9.2.4 where it is shown that the optimal (Bayes) solution leads to exploding computational requirements. Two suboptimal techniques are presented to overcome this problem.

The section concluded with a technique for feature ordering and selection in Section 9.2.5. The technique uses the discrete Karhunen–Loéve transform in selecting the features.

As has been the case in the earlier chapters, we have not considered the use of nonparametric techniques in pattern recognition. The nearest neighbor technique is an important classification technique which uses distance functions to effect classification [19,20]. Other such approaches are the stochastic automaton [3] and the stochastic approximation methods [21]. Also, the techniques of this chapter can easily be modified to obtain sequential algorithms for pattern recognition. These follow along the lines of our discussion in Chapter 3. Details can be found in [2,22].

9.3. APPLICATION TO SYSTEM IDENTIFICATION

The design of any system requires a knowledge of its dynamic characteristics. While a simple system may be built with only an intuitive feeling about its behavior, its proper operation depends upon a choice of parameters suitable to the characteristics of the particular system. These characteristics must usually be determined by measurements on the system inputs and outputs. In this section, we consider the application of the techniques discussed earlier, to the identification of systems with random inputs.

A system can be considered to be a transformation of an input into an output. System identification then implies the determination of this transformation based on measurements of the input and the output. The first step is to construct a mathematical model within a specified class of models which describes the physical transformation occuring in the system. To make this meaningful, we select a mathematical error criterion and choose the model within the specified class which minimizes the criterion [23]. While several criteria can be used, as in earlier chapters, we will for the most part restrict ourselves to the minimum mean square error criterion.

Mathematical models can be linear or nonlinear, continuous or discrete-time, deterministic or stochastic, with finite or infinite memory, and so on. The choice of the model used in a particular case depends on the available data, the purpose to which the model is to be put, and the complexity of the algorithms involved. Quite often, the model has little connection with the actual system. For example, a linear time-invariant model is often a sufficiently accurate representation of a nonlinear time-varying system. Once the model is chosen, it can be characterized in terms of suitable descriptions such as a differential equation, transfer function, impulse response, power series expansion. The characterization may be parametric or nonparametric. Parametric characterizations assume a functional form (for example, a differential equation) for the input–output relation, whereas nonparametric characterizations (for example, impulse responses) do not. The former are easier to determine since all that is required is the estimation of certain parameter values from measured data, so that the performance criterion is minimized.

The choice of the mathematical model is neither simple nor straightforward in a general situation. Questions such as what models to select or indeed what models are available or whether the solution to the identification problem is unique are not easy to answer. However, for linear and time-invariant systems these problems are fairly easily answered. The characterization is well established and the order of the system determines the maximum number of parameters to be identified. From a practical point of view, the assumption of a linear time-invariant model for the system considerably reduces the complexity of the identification problem. For such an assumption to be valid, the system parameters must vary slowly in comparison to the signals. The linearity assumption is justified if the system is close to normal operating conditions and we are interested in determining changes away from normal.

In this section we are primarily concerned with the identification of linear time-invariant system with random inputs, on the basis of normal operating records. We assume a characterization of the system in terms of either differential equations or transfer functions. Thus the characterization is parametric so that the problem reduces to one of identifying certain parameters in the system. We could, therefore, use the techniques discussed in Chapter 5 for the estimation of random parameters. However in most applications, the dependence of the output signal on the parameters to be estimated is only indirect and the parameters cannot be obtained in a straightforward manner. We discuss some techniques for obtaining such estimates from measurements of the output.

Our interest is primarily in discussing identification algorithms which can be easily processed by a digital computer. We therefore use exclusively discrete-time models throughout this section. Since the literature in the area of system identification represents more a collection of individual techniques rather than a unified theory, we content ourselves with presenting a few basic techniques and not attempt to discuss all available methods. We refer the reader to the recent books on identification [23,24] for a discussion of these techniques.

A problem closely related to the identification of stochastic models is that of estimating the power spectrum of a stationary, stochastic process. In many of our discussions in the previous chapters, we have assumed that the statistics of the underlying distributions are known, at least to second order. In a practical situation, these statistics, if not assumed a priori, will have to be determined on the basis of records of typical data. We have discussed earlier the estimation of the mean and the variance of a random variable on the basis of observed values. More generally, given sample values of a stationary process, we may be interested in estimating the autocorrelation function or power spectrum of the process. If the underlying process is continuous-time, the sample values must be obtained by sampling the signal. If the signal is band limited, the sampling frequency must be chosen to be at least twice as large as the bandwidth of the signal (the Nyquist rate). If the signal is not band limited, to prevent aliasing, frequently the signal is first lowpass filtered before sampling.

There are two approaches that can be used in estimating the correlation function $\phi_x(k)$. In a nonparametric approach, we try to obtain a numerical estimate for each of the M numbers $\phi_x(0), \phi_x(1), \ldots, \phi_x(M-1)$ for some choice of M. In the parametric approach, we might assume a functional form for the autocorrelation function and determine certain associated (unknown) parameters. For example we may assume that

$$\phi_x(k) = \phi_x(0)\alpha^{|k|} \qquad |\alpha| < 1$$

and estimate the parameters α and $\phi_x(0)$. We note that the signal $x(k)$ can be considered to be the output of a first-order all-pole filter driven by white noise. The problem of spectral estimation then reduces to one of estimating the parameters of the corresponding system model. Thus the approaches discussed in this section can also be used for spectral estimation. The reader is referred to [25–29] for a discussion of power spectrum estimation by various methods.

9.3.1 Identification of ARMA Models

We recall from our discussions in Chapter 2 that a stationary stochastic process can be considered to be the output of a linear time-invariant system driven by white noise. For example, for the process with autocorrelation sequence

$$\phi_x(k) = \phi_x(0)\alpha^{|k|} \tag{9.3.1}$$

the power spectrum in the Z-domain is given by

$$\Phi_x(z) = \frac{\phi_x(0)(1-\alpha^2)}{(z-\alpha)(z^{-1}-\alpha)} \tag{9.3.2}$$

If we let $\phi_x(0)(1-\alpha^2) = G^2$, we can write the preceding as

$$\Phi_x(z) = \frac{G}{z-\alpha}\frac{G}{z^{-1}-\alpha} \tag{9.3.3}$$

It then follows that $\Phi_x(z)$ can be considered to be the spectrum of the output of a linear discrete-time system with transfer function

$$H(z) = \frac{G}{z-\alpha} \tag{9.3.4}$$

driven by a zero mean white sequence $w(k)$ with unity variance. We can therefore write

$$X(z) = \frac{G}{z-\alpha}W(z) \tag{9.3.5}$$

where $X(z)$ and $W(z)$ represent the transforms of $x(k)$ and $w(k)$.

The corresponding difference equation describing the process $x(k)$ is

$$x(k+1) - \alpha x(k) = Gw(k) \tag{9.3.6}$$

A process defined by an equation such as (9.3.6) is said to be an *autoregressive* (AR) process. The spectrum estimation problem thus reduces to one of identification of the parameters of the autoregressive model of the process. In the sequel, we assume that $x(\cdot)$ is zero mean.

The process described by Eq. (9.3.6) is a first-order process. We also note from the system transfer function Eq. (9.3.4) that $H(z)$ has no finite zeros. We can generalize the model by assuming that $H(z)$ is a general transfer function of the form

$$H(z) = \frac{G(1+\beta_1'z^{-1}+\cdots\beta_M'z^{-M})}{1+\alpha_1 z^{-1}+\cdots+\alpha_N z^{-N}} \tag{9.3.7}$$

so that the system has M zeros and N poles.

Let

$$G\beta_i' = \beta_i \quad \text{for } i = 1 \ldots M \tag{9.3.8}$$

Then

$$H(z) = \frac{G + \sum_{i=1}^{M} \beta_i z^{-i}}{1 + \sum_{i=1}^{N} \alpha_i z^{-i}} \tag{9.3.9}$$

The output $X(z)$ is given by

$$X(z) = H(z)W(z) = \frac{G + \sum_{i=1}^{M} \beta_i z^{-i}}{1 + \sum_{i=1}^{N} \alpha_i z^{-i}} W(z) \tag{9.3.10}$$

The corresponding difference equation can be written as

$$x(k) + \sum_{i=1}^{N} \alpha_i x(k-i) = Gw(k) + \sum_{i=1}^{M} \beta_i w(k-i) \tag{9.3.11}$$

This equation represents an *autoregressive moving-average* (*ARMA*) process. We note that if $\beta_i = 0$, $i = 1 \ldots M$, we obtain the autoregressive process defined earlier. If $\alpha_i = 0$, $i = 1 \ldots N$, then $x(k)$ is a *moving-average* (*MA*) process. In thc former case $H(z)$ will be an all-pole function with no finite zeros. For the moving-average process, $H(z)$ represents an all zero system.

Let us denote the estimates of G, α_i, and β_i as $\hat{G}$, $\hat{\alpha}_i$, and $\hat{\beta}_i$, respectively. We then define a *residual* or *error* as

$$\epsilon(k) = x(k) + \sum_{i=1}^{N} \hat{\alpha}_i x(k-i) - \sum_{i=1}^{M} \hat{\beta}_i w(k-i) \tag{9.3.12}$$

Following our approach in earlier chapters, we will seek our estimates $\hat{\alpha}_i$ and $\hat{\beta}_i$ such that the mean square error $E\{\epsilon^2(k)\}$ is minimized. The model so obtained corresponds to the best linear mean-square fit to the observed data. That is, the model yields the best (in the mean-square sense) predicted value of $x(k)$ based on past observations and inputs.

We now discuss several techniques for the identification of the parameters of the ARMA process defined previously. We first discuss the identification of autoregressive parameters by observing $x(k)$. Later we will extend our results to ARMA models.

Autoregressive Parameter Identification

Let us for the moment assume that $\beta_i = 0$ so that $x(k)$ is an autoregressive process. For this case Eq. (9.3.11) becomes

$$x(k) + \sum_{i=1}^{N} \alpha_i x(k-i) = Gw(k) \tag{9.3.13}$$

The problem can now be stated: Given measurements of the signal $x(k)$, determine the parameters α_i, $i = 1 \ldots N$. If $\hat{\alpha}_i$ represents the estimate of α_i, the residual is given by

$$\varepsilon(k) = x(k) + \sum_{i=1}^{N} \hat{\alpha}_i x(k-i) \tag{9.3.14}$$

We now determine the estimates by minimizing the mean square error

$$J = \mathrm{E}\left[\varepsilon^2(k)\right] = \mathrm{E}\left[\left\{x(k) + \sum_{i=1}^{N} \hat{\alpha}_i x(k-i)\right\}^2\right] \tag{9.3.15}$$

Setting the partial derivatives of J with respect to $\hat{\alpha}_j$ equal to zero for $j = 1 \ldots N$, we get

$$\mathrm{E}\left\{\left[x(k) + \sum_{i=1}^{N} \hat{\alpha}_i x(k-i)\right] x(k-j)\right\} = 0 \qquad j = 1 \ldots N \tag{9.3.16}$$

That is, $\varepsilon(k)$ is orthogonal to $x(k-j)$ for $j = 1 \ldots N$. Equation (9.3.16) can be rearranged to give

$$\sum_{i=1}^{N} \hat{\alpha}_i \phi_x(j-i) = -\phi_x(j) \qquad j = 1 \ldots N \tag{9.3.17}$$

where

$$\phi_x(j-i) = \mathrm{E}\{x(k-i)x(k-j)\} = \mathrm{E}\{x(j)x(i)\} \tag{9.3.18}$$

Equation (9.3.18) follows since the process $x(k)$ is stationary. We can write the set of equations (9.3.17) in compact form by using matrix notation. We then obtain

$$\begin{bmatrix} \phi_x(0) & \phi_x(1) & \cdots & \phi_x(N-1) \\ \phi_x(1) & \phi_x(0) & \cdots & \phi_x(N-2) \\ \phi_x(2) & & & \\ \vdots & \vdots & & \vdots \\ \phi_x(N-1) & & \cdots & \phi_x(0) \end{bmatrix} \begin{bmatrix} \hat{\alpha}_1 \\ \hat{\alpha}_2 \\ \\ \vdots \\ \hat{\alpha}_N \end{bmatrix} = - \begin{bmatrix} \phi_x(1) \\ \phi_x(2) \\ \\ \vdots \\ \phi_x(N) \end{bmatrix} \quad \textbf{(9.3.19)}$$

These equations are generally known as the Yule–Walker equations. The matrix on the left side of Eq. (9.3.19) is special kind of a matrix known as a Toeplitz matrix [32]. As can be seen, the elements along any diagonal of this matrix are the same. As such the matrix possesses certain useful properties. The matrix will be positive definite and hence invertible. Furthermore the estimates $\hat{\alpha}_i$ will form a stable set. That is, the polynomial

$$1 + \sum_{i=1}^{N} \hat{\alpha}_i z^{-i} = 0 \quad \textbf{(9.3.20)}$$

has all its zeros inside the unit circle in the Z-plane. This guarantees that the autoregressive model is stable [32]. Let us define the matrix $\boldsymbol{\phi}^N$ and the vectors $\hat{\boldsymbol{\alpha}}^N$ and $\boldsymbol{\rho}^N$ as

$$\boldsymbol{\phi}^N = \begin{bmatrix} \phi_x(0) & \cdots & \phi_x(N-1) \\ \vdots & & \vdots \\ \phi_x(N-1) & \cdots & \phi_x(0) \end{bmatrix}, \hat{\boldsymbol{\alpha}}^N = \begin{bmatrix} \hat{\alpha}_1 \\ \vdots \\ \hat{\alpha}_N \end{bmatrix}, \text{ and } \boldsymbol{\rho}^N = \begin{bmatrix} \phi_x(1) \\ \vdots \\ \phi_x(N) \end{bmatrix} \quad \textbf{(9.3.21)}$$

The solution to the set of Eqs. (9.3.19) can then be written as

$$\hat{\boldsymbol{\alpha}}^N = -\left[\boldsymbol{\phi}^N\right]^{-1} \boldsymbol{\rho}^N \quad \textbf{(9.3.22)}$$

The minimum value of J obtained by using this choice of $\hat{\boldsymbol{\alpha}}$ is easily seen to be

$$J = \mathrm{E}\{\varepsilon(k)x(k)\} = \phi_x(0) + \sum_{i=1}^{N} \hat{\alpha}_i \phi_x(i) \quad \textbf{(9.3.23)}$$

To compute the gain G, we note that if the parameters have been identified correctly $\varepsilon(k) = Gw(k)$ and hence

$$\hat{G}^2 \mathrm{E}\{w^2(k)\} = \hat{G}^2 = \mathrm{E}\{\varepsilon^2(k)\} = \phi_x(0) + \sum_{i=1}^{N} \hat{\alpha}_i \phi_x(i) \tag{9.3.24}$$

It only remains to determine the autocorrelation coefficients $\phi_x(k)$ for $k = 1 \ldots N-1$. These coefficients may be estimated using any of the techniques discussed in [25–27]. A commonly used estimate is

$$\hat{R}_x(k) = \frac{1}{L} \sum_{n=0}^{L-|k|-1} x(n)x(n+k) \tag{9.3.25}$$

where L is the length of the data sequence.

Example 9.3.1 We can relate our results on the estimation of parameters of the AR process to the ML estimates of Chapter 5. Suppose we are given measurements $x(1), x(2), \ldots, x(L)$. We recall that the maximum likelihood estimate $\hat{\boldsymbol{\alpha}}_{ml}$ is obtained by solving the likelihood equation

$$\left. \frac{\partial \ln p(x(1), x(2), \ldots, x(L)|\boldsymbol{\alpha})}{\partial \boldsymbol{\alpha}} \right|_{\boldsymbol{\alpha} = \hat{\boldsymbol{\alpha}}_{ml}} = 0$$

Unfortunately, the determination of this density is not easy in most cases. We can obtain an approximate expression for the density in the case where $w(k)$ is a Gaussian sequence. In order to do so, let us denote the set $x(1), x(2) \ldots x(R) = \mathbf{x}_R$. Then

$$\begin{aligned} p(x(1), x(2) \ldots x(L)|\alpha) &= p(\mathbf{x}_L|\alpha, G) \\ &= p(x(N+1) \ldots x(L)|\mathbf{x}_N, \alpha)p(\mathbf{x}_N|\alpha) \end{aligned}$$

When L is large compared to N, the contribution from the term $p(x(1), x(2), \ldots, x(N)|\boldsymbol{\alpha})$ can be neglected. Since $w(\cdot)$ is zero-mean, Gaussian, white with unity covariance

$$p(w(N+1), w(N+2), \ldots, w(L)) = \frac{1}{(\sqrt{2\pi})^{L-N}} \exp\left\{-\frac{1}{2} \sum_{k=N+1}^{L} w_k^2\right\}$$

It then follows from Eq. (9.3.13) that

$$\begin{aligned} &p(x(N+1)x(N+2), \ldots, x(L)|\boldsymbol{\alpha}, G) \\ &= \frac{1}{(\sqrt{2\pi}\, G)^{L-N}} \exp\left\{-\frac{1}{2G^2} \sum_{k=N+1}^{L} \left[x(k) + \sum_{i=1}^{N} \alpha_i x(k-i)\right]^2\right\} \end{aligned}$$

The log-likelihood function is then approximately given by

$$\ln p(x(1), x(2), \ldots, x(L)|\boldsymbol{\alpha}, G)$$
$$\simeq -(L-N)(\sqrt{2\pi}\, G)$$
$$-\frac{1}{2G^2}\sum_{k=N+1}^{L}\left[x(k)+\sum_{k=1}^{N}\alpha_i x(k-i)\right]^2$$

The ML estimates $\hat{\boldsymbol{\alpha}}_{ml}$ are obtained by solving the equations

$$\sum_{k=N+1}^{L}\left[x(k)+\sum_{i=1}^{N}\hat{\alpha}_i x(k-i)\right]x(k-j)=0 \qquad \text{for} \quad j=1,2,\ldots,N$$

or equivalently

$$\sum_{i=1}^{N}\hat{\alpha}_i\left[\sum_{k=N+1}^{L} x(k-i)x(k-j)\right]=-\sum_{k=N+1}^{L} x(k)x(k-j) \qquad j=1,2,\ldots,N$$

It easily follows from Eq. (9.3.25) that

$$\sum_{k=N+1}^{L} x(k-i)x(k-j)=L\hat{R}_x(|j-i|)$$

so that

$$\sum_{i=1}^{N}\hat{\alpha}_i\hat{R}_x(|j-i|)=-\hat{R}_x(j)$$

which is the same as Eq. (9.3.17) with $\hat{R}_x(\cdot)$ replacing $\phi(\cdot)$.

Example 9.3.2 We consider the indentification of the parameters of a third-order autoregressive model by using Eq. (9.3.22) based on 1024 points of the data sequence. The data were generated by simulating the process on a digital computer. The model was assumed to be

$$x(k)+0.9x(k-1)+0.36x(k-2)+0.054x(k-3)=w(k)$$

so that the actual parameters are

$$\alpha_1=0.9,\ \alpha_2=0.36, \qquad \text{and} \quad \alpha_3=0.054$$

The input $w(k)$ is a stationary, zero-mean, white sequence with unity variance. The four values of the autocorrelation coefficients needed to obtain the estimates were determined from Eq. (9.3.25) as

$$\hat{R}_x(0) = 2.0307,\ \hat{R}_x(1) = -1.3893,\ \hat{R}_x(2) = 0.6778,$$
$$\text{and} \quad \hat{R}_x(3) = -0.3126$$

The estimates as obtained from Eq. (9.3.22) are

$$\hat{\alpha}_1 = 0.8932,\ \hat{\alpha}_2 = 0.3222,\ \hat{\alpha}_0 = 0.0815$$

There are some disadvantages to determining the estimate $\hat{\alpha}^N$ from Eq. (9.3.22). First, the order N of the model must be known (or at least fixed) a priori. Also in order to compute the correlation coefficients, the data sequence must be processed all at once. Hence the technique is a *batch-processing* technique. The computation of the correlation coefficients is time-consuming. Finally, Eq. (9.3.22) involves inverting the matrix $\boldsymbol{\phi}^N$. The last can be overcome by using a trangularization procedure. The other two disadvantages, however, still remain.

We can try to overcome the first one by computing the estimates $\hat{\alpha}^N$ for various values of N and accepting as the true model order the one which yields the least value for J. This, however, requires that Eq. (9.3.22) be solved repeatedly for various N. We now briefly discuss a technique which avoids this problem.

9.3.2 The Levinson Algorithm [31,32]

Let us assume that the model order is n and let the corresponding estimates be denoted by $\hat{\alpha}_1^n, \hat{\alpha}_2^n, \ldots, \hat{\alpha}_n^n$. Let

$$\hat{\boldsymbol{\alpha}}^n = [\hat{\alpha}_1^n \ldots \hat{\alpha}_n^n]^T \tag{9.3.26}$$

(We use the superscript n in denoting the estimates since the model order is assumed to be n.) Then from Eq. (9.3.19) the estimate $\hat{\boldsymbol{\alpha}}^n$ satisfies the equation

$$\begin{bmatrix} \phi_x(0) & \ldots & \phi_x(n-1) \\ \vdots & & \vdots \\ \phi_x(n-1) & \ldots & \phi_x(0) \end{bmatrix} \begin{bmatrix} \hat{\alpha}_1^n \\ \vdots \\ \hat{\alpha}_n^n \end{bmatrix} = -\begin{bmatrix} \phi_x(1) \\ \vdots \\ \phi_x(n) \end{bmatrix} \tag{9.3.27}$$

By reversing the order of the equations, Eq. (9.3.27) can also be written as

$$\begin{bmatrix} \phi_x(0) & \cdots & \phi_x(n-1) \\ \vdots & & \vdots \\ \phi_x(n-1) & \cdots & \phi_x(0) \end{bmatrix} \begin{bmatrix} \hat{\alpha}_n^n \\ \vdots \\ \hat{\alpha}_1^n \end{bmatrix} = - \begin{bmatrix} \phi_x(n) \\ \vdots \\ \phi_x(1) \end{bmatrix} \tag{9.3.28}$$

Let us now increase the model order by one and let the corresponding estimate be

$$\hat{\boldsymbol{\alpha}}^{n+1} = \left[\hat{\alpha}_1^{n+1} \ldots \hat{\alpha}_{n+1}^{n+1} \right]^T$$

Then $\hat{\boldsymbol{\alpha}}^{n+1}$ is determined by the set of equations

$$\left[\begin{array}{ccc|c} \phi_x(0) & \cdots & \phi_x(n-1) & \phi_x(n) \\ \vdots & & \vdots & \vdots \\ \phi_x(n-1) & \cdots & \phi_x(0) & \phi_x(1) \\ \hline \phi_x(n) & \cdots & \phi_x(1) & \phi_x(0) \end{array}\right] \left[\begin{array}{c} \hat{\alpha}_1^{n+1} \\ \vdots \\ \hat{\alpha}_n^{n+1} \\ \hline \hat{\alpha}_{n+1}^{n+1} \end{array}\right]$$

$$= - \left[\begin{array}{c} \phi_x(1) \\ \vdots \\ \phi_x(n) \\ \hline \phi_x(n+1) \end{array}\right] \tag{9.3.29}$$

By partitioning the matrices and vectors as shown, it follows that Eq. (10.3.29) can be written as

$$\begin{bmatrix} \phi_x(0) & \cdots & \phi_x(n-1) \\ \vdots & & \vdots \\ \phi_x(n-1) & \cdots & \phi_x(0) \end{bmatrix} \begin{bmatrix} \hat{\alpha}_1^{n+1} \\ \vdots \\ \hat{\alpha}_n^{n+1} \end{bmatrix}$$

$$+ \hat{\alpha}_{n+1}^{n+1} \begin{bmatrix} \phi_x(n) \\ \vdots \\ \phi_x(1) \end{bmatrix} = - \begin{bmatrix} \phi_x(1) \\ \vdots \\ \phi_x(n) \end{bmatrix} \tag{9.3.30}$$

and

$$\sum_{j=1}^{n} \hat{\alpha}_i^{n+1} \phi_x(n+1-i) + \hat{\alpha}_{n+1}^{n+1} \phi_x(0) = -\phi_x(n+1) \tag{9.3.31}$$

Premultiplying Eq. (9.3.30) by the inverse of the correlation matrix ϕ^N and using Eq. (9.3.27) and (9.3.28), we can write

$$\begin{bmatrix} \hat{\alpha}_1^{n+1} \\ \vdots \\ \hat{\alpha}_n^{n+1} \end{bmatrix} - \hat{\alpha}_{n+1}^{n+1} \begin{bmatrix} \hat{\alpha}_n^n \\ \vdots \\ \hat{\alpha}_1^n \end{bmatrix} = \begin{bmatrix} \hat{\alpha}_1^n \\ \vdots \\ \hat{\alpha}_n^n \end{bmatrix} \tag{9.3.32}$$

Let us now define

$$\hat{\alpha}_0^n = \hat{\alpha}_0^{n+1} = 1 \tag{9.3.33}$$

Equations (9.3.32) and (9.3.33) constitute a set of equations for recursively computing the estimates for any order $(n + 1)$ given the estimates corresponding to model order n. This can be seen by explicitly rewriting these equations as follows. Let us denote $\hat{\alpha}_{n+1}^{n+1}$ by p_{n+1}. We can then write

$$\begin{aligned} \hat{\alpha}_0^{n+1} &= \hat{\alpha}_0^n \\ \hat{\alpha}_i^{n+1} &= \hat{\alpha}_i^n + p_{n+1}\hat{\alpha}_{n+1-i}^n \qquad i = 1, \ldots, n \end{aligned} \tag{9.3.34}$$

where

$$p_{n+1} = \hat{\alpha}_{n+1}^{n+1} \tag{9.3.35}$$

The parameters p_i are known as *partial correlation coefficients*. It is clear from Eq. (9.3.34) that the estimates $\hat{\alpha}_i^n$ for any n can be computed if we know $p_1, p_2, \ldots, p_n$. We now consider the determination of these coefficients. Let us multiply $\hat{\alpha}_i^{n+1}$ by z^{-i}, sum over i and use Eqs. (9.3.34) and (9.3.35) to obtain

$$\sum_{i=0}^{n+1} \hat{\alpha}_i^{n+1} z^{-i} = \sum_{i=0}^{n} \hat{\alpha}_i^n z^{-i} + p_{n+1} z^{-1} \left[\sum_{i=1}^{n} \hat{\alpha}_{n+1-i}^n z^{-(i-1)} + z^{-n} \right] \tag{9.3.36}$$

Let us define the polynomials

$$F_n(z^{-1}) = \sum_{i=0}^{n} \hat{\alpha}_i^n z^{-i} \tag{9.3.37}$$

and

$$G_n(z^{-1}) = z^{-1} \left[\sum_{i=0}^{n-1} \hat{\alpha}_{n-i}^n z^{-i} + z^{-n} \right] \tag{9.3.38}$$

In terms of these polynomials, Eq. (9.3.36) can be written as

$$F_{n+1}(z^{-1}) = F_n(z^{-1}) + p_{n+1}G_n(z^{-1}) \tag{9.3.39}$$

We now determine a relation for $G_{n+1}(z^{-1})$ in terms of $F_n(z^{-1})$ and $G_n(z^{-1})$. In order to do so, we rewrite Eq. (9.3.34) (by replacing i by $n+1-i$) to get

$$\hat{\alpha}^{n+1}_{n+1-i} = \hat{\alpha}^{n}_{n+1-i} + p_{n+1}\hat{\alpha}^n_i \qquad i = 1, \ldots, n \tag{9.3.40}$$

We also have from Eq. (9.3.38)

$$\begin{aligned} zG_{n+1}(z^{-1}) &= \sum_{i=0}^{n} \hat{\alpha}^{n+1}_{n+1-i} z^{-i} + z^{-(n+1)} \\ &= \hat{\alpha}^{n+1}_{n+1} + \sum_{i=1}^{n} \hat{\alpha}^{n+1}_{n+1-i} z^{-i} + z^{-(n+1)} \end{aligned} \tag{9.3.41}$$

Substitution of equations (9.3.35) and (9.3.40) in Eq. (9.3.41) yields

$$\begin{aligned} zG_{n+1}(z^{-1}) &= \sum_{i=1}^{n} \left[\hat{\alpha}^{n}_{n+1-i} + p_{n+1}\hat{\alpha}^n_i\right] z^{-i} + z^{-(n+1)} + p_{n+1} \\ &= z^{-1} \sum_{i=0}^{n} \hat{\alpha}^{n}_{n-i} z^{-i} + z^{-n} + p_{n+1}\left[\sum_{i=0}^{n} \hat{\alpha}^n_i z^{-i}\right] \end{aligned} \tag{9.3.42}$$

We now use Eqs. (9.3.37) and (9.3.38) to write

$$zG_{n+1}(z^{-1}) = G_n(z^{-1}) + p_{n+1}F_n(z^{-1}) \tag{9.3.43}$$

We can use Eqs. (9.3.39) and (9.3.43) to recursively compute the estimates of p_i. Equation (9.3.34) can then be used to obtain the estimates $\hat{\alpha}^n_i$. The initial conditions for the recursion are obtained from (9.3.37) and (9.3.38) as

$$F_0(z^{-1}) = 1 \quad \text{and} \quad G_0(z^{-1}) = z^{-1} \tag{9.3.44}$$

Starting with $F_0(z^{-1})X(z) = X(z)$ and $G_0(z^{-1})X(z) = z^{-1}X(z)$ and using Eqs. (9.3.39) and (9.3.45) we obtain the lattice structure shown in Fig. (9.3.1). We note from Fig. 9.3.1 that $\varepsilon^n_f(k)$, the output at the nth summer along the upper leg of the lattice is the residual obtained by assuming a model of order n. Thus we can minimize $\mathrm{E}\{\varepsilon^n_f(k)^2\}$ to obtain the parameter estimates $\hat{\alpha}^n_1 \ldots \hat{\alpha}^n_n$.

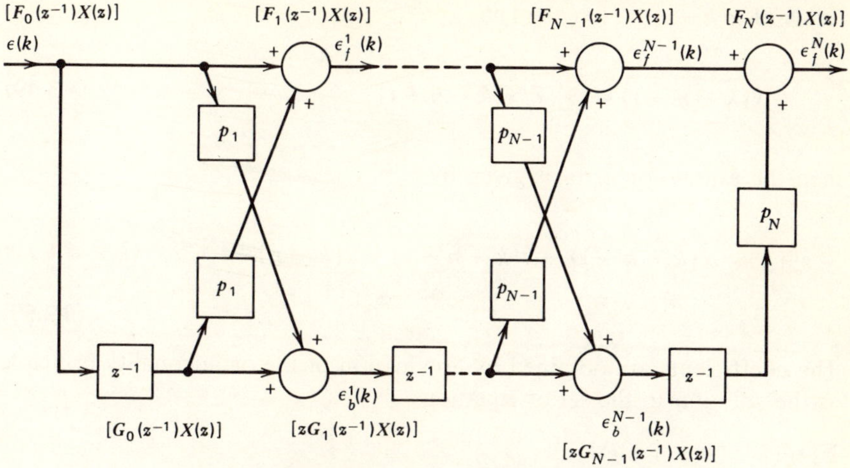

Fig. 9.3.1 Identification using the Levinson algorithm.

As indicated earlier, the structure so obtained yields the best linear mean square estimate of $x(k)$ in terms of its past values $x(k-1)\ldots x(k-n)$ [31]. In order to see this, let us define the estimate as

$$\hat{x}(k) = -\sum_{i=1}^{n} \hat{\gamma}_i^n x(k-i) \tag{9.3.45}$$

Then by the orthogonality principle, the coefficients $\hat{\gamma}_i^n$ must be such that the estimation error

$$\varepsilon_f^n(k) = x(k) - \hat{x}(k) = x(k) + \sum_{i=1}^{n} \hat{\gamma}_i^n x(k-i) \tag{9.3.46}$$

is orthogonal to the set $x(k-j)$ for $i = 1 \ldots n$.

Hence

$$\mathrm{E}\left\{\left[x(k) + \sum_{i=1}^{n} \hat{\gamma}_i^n x(k-i)\right] x(k-j)\right\} = 0; \qquad j = 1 \ldots n \tag{9.3.47}$$

Comparison with Eq. (9.3.27) shows that

$$\hat{\gamma}_i^n = \hat{\alpha}_n \qquad \text{for} \quad i = 1 \ldots n \tag{9.3.48}$$

It can similarly be shown that if we write the estimate of $x(k-n-1)$ in terms of $x(k-n)\ldots x(k-1)$ as

$$\hat{x}(k-n-1)=\sum_{i=0}^{n-1}\hat{\delta}_i^n x(k-n+i) \tag{9.3.49}$$

then the estimation error is given by

$$\varepsilon_b^n(k)=x(k-n-1)-\hat{x}(k-n-1)=x(k-n)+\sum_{i=0}^{n-1}\hat{\delta}_i^n x(k-n+i) \tag{9.3.50}$$

The coefficients are obtained by application of the orthogonality principle as the solution to the set of equations

$$\mathrm{E}\{\varepsilon_b^n(k)x(k-n+j)\} = \mathrm{E}\left\{\left[x(k-n)+\sum_{i=0}^{n-1}\hat{\delta}_i^n x(k-n+i)\right]x(k-n+j)\right\}=0 \qquad j=1\ldots n \tag{9.3.51}$$

Comparison with Eq. (9.3.28) shows that

$$\hat{\delta}_i^n=\hat{\alpha}_{n-i+1}^n \qquad \text{for} \quad i=1\ldots n \tag{9.3.52}$$

It follows from Eqs. (9.3.46), (9.3.48), and (9.3.37) that

$$\varepsilon_f^n(z)=F_n(z^{-1})X(z) \tag{9.3.53}$$

Similarly Eqs. (9.3.50), (9.3.52), and (9.3.38) yield

$$\varepsilon_b^n(z)=zG_n(z^{-1})X(z) \tag{9.3.54}$$

It can be seen from Fig. (9.3.1) that the residuals $\varepsilon_f^n(k)$ and $\varepsilon_b^n(k)$ are defined recursively as follows.

$$\varepsilon_f^{n+1}(k)=\varepsilon_f^n(k)+p_{n+1}\varepsilon_b^n(k-1) \tag{9.3.55}$$

$$\varepsilon_b^{n+1}(k)=\varepsilon_b^n(k-1)+p_{n+1}\varepsilon_f^n(k) \tag{9.3.56}$$

We can compute p_{n+1} by minimizing either $\mathrm{E}\{\varepsilon_f^{n+1}(k)^2\}$ or $\mathrm{E}\{\varepsilon_b^{n+1}(k)^2\}$. It can easily be verified that the two estimates for p_{n+1} follow as

$$\hat{p}_{n+1}^f=-\left\{\frac{\mathrm{E}\{\varepsilon_f^n(k)\varepsilon_b^n(k-1)\}}{\mathrm{E}\{[\varepsilon_b^n(k-1)]^2\}}\right\} \tag{9.3.57}$$

and

$$\hat{p}^{b}_{n+1} = -\left\{\frac{\mathrm{E}\{\varepsilon_f^n(k)\varepsilon_b^n(k-1)\}}{\mathrm{E}\{[\varepsilon_b^n(k)]^2\}}\right\} \tag{9.3.58}$$

Since $x(k)$ is stationary, it can be shown by direct computation that

$$\mathrm{E}\{[\varepsilon_b^n(k-1)]^2\} = \mathrm{E}\{[\varepsilon_f^n(k)]^2\} \tag{9.3.59}$$

We can therefore combine Eqs. (9.3.57) and (9.3.58) to obtain as our estimate

$$\hat{p}_{n+1} = -\frac{2\mathrm{E}\{\varepsilon_f^n(k)\varepsilon_b^n(k-1)\}}{\left[\mathrm{E}\{\varepsilon_b^n(k-1)^2\} + \mathrm{E}\{\varepsilon_f^n(k)^2\}\right]} \tag{9.3.60}$$

The crosscorrelations in Eq. (9.3.60) can be computed in a manner similar to Eq. (9.3.25).

We now briefly describe the algorithm. Given the data sequence $x_0, x_1, \ldots, x_{L-1}$, we form the two sequences

$$\varepsilon_f^0(k) = x(k) \qquad \text{and} \quad \varepsilon_b^0(k) = x(k)$$

We now compute $\hat{p}_1$ from Eq. (9.3.59). We can then use Eqs. (9.3.55) and (9.3.56) to determine $\varepsilon_f^2(k)$ and $\varepsilon_b^2(k)$ for $k = 0, 1, \ldots, L-1$. We form the cross-correlations needed in Eq. (9.3.60) and compute $\hat{p}_2$. This procedure can be carried as far as needed.

Since p_{n+1} depends only on $\varepsilon_b^n(k)$, it is clear that p_{n+1} is independent of p_j for $j > n+1$. This means that in contrast to the computation of the coefficients α_i, we need not recompute p_i when the assumed model order is increased. In fact the algorithm also gives us a means of determining the model order N. As mentioned earlier, if the parameters are correctly identified, the residual will be white. Let us therefore compute the mean square value of the residual at each stage of the recursion as

$$J_n = \mathrm{E}\{[\varepsilon_f^n(k)]^2\} \tag{9.3.61}$$

and let us form the difference

$$J_{n+1} - J_n \tag{9.3.62}$$

for each n. If this quantity attains a minimum or becomes essentially constant, this implies that the residuals cannot be reduced by further processing. The value of n at which this happens should thus correspond to the model order. The coefficients $\hat{\alpha}_i$ can be obtained from $\hat{p}_i$ by use of Eq. (9.3.34). For example, for $n = 1$, we get

$$\hat{\alpha}_1^1 = \hat{p}_1$$

and for $n = 2$, we have

$$\hat{\alpha}_1^2 = \hat{\alpha}_1^1 + \hat{p}_2\hat{\alpha}_1^2 = \hat{p}_1 + \hat{p}_1\hat{p}_2$$

$$\hat{\alpha}_2^2 = \hat{p}_2$$

A fast computer program for determining the coefficients $\hat{p}_i$ and hence $\hat{\alpha}_i^n$ is also available [32].

The Levinson algorithm avoids the need for predetermining the model order N. It is, however, a batch processing technique and involves the computation of correlation coefficients. We now discuss several sequential techniques for the identification of AR parameters.

9.3.3 Sequential Identification of Partial Correlation Coefficients [37]

We can obtain a sequential variation of the Levinson algorithm by replacing the expectation operation in the previous sections by the time average. Thus given a signal $y(k)$, we replace $\mathrm{E}\{y^2(k)\}$ by

$$\frac{1}{k}\sum_{i=1}^{k} y^2(i) \tag{9.3.63}$$

It follows that we can obtain an equation analogous to Eq. (9.3.60) for the estimate $\hat{p}_{n+1}$ as

$$\hat{p}_{n+1}(k) = -\frac{2\sum_{i=1}^{k} \varepsilon_f^n(i)\varepsilon_b^n(i-1)}{\sum_{i=1}^{k}\left\{\left[\varepsilon_b^n(i-1)\right]^2 + \left[\varepsilon_f^n(i)\right]^2\right\}} \tag{9.3.64}$$

where $\hat{p}_{n+1}(k)$ denotes the estimate obtained at time k. We can obtain a sequential algorithm for estimating $\hat{p}_{n+1}$ by writing Eq. (9.3.64) as

$$\hat{p}_{n+1}(k) = -\frac{2\sum_{i=1}^{k} \varepsilon_f^n(i)\varepsilon_b^n(i-1)}{\left[\sum_{i=1}^{k}\left[\varepsilon_b^n(i-1)\right]^2 + \sum_{i=1}^{k}\left[\varepsilon_f^n(i)\right]^2\right]} - \frac{2\varepsilon_f^n(k)\varepsilon_b^n(k-1)}{\left[\sum_{i=1}^{k}\left[\varepsilon_b^n(i-1)\right]^2 + \sum_{i=1}^{k}\left[\varepsilon_f^n(i)\right]^2\right]} \tag{9.3.65}$$

This can be written as

$$\hat{p}_{n+1}(k) = \hat{p}_{n+1}(k-1)\frac{\left[\sum_{i=1}^{k-1}[\varepsilon_b^n(i-1)]^2 + \sum_{i=1}^{k-1}[\varepsilon_f^n(i)]^2\right]}{\left[\sum_{i=1}^{k}[\varepsilon_b^n(i-1)]^2 + \sum_{i=1}^{k}[\varepsilon_f^n(i)]^2\right]}$$

$$-\frac{2\varepsilon_f^n(k)\varepsilon_b^n(k-1)}{\left[\sum_{i=1}^{k}[\varepsilon_b^n(i-1)]^2 + \sum_{i=1}^{k}[\varepsilon_f^n(i)]^2\right]} \qquad \textbf{(9.3.66)}$$

By adding and subtracting $\{[\varepsilon_b^n(k-1)]^2 + [\varepsilon_f^n(k)]^2\}$ in the numerator of the first term of the right-hand side of Eq. (9.3.66), it follows that

$$\hat{p}_{n+1}(k) = \hat{p}_{n+1}(k-1) - \frac{1}{\sum_{i=1}^{k}\left\{[\varepsilon_f^n(i)]^2 + [\varepsilon_b^n(i-1)]^2\right\}}$$

$$\times\left[\hat{p}_{n+1}(k-1)\left\{(\varepsilon_f^n(k))^2 + (\varepsilon_b^n(k-1))^2\right\} + 2\varepsilon_f^n(k)\varepsilon_b^n(k-1)\right] \qquad \textbf{(9.3.67)}$$

Equations (9.3.67) together with (9.3.55) and (9.3.56) constitute a complete set of algorithms for sequential identification of the parameters $\hat{p}_i$. The parameters $\hat{\alpha}_i^n$ can again be computed by use of Eq. (9.3.34). The algorithms are summarized in Table 9.3.1.

Example 9.3.3 We use the sequential algorithms of Table 9.3.1 to estimate the parameters of a sixth order AR model. The data was generated by simulating the system on a digital computer. The input $w(k)$ to the model was a zero-mean, white, Gaussian sequence, with unity variance.

The actual parameter values and the identified values at the end of 250 samples for these two cases is shown in Table 9.3.2a. For purposes of determining the accuracy of the estimates the squared error $e_i = \sum_{k=1}^{k_f} [p_i(k) - \hat{p}_i(k)]^2$; $i = 1 \ldots N$, was computed for each i for $k_f = 250$, and is also tabulated in the table.

The differences of the squared sum of the residuals J_n calculated for various values of n are tabulated in Table 9.3.2b. The quantity J_n attains a minimum for $n = 6$.

TABLE 9.3.1 SEQUENTIAL IDENTIFICATION OF PARTIAL CORRELATION COEFFICIENTS

Parameter Identification Algorithm:

$$\hat{p}_{n+1}(k) = \hat{p}_{n+1}(k-1) - \frac{\left[\hat{p}_{n+1}(k-1)\left\{(\varepsilon_f^n(k))^2 + (\varepsilon_b^n(k-1))^2\right\} + 2\varepsilon_f^n(k)\varepsilon_b^n(k-1)\right]}{\sum_{i=1}^{k}\left[(\varepsilon_f^n(i))^2 + (\varepsilon_b^n(i-1))^2\right]} \tag{9.3.67}$$

$$\varepsilon_f^{j+1}(k) = \varepsilon_f^j(k) + \hat{p}_{j+1}(k)\varepsilon_b^j(k-1) \tag{9.3.55}$$

$$\varepsilon_b^{j+1}(k) = \varepsilon_b^j(k-1) + \hat{p}_{j+1}(k)\varepsilon_f^j(k) \tag{9.3.56}$$

Initial Conditions:

$\varepsilon_b^0(k) = \varepsilon_f^0(k) = x(k)$ Observations.

Conversion of $\hat{p}_n$ to $\hat{\alpha}_i$: For $n = 1, 2, \ldots, N$

$$\hat{\alpha}_i^{n+1} = \hat{\alpha}_i^n + \hat{p}_n\hat{\alpha}_{n+1-i}^n \tag{9.3.34}$$

$$\hat{\alpha}_{n+1}^{n+1} = \hat{p}_{n+1} \tag{9.3.35}$$

9.3.4 Sequential Identification of AR Parameters

We now consider the sequential identification of the parameters α_i. The algorithm that we obtain will be similar to the Kalman filter algorithm that we studied in Chapter 6.

Let us define the parameter vector $\boldsymbol{\alpha}$ as

$$\boldsymbol{\alpha}^T = [\alpha_1 \ldots \alpha_N] \tag{9.3.68}$$

Since $\boldsymbol{\alpha}$ is a constant vector, we can model it by the equation

$$\boldsymbol{\alpha}(k+1) = \boldsymbol{\alpha}(k) \tag{9.3.69}$$

Our problem is to estimate $\boldsymbol{\alpha}(k)$ on the basis of the sequence of observations $x(1)$, $x(2)$, and so on. The observation model can be obtained from Eq. (9.3.13) as

$$z(k) = x(k) = -\sum_{i=1}^{N} \alpha_i x(k-i) + Gw(k) \tag{9.3.70}$$

TABLE 9.3.2a SEQUENTIAL IDENTIFICATION OF PARTIAL CORRELATION COEFFICIENTS IN EXAMPLE 9.3.2.

INDEX i	1	2	3	4	5	6
Actual Parameter Values p_i	0.885	−0.554	0.754	0.884	−0.638	0.385
Identified Parameter Values $\hat{p}_i$	0.9185	−0.6132	0.8475	0.7634	−0.6327	0.4622
Initial Mean Square Error e_i	0.7832	0.3069	0.5685	0.7815	0.4070	0.1482
Final Mean Square Error e_i	$.1125 \times 10^{-2}$	$.3505 \times 10^{-2}$	$.8733 \times 10^{-2}$	$.1455 \times 10^{-1}$	$.2767 \times 10^{-4}$	$.5955 \times 10^{-2}$

TABLE 9.3.2b COST AS A FUNCTION OF MODEL ORDER

SYSTEM ORDER: n	1	2	3	4	5	6	7	8
J_n	$.1142 \times 10^7$	$.1114 \times 10^6$	$.1482 \times 10^6$	$.3270 \times 10^5$	$.7112 \times 10^4$	$.2280 \times 10^4$	$.2782 \times 10^4$	$.2644 \times 10^4$

Let us define the matrix $\mathbf{C}(k)$ as

$$\mathbf{C}(k-1) = [-x(k-1) \ldots -x(k-N)]$$

Then the observation equation can be written as

$$z(k) = \mathbf{C}(k-1)\boldsymbol{\alpha}(k) + v(k) \tag{9.3.71}$$

where we have set

$$Gw(k) = v(k) \tag{9.3.72}$$

so that $v(k)$ is white with covariance kernel G^2. Comparison of Eqs. (9.3.69) and (9.3.71) with the message and observation models of the Kalman filter discussed in Chapter 6 shows the equivalence between the two problems. To complete the equivalence, let us assume that $\boldsymbol{\alpha}(0)$ is a random vector with known mean and covariance. The equations for the estimate $\hat{\boldsymbol{\alpha}}(k)$ directly follow from Table 6.4.1. as

$$\hat{\boldsymbol{\alpha}}(k+1) = \hat{\boldsymbol{\alpha}}(k) + \mathbf{K}(k+1)[x(k+1) - \mathbf{C}(k)\hat{\boldsymbol{\alpha}}(k)] \tag{9.3.73}$$

The gain vector $\mathbf{K}(k+1)$ is given

$$\mathbf{K}(k+1) = \frac{\mathbf{V}_{\tilde{\alpha}}(k)\mathbf{C}^T(k)}{G^2 + \mathbf{C}(k)\mathbf{V}_{\tilde{\alpha}}(k)\mathbf{C}^T(k)} \tag{9.3.74}$$

and the error variance matrix $\mathbf{V}_{\tilde{\alpha}}(k)$ satisfies the difference equation

$$\mathbf{V}_{\tilde{\alpha}}(k+1) = \mathbf{V}_{\tilde{\alpha}}(k) - \frac{\mathbf{V}_{\tilde{\alpha}}(k)\mathbf{C}^T(k)\mathbf{C}(k)\mathbf{V}_{\tilde{\alpha}}(k)}{G^2 + \mathbf{C}(k)\mathbf{V}_{\tilde{\alpha}}(k)\mathbf{C}^T(k)} \tag{9.3.75}$$

The technique is therefore sequential in nature. The model order N, however, must be fixed a priori. We need to specify the gain G, the initial value for the estimate $\hat{\boldsymbol{\alpha}}(0)$ and the initial error variance $\mathbf{V}_{\tilde{\alpha}}(0)$. The convergence of the algorithm is clearly determined by the rate at which $\mathbf{K}(k)$ tends to zero. From Eq. (9.3.74), it can be seen that this is influenced by the choice of G. To see this, we note that Eq. (9.3.75) can be written as

$$\mathbf{V}_{\tilde{\alpha}}(k+1) = [\mathbf{I} - \mathbf{K}(k+1)\mathbf{C}(k)]\mathbf{V}_{\tilde{\alpha}}(k) \tag{9.3.76}$$

For small G, the product $\mathbf{K}(k+1)\mathbf{C}(k)$ is approximately equal to the identity matrix so that $\mathbf{V}_{\tilde{\alpha}}(k)$ tends to zero after only a few samples have been processed. Conversely if G is large, the estimator gain $\mathbf{K}(k)$ tends to go to zero very slowly.

While it may thus appear to be desirable to choose the value of G small so that convergence is accelerated, such a procedure could cause the algorithm to converge to a value which is not the correct value of the parameter. This is related to the phenomemon of divergence in the Kalman filter. (See [34] for a discussion of divergence.)

The choice of the initial values of $\hat{\boldsymbol{\alpha}}$ and $\mathbf{V}_{\tilde{\boldsymbol{\alpha}}}$ are intertwined in that the choice of $\mathbf{V}_{\tilde{\boldsymbol{\alpha}}}$ should reflect the degree of confidence placed in the initial choice of $\hat{\boldsymbol{\alpha}}$. If this initial choice is thought to be good, $\mathbf{V}_{\tilde{\boldsymbol{\alpha}}}(0)$ should be small so that the estimates change very little. If the initial choice is thought to be poor (or uncertain), $\mathbf{V}_{\tilde{\boldsymbol{\alpha}}}(0)$ should be chosen large.

An alternative way of accounting for the uncertainty in the choice of the initial value for the parameter vector is to model $\boldsymbol{\alpha}(k)$ not as a constant vector but as a Markov process described by the equation

$$\boldsymbol{\alpha}(k+1) = \boldsymbol{\alpha}(k) + \mathbf{u}(k) \tag{9.3.77}$$

where $\mathbf{u}(k)$ is a stationary white sequence. Equation (9.3.76) then becomes

$$\mathbf{V}_{\tilde{\boldsymbol{\alpha}}}(k+1) = [\mathbf{I} - \mathbf{K}(k+1)\mathbf{C}(k)]\mathbf{V}_{\tilde{\boldsymbol{\alpha}}}(k) + \mathbf{V}_{\mathbf{u}} \tag{9.3.78}$$

where $\mathbf{V}_{\mathbf{u}}$ is the variance of the process $\mathbf{u}(k)$. The choice of $\mathbf{V}_{\mathbf{u}}$ will then determine the rate of convergence of the algorithm. Since $\mathbf{V}_{\tilde{\boldsymbol{\alpha}}}(k)$ is now bounded below by $\mathbf{V}_{\mathbf{u}}$ the problem of converging to a wrong value because of the gain vector going to zero too soon is obviated.

Example 9.3.4 We consider the identification of a sixth order model using the sequential algorithms of Eqs. (9.3.73)–(9.3.75). The system was simulated on a digital computer and the results at the end of 200 iterations are shown in Table 9.3.3. The simulations were carried out using two different

TABLE 9.3.3 SEQUENTIAL IDENTIFICATION OF AUTOREGRESSIVE PARAMETERS.

PARAMETER:	α_1	α_2	α_3	α_4	α_5	α_6	ERROR
True Value	−1.052	0.1071	.5371	−.53	−1.094	−0.529	—
Estimate for $G^2 = 3$	−1.0602	0.1477	.4914	−.4292	−1.069	−.469	0.0108
Estimate for $G^2 = 100$	−1.008	0.1772	.4375	−.5593	−1.026	−.443	0.0296

values of G^2. The mean square error obtained in each case, $\sum_{i=1}^{6} (\alpha_i - \hat{\alpha}_i)^2$ is also shown in the table. The initial values for the parameters were chosen to be zero in both cases. Convergence as indicated by the mean square error seems to be faster for the lower value of G^2. This is in conformity with the earlier discussion on the choice of G^2.

As stated earlier, one of the applications of determining the parameters of an autoregressive model of a stationary stochastic process is in the computation of the power spectrum. It is thus instructive to see how well the power spectrum is identified by this technique. Figure 9.3.2 shows the true spectrum and the spectrum obtained from the autoregressive model. As can be seen, the true spectrum has peaks at 500, 1800, and 2200 Hz. The AR model does identify the location of all three spectral peaks fairly accurately. In general, the AR model fits the true spectrum better at spectral peaks than at troughs. This is particularly useful in applications such as speech processing, since it has been determined that the ear is relatively sensitive to the location of the peaks of the speech spectrum.

9.3.5 Extensions to ARMA Processes

We will now extend the algorithms of the previous sections to the identification of the parameters of an ARMA process described in terms of either the difference equation Eq. (9.3.11) or in terms of the Z-transform description Eq. (9.3.10).

$$\frac{X(z)}{W(z)} = \frac{G(1 + \beta_1 z^{-1} + \cdots + \beta_M z^{-M})}{1 + \alpha_1 z^{-1} + \cdots + \alpha_N z^{-N}} \tag{9.3.79}$$

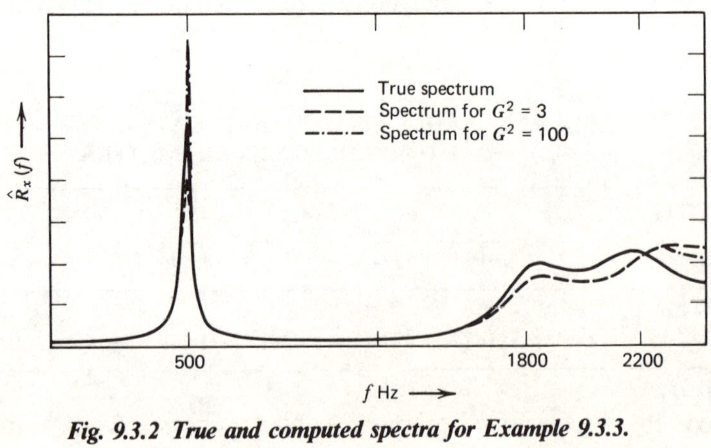

Fig. 9.3.2 True and computed spectra for Example 9.3.3.

We can approximate the transfer function on the right-hand side of Eq. (9.3.79) by an all-pole transfer function by replacing each zero by a sufficient number of poles [35]. This can be seen by writing a zero as

$$1+\beta z^{-1}=\frac{1}{1-\beta z^{-1}-\beta z^{-2}-\beta z^{-3}-\cdots} \tag{9.3.80}$$

In most applications it suffices to use just the first few terms in the series. We can therefore approximate the ARMA process $x(k)$ by a purely AR process by writing the right side of Eq. (9.3.79) as

$$\frac{X(Z)}{W(Z)}=\frac{G}{1+r_1 z^{-1}+r_2 z^{-2}+\cdots+r_{N'} z^{-N'}} \tag{9.3.81}$$

While $N'=\infty$ yields an exact equivalence, the approximation is fairly good for finite N'. The value of N' is generally greater than $N+M$.

In order to determine the relation between the parameters $\{r_i\}$ of the approximate AR model and $\{\alpha_i\}$ and $\{\beta_i\}$ corresponding to the ARMA model, we equate the right-hand sides of Eqs. (9.3.79) and (9.3.81) so that we have

$$\frac{G\left(1+\beta_1 z^{-1}+\cdots+\beta_M z^{-M}\right)}{1+\alpha_1 z^{-1}+\cdots+\alpha_N z^{-N}}=\frac{G}{1+r_1 z^{-1}+\cdots+r_{N'} z^{-N'}} \tag{9.3.82}$$

Cross multiplying yields

$$\begin{aligned}\left(1+\beta_1 z^{-1}+\cdots+\beta_M z^{-M}\right)&\left(1+r_1 z^{-1}+\cdots+r_{N'} z^{-N'}\right)\\ &=1+\alpha_1 z^{-1}+\cdots+\alpha_N z^{-N}\end{aligned} \tag{9.3.83}$$

Equating coefficients of like powers of z^{-1} gives us the following set of equations

$$\begin{aligned}\alpha_1&=\beta_1+r_1\\ \alpha_2&=\beta_2+\beta_1 r_1+r_2\\ \vdots&=\\ \alpha_N&=\beta_N+\beta_{N-1}r_1+\cdots+r_N\\ 0&=\beta_N r_i+\beta_{N-1}r_{i+1}+\cdots\beta_1 r_{N+i-1}+r_{N+i}\qquad i=1,2,\ldots N'\end{aligned} \tag{9.3.84}$$

where $\beta_j = 0$ for $j = M+1, M+2, \ldots, N$. These equations can be put in matrix form as

$$\begin{bmatrix} r_N & r_{N-1} & \cdots & r_{N-M+1} \\ r_{N+1} & r_N & \cdots & r_{N-M+2} \\ \vdots & & & \vdots \\ r_{N'-1} & r_{N'-2} & \cdots & r_{N'-M} \end{bmatrix} \begin{bmatrix} \beta_1 \\ \beta_2 \\ \vdots \\ \beta_M \end{bmatrix} = - \begin{bmatrix} r_{N+1} \\ r_{N+2} \\ \vdots \\ r_{N'} \end{bmatrix} \tag{9.3.85}$$

and

$$\begin{bmatrix} \alpha_1 \\ \alpha_2 \\ \alpha_3 \\ \vdots \\ \alpha_N \end{bmatrix} = \begin{bmatrix} \beta_1 \\ \beta_2 \\ \beta_3 \\ \vdots \\ \beta_N \end{bmatrix} + \begin{bmatrix} 1 & 0 & 0 & \cdots & 0 \\ \beta_1 & 1 & 0 & \cdots & 0 \\ \beta_2 & \beta_1 & 1 & \cdots & 0 \\ \vdots & & & & \\ \beta_{N-1} & \beta_{N-2} & \beta_{N-3} & & 1 \end{bmatrix} \begin{bmatrix} r_1 \\ r_2 \\ r_3 \\ \vdots \\ r_N \end{bmatrix} \tag{9.3.86}$$

Equation (9.3.85) represents a set of $(N' - N)$ equations in M unknowns. In order to obtain a solution, we must therefore choose $N' = N + M$. Then once the parameters $r_1 \ldots r_{N'}$ have been identified, we can use Eqs. (9.3.85) and (9.3.86) to determine the parameter sets $\{\alpha_i\}$ and $\{\beta_i\}$. The parameters r_i can be identified using either of the techniques discussed in the previous section. Identification of the partial correlation coefficients associated with $\{r_i\}$ has the advantage that the model order N' need not be fixed a priori but can be determined from the computational scheme. On the other hand, in determining the parameters $\{r_i\}$ using the Kalman-type algorithm, the model order N' must be predetermined.

With either scheme, there still remains the problem of fixing the relative orders N and M. Let $\hat{N}$ and $\hat{M}$ denote assumed values of N and M. Denoting the square matrix on the left-hand side of Eq. (9.3.85) by $A_{\hat{N},\hat{M}}$, we can test the rank of $A_{\hat{N},\hat{M}}$ for values of $\hat{N}$, $\hat{M}$ such that $\hat{N} + \hat{M} = \hat{N}'$ until for some M and N

$$\text{Rank}\{A_{\hat{N},\hat{M}}\} = M_1 \qquad \text{for} \quad \hat{N} > \hat{N}_1,\ \hat{M} > \hat{M}_1$$

or alternatively

$$\text{Det}\, A_{\hat{N},\hat{M}} = 0 \qquad \text{for} \quad \hat{N} > \hat{N}_1,\ \hat{M} > \hat{M}_1 \tag{9.3.87}$$

In practice, because of round-off and truncation errors, the determinant in

Eq. (9.3.87) is rarely zero. We can therefore use as our criterion

$$\text{Det } A_{\hat{N}, \hat{M}} \leqslant \varepsilon \qquad \text{for some} \quad \varepsilon \geqslant 0$$

If M and N are estimated on the low side, there is the possibility of omitting the dynamics essential in signal description. On the other hand, if M and N are overestimated, some negligibly small coefficients are introduced into the model. These correspond to the addition of small-magnitude high-frequency terms and should not cause difficulties in most applications.

Example 9.3.5 We consider the identification of the ARMA model

$$x(k) + 1.5x(k-1) + 0.625x(k-2) = w(k) - 0.5w(k-1)$$

where $w(k)$ is a zero mean, unity variance, white, Gaussian sequence. The data was again generated by simulating the system on a digital computer. Table 9.3.4a shows the cost for various values of N'. As can be seen from the table, the cost becomes essentially constant after $N' = 3$. We thus choose $N' = 3$. This is acceptable since $M + N = 3$ for this example. The true values of the parameters and the estimates obtained on the basis of 250 samples are shown in Table 9.3.4b.

TABLE 9.3.4a DETERMINATION OF MODEL ORDER FOR EXAMPLE 9.3.4

SYSTEM ORDER n'	$n' = 1$	$n' = 2$	$n' = 3$	$n' = 4$	$n' = 5$
$J_{n'}$	1.976	1.871	0.1362	0.1218	0.1039

TABLE 9.3.4b IDENTIFICATION OF AR AND MA PARAMETERS FOR THE ARMA MODEL OF EXAMPLE 9.3.4

True Values	$\alpha_1 = 1.5$	$\alpha_2 = .625$	$\beta_1 = -.5$
Estimated Values	$\hat{\alpha}_1 = 1.509$	$\hat{\alpha}_2 = .703$	$\hat{\beta}_1 = -.445$

9.3.6 Identification of ARMA Models: Direct Determination

We can set up an algorithm for simultaneously identifying the $\{\alpha_i\}$ and $\{\beta_i\}$ parameters in the ARMA model using a procedure similar to the one for identifying the $\{\alpha_i\}$ parameters in the AR model using the Kalman algorithm [36]. Let $\boldsymbol{\theta}$ denote the $(N+M)$ column vector

$$\boldsymbol{\theta} = [\alpha_1 \ldots \alpha_N, \beta_1 \ldots \beta_M]^T \tag{9.3.88}$$

Since $\boldsymbol{\theta}$ is a constant vector, it can be modeled by the equation

$$\boldsymbol{\theta}(k+1) = \boldsymbol{\theta}(k) \tag{9.3.89}$$

The observation model is obtained from Eq. (9.3.11) as

$$z_k = x_k = -\sum_{i=1}^{N} \alpha_i x(k-i) + \sum_{i=1}^{M} \beta_i w(k-i) + Gw(k) \tag{9.3.90}$$

If the input $w(k)$ were available for measurement we can form the row vector $\mathbf{C}(k)$

$$\mathbf{C}(k-1) = [-x(k-1), -x(k-2) \ldots -x(k-N), w(k-1) \ldots w(k-M)] \tag{9.3.91}$$

and write the observation equation as

$$z(k) = \mathbf{C}(k-1)\boldsymbol{\theta}(k) + v(k) \tag{9.3.92}$$

where again we have defined

$$v(k) = Gw(k) \tag{9.3.93}$$

Equations (9.3.89) and (9.3.92) are similar to Eqs. (9.3.69) and (9.3.71) and hence the Kalman algorithm may be applied to estimate the parameter vector $\boldsymbol{\theta}$. Since, however, in most applications the input $w(k)$ is not available, we may seek to estimate it at any stage by using Eq. (9.3.90).

We thus write

$$\hat{G}(k)\hat{w}(k) = x_k + \sum_{i=1}^{N} \hat{\alpha}_i(k)x(k-i) - \sum_{i=1}^{M} \hat{\beta}_i \hat{w}(k-i) \tag{9.3.94}$$

From Eq. (9.3.12), we see that the quantity on the right-hand side of Eq. (9.3.94) represents the residual $\varepsilon(k)$ so that

$$\hat{w}(k) = \frac{1}{\hat{G}(k)} \epsilon(k) \tag{9.3.95}$$

where

$$\epsilon(k) = x(k) + \sum_{i=1}^{N} \hat{\alpha}_i(k)x(k-i) - \sum_{i=1}^{M} \hat{\beta}_i \hat{w}(k-i) \tag{9.3.96}$$

To evaluate $\hat{G}(k)$ from Eq. (9.3.95) we use the fact that $w(k)$ is white with unity spectral height to obtain

$$E\{[\hat{G}(k)\hat{w}(k)]^2\} = \hat{G}^2(k) = E\{\epsilon^2(k)\} \tag{9.3.97}$$

where $E\{\epsilon^2(k)\}$ can be computed as

$$E\{\epsilon^2(k)\} = \frac{1}{k}\sum_{i=1}^{k}\epsilon^2(i)$$

If we now replace $w(k-i)$, $i = 1, 2, \ldots, M$ in Eq. (9.3.91) by $\hat{w}(k-i)$ obtained from Eq. (9.3.95) we have

$$\mathbf{C}(k-1) = [-x(k-1)\ldots - x(k-N), \hat{w}(k-1)\ldots \hat{w}(k-M)] \tag{9.3.98}$$

The identification algorithm becomes

$$\hat{\boldsymbol{\theta}}(k+1) = \hat{\boldsymbol{\theta}}(k) + \mathbf{K}(k+1)[x(k+1) - \mathbf{C}(k)\hat{\boldsymbol{\theta}}(k)] \tag{9.3.99}$$

$$\mathbf{K}(k+1) = \mathbf{V}_{\tilde{\theta}}(k)\mathbf{C}^T(k)[G^2(k) + \mathbf{C}(k)\mathbf{V}_{\tilde{\theta}}(k)\mathbf{C}^T(k)] \tag{9.3.100}$$

and

$$\mathbf{V}_{\tilde{\theta}}(k+1) = [\mathbf{I} - \mathbf{K}(k+1)\mathbf{C}(k)]\mathbf{V}_{\tilde{\theta}}(k) \tag{9.3.101}$$

Equations (9.3.94)–(9.3.101) constitute a complete set of algorithms for identifying the parameters of the ARMA model of (9.3.11). It has been shown that under certain circumstances the estimate $\hat{\boldsymbol{\theta}}(k)$ converges to the true value with probability 1 [36].

Example 9.3.6 We consider the identification of the parameters of the ARMA model of Example (9.3.4) using the direct method. Table 9.3.5 shows the true parameter values, and estimates obtained on the basis of 250 data samples and 1000 samples, respectively. The identification of both the AR and MA parameters seems to be slightly better in this example than in Example 9.3.4. However, as pointed out earlier, we need to assume the model order a priori in this technique.

TABLE 9.3.5 DIRECT DETERMINATION OF ARMA PARAMETERS

True Values	$\alpha_1 = 1.5$	$\alpha_2 = .625$	$\beta_1 = -.5$
Estimated Values, $k_f = 1000$	$\hat{\alpha}_1 = 1.49$	$\hat{\alpha}_2 = .604$	$\hat{\beta}_1 = -.518$
Estimated Values $k_f = 250$	$\hat{\alpha}_1 = 1.463$	$\hat{\alpha}_2 = .5825$	$\hat{\beta}_1 = -.505$

9.3.7 Estimation in Nonlinear Models: Extended Kalman Filter

We have so far considered the identification of the parameters of a linear time-invariant model with stationary white noise input. We now consider a general nonlinear model and present a technique for identifying the parameters of such a model.

Let $\boldsymbol{\theta}$ represent the vector of unknown parameters in the system. Let us consider the system described by

$$\mathbf{x}(k+1) = \mathbf{f}(\mathbf{x}, \boldsymbol{\theta}, k) + \mathbf{g}(\mathbf{x}, \boldsymbol{\theta}, k)\mathbf{w}(k) \tag{9.3.102}$$

where $\mathbf{x}(k)$ is an n-vector describing the state of the system and where $\mathbf{w}(k)$ is a zero mean, white input sequence with covariance

$$\text{cov}\,\{\mathbf{w}(k)\mathbf{w}(j)\} = \mathbf{V}_{\mathbf{w}}(k)\delta_{kj} \tag{9.3.103}$$

The observation vector $z(k)$ is

$$\mathbf{z}(k) = \mathbf{h}(\mathbf{x}, \boldsymbol{\theta}, k) + \mathbf{v}(k) \tag{9.3.104}$$

with the observation noise $\mathbf{v}(k)$ being zero mean, white, with covariance

$$\text{cov}\,\{\mathbf{v}(k)\mathbf{v}(j)\} = \mathbf{V}_{\mathbf{v}}(k)\delta_{kj} \tag{9.3.105}$$

and independent of $\mathbf{w}(k)$.

Assume that $\mathbf{x}(k_0)$ is a random variable with known mean and covariance and independent of $\mathbf{w}(k)$ and $\mathbf{v}(k)$ for all $k > k_0$, so that

$$E\{\mathbf{x}(k_0)\} = \boldsymbol{\mu}_{\mathbf{x}_0}, \quad \text{var}\,\{\mathbf{x}(k_0)\} = \mathbf{V}_{\mathbf{x}_0}$$

$$\text{cov}\,\{\mathbf{x}(k_0), \mathbf{w}(k)\} = \text{cov}\,\{\mathbf{x}(k_0), \mathbf{v}(k)\} = \text{cov}\,\{\mathbf{w}(k), \mathbf{v}(j)\} = 0 \tag{9.3.106}$$

If the time evolution of the parameter vector $\boldsymbol{\theta}$ can be described by a difference equation of the form

$$\boldsymbol{\theta}(k+1) = \boldsymbol{\psi}(\boldsymbol{\theta}, k) \tag{9.3.107}$$

we can define an augmented state vector $\mathbf{X}$ as

$$\mathbf{X} = \left[\frac{\mathbf{x}}{\boldsymbol{\theta}}\right]$$

From Eqs. (9.3.101) and (9.3.106) we therefore have

$$\mathbf{X}(k+1) = \left[\frac{\mathbf{x}(k+1)}{\boldsymbol{\theta}(k+1)}\right] = \left[\frac{\mathbf{f}(\mathbf{x}, \boldsymbol{\theta}, k) + \mathbf{g}(\mathbf{x}, \boldsymbol{\theta}, k)\mathbf{w}(k)}{\boldsymbol{\psi}(\boldsymbol{\theta}, k)}\right]$$

$$= \boldsymbol{\phi}(\mathbf{X}, k) + \boldsymbol{\Gamma}(\mathbf{X}, k)\mathbf{w}(k) \tag{9.3.108}$$

Similarly the observation model can be written in terms of the augmented state vector as

$$\mathbf{Z}(k) = \mathbf{h}(\mathbf{X}, k) + \mathbf{v}(k) \tag{9.3.109}$$

The problem of identification of the parameter vector now reduces to one of estimating the states $\mathbf{X}(k)$ of the augmented system and observation models Eqs. (9.3.110) and (9.3.111). Since the models are nonlinear, we have to use an approximate algorithm for determining the estimates. In fact, we have already seen an application of this technique in Chapter 7 in our discussion of equalizers for intersymbol interference. As in Chapter 7, we may use the extended Kalman filter algorithms of Section 6.6 to obtain estimates of the augmented state $\mathbf{X}(k)$ for the message and observation models of Eqs. (9.3.108) and (9.3.109). The algorithms are:

$$\hat{\mathbf{X}}(k+1) = \hat{\mathbf{X}}(k+1|k) + \mathbf{K}(k+1)\{\mathbf{Z}(k+1) - \mathbf{h}[\hat{\mathbf{X}}(k+1|k), k+1]\} \tag{9.3.110a}$$

$$\hat{\mathbf{X}}(k+1|k) = \boldsymbol{\phi}[\hat{\mathbf{X}}(k|k), k] \tag{9.3.110b}$$

$$\mathbf{K}(k+1) = \mathbf{V}_{\tilde{\mathbf{X}}}(k+1|k+1)\frac{\partial \mathbf{h}^T[\hat{\mathbf{X}}(k+1|k), k+1]}{\partial \hat{\mathbf{X}}(k+1|k)}\mathbf{V}_{\mathbf{v}}^{-1}(k+1) \tag{9.3.110c}$$

$$\mathbf{V}_{\tilde{\mathbf{X}}}(k+1|k) = \frac{\partial \boldsymbol{\phi}[\hat{\mathbf{X}}(k|k), k]}{\partial \hat{\mathbf{X}}(k|k)}\mathbf{V}_{\tilde{\mathbf{X}}}(k|k)\frac{\partial \boldsymbol{\phi}^T[\hat{\mathbf{X}}(k|k), k]}{\partial \hat{\mathbf{X}}(k|k)} + \boldsymbol{\Gamma}[\hat{\mathbf{X}}(k|k), k]\mathbf{V}_{\mathbf{w}}(k)\boldsymbol{\Gamma}^T[\hat{\mathbf{X}}(k|k), k] \tag{9.3.110d}$$

$$\begin{aligned}\mathbf{V}_{\tilde{\mathbf{X}}}(k+1|k+1) = \mathbf{V}_{\tilde{\mathbf{X}}}(k+1|k) - \mathbf{V}_{\tilde{\mathbf{X}}}(k+1|k)\frac{\partial \mathbf{h}^T[\hat{\mathbf{X}}(k+1|k), k+1]}{\partial \hat{\mathbf{X}}(k+1|k)} \\ \times \left\{\frac{\partial \mathbf{h}[\hat{\mathbf{X}}(k+1|k), k+1]}{\partial \hat{\mathbf{X}}(k+1|k)}\mathbf{V}_{\tilde{\mathbf{X}}}(k+1|k)\right. \\ \left. \times \frac{\partial \mathbf{h}^T[\hat{\mathbf{X}}(k+1|k), k+1]}{\partial \hat{\mathbf{X}}(k+1|k)} + \mathbf{V}_{\mathbf{v}}(k+1)\right\} \\ \times \frac{\partial \mathbf{h}[\hat{\mathbf{X}}(k+1|k), k+1]}{\partial \hat{\mathbf{X}}(k+1|k)}\mathbf{V}_{\tilde{\mathbf{X}}}(k+1|k)\end{aligned} \tag{9.3.110e}$$

We note that the algorithms simultaneously estimate the states of the system and the unknown parameters.

Example 9.3.7 We consider the problem of identifying the time constant in a first-order system described by

$$x(k+1) = ax(k) + w(k)$$

The observations are given by

$$z(k) = x(k) + v(k)$$

Since a is a constant parameter, it can be modeled as

$$a(k+1) = a(k)$$

Let us define the states of the augmented model as

$$X_1(k) = x(k)$$
$$X_2(k) = a(k)$$

Then

$$X_1(k+1) = X_1(k)X_2(k) + w(k)$$
$$X_2(k+1) = X_2(k)$$

and

$$Z(k) = X_1(k) + v(k)$$

We note that even though the system was linear, the augmented message model is nonlinear.

The results of a digital computer simulation of the algorithms are shown in Fig. 9.3.3. Three simulations corresponding to different values of the input covariance kernel $V_w(k)$ and the measurement noise covariance kernel $V_v(k)$ are shown. The initial conditions for the simulations were chosen as $\mathbf{X}(0) = \begin{bmatrix} 0 \\ 1 \end{bmatrix}$ and $V_x(0) = I$. The results indicate that the identification of the unknown parameter is better at the lower values of V_w and V_v.

9.3.8 Maximum Likelihood Identification [23,37,41]

In our discussions in Chapter 5, we saw that the maximum likelihood technique plays an important role in the identification of nonrandom

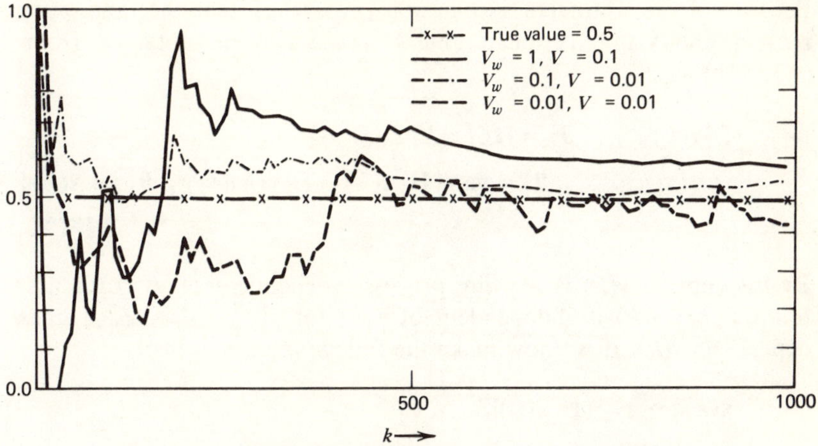

Fig. 9.3.3 Identification using the extended Kalman filter.

parameters. In this section, we consider the identification of the parameters of linear time-invariant systems using the maximum likelihood approach. We assume that the system is described by the state-space model

$$\mathbf{x}(k+1) = \boldsymbol{\phi}\mathbf{x}(k) + \boldsymbol{\Gamma}\mathbf{w}(k) + \mathbf{B}\mathbf{u}(k) \tag{9.3.111a}$$

$$\mathbf{z}(k) = \mathbf{H}\mathbf{x}(k) + \mathbf{v}(k) \tag{9.3.111b}$$

where the state vector $\mathbf{x}(k)$ is of dimension n, $\mathbf{u}(k)$ is a p-vector of deterministic inputs, $\mathbf{z}(k)$ is a q-vector of observations. $\mathbf{w}(k)$ and $\mathbf{v}(k)$ are assumed to be zero-mean, stationary, Gaussian, white processes, independent of each other and with covariance kernels $\mathbf{V}_\mathbf{w}$ and $\mathbf{V}_\mathbf{v}$, respectively. $\boldsymbol{\phi}$, $\boldsymbol{\Gamma}$, $\mathbf{B}$ and $\mathbf{H}$ are constant matrices of appropriate order.

The most general identification problem arises when some or all of the elements of the matrices $\boldsymbol{\phi}$, $\boldsymbol{\Gamma}$, $\mathbf{B}$, $\mathbf{H}$, $\mathbf{V}_\mathbf{w}$, and $\mathbf{V}_\mathbf{v}$ are unknown. Let us denote by $\boldsymbol{\theta}$ the set of unknown parameters in the system and let $\mathbf{Z}_N$ denote the set of observations $\mathbf{z}(1), \mathbf{z}(2), \ldots, \mathbf{z}(N)$. The ML estimate of $\boldsymbol{\theta}$ is then obtained by maximizing $p(\mathbf{Z}_N|\boldsymbol{\theta})$. We note that

$$\begin{aligned} p(\mathbf{Z}_N|\boldsymbol{\theta}) &= p(\mathbf{z}(N)|\mathbf{Z}_{N-1}, \boldsymbol{\theta})p(\mathbf{Z}_{N-1}|\boldsymbol{\theta}) \\ &= p(\mathbf{z}(N)|\mathbf{Z}_{N-1}, \boldsymbol{\theta})p(\mathbf{z}(N-1)|\mathbf{Z}_{N-2}, \boldsymbol{\theta}) \ldots p(\mathbf{z}(0)) \end{aligned} \tag{9.3.112}$$

so that

$$\ln p(\mathbf{Z}_N|\boldsymbol{\theta}) = \ln \sum_{k=1}^{N} p(\mathbf{z}(k)|\mathbf{Z}_{k-1}, \boldsymbol{\theta}) \tag{9.3.113}$$

Because of our Gaussian assumptions, it is clear that the random vector $\mathbf{z}(k)$ is Gaussian. Its mean and variance can be obtained from Eq. (9.3.111b) as

$$
\begin{aligned}
&E\{\mathbf{z}(k)|\mathbf{Z}_{k-1}, \boldsymbol{\theta}\} = \mathbf{H}E\{\mathbf{x}(k)|\mathbf{Z}_{k-1}, \boldsymbol{\theta}\} \\
&\text{var}\,\{\mathbf{z}(k)|\mathbf{Z}_{k-1}, \boldsymbol{\theta}\} = \text{var}\,\{\mathbf{H}[\mathbf{x}(k) - E\{\mathbf{x}(k)|\mathbf{Z}_{k-1}, \boldsymbol{\theta}\}] + \mathbf{v}(k)\}
\end{aligned} \tag{9.3.114}
$$

By assumption $\mathbf{v}(k)$ is a white process independent of $\mathbf{w}(k)$ for all k. It follows that $\mathbf{v}(k)$ is independent of $\mathbf{x}(k)$ for all k. Also, $\mathbf{Z}_{k-1}$ does not depend on $\mathbf{v}(k)$. If we now make the following definitions:

$$\hat{\mathbf{x}}(k|k-1) = E\{\mathbf{x}(k)|\mathbf{Z}_{k-1}, \boldsymbol{\theta}\}$$

$$\tilde{\mathbf{x}}(k|k-1) = \mathbf{x}(k) - \hat{\mathbf{x}}(k|k-1)$$

and

$$\mathbf{V}_{\tilde{\mathbf{x}}}(k|k-1) = \text{var}\,\{\tilde{\mathbf{x}}(k|k-1)\} \tag{9.3.115}$$

it follows that the mean and variance of $\mathbf{z}(k)$ can be written from Eqs. (9.3.114) as

$$
\begin{aligned}
&E\{\mathbf{z}(k)|\mathbf{Z}_{k-1}, \boldsymbol{\theta}\} = \mathbf{H}\hat{\mathbf{x}}(k|k-1) \\
&\text{var}\,\{\mathbf{z}(k)|\mathbf{Z}_{k-1}, \boldsymbol{\theta}\} = \mathbf{H}\mathbf{V}_{\tilde{\mathbf{x}}}(k|k-1)\mathbf{H}^T + \mathbf{V}_{\mathbf{v}}
\end{aligned} \tag{9.3.116}
$$

Substitution in Eq. (9.5.3) then yields the log-likelihood function as

$$
\begin{aligned}
\ln p(\mathbf{Z}_N|\boldsymbol{\theta}) = &-\frac{1}{2}\sum_{k=1}^{N} \ln\,\{|\mathbf{H}\mathbf{V}_{\tilde{\mathbf{x}}}(k|k-1)\mathbf{H}^T + \mathbf{V}_{\mathbf{v}}|\} - \frac{Nq}{2}\ln\,(2\pi) \\
&-\frac{1}{2}\sum_{k=1}^{N} [\mathbf{z}(k) - \mathbf{H}\hat{\mathbf{x}}(k|k-1)]^T \\
&\times [\mathbf{H}\mathbf{V}_{\tilde{\mathbf{x}}}(k|k-1)\mathbf{H}^T + \mathbf{V}_{\mathbf{v}}]^{-1}[\mathbf{z}(k) - \mathbf{H}\hat{\mathbf{x}}(k|k-1)]
\end{aligned} \tag{9.3.117}
$$

It only remains to obtain the quantities $\hat{\mathbf{x}}(k|k-1)$ and $\mathbf{V}_{\tilde{\mathbf{x}}}(k|k-1)$. Because of the form of the system model Eq. (9.3.111), it is clear that these quantities can be obtained by using the discrete Kalman filter algorithms of Chapter 6. The algorithms are as under:

$$\hat{\mathbf{x}}(k+1|k) = \boldsymbol{\phi}\hat{\mathbf{x}}(k|k-1) + \mathbf{K}(k+1, k)\left[\mathbf{z}(k) - \mathbf{H}\hat{\mathbf{x}}(k|k-1)\right] + \mathbf{B}\mathbf{u}(k) \tag{9.3.118a}$$

$$\mathbf{K}(k+1|k) = \boldsymbol{\phi}\mathbf{V}_{\tilde{\mathbf{x}}}(k|k-1)\mathbf{H}^T\left[\mathbf{H}\mathbf{V}_{\tilde{\mathbf{x}}}(k|k-1)\mathbf{H}^T + \mathbf{V}_{\mathbf{v}}\right]^{-1} \tag{9.3.118b}$$

$$\begin{aligned}\mathbf{V}_{\tilde{\mathbf{x}}}(k+1|k) &= \boldsymbol{\phi}\mathbf{V}_{\tilde{\mathbf{x}}}(k|k-1)\boldsymbol{\phi}^T + \boldsymbol{\Gamma}\mathbf{V}_{\mathbf{w}}\boldsymbol{\Gamma}^T \\ &\quad - \boldsymbol{\phi}\mathbf{V}_{\tilde{\mathbf{x}}}(k|k-1)\mathbf{H}^T\left[\mathbf{H}\mathbf{V}_{\tilde{\mathbf{x}}}(k|k-1)\mathbf{H}^T + \mathbf{V}_{\mathbf{v}}\right]^{-1}\mathbf{H}\mathbf{V}_{\tilde{\mathbf{x}}}(k|k-1)\boldsymbol{\phi}^T\end{aligned} \tag{9.3.118c}$$

These equations differ from the usual Kalman algorithms in that the quantity $\hat{\mathbf{x}}(k|k-1)$ is the one-stage predicted estimate of $\mathbf{x}(k)$ conditioned on the parameter vector $\boldsymbol{\theta}$. Also Eq. (9.3.118a) has been modified to take into account the nonrandom input $\mathbf{u}(k)$.

While equations (9.3.117) and (9.3.118) give us a means of determining the log-likelihood function, there still remains the problem of finding the estimate $\hat{\boldsymbol{\theta}}$ by maximizing $\ln p(\mathbf{Z}_N|\boldsymbol{\theta})$. In general this is a fairly difficult task. We can consider an iterative approach in which the estimate $\boldsymbol{\theta}^j$ at the jth iteration is obtained from the estimate at the previous iteration by using an algorithm of the type

$$\hat{\boldsymbol{\theta}}^j = \hat{\boldsymbol{\theta}}^{j-1} + \Delta\hat{\boldsymbol{\theta}} \tag{9.3.119}$$

where $\Delta\hat{\boldsymbol{\theta}}$ represents a correction term. Usually, $\Delta\hat{\boldsymbol{\theta}}$ is chosen to be of the form

$$\Delta\hat{\boldsymbol{\theta}} = -\mathbf{K}(j)\left.\frac{\partial \ln p(\mathbf{Z}_N|\boldsymbol{\theta})}{\partial \boldsymbol{\theta}}\right|_{\boldsymbol{\theta}=\hat{\boldsymbol{\theta}}^{j-1}} \tag{9.3.120}$$

where $\mathbf{K}(j)$ represents a suitably chosen matrix. The solution is an off-line solution and the data set $\mathbf{Z}_N$ may have to be used over several iterations to obtain reasonably converging solutions. In general, the computation of $\Delta\hat{\boldsymbol{\theta}}$ can be quite tedious.

If the dimension of the parameter vector $\boldsymbol{\theta}$ is small and if the range of values which $\boldsymbol{\theta}$ takes are known or can be reasonably guessed, we can simply evaluate $\ln p(\mathbf{Z}_N|\boldsymbol{\theta})$ for several values of $\boldsymbol{\theta}$ over its range and choose as our estimate, that value which yields the largest value of $\ln p(\mathbf{Z}_N|\boldsymbol{\theta})$.

While the solution of the likelihood equations is, in general, difficult, we can obtain simple expressions for the estimates in some special cases. We illustrate this by means of two examples.

Example 9.3.8 Consider the scalar model

$$x(k+1) = \phi x(k) + w(k) + u(k) \tag{9.3.121a}$$

$$z(k) = x(k) \tag{9.3.121b}$$

Let us assume that we want to identify ϕ and V_w so that $\boldsymbol{\theta} = [\phi \; V_w]^T$. It follows from Eq. (9.3.117) that

$$\ln p(\mathbf{Z}_N|\boldsymbol{\theta}) = -\frac{1}{2}\sum_{k=1}^{N} \ln V_{\tilde{x}}(k|k-1) - \frac{N}{2}\ln(2\pi) - \frac{1}{2V_{\tilde{x}}(k|k-1)}\sum_{k=1}^{N}[z(k) - \hat{x}(k|k-1)]^2 \tag{9.3.122}$$

Now

$$\begin{aligned}\hat{x}(k|k-1) &= \mathrm{E}\{x(k)|\mathbf{Z}_{k-1}, \boldsymbol{\theta}\} \\ &= \mathrm{E}\{\phi x(k-1) + w(k-1) + u(k-1)|\mathbf{Z}_{k-1}, \boldsymbol{\theta}\} \\ &= \phi z(k-1) + u(k-1)\end{aligned} \tag{9.3.123}$$

so that

$$\begin{aligned}\tilde{x}(k|k-1) &= x(k) - \hat{x}(k|k-1) \\ &= z(k) - \phi z(k-1) - u(k-1)\end{aligned}$$

and

$$V_{\tilde{x}}(k|k-1) = [z(k) - \phi z(k-1) - u(k-1)]^2 \tag{9.3.124}$$

Furthermore, we can write

$$\begin{aligned}V_w &= \mathrm{E}\{w^2(k)\} = \mathrm{E}\{[x(k+1) - \phi x(k) - u(k)]^2\} \\ &= [z(k+1) - \phi z(k) - u(k)]^2 \\ &= V_{\tilde{x}}(k|k-1)\end{aligned} \tag{9.3.125}$$

Substitution of Eqs. (9.3.123) and (9.3.125) in Eq. (9.3.122) yields the log-likelihood function as

$$\ln p(\mathbf{Z}_N|\boldsymbol{\theta}) = -\frac{N}{2}\ln V_w - \frac{N}{2}\ln(2\pi) - V_w^{-1}\sum_{k=1}^{N}[z(k) - \phi z(k-1) - u(k-1)]^2$$

The ML estimates obtained by setting the partial derivatives of $\ln p(\mathbf{Z}_N|\boldsymbol{\theta})$ with respect to ϕ and V_w equal to zero are

$$\hat{\phi} = \frac{\sum_{k=1}^{N} z(k-1)[z(k) - u(k-1)]}{\sum_{k=1}^{N} z^2(k-1)}$$

and

$$\hat{V}_w = \frac{1}{N} \sum_{k=1}^{N} [z(k) - \hat{\phi} z(k-1) - u(k-1)]^2$$

Example 9.3.9 We consider the scalar autoregressive model

$$x(k) + \sum_{k=1}^{N} \alpha_i x(k-i) = w(k) + u(k)$$

$$z(k) = x(k)$$

where $w(k)$ is zero mean, white, Gaussian with variance V_w.

We want to identify the coefficients α_i and V_w. The parameter vector, therefore, is

$$\boldsymbol{\theta} = [\alpha_1\ \alpha_2 \ldots \alpha_K\ V_w]^T$$

Because the observations are noisefree, it easily follows from our derivation in the previous example that

$$\hat{x}(k|k-1) = -\sum_{i=1}^{K} \alpha_i z(k-i) + u(k)$$

and

$$V_w = V_{\tilde{x}}(k|k-1) = \left[z(k) + \sum_{i=1}^{K} \alpha_i z(k-i) - u(k) \right]^2$$

The log-likelihood function, therefore, is

$$\ln p(\mathbf{Z}_N|\boldsymbol{\theta}) = -\frac{N}{2} \ln V_w - \frac{N}{2} \ln(2\pi) - V_w^{-1} \sum_{k=1}^{K} \left[z(k) + \sum_{i=1}^{K} \alpha_i z(k-i) - u(k) \right]^2$$

It follows that the estimates can be obtained by solving the set of equations

$$\sum_{i=1}^{K} \hat{\alpha}_i \left[\sum_{k=1}^{K} z(k-i)z(k-j) \right] = \sum_{k=1}^{K} [u(k) - z(k)]z(k-j),$$

$$j = 1, 2, \ldots, K$$

$$\hat{V}_w = \frac{1}{N} \sum_{k=1}^{N} \left[z(k) + \sum_{i=1}^{K} \hat{\alpha}_i z(k-i) - u(k) \right]^2$$

We conclude this section by noting that while our results were derived under Gaussian assumptions, as the number of observations become very large, these assumptions are not crucial to our results. A lower bound on the maximum achievable accuracy using the ML technique can be obtained from the Cramér–Rao bound discussed in Chapter 5.

9.3.9 Section Summary

In this section we have considered the application of the concepts of estimation theory to the identification of linear time-invariant systems. We assumed that the system was of known structure so that the problem is essentially one of estimating certain parameters associated with the model. The specific models that we were mostly concerned with were the autoregressive-moving average model and the state space models.

In Section 9.3 we considered the estimation of parameters in an ARMA model representation of a stationary stochastic process. We first derived the Yule–Walker equations for the parameter set and subsequently in Section 9.3.2 derived the Levinson algorithm for solving these equations. We also derived a sequential version of the Levinson algorithm in Section 9.3.3. In Section 9.3.4 we derived an alternate sequential identification algorithm similar to the Kalman filter algorithms of Chapter 6. Section 9.3.5 considered the extension of some of these techniques for the identification of ARMA models.

Identification of parameters in a state variable representation of the system was considered in Section 9.3.6. By treating the parameters as states, we can pose the problem as one of estimation in nonlinear state variable models. Approximate algorithms for the parameter estimates were obtained by using the extended Kalman filter algorithms of Chapter 6.

For nonrandom parameter estimation, ML estimates are very attractive because of their properties. For state variable models, expressions for the likelihood function were derived in Section 9.3.7 under Gaussian assumptions. The resulting equations for the estimates are usually difficult to solve. In most cases an off-line iterative technique must be resorted to. For some special cases, the algorithms take simple forms.

The results of Section 9.3.1 can be extended to vector ARMA processes [38]. Some recent results are contained in [39]. The choice of model order using an information theoretic criterion is discussed in [40] (see also [41]).

An interesting topic is the choice of input signals in the model of Section 9.3.7. If we are free to choose the input signal, then we could conceivably design the signal so as to aid us in the identification of the model parameters. The problem is related to that of signal design in the detection of known waveforms that we encountered in Chapter 4. The interested reader is referred to [43] for a survey of recent results in this area.

EXERCISES

9.1 In a two-category classification problem, the class-conditional densities satisfy a Cauchy distribution

$$p(x|H_i) = \frac{1}{\pi\alpha} \cdot \frac{1}{1 + \left(\dfrac{x - \beta_i}{\alpha}\right)} \qquad i = 1, 2$$

Assuming equal prior probabilities for the two classes, determine the discriminant function. What is the probability of misclassification?

9.2 The Bhattacharya coefficient is defined as

$$\rho = \int [p(\mathbf{x}|H_1)p(\mathbf{x}|H_2)]^{1/2}\, d\mathbf{x}$$

Show that the error probability for two-class problem satisfies

$$P_e \leqslant \sqrt{P_1 P_2}\, \rho \leqslant \frac{1}{2}\rho$$

9.3 In the two-category classification problem, instead of assigning a pattern $\mathbf{x}$ to one of the two classes, we may, under certain circumstances, reject it as being unrecognizable. If we consider rejection as assigning a pattern to class H_3, we can formulate the problem as a ternary hypothesis testing problem. We make the following assignment of costs:

$$c_{ii} = 0 \qquad c_{ij} = c_s \quad i, j = 1, 2 \qquad c_{i3} = c_r \quad i = 1, 2$$

Show that the optimal classification rule is to assign **x** to class i if

$$P(H_i|\mathbf{x}) \geqslant P(H_j|\mathbf{x}) \quad \text{and} \quad P(H_i|\mathbf{x}) \geqslant 1 - \frac{c_r}{c_s} \qquad i, j = 1, 2$$

and reject otherwise and that the corresponding discriminant functions are given by

$$f_i(x) = \begin{cases} p(\mathbf{x}|H_i)P(H_i) & i = 1, 2 \\ \dfrac{c_a - c_s}{c_s} \displaystyle\sum_{j=1}^{2} p(x|H_j)P(H_j) & i = 3 \end{cases}$$

9.4 In a binary scalar pattern classification problem, the distributions are Gaussian as follows:

$$p(x|H_i) = \frac{1}{\sqrt{2\pi}\, V_i} \exp\left[-\frac{1}{2} \frac{(x - m_i)^2}{V_i^2} \right], \qquad i = 1, 2$$

Assume $P_1 = P_2 = \frac{1}{2}$. Generate a set of samples corresponding to each class and use the decision rules of Section 9.3 to classify them. Obtain an estimate of the probability of error by taking the ratio of the number of correctly classified samples to the total number of samples and compare with the theoretical value. Consider the cases

(a) $V_i = 1$ and $m_i = (-1)^i, \quad i = 1, 2$

(b) $V_1 = \frac{1}{2}, V_2 = 4$ and $m_i = 0, \quad i = 1, 2$

(c) $V_1 = \frac{1}{2}, V_2 = 4$ and $m_1 = -1$ and $m_2 = 1$.

9.5 In a supervised learning problem, the class conditional densities are given as

$$p(x|\theta_i) = \begin{cases} \theta_i e^{-\theta_i x} & x \geqslant 0 \\ 0 & \text{otherwise} \end{cases}$$

where θ_i are nonrandom. Given N_1 samples from class 1 and N_2 samples from class 2, set up the learning algorithm for estimating θ_i. What is the discriminant function?

9.6 In Exercise 9.5, assume that θ_i are random parameters. Find the algorithms for learning θ_i in this case by choosing a suitable reproducing form for the prior density of θ_i. (Refer to Section 5.7.)

9.7 Repeat Exercise 9.6 if the class conditional densities are

$$P(x|\theta_i) = \theta_i^{x_i}(1 - \theta_i)^{x_i}$$

where θ_i is the probability that $x_i = 1$.

9.8 Obtain the learning algorithms in Example 9.2.2.

9.9 In a two-category unsupervised learning problem, the class-conditional densities are

$$P(\mathbf{x}|H_i, \boldsymbol{\theta}_i) = \prod_{j=1}^{n} \theta_{ij}^{x_j}(1 - \theta_{ij})^{1-x_j} \qquad i = 1, 2$$

so that the mixture density for $\mathbf{x}$ is

$$P(\mathbf{x}|\boldsymbol{\theta}) = \sum_{i=1}^{2} P(\mathbf{x}|H_i, \boldsymbol{\theta}_i)P(H_i)$$

Find the maximum likelihood estimate $\hat{\boldsymbol{\theta}}_i$ of $\boldsymbol{\theta}_i$.

9.10 Formulate the nonsupervised learning pattern classification problem for the data in Exercise 9.4, assuming that the mean of class H_1 is not known. Simulate the probabilistic teacher and decision directed schemes for the problem. Plot the curve of estimated mean versus number of learning samples and the graph of an estimate of the probability versus number of total test samples for both the schemes.

9.11 Entropy is a statistical measure of uncertainty. Features that reduce the uncertainty within a pattern vector can be considered to be more informative than those that do not. A technique for feature selection therefore is to minimize the entropy of the pattern classes. The entropy of the ith class is defined as

$$H_i = -\int_x p(\mathbf{x}|H_i) \ln p(\mathbf{x}|H_i)\, d\mathbf{x}$$

We seek an $r \times n$ matrix $\mathbf{T}$ which operating on the n-dimensional vector $\mathbf{x}(n)$ will yield a feature vector of reduced dimension $\mathbf{x}(r)$, $r < n$

$$\mathbf{x}(r) = \mathbf{T}\mathbf{x}(n)$$

Let us consider the two-category problem in which the class conditional densities are Gaussian with equal covariances $\mathbf{V}$ but unequal means. Show that to minimize the entropy, the rows of $\mathbf{T}$ must be chosen as those eigenvectors of $\mathbf{V}$ corresponding to the r smallest eigenvalues [43].

9.12 The estimate of the autocorrelation sequence of a data sequence of 1000 samples for the first 10 lags is as follows:

$$\hat{R}_x(0) = 1,\ \hat{R}_x(1) = .05,\ \hat{R}_x(2) = .25,\ \hat{R}_x(3) = .025$$
$$\hat{R}_x(4) = .075,\ \hat{R}_x(5) = .25,\ \hat{R}_x(6) = .125,\ \hat{R}_x(7) = .025,$$
$$\hat{R}_x(8) = .125,\ \hat{R}_x(9) = -.0125,\ \hat{R}_x(10) = 0$$

Assume that we want to fit an expression of the form

$$\hat{R}_x(k) = \hat{R}_x(0)\alpha^{|k|}$$

to this data. Find the maximum likelihood estimate of α.

9.13 Establish Eq. (9.3.59).

9.14 Consider the system

$$\dot{x}(t) = .5x(t) + bw(t)$$

where $w(t)$ is zero mean, white noise with unity variance. The observations are

$$z(t) = x(t) + v(t)$$

where $v(t)$ is zero mean, white with covariance kernel 4, uncorrelated with $w(t)$. We wish to identify the unknown parameter b.

(a) Set up the extended Kalman algorithms for this problem. Can you identify b uniquely?

(b) Define the variable $y(t)$ as $y(t) = (1/b)x(t)$. Set up the problem in terms of the variable $y(t)$ and obtain the equations for the identification of the unknown parameter. Is this scheme better or worse than the scheme in part (a)?

9.15 Consider the system

$$\mathbf{x}(k+1) = \boldsymbol{\phi}\mathbf{x}(k) + \boldsymbol{\Gamma}\mathbf{w}(k)$$
$$\mathbf{z}(k) = \mathbf{H}\mathbf{x}(k) + \mathbf{v}(k)$$

with the usual assumptions on $\mathbf{w}$ and $\mathbf{v}$.

(a) Obtain the ML estimate of $\mathbf{H}$ assuming $\boldsymbol{\Gamma}$ is known.

(b) Obtain the ML estimate of $\boldsymbol{\Gamma}$ assuming $\mathbf{H}$ is known.

9.16 Use the state-augmentation method in Exercise 9.15 and obtain algorithms for estimating $\mathbf{H}$ and $\mathbf{B}$.

9.17 Derive the algorithms for estimating $\mathbf{B}$ in the model of Eq. (9.3.111) using the state-augmentation technique. Simplify the algorithms when we have perfect observation of the state so that

$$\mathbf{z}(k) = \mathbf{x}(k)$$

9.18 Consider ML estimation in the model of Eq. (9.3.111) when the observation noise is zero. Assume that the unknown parameters are the elements of $\boldsymbol{\phi}$, $\mathbf{B}$, $\mathbf{V}_w$.

(a) Derive the equations for obtaining $\hat{\boldsymbol{\theta}}$. Assume $\mathbf{H} = \mathbf{I}$.

(b) Do these algorithms specialize to the algorithms of Example 9.5.1?

(c) Can you write the algorithms in sequential form?

9.19 Set up the ML estimation algorithms for the model of Eq. (9.3.112) if $\boldsymbol{\Gamma} = \mathbf{0}$. Assume that $\mathbf{V}_v$ and $\mathbf{H}$ are unknown.

9.20 In the problem of Example 9.3.6, assume that only ϕ is unknown. Find $\hat{\phi}_{\text{ML}}$. Is it an unbiased estimate? Find a lower limit on the achievable accuracy in identification. (Use the Cramér–Rao bound.)

9.21 Consider the general ARMA model of the form

$$x(k) + \sum_{i=1}^{M} \alpha_i x(k-i) = \sum_{i=1}^{M} \beta_i u(k-i) + w(k)$$

where $u(\cdot)$ is a known input and $w(k)$ is a zero mean, Gaussian white noise with $\mathrm{E}\{w(k)w^T(j)\} = V_w \delta_{kj}$. Derive the ML estimates $\hat{\alpha}_i$, $\hat{\beta}_j$, $i = 1 \ldots N, j = 1 \ldots M$. Assume $z(k) = x(k)$.

9.22 In Exercise (9.21), assume that $w(k)$ is colored with

$$\mathrm{E}\{w(k)w(k-j)\} = \begin{cases} V_j & j = 0, 1, \ldots, N \\ 0 & \text{otherwise} \end{cases}$$

with observations

$$z(k) = x(k) + v(k)$$

Find the ML estimates of $\hat{\alpha}_i$, $\hat{\beta}_i$, V_j [40].

9.23 In Section 9.38, we derived the ML estimates of the parameters in a state variable model of a linear time-invariant system under the assumption that the parameters were nonrandom. Derive MAP algorithms for the model of Eq. (9.3.111) if the unknown parameter vector $\boldsymbol{\theta}$ is Gaussian with mean $\boldsymbol{\mu}_\theta$ and variance $\mathbf{V}_\theta$.

9.24 [45,46] Let $x(\cdot)$ be a zero mean, stationary, Gaussian process whose correlation coefficients for the first N lags $r(0)$, $r(1)$, $r(2), \ldots,$ $r(N-1)$ are known. Suppose we now want to determine $r(N)$. One way of doing this is to maximize the entropy of the process defined as

$$H = -\int p(x) \ln p(x)\, dx$$

where $p(x)$ denotes the multivariate density function of the process. For the N-variate Gaussian density function with covariance matrix

R_N where

$$R_N = \begin{bmatrix} r_0 & r_1 & r_2 & \cdots & r_N \\ r_1 & r_0 & r_1 & \cdots & r_{N-1} \\ \vdots & & & & \vdots \\ r_N & r_{N-1} & r_{N-2} & \cdots & r_0 \end{bmatrix}$$

show that maximizing the entropy is equivalent to maximizing

$$\ln (2\pi)^{N/2} \det [R_N]^{1/2}$$

Hence determine the value of r_N which maximizes the entropy of the process. Can you interpret your results in terms of the autoregressive models discussed in Section 9.3.1?

REFERENCES

1. G. Nagy, State of the Art in Pattern Recognition, *Proc. IEEE*, **56**, No. 5, May, 836–862 (1968).
2. K. S. Fu, *Sequential Methods in Pattern Recognition and Machine Learning*, Academic Press, New York, 1968.
3. J. M. Mendel and K. S. Fu, *Adaptive Learning and Pattern Recognition Systems: Theory and Applications*, Academic Press, New York, 1970.
4. K. Fukunaga, *Introduction to Statistical Pattern Recognition*, Academic Press, New York, 1972.
5. H. C. Andrews, *Introduction to Mathematical Techniques in Pattern Recognition*, John Wiley and Sons, New York, 1972.
6. R. O. Duda and P. E. Hart, *Pattern Classification and Scene Analysis*. Wiley-Interscience, New York, 1973.
7. J. T. Tou and R. C. Gonzalez, *Pattern Recognition Principles*, Addison-Wesley Publishing Co., Reading, Mass., 1974.
8. D. J. Rogers, and T. T. Tanimoto, A Computer Program for Classifying Plants, *Science*, **132**, Oct., 1115–1118 (1960).
9. J. Spragins, Reproducing Distributions for Machine Learning, *Tech. Rep. 6103-7*, Stanford University, Palo Alto, Calif., Nov. 1963.
10. N. Abramson, and D. Braverman, Learning to Recognize Patterns in a Random Environment, *IEEE Trans. Info. Theory*, **IT-8**, No. 5, Sept., 58–63 (1962).
11. D. G. Keehn, A Note on Learning Gaussian Properties, *IEEE Trans. Info. Theory*, **IT-11**, No. 1, Jan., 126–132 (1965).
12. H. Teicher, Identifiability of Mixtures, *Ann. Math. Stat.*, **32**, Mar., 244–248 (1961).

13. S. J. Yakowitz and J. D. Spragins, On the Identifiability of Finite Mixtures, *Ann. Math. Stat.*, **39**, Feb., 209–214 (1968).
14. S. J. Yakowitz, Unsupervised Learning and the Identification of Finite Mixtures, *IEEE Trans. Info. Theory*, **IT-16**, No. 3, May, 330–338 (1970).
15. S. C. Fralick, Learning to Recognize Patterns without a Teacher, *IEEE Trans. Info. Theory*, **IT-13**, No. 1, Jan., 544–549 (1967).
16. R. F. Daly, The Adaptive Binary Detection Problem on the Real Line, *Tech. Report 2003-3*, Stanford University, Palo Alto, Calif., Feb. 1962.
17. H. J. Scudder III, Adaptive Communication Receivers, *IEEE Trans. Info. Theory*, **IT-14**, No. 5, Oct., 544–549 (1965).
18. A. K. Agrawala, Learning with a Probabilistic Teacher, *IEEE Trans. Info. Theory*, **IT-16**, No. 4, July, 373–379 (1970).
19. E. A. Patrick, *Fundamentals of Pattern Recognition*, Prentice-Hall, Englewood Cliffs, N. J., 1972.
20. T. M. Cover, and P. E. Hart, Nearest Neighbor Pattern Classification, *IEEE Trans. Info. Theory*, **IT-13**, Jan., 21–27 (1967).
21. Ya. Z. Tsypkin, *Adaptation and Learning in Automatic Systems*, Translated by Z. J. Nikolic, Academic Press, New York, 1971.
22. Y. T. Chien and K. S. Fu, Sequential Recognition Using a Non-Parametric Ranking Procedure, *IEEE Trans. Info. Theory*, **IT-13**, No. 4, July, 484–492 (1967).
23. P. Eykhoff, *System Identification*, John Wiley and Sons, New York, 1974.
24. A. P. Sage and J. L. Melsa, *System Identification*, Academic Press, New York, 1971.
25. R. B. Blackman and J. W. Tukey, *The Measurement of Power Spectra*, Dover Publications, New York, 1958.
26. E. J. Hannan, *Time Series Analysis*, Metheun and Co., London, 1960.
27. G. M. Jenkins and D. G. Watts, *Spectrum Analysis and Its Applications*, Holden-Day, San Francisco, 1968.
28. E. J. Hannan, Time Series Analysis, *IEEE Trans. Automatic Control*, **AC-19**, No. 6, Dec., 706–715 (1974).
29. E. Parzen, Some Recent Advances in Time Series Modeling, *IEEE Trans. Automatic Control*, **AC-19**, No. 6, Dec, 723–730 (1974).
30. U. Grenander and G. Szego, *Toeplitz Forms and Their Applications*, University of California Press, Berkley, 1958.
31. N. Levinson, The Wiener RMS (root mean square) Error Criterion in Filter Design and Prediction, *J. Math. Phys.*, **25**, Jan., 261–278 (1947).
32. E. A. Robinson, *Multichannel Time Series Analysis with Digital Computer Programs*, Holden-Day, San Francisco, Calif., 1967.
33. M. D. Srinath and M. M. Viswanathan, Sequential Algorithm for Identification of Parameters of an Autoregressive Process, *IEEE Trans. Automatic Control*, **AC-20**, No. 4, Aug., 542–546 (1975).
34. P. C. Young, An Instrument Variable Method for Real-Time Identification of a Noisy Process, *Automatica*, **6**, 271–287 (1970).
35. A. P. Sage and J. L. Melsa, *Estimation Theory with Applications to Communications and Control*, McGraw-Hill Book Co., New York, 1971.

36. D. Graupe, D. J. Krause, and J. B. Moore, Identification of Autoregressive Moving-Average Parameters of Time Series, *IEEE Trans. Automatic Control*, **AC-20**, No. 1, Feb., 104–107 (1975).
37. R. L. Kashyap, Estimation of Parameters in a Partially Whitened Representation of Stochastic Process, *IEEE Trans. Automatic Control*, **AC-19**, No. 1, Feb., 13–21 (1974).
38. F. C. Schweppe, *Uncertain Dynamical Systems*, Prentice-Hall, Englewood Cliffs, N. J., 1973.
39. R. L. Kashyap, Maximum Likelihood Identification of Stochastic Linear Systems, *IEEE Trans. Automatic Control*, **AC-15**, No. 1, Feb., 25–34 (1970).
40. R. L. Kashyap and R. E. Nasburg, Parameter Estimation in Multivariate Stochastic Difference Equations, *IEEE Trans. Automatic Control*, **AC-19**, No. 6, Dec., 784–797 (1974).
41. H. Akaike, Use of an Information Theoretic Quantity for Statistical Model Identification, *Proc. 5th Hawaii Intl. Confce. System Sciences*, 1972, 249–250.
42. H. Akaike, A New Look at the Statistical Model Identification, *IEEE Trans. Automatic Control*, **AC-19**, No. 6, Dec., 716–723 (1974).
43. R. K. Mehra, Optimal Input Signals for Parameter Estimation in Dynamic Systems-Survey and New Results, *IEEE Trans. Automatic Control*, **AC-19**, No. 6, Dec., 753–768 (1974).
44. J. T. Tou and R. P. Heydorn, Some Approaches to Optimum Feature Extraction, in *Computer and Information Sciences—II* (J. T. Tou, Ed.), Academic Press, New York, 1967.
45. A. van den Bos, Alternative Interpretation of Maximum Entropy Spectral Analysis, *IEEE Trans. Info. Theory*, **IT-17**, No. 4, July, 493–494 (1971).
46. J. A. Edward and M. M. Fitelson, Notes on Maximum-Entropy Processing, *IEEE Trans. Info. Theory*, **IT-19**, No. 2, Mar., 232–234 (1973).

APPENDIX A

BILATERAL TRANSFORMS

We have used the bilateral or two sided Laplace and Z-transforms in the text in our discussions of power spectra of stationary stochastic processes. We recall that the unilateral Laplace transform of a causal function $f(t)$ is defined as

$$\mathfrak{L}\{f(t)\} = F(s) = \int_0^{\infty} F(t)e^{-st}\, dt \tag{A.1}$$

where s is a complex variable. For this integral to exist, there must be a real constant α such that

$$|f(t)| < k_1 e^{\alpha t}$$

The integral is then absolutely convergent for Re $s > \alpha$, so that the region of convergence is to the right of a vertical line in the s-plane through the point Re $s = \alpha$.

For functions defined for negative values of t, that is for anticausal functions, the one-sided Laplace transform is given by

$$\mathfrak{L}\{f(t)\} = F(s) = \int_{-\infty}^{0} f(t)e^{-st}\, dt \tag{A.2}$$

If $|f(t)| < k_2 e^{\beta t}$ for some constant $\beta > 0$, the integral exists for values for s such that Re $s = \sigma < \beta$. The region of convergence is therefore to the left of a vertical line through β.

We note that replacing t with $-t$ in Eq. (A.2) yields

$$F(s) = \int_0^{\infty} f(-t)e^{st}\, dt \tag{A.3}$$

If we therefore define the causal function $f_1(t)$ as

$$f_1(t) = f(-t) \qquad t \geqslant 0,$$

we have

$$F(s) = F_1(-s)$$

Thus the properties of the $\mathcal{L}$-transform of the anticausal function $f(t)$ can be determined in terms of the one-sided transform of the causal function $f_1(t)$.

Let $f(t)$ be a function defined over the entire time axis such that

$$|f(t)| < e^{\beta t} \qquad \text{for} \quad t < 0$$

and

$$|f(t)| < e^{\alpha t} \qquad \text{for} \quad t \geqslant 0$$

We define the bilateral transform of $f(t)$ as

$$F(s) = \int_{-\infty}^{\infty} f(t) e^{-st}\, dt \tag{A.4}$$

The region of convergence of $F(s)$ is the vertical strip in the s-plane defined by $\alpha < \text{Re}\, s < \beta$. It follows that for the bilateral Laplace transform of a function to exist, β must be greater than α.

The time function $f(t)$ corresponding to a transform $F(s)$ can be obtained as

$$f(t) = \frac{1}{2\pi j} \int_{c-j\infty}^{c+j\infty} F(s) e^{st}\, ds \tag{A.5}$$

Here the integration is performed along a vertical line in the s-plane corresponding to $\text{Re}\, s = c$ as $\text{Im}\, s$ varies from $-\infty$ to ∞. The line is chosen to lie in the region of convergence of $F(s)$.

In evaluating the integral in Eq. (A.5), we note that if the function $f(t)$ is causal, the region of convergence will be to the right of all the singularities of $F(s)$. If $f(t)$ is anticausal, the region of convergence will be to the left of the singularities of $F(s)$. If $f(t)$ has both a causal and an anticausal part, the singularities of $F(s)$ to the right of the region of convergence will correspond to the anticausal part of $f(t)$ while those to the left will correspond to the causal part. If $F(s)$ vanishes as $s \to \infty$, the integral can be most easily evaluated by closing the contour of integration with a semicircle of infinite radius and using the residue theorem. If the contour is closed in the left half-plane we obtain the causal part of $f(t)$.

For discrete sequences, we can similarly define a bilateral Z-transform. We recall that given the causal sequence $f(k)$ defined for nonnegative integer values of k, we define the Z-transform of the sequence as

$$Z\{f(k)\} = F(z) = \sum_{k=0}^{\infty} f(k)z^{-k} \tag{A.6}$$

The series is absolutely convergent for all z such that $|z| > R_{c^+}$ where

$$R_{c^+} = \overline{\lim}\,|f(k)|^{1/k}$$

Here, $\overline{\lim}$ refers to the largest of the limit points of $|f(k)|^{1/k}$ as $k \to \infty$. For sequences defined for all integer values of k, we define a bilateral or two-sided Z-transform

$$F(z) = \sum_{k=-\infty}^{\infty} f(k)z^{-k} \tag{A.7}$$

Let R_{c^+} and R_{c^-} be the largest and smallest of the limit points to which $|f(k)|^{1/k}$ tends as $k \to \infty$ and $k \to -\infty$, respectively. Then $F(z)$ converges absolutely for all z such that $R_{c^+} < |z| < R_{c^-}$. It follows, therefore, that for the bilateral transform to exist, we must have $R_{c^+} < R_{c^-}$. If $f(k)$ is a causal sequence, its inverse transform is given by

$$f(k) = \frac{1}{2\pi j}\oint_\Gamma z^{k-1}F(z)\,dz \tag{A.8}$$

where Γ is any closed contour in the region of convergence of $F(z)$. If $f(k)$ is an anticausal sequence, the inverse transform is given by

$$f(k) = \frac{1}{2\pi j}\oint_\Gamma F(z)z^{-(k+1)}\,dz \qquad \text{for} \quad k < 0 \tag{A.9}$$

REFERENCES

1. H. Freeman, *Discrete-Time Systems: An Introduction to the Theory*, John Wiley and Sons, New York, 1965.
2. G. Doetsch, *Guide to the Applications of the Laplace Transformation*, Van Nostrand, Princeton, N. J., 1961.

APPENDIX B

CALCULUS OF EXTREMA

We consider the problem of optimizing a function $\phi(x_1, x_2, \ldots, x_n)$ of n variables $x_1, x_2, \ldots, x_n$. A necessary condition that the function have a stationary value is that

$$d\phi = \frac{\partial\phi}{\partial x_1} dx_1 + \frac{\partial\phi}{\partial x_1} dx_2 + \cdots + \frac{\partial\phi}{\partial x_n} dx_n = 0 \tag{B.1}$$

for all permissible values of the differentials $dx_1, \ldots, dx_n$. If the n variables are all independent, this is equivalent to the n conditions

$$\frac{\partial\phi}{\partial x_1} = \frac{\partial\phi}{\partial x_2} = \cdots = \frac{\partial\phi}{\partial x_n} = 0 \tag{B.2}$$

If the n variables are dependent, then the method of *Lagrange multipliers* is most useful in obtaining necessary conditions for a stationary point. Let the dependence between the variables be expressed in terms of the *constraint* equations

$$f_i(x_1, x_2, \ldots, x_n) = 0 \qquad i = 1, \ldots, N \tag{B.3}$$

We form the augmented function

$$\phi^1(x_1, x_2, \ldots, x_n;\ \lambda_1 \ldots \lambda_N) = \phi(x_1, x_2, \ldots, x_n) + \sum_{i=1}^{N} \lambda_i f_i(x_1 \ldots x_n) \tag{B.4}$$

The problem of optimizing the function $\phi(\cdot)$ subject to the constraints, Eq. (B.3) is then equivalent to optimizing the function $\phi^1(\cdot)$ in which the variables are treated as being independent. The constants λ_i are known as Lagrange multipliers.

We can therefore obtain necessary conditions similar to Eq. (B.1) as

$$\frac{\partial \phi^1}{\partial x_i} = \frac{\partial \phi}{\partial x_i} + \sum_{i=1}^{N} \lambda_i \frac{\partial f_i}{\partial x_i} = 0 \qquad i = 1, 2, \ldots, n \tag{B.5}$$

The set of Eqs. (B.3) and (B.5) constitute a set of $n + N$ equations in the $(n + N)$ variables $\{x_1, \ldots, x_n; \lambda_1 \ldots \lambda_N\}$ and can be solved in any suitable manner.

CALCULUS OF VARIATIONS

In many problems the quantities x_i are dependent on another variable (for example time). The function to be optimized (cost function) is an integral of the form

$$I = \int_{t=a}^{t=b} \phi(x_1, \ldots, x_n, \dot{x}_1, \ldots, \dot{x}_n, t)\, dt \tag{B.6}$$

where

$$\dot{x}_i = \frac{dx_i}{dt}$$

We confine ourselves to the case of a single variable x, so that Eq. (B.6) becomes

$$I = \int_{t=a}^{t=b} \phi(x, \dot{x}, t)\, dt \tag{B.7}$$

In order to optimize I, we consider a variation δx in x at a fixed value of t of the form

$$\delta x(t) = \varepsilon \eta(t) \tag{B.8}$$

where ε is a fixed scalar and $\eta(\cdot)$ is an arbitrary function. The corresponding change in the functional ϕ is then given by

$$\begin{aligned} \Delta\phi &= \phi(x + \delta x, \dot{x} + \delta\dot{x}, t) - \phi(x, \dot{x}, t) \\ &= \frac{\partial \phi}{\partial x}\delta x + \frac{\partial \phi}{\partial \dot{x}}\delta \dot{x} + \text{(higher-order terms)} \end{aligned}$$

The first order variation in ϕ is then defined to be

$$\delta\phi = \frac{\partial\phi}{\partial x}\,\delta x + \frac{\partial\phi}{\partial \dot{x}}\,\delta\dot{x} \tag{B.9}$$

and represents a first-order approximation to the change in ϕ as the locus of x changes from one curve to another. The laws of variation of sums, products, ratios, etc. are analogous to the corresponding laws of differentiation. For example,

$$\delta(\phi_1\phi_2) = (\delta\phi_1)\phi_2 + \phi_1(\delta\phi_2) \tag{B.10a}$$

$$\delta(\phi_1/\phi_2) = \frac{\phi_2\delta\phi_1 - \phi_1\delta\phi_2}{\phi_2^2} \tag{B.10b}$$

$$\delta\int_a^b \phi\, dt = \int_a^b \delta\phi\, dt \tag{B.10c}$$

In terms of variations, necessary condition that I should be stationary is that δI should be zero, which yields

$$\delta I = \delta\int_a^b \phi(x, \dot{x}, t)\, dt = \int_a^b \delta\phi(x, \dot{x}, t)\, dt = \int_a^b \left(\frac{\partial\phi}{\partial x}\,\delta x + \frac{\partial\phi}{\partial\dot{x}}\,\partial\dot{x}\right) dt \tag{B.11}$$

By integrating the second term by parts, we obtain as a necessary condition for an optimum, the Euler–Lagrange equation

$$\frac{\partial\phi}{\partial x} - \frac{d}{dt}\left(\frac{\partial\phi}{\partial\dot{x}}\right) = 0 \tag{B.12a}$$

and the transversality conditions

$$\frac{\partial\phi}{\partial\dot{x}}\,\delta x(t)\bigg|_a^b = 0 \tag{B.12b}$$

The transversality conditions together with any specified boundary conditions on $x(t)$ or $\dot{x}(t)$ provide necessary boundary conditions for solving Eq. (B.12).

The extension to several variables follows easily. The conditions for stationarity of the integral in Eq. (B.6) are

$$\frac{\partial\phi}{\partial x_i} - \frac{d}{dt}\left(\frac{\partial\phi}{\partial\dot{x}_i}\right) = 0 \tag{B.13}$$

with the transversality conditions

$$\sum_{i=1}^{n} \frac{\partial \phi}{\partial \dot{x}_i} \delta x_i(t) \bigg|_a^b = 0 \tag{B.14}$$

If there are additional constraints on x or $\dot{x}$ of the form

$$f(x, \dot{x}, t) = 0 \tag{B.15}$$

we form the integral

$$\begin{aligned} I' &= \int_a^b [\phi(x, \dot{x}, t) + \lambda f(x, \dot{x}, t)]\, dt \\ &= \int_a^b \phi'(x, \dot{x}, \lambda, t)\, dt \end{aligned} \tag{B.16}$$

where $\phi'(\cdot)$ has the obvious definition. We can then treat the problem as one of finding the optimum of I' with respect to the functions x and λ. The necessary conditions for stationarity of I' follow from Eq. (B.11) as

$$\frac{\partial \phi'}{\partial x} - \frac{d}{dt}\left(\frac{\partial \phi'}{\partial \dot{x}}\right) = 0 \tag{B.17a}$$

and

$$\frac{\partial \phi'}{\partial \lambda} - \frac{d}{dt}\left(\frac{\partial \phi'}{\partial \dot{\lambda}}\right) = 0 \tag{B.17b}$$

Explicitly these equations can be written as

$$\frac{d}{dt}\left[\frac{\partial \phi}{\partial \dot{x}} + \lambda f\right] = \frac{\partial \phi}{\partial \dot{x}} \tag{B.18a}$$

$$f(x, \dot{x}, t) = 0 \tag{B.18b}$$

The transversality conditions obtained from Eq. (B.12b) are

$$\left[\frac{\partial \phi}{\partial \dot{x}} + \lambda \frac{\partial f}{\partial \dot{x}}\right] \delta x \bigg|_a^b = 0 \tag{B.19}$$

These equations together form a set of equations for solving for x and λ. The extension to the n-variable case is straightforward.

REFERENCES

1. F. B. Hildebrand, Methods of Applied Mathematics, Prentice-Hall, Englewood Cliffs, N. J., 1952.

APPENDIX C

VECTORS AND MATRICES

We list here some definitions and properties which are useful. Details may be found in the references at the end of the appendix.

VECTOR SPACES

For a given integer n, the set of all ordered n-tuples is an n-dimensional *vector space*.

A set of k vectors $\mathbf{x}_1, \mathbf{x}_2, \ldots, \mathbf{x}_k$ is said to be *linearly independent* if there exists scalars $c_1, c_2, \ldots, c_k$, not all zero such that

$$c_1\mathbf{x}_1 + c_2\mathbf{x}_2 + \cdots + c_k\mathbf{x}_k = 0 \tag{C.1}$$

If no such set of scalars exists, the vectors are independent. A set of vectors $\mathbf{x}_1, \mathbf{x}_2, \ldots, \mathbf{x}_k$ is said to *span* a vector space V if every vector $\mathbf{x}$ in V can be expressed as a linear combination of the vectors $\mathbf{x}_i$; that is

$$\mathbf{x} = \sum_{i=1}^{k} c_i\mathbf{x}_i \tag{C.2}$$

A set of linearly independent vectors which span a space V form a *basis* for V. Given a basis for V, any vector in V is expressible *uniquely* as a linear combination of the members of the basis.

MATRICES

A matrix of order $m \times n$ is a rectangular array of mn quantities with m rows and n columns. The matrix is square if $m = n$.

A matrix with only one column (row) is a column (row) vector. The *transpose* of a matrix $\mathbf{A}$ obtained by interchanging rows and columns is denoted by $\mathbf{A}^T$. The transpose of the product of two matrices is $(\mathbf{AB})^T = \mathbf{B}^T\mathbf{A}^T$. A square matrix $\mathbf{A}$ has an *inverse* $\mathbf{A}^{-1}$

$$\mathbf{AA}^{-1} = \mathbf{A}^{-1}\mathbf{A} = \mathbf{I} \qquad \text{(identity matrix)} \tag{C.3}$$

if and only if its determinant is nonzero. $\mathbf{A}$ is then said to be *nonsingular*. The inverse of the product of two nonsingular matrices $\mathbf{A}$ and $\mathbf{B}$ is

$$(\mathbf{AB})^{-1} = \mathbf{B}^{-1}\mathbf{A}^{-1} \tag{C.4}$$

A matrix for which $\mathbf{A}^T = \mathbf{A}^{-1}$ is *orthogonal*. If $\mathbf{A}^T = \mathbf{A}$, it is *symmetric* and if $\mathbf{A}^T = -\mathbf{A}$, it is *skew-symmetric*. Any arbitrary matrix can be expressed as the sum of a symmetric matrix and a skew-symmetric matrix as

$$\mathbf{A} = \frac{\mathbf{A} + \mathbf{A}^T}{2} + \frac{\mathbf{A} - \mathbf{A}^T}{2} \tag{C.5}$$

If $\mathbf{A}$ is an $m \times n$ matrix, the expression

$$\mathbf{y} = \mathbf{Ax}$$

represents a linear transformation which maps an element $\mathbf{x}$ of the vector space V_x into an element $\mathbf{y}$ of the vector space V_y. If $\mathbf{A}$ is a square matrix of order n, $\mathbf{A}$ transforms a vector space into itself. In this case, there exist certain vectors whose direction remains unchanged under the transformation. Such vectors are the *eigenvectors* of the transformation (matrix) $\mathbf{A}$ and satisfy the relation

$$\mathbf{Ax} = \lambda\mathbf{x} \tag{C.6}$$

for some scalar λ. The scalar λ is the *eigenvalue* associated with the eigenvector $\mathbf{x}$. Thus any nonzero vector satisfying

$$[\mathbf{A} - \lambda\mathbf{I}]\mathbf{x} = 0 \tag{C.7}$$

is an eigenvector of $\mathbf{A}$. The values of λ for which this equation has a nonzero solution are determined by solving the *characteristic equation*

$$\det\left[\mathbf{A}-\lambda\mathbf{I}\right]=0 \tag{C.8}$$

The nth degree polynomial, det $(\mathbf{A}-\lambda\mathbf{I})$, is the *characteristic polynomial* of **A**. Thus the matrix, **A** has n characteristic or eigenvalues. A square matrix satifies its own characteristic equation (Cayley–Hamilton theorem). That is, if the characteristic equation of **A** is

$$\sum_{i=0}^{n} a_i\lambda^i=0 \tag{C.9}$$

then

$$\sum_{i=0}^{n} a_i\mathbf{A}^i=\mathbf{0} \tag{C.10}$$

An $n\times n$ matrix **A** is *similar* to an $n\times n$ matrix **B** if there exists a nonsingular matrix **P** such that

$$\mathbf{B}=\mathbf{P}^{-1}\mathbf{A}\mathbf{P}. \tag{C.11}$$

P is then said to be a *similarity* transformation.

If there are n independent eigenvectors $\mathbf{x}_1 \ldots \mathbf{x}_n$, of the matrix **A**, then **A** is similar to a diagonal matrix of the form

$$\mathbf{A}_D=\begin{bmatrix}\lambda_1 & & & 0\\ & \lambda_2 & & \\ & & \ddots & \\ 0 & & & \lambda_n\end{bmatrix} \tag{C.12}$$

where λ_i, $i=1\ldots n$ are the eigenvalues of **A**. The eigenvalues may or may not be distinct. The corresponding similarity transformation **P** is

$$\mathbf{P}=\left[\mathbf{x}_1 \,\vdots\, \mathbf{x}_2 \,\vdots\, \ldots \mathbf{x}_n\right] \tag{C.13}$$

A *quadratic form* $Q(\mathbf{x})$ associated with a square matrix **A** is

$$Q(\mathbf{x})=\mathbf{x}^T\mathbf{A}\mathbf{x}=\sum_{i=1}^{n}\sum_{j=1}^{n} a_{ij}x_ix_j \tag{C.14}$$

The quadratic form associated with a matrix **A** is identical to that associated with its symmetric part.

A square matrix **A** is *positive definite* if the associated quadratic form $Q(\mathbf{x})$ positive for all $\mathbf{x} \neq 0$. It is positive semidefinite if $Q(\mathbf{x})$ is only nonnegative for all **x**.

The following results hold for real, symmetric matrices:

The eigenvalues of any real symmetric matrix are real.

A real symmetric matrix is positive definite if and only if all eigenvalues are positive. Thus any real, symmetric positive definite matrix is nonsingular and its inverse is also positive definite.

A real symmetric matrix is positive definite if and only if its principal minors are positive (Sylvester test).

A real symmetric matrix **A** is positive definite if and only if it can be expressed in the form

$$\mathbf{A} = \mathbf{M}\mathbf{M}^T \tag{C.15}$$

where **M** is real and nonsingular.

If **A** is real, symmetric and positive definite then there exist matrices $\mathbf{A}^{1/2}$ and $\mathbf{A}^{-1/2}$ such that

$$[\mathbf{A}^{1/2}]^2 = \mathbf{A} \tag{C.16a}$$

and

$$[\mathbf{A}^{-1/2}]^2 = \mathbf{A}^{-1} \tag{C.16b}$$

Thus a real, symmetric positive definite matrix possesses a *square root* which is also positive definite.

A real symmetric matrix **A** can always be transformed into a diagonal matrix of its eigenvalues, $\mathbf{\Lambda}$. The transformation matrix **P** is also symmetric, and orthogonal ($\mathbf{P}^T = \mathbf{P}^{-1}$):

$$\mathbf{P}^T\mathbf{A}\mathbf{P} = \mathbf{\Lambda} \tag{C.17}$$

SIMULTANEOUS DIAGONALIZATION OF TWO MATRICES

Let $\mathbf{A}_1$ and $\mathbf{A}_2$ be two real symmetric positive definite matrices. Let $\mathbf{\Lambda}_1$ and $\mathbf{\Lambda}_2$ be diagonal matrices of eigenvalues of $\mathbf{A}_1$ and $\mathbf{A}_2$ respectively. We seek a transformation matrix $\mathbf{P}$ such that

$$\mathbf{P}^T\mathbf{A}_1\mathbf{P} = \mathbf{I} \quad \text{and} \quad \mathbf{P}^T\mathbf{A}_2\mathbf{P} = \mathbf{\Lambda}_2 \tag{C.18}$$

We determine $\mathbf{P}$ as follows. Let $\mathbf{P}_i$ be the matrix of eigenvectors of $\mathbf{A}_i$, $i = 1, 2$. Then, from Eq. (C.17) we have

$$\mathbf{P}_1^T\mathbf{A}_1\mathbf{P}_1 = \mathbf{\Lambda}_1 \tag{C.19}$$

Pre- and postmultiplying Eq. (C.19) by $\mathbf{\Lambda}_1^{-1/2}$, we obtain

$$\mathbf{\Lambda}_1^{-1/2}\mathbf{P}_1^T\mathbf{A}_1\mathbf{P}_1\mathbf{\Lambda}_1^{-1/2} = \mathbf{I} \tag{C.20}$$

Let us define the following matrix:

$$\mathbf{K} = \mathbf{\Lambda}_1^{-1/2}\mathbf{P}_1^T\mathbf{A}_2\mathbf{P}_1\mathbf{\Lambda}_1^{-1/2} \tag{C.21}$$

Let $\mathbf{P}$ denote the eigenvector matrix of $\mathbf{K}$. The required transformation is then given by

$$\mathbf{P} = \mathbf{P}_1\mathbf{\Lambda}_1^{-1/2}\mathbf{P}_2 \tag{C.22}$$

This follows since

$$\begin{aligned}\mathbf{P}^T\mathbf{A}_2\mathbf{P} &= \mathbf{P}_2^T\mathbf{\Lambda}_1^{-1/2}\mathbf{P}_1^T\mathbf{A}_2\mathbf{P}_1\mathbf{\Lambda}_1^{-1/2}\mathbf{P}_2\\ &= \mathbf{P}_2^T\mathbf{K}\mathbf{P}_2 \qquad \text{(from (C.21))}\\ &= \mathbf{\Lambda}_2\end{aligned}$$

The last step follows since the eigenvalues of $\mathbf{A}_2$ and $\mathbf{K}$ are the same. Furthermore, since $\mathbf{K}$ is symmetric, we have $\mathbf{P}_2^T\mathbf{P} = \mathbf{I}$. Premultiplying Eq. (C.20) by $\mathbf{P}_2^T$ and postmultiplying by $\mathbf{P}_2$ then yields

$$\mathbf{P}^T\mathbf{A}_1\mathbf{P} = \mathbf{P}_2^T\mathbf{P} = \mathbf{I}$$

The transformation $\mathbf{P}_1\mathbf{\Lambda}_1^{-1/2}$ is usually referred to as the *whitening transformation* of the matrix $\mathbf{A}_1$.

THE MATRIX EXPONENTIAL

If $\mathbf{A}$ is an $n \times n$ matrix, the exponential $e^{\mathbf{A}}$ is defined by the convergent series

$$e^{\mathbf{A}} = \mathbf{I} + \sum_{n=1}^{\infty} \frac{\mathbf{A}^n}{n!} \tag{C.23}$$

Some of its properties are

$$e^{\mathbf{PAP}^{-1}} = \mathbf{P}e^{\mathbf{A}}\mathbf{P}^{-1}$$

$$e^{\mathbf{A}+\mathbf{B}} = e^{\mathbf{A}} \cdot e^{\mathbf{B}} \qquad \text{if } \mathbf{AB} = \mathbf{BA}$$

$$\det e^{\mathbf{A}} = e^{\operatorname{tr} \mathbf{A}} \qquad \text{where tr } \mathbf{A} = \text{trace of } \mathbf{A} = \sum_{i=1}^{n} a_{ii}$$

$$(e^{\mathbf{A}})^{-1} = e^{-\mathbf{A}}$$

$$e^{\mathbf{0}} = \mathbf{I} \tag{C.24}$$

DIFFERENTIATION OF MATRICES AND VECTORS

If $\mathbf{A} = \mathbf{A}(t)$

$$\frac{d\mathbf{A}}{dt} = \begin{bmatrix} \dfrac{da_{11}}{dt} & \cdots & \dfrac{da_{1n}}{dt} \\ \vdots & & \vdots \\ \dfrac{da_{n1}}{dt} & \cdots & \dfrac{da_{nn}}{dt} \end{bmatrix} \tag{C.25}$$

If $\mathbf{x}$ is a vector and $f(\mathbf{x})$ is a scalar function of $\mathbf{x}$, then the gradient of $f(\cdot)$ with respect to $\mathbf{x}$ is

$$\frac{df}{d\mathbf{x}} = \nabla_{\mathbf{x}} f = \left[\frac{df}{dx_1} \cdots \frac{df}{dx_n} \right]^T \tag{C.26}$$

If $\mathbf{g}(\mathbf{x})$ is a vector function of the vector $\mathbf{x}$, then the Jacobian is

$$\frac{\partial \mathbf{g}}{\partial \mathbf{x}} = \begin{bmatrix} \dfrac{\partial g_1}{\partial x_1} & \dfrac{\partial g_1}{\partial x_2} & \cdots & \dfrac{\partial g_1}{\partial x_n} \\ \vdots & & & \vdots \\ \dfrac{\partial g_m}{\partial x_1} & & \cdots & \dfrac{\partial g_m}{\partial x_n} \end{bmatrix} \tag{C.27}$$

The derivative of a matrix function $\mathbf{F}(\mathbf{x})$ with respect to the vector $\mathbf{x}$ is the partitioned matrix

$$\frac{d\mathbf{F}}{d\mathbf{x}} = \left[\frac{d\mathbf{F}}{dx_1} \cdots \frac{d\mathbf{F}}{dx_n} \right]^T \tag{C.28}$$

NORMS OF VECTORS AND MATRICES

A norm of a vector $\mathbf{x}$ or matrix $\mathbf{A}$ is a generalization of the concept of length.

For a vector $\mathbf{x}$ the norm is a real nonnegative number $|\mathbf{x}|$ satisfying the following relations:

(i) $|\mathbf{x}| > 0$ for $|\mathbf{x}| \neq 0$; $|\mathbf{x}| = 0$ if $\mathbf{x} = 0$
(ii) For any scalar α
$|\alpha \mathbf{x}| = |\alpha||\mathbf{x}|$ where $|\alpha|$ is the absolute value
(iii) $|\mathbf{x}_1 + \mathbf{x}_2| \leqslant |\mathbf{x}_1| + |\mathbf{x}_2|$ for all $\mathbf{x}_1$ and $\mathbf{x}_2$ (triangle inequality)

The l_p norm of a vector is defined as

$$|\mathbf{x}|_p = \left(\sum_{i=1}^{n} |x_i|^p \right)^{1/p}$$

The l^2 norm is the Euclidean norm. It can be shown that for $p = \infty$

$$|\mathbf{x}|_\infty = \max_i |x_i|$$

The norm of a matrix $\mathbf{A}$ is a nonnegative number $|\mathbf{A}|$ satisfying

(i) $|\mathbf{A}| > 0$ for $\mathbf{A} \neq 0$, $|\mathbf{A}| = 0$ if $\mathbf{A} = 0$
(ii) $|\alpha \mathbf{A}| = |\alpha||\mathbf{A}|$ for all scalar α
(iii) $|\mathbf{A} + \mathbf{B}| \leqslant |\mathbf{A}| + |\mathbf{B}|$ for all $\mathbf{A}$ and $\mathbf{B}$
(iv) $|\mathbf{AB}| < |\mathbf{A}| \cdot |\mathbf{B}|$ for all $\mathbf{A}$ and $\mathbf{B}$

Typical norms are

$$|\mathbf{A}|_1 = \max_j \sum_{i=1}^{n} |a_{ij}|, \qquad \text{the } l_1 \text{ norm}$$

$$|\mathbf{A}|_\infty = \max_i \sum_{j=1}^{n} |a_{ij}|, \qquad \text{the } l_\infty \text{ norm}$$

The definitions can be extended to the case where the elements of a vector or matrix are functions of time. For example for a vector, $\|\mathbf{x}\|_p = \left[\int_{-\infty}^{\infty} (|\mathbf{x}(t)|_p)^p \, dt \right]^{1/p}$ defines the L_p norm of the vector $\mathbf{x}(t)$. For $p = \infty$, this becomes

$$\|\mathbf{x}\|_\infty = \sup_{-\infty < t < \infty} |\mathbf{x}(t)|_\infty$$

A MATRIX LEMMA

We now prove the following useful lemma: *Let* $\mathbf{A}$ *be a nonsingular matrix and let* $\mathbf{B}$ *and* $\mathbf{C}$ *be two matrices such that* $\mathbf{A}+\mathbf{BC}$ *and* $\mathbf{I}+\mathbf{CA}^{-1}\mathbf{B}$ *are nonsingular*. *Then*

$$(\mathbf{A}+\mathbf{BC})^{-1}=\mathbf{A}^{-1}-\mathbf{A}^{-1}\mathbf{B}(\mathbf{I}+\mathbf{CA}^{-1}\mathbf{B})^{-1}\mathbf{CA}^{-1} \tag{C.29}$$

Proof: Let $(\mathbf{A}+\mathbf{BC})^{-1}=\mathbf{M}^{-1}$ Then

$$\mathbf{I}=\mathbf{M}^{-1}\mathbf{M}=\mathbf{M}^{-1}(\mathbf{A}+\mathbf{BC})=\mathbf{M}^{-1}\mathbf{A}+\mathbf{M}^{-1}\mathbf{BC} \tag{C.30}$$

Postmultiply by $\mathbf{A}^{-1}$ and rearrange to get

$$\mathbf{M}^{-1}\mathbf{BCA}^{-1}=\mathbf{A}^{-1}-\mathbf{M}^{-1} \tag{C.31}$$

Postmultiplying (C.31) by $\mathbf{B}$ and rearranging yields

$$\mathbf{M}^{-1}\mathbf{BCA}^{-1}\mathbf{B}+\mathbf{M}^{-1}\mathbf{B}=\mathbf{M}^{-1}\mathbf{B}(\mathbf{I}+\mathbf{CA}^{-1}\mathbf{B})=\mathbf{A}^{-1}\mathbf{B} \tag{C.32}$$

Postmultiply by $(\mathbf{I}+\mathbf{CA}^{-1}\mathbf{B})^{-1}\mathbf{CA}^{-1}$ to get

$$\mathbf{M}^{-1}\mathbf{BCA}^{-1}=\mathbf{A}^{-1}\mathbf{B}(\mathbf{I}+\mathbf{CA}^{-1}\mathbf{B})\mathbf{CA}^{-1} \tag{C.33}$$

which together with (C.31) yields (C.29). □

REFERENCES

1. R. Bellman, *Introduction to Matrix Analysis*, McGraw-Hill Book Co., New York, 1960.
2. F. R. Gantmacher, *The Theory of Matrices*, Vol. I and II, Chelsea Publishing Co., New York, 1959.
3. P. R. Halmos, Finite Dimensional Vector Spaces, *Annals of Mathematics, Study No. 7*, Princeton University Press, Princeton, N. J., 1948.
4. T. M. Apostol, *Mathematical Analysis*, Addison-Wesley, Reading, Mass., 1957.
5. K. Hoffman and R. Kunze, *Linear Algebra*, Prentice-Hall, Englewood Cliffs, N. J., 1965.

SUPPLEMENTARY BIBLIOGRAPHY

In addition to the references provided at the end of each chapter, we provide in the following a list of other pertinent books, reports and papers which relate to the material discussed in the test. The list, which is evidently not exhaustive, is arranged alphabetically. The numbers in parentheses at the ends of the references refer to pertinent chapters of this book.

H. B. Aasnaes and T. Kailath, An Innovations Approach to Least-Squares Estimation—Part VII: Some Applications of Vector Autoregressive—Moving Average Models, *IEEE Trans. Automatic Control*, **AC-18**, No. 6, Dec. 1973, pp. 601–607. *(6, 9)*

H. B. Aasnaes and T. Kailath, Initial-Condition Robustness of Linear Least Squares Filtering Algorithms, *IEEE Trans. Automatic Control*, **AC-19**, No. 4 August 1974, pp. 393–397. *(6, 9)*

M. Abramowitz and I. A. Stegun, *Handbook of Mathematical Functions*, Nat. Bur. Std. Appl. Math. Ser. *55*, August 1966. *(3, 4, 7, 8)*

N. Abramson and D. Braverman, Learning to Recognize Patterns in a Random Environment, *IRE Trans. Inform. Theory*, **IT-8**, No. 5, Sept. 1962, pp. 558–563. *(9)*

V. R. Algazi and R. M. Lerner, Optimum Binary Detection in Non-Gaussian Noise, *IEEE Trans. Info. Theory*, **IT-12**, No. 2, March 1966, p. 269. *(4)*

B. D. O. Anderson and J. B. Moore, State Estimation via the Whitening Filter, *Proc. Joint Autom. Control. Conf. June 1968, Ann Arbor, Mich*, pp. 123–129. *(6)*

B. D. O. Anderson and J. B. Moore, The Kalman-Bucy Filter as a True Time Varying Wiener Filter, *IEEE Trans. Sys. Man., Cyb.*, **SMC-1**, No. 2, April 1971, pp. 119–128. *(6)*

M. Aoki, *Optimization of Stochastic Systems: Topics In Discrete Systems*, Academic Press, New York, 1967. *(9)*

K. Arbenz, Optimum Cross Coupled Tracker for Pulse Doppler Radar, *IEEE Trans. Aerospace Elect. Syst.*, **AES-9**, No. 4, July 1973, pp. 496–498. *(8)*

K. J. Astrom, On the Achievable Accuracy in Identification Problems, *IFAC Symp. —Identification in Automatic Control Systems*, Prague, 1967. *(9)*

K. J. Astrom and P. Eykhoff, System Identification—A Survey, *2nd IFAC Symp. Identification and Process Parameter Estimation*, Prague, 1970; also in *Automatica*, **7**, pp. 123–162. *(9)*

M. Athans and E. Tse, A Direct Derivation of the Optimal Linear Filter Using the Maximum Principle, *IEEE Trans. Automatic Control*, **AC-12**, No. 6, Dec. 1967, pp. 690–698. *(6)*

M. Athans, The Relationship of Alternate State-Space Representations in Linear Filtering Problems, *IEEE Trans. Automatic Control*, **AC-12**, No. 6, Dec. 1967, pp. 775–776. *(6)*

M. Athans, R. P. Wishner, and A. Bertolini, Suboptimal State Estimation for Continuous-Time Nonlinear Systems from Discrete Noise Measurements, *IEEE Trans. Automatic Control*, **AC-13**, No. 5, Oct. 1968, pp. 504–514. *(6)*

C. C. Babu, On The Application of Probabilistic Distance Measures for the Extraction of Features from Imperfectly Labeled Patterns, *Internat. J. Computer Info. Sci.*, **2**, No. 2, 1973, pp. 103–114. *(9)*

A. V. Balakrishnan, *Communication Theory*, McGraw-Hill Book Co., New York, 1968. *(7)*

Y. Bar-Shalom, Optimal Simultaneous State Estimation and Parameter Identification in Linear Discrete-Time Systems, *IEEE Trans. Automatic Control*, **AC-17**, No. 3, June 1972, pp. 308–319. *(9)*

H. M. Beisner, A Recursive Bayesian Approach to Pattern Recognition, *Pattern Recognition*, **1**, July 1968, pp. 13–31. *(9)*

P. A. Bello, Binary Error Probabilities Over Selectively Fading Channels Containing Specular Components, *IEEE Trans. Commun. Technol.*, **COM-14**, No. 4., Aug. 1966, pp. 400–406. *(7)*

N. J. Bershad and P. A. Vena, Eliminating Intersymbol Interference—A State-Space Approach, *IEEE Trans. Info. Theory*, **IT-18**, No. 2, Mar. 1972, pp. 275–281. *(7)*

A. Bhattacharya, On a Measure of Divergence between Two Statistical Populations Defined by Their Probability Distributions, *Bull. Calcutta Math. Soc.*, **35**, No. 3., 1973, pp. 99–110. *(9)*

G. J. Bierman, A Comparison of Discrete Linear Filtering Algorithms, *IEEE Trans. Aerospace Elect. Syst.*, **AES-9**, No. 1, Jan. 1973, pp. 28–37. *(6)*

L. E. Brennan, I. S. Reed, and W. Sollfrey, A Comparison of Average Likelihood and Maximum Likelihood Ratio Tests for Detecting Radar Targets of Unknown Doppler Frequency, *IEEE Trans. Info. Theory*, **IT-4**, No. 1, Jan. 1968, pp. 104–110. Also *Rand Corp. Rep.*, **RM-5198-PR**, AD 646366, Dec. 1966. *(8)*

A. E. Bryson and D. E. Johansen, Linear Filtering for Time-Varying Systems Using Measurements Containing Colored Noise, *IEEE Trans. Automatic Control*, **AC-10**, No. 1, Jan. 1965, pp. 4–10. *(6)*

R. S. Bucy, Optimal Filtering for Correlated Noise, *J. Math. Anal. Appln.*, **20**, 1967, pp. 1–8. *(6)*

P. Caines, *Maximum Likelihood Identification of Multivariable Systems*, Ph. D. Thesis, Imperial College, London, 1970. *(9)*

J. Capon, Radon-Nikodym Derivatives of Stationary Gaussian Measures, *Ann. Math. Stat.*, **35**, June 1963, pp. 517–531. *(4)*

J. Capon, Hilbert Space Methods for Detection Theory and Pattern Recognition, *IEEE Trans. Inform. Theory*, **IT-11**, No. 2, April 1965, pp. 247–259. *(4, 9)*

B. Chandrasekaran, Independence of Measurements and the Mean Recognition Accuracy, *IEEE Trans. Info. Theory*, **IT-17**, No. 4, July 1971, pp. 452–456. *(9)*

D. Chase, Digital Signal Design for a Time-Varying Rician Channel, *IEEE Trans. Commun.*, **COM-24**, No. 2, Feb. 1976, pp. 164–172. *(7)*

C. H. Chen, A Theory of Bayesian Learning Systems, *IEEE Trans. Sys. Sci. Cyb.*, **SSC-5**, Jan. 1969, pp. 30–37. *(9)*

Y. T. Chien and K. S. Fu, A Modified Sequential Recognition Machine Using Time-Varying Stopping Boundaries, *IEEE Trans. Inform. Theory*, **IT-12**, No. 2, April 1966, pp. 206–214. *(9)*

Y. T. Chien and K. S. Fu, On the Generalized Karhunen–Loéve Expansion, *IEEE Trans. Inform. Theory*, **IT-13**, No. 3, June 1967, pp. 518–520. *(4)*

L. R. Chow, H. A. Owen, and P. P. Wang, An Adaptive Estimation Algorithm for Time-Varying Bit Synchronizers, *IEEE Trans. Aerospace Elect. Syst.*, **AES-9**, No. 1, January 1973, pp. 76–83. *(7)*

J. T. Chu and J. C. Chuech, Error Probability in Decision Functions for Character Recognition, *J. ACM*, **14**, April 1967, pp. 273–280. *(9)*

D. B. Cooper and P. W. Cooper, Nonsupervised Adaptive Signal Detection and Pattern Recognition, *Info. Control*, **7**, 1964, pp. 416–444. *(9)*

P. W. Cooper, Nonsupervised Learning in Statistical Pattern Recognition in *Methodologies of Pattern Recognition*, S. Watanabe, Ed., Academic Press, New York, 1969, pp. 97–109. *(9)*

T. M. Cover, Learning in Pattern Recognition, in *Methodologies of Pattern Recognition*. S. Watanabe, Ed., Academic Press, New York, 1969, pp. 111–132. *(9)*

H. Cox, On the Estimation of State Variables and Parameters for Noisy Dynamic Systems, *IEEE Trans. Automatic Control*, **AC-9**, No. 1, Jan. 1964, pp. 5–12. *(6)*

M. Cuenod, and A. P. Sage. Comparison of Some Methods Used for Process Identification, *IFAC Symp. Identification in Autom. Control Systems*, Prague, 1967; also in *Automatica*, **4**, 1968, pp. 235–269. *(9)*

M. C. Davis, Factoring the Spectral Matrix, *IEEE Trans. Automatic Control*, **AC-8**, No. 4, Oct. 1963, pp. 296–305. *(6)*

J. B. Denuzzo and P. N. Robinson, Target Spectra Feature Selection Using A Priori Information, *IEEE Trans. Aerospace Elect. Syst.*, **AES-11**, No. 4, July 1975, pp. 650–654. *(8)*

D. M. Detchmendy and R. Sridhar, Sequential Estimation of States and Parameters in Noisy Nonlinear Dynamical Systems, *J. Basic Eng., (Trans. ASME)*, **88D**, 1966, pp. 362–368. *(6)*

T. E. Duncan, Evaluation of Likelihood Ratios, *Info. Control*, **13**, 1968, pp. 62–74. *(4)*

R. E. Eaves and A. H. Levesque, Probability of Block Error for Very Slow Rayleigh Fading in Gaussian Noise, *IEEE Trans. Commun.*, **COM-25**, No. 3, March 1977, pp. 368–374. *(7)*

R. Esposito, On a Relation Between Detection and Estimation in Decision Theory, *Info. Control*, **12**, 1968, pp. 116–120. *(4, 6)*

D. D. Falconer and J. Salz, Optimal Reception of Digital Data Over the Gaussian Channel with Unknown Delay and Phase Filter, *IEEE Trans. Info. Theory*, **23**, No. 1, Jan. 1977, pp. 117–126. *(7)*

P. L. Feintuch and N. J. Bershad, Envelope Detector Performance for a Partially Coherent Fading Sine Wave in Noise, *IEEE Trans. Commun.*, **COM-24**, No. 8, Aug. 1976, pp. 913–917. *(7)*

G. D. Forney Jr., The Viterbi Algorithm, *Proc. IEEE*, **61**, No. 3, March 1973, pp. 268–278. *(7)*

G. J. Foschini, R. D. Gitlin, and S. B. Weinstein, Optimum Detection of Quantized PAM Data Signals, *IEEE Trans. Commun.*, **COM-24**, No. 12, Dec. 1976, pp. 1301–1309. *(7)*

B. D. Fritchman, C. D. Hafer, J. C. Mixsell and M. A. Sattar, Approximations to a Joint Detection/Decoding Algorithm, *IEEE Trans. Commun.*, **COM-25**, No. 2, Feb. 1977, pp. 271–278. *(7)*

M. A. Gallop and L. W. Nolte, Bayesian Detection of Targets of Unknown Location, *IEEE Trans. Aerospace Elect. Syst.*, **AES-10**, No. 4, July 1974, pp. 429–435. *(8)*

A. A. Galvin, A Sequential Detection System for the Processing of Radar Returns, *Proc. IRE*, **49**, No. 9, Sept. 1961, pp. 1417–1423. *(8)*

W. A. Gardner, Structurally Constrained Receivers for Signal Detection and Estimation, *IEEE Trans. Commun.*, **COM-24**, No. 6, June 1976, pp. 578–592. *(7)*

A. Gelb, *Applied Optimal Estimation*, MIT Press, 1974. *(6)*

A. Gersho, Adaptive Equalization of Highly Dispersive Channels for Data Transmission, *Bell System Tech. J.* ***48**, Jan. 1969, pp. 55–70. (7)*

M. R. Gevers and T. Kailath, An Innovations Approach to Least-Squares Estimation, Part VI: Discrete-Time Innovations Representations and Recursive Estimation, *IEEE Trans. Automatic Control*, **AC-18**, No. 6, Dec. 1973, pp. 588–600. *(6)*

J. D. Gibbons, *Non-Parametric Statistical Inference*, McGraw-Hill, New York, 1971. *(3)*

I. I. Gikhman and A. V. Skorokhod, *Introduction to the Theory of Random Processes*, Saunders, Philadelphia, Penna., 1969. *(2, 6)*

D. Graupe, *Identification of Systems*, Van Nostrand, New York, 1972. *(9)*

W. D. Gregg and J. C. Hancock, An Optimum Decision Directed Scheme for Gaussian Mixtures, *IEEE Trans. Info. Theory*, **IT-14**, No. 3, May 1968, pp. 451–461. *(4)*

T. L. Grettenberg, Signal Selection in Communication and Radar Systems, *IEEE Trans. Info. Theory*, **IT-9**, No. 4, Oct. 1963, pp. 265–275. *(7, 8)*

P. F. Guarguaslini and F. Marcoz, The DFTSD: A Sequential Suboptimum Processor for Multiple Range-Bin Radar Systems, *IEEE Trans. Aerospace Elect. Syst.*, **AES-10**, No. 2, March 1974, pp. 193–203. *(8)*

C. W. Helstrom, Solution of the Detection Integral Equation for Stationary Filtered White Noise, *IEEE Trans. Info. Theory*, **IT-11**, No. 3, July 1965, pp. 335–339. *(4)*

C. G. Hilborn and D. G. Lainiotis, Optimal Estimation in the Presence of Unknown Parameters, *IEEE Trans. Systems Sci. Cyb.*, **SSC-5**, No. 2, April 1969, pp. 38–43. *(6)*

Y. C. Ho and A. K. Agrawala, On Pattern Classification Algorithms: Introduction and Survey, *Proc. IEEE*, **56**, No. 12, Dec. 1968, pp. 2101–2114. *(9)*

W. S. Hodgkiss and L. W. Nolte, Bayes Optimum Likelihood Approaches in Array Processing Problem, *IEEE Trans. Aerospace Elect. Syst.*, **AES-11**, No. 5, Sept. 1975, pp. 913–916. *(8)*

W. S. Hodgkiss and L. W. Nolte, A Note on the Calculation of Performance for Likelihood Ratio Processors via Computer Simulation, *IEEE Trans. Info. Theory*, **21**, No. 6, Nov. 1975, pp. 695–698. *(4)*

C. E. Hutchinson, J. A. D'Apollito, and K. J. Roy, Applications of Minimum Variance Reduced State Estimators, *IEEE Trans. Aerospace Elect. Syst.*, **AES-11**, No. 5, Sept. 1975, pp. 785–794. *(6)*

J. R. Huynen, F. McNulty, and E. Hansen, Component Distributions for Fluctuating Radar Targets, *IEEE Trans. Aerospace Elect. Syst.*, **AES-11**, No. 6, Nov. 1975, pp. 1316–1331. *(8)*

International Cumulative Index on Radar Systems, *IEEE Trans. Aerospace Elect. Syst.*, **AES-11**, No. 3, May 1975, p. 429. *(8)*

I. M. Jacobs, Comparison of *M*-ary Modulation Systems, *Bell System Tech. J.*, May–June 1967. *(4, 7)*

I. M. Jacobs, Probability-of-Error Bounds for Binary Transmission on the Slowly Fading Rician Channel, *IEEE Trans. Info. Theory*, **IT-12**, No. 4, Oct. 1966, pp. 431–441. *(4, 7)*

D. E. Johansen, Solution of a Linear Mean Square Estimation Problem when Process Statistics Are Undefined, *IEEE Trans. Automatic Control*, **AC-11**, No. 1, Jan. 1966, pp. 20–30. *(6)*

T. T. Kadota, Optimum Reception of M-ary Gaussian Signals in Gaussian Noise, *Bell System Tech. J.*, Nov. 1965, pp. 2187–2197. *(4)*

T. Kailath, Some Integral Equations with Nonrational Kernels, *IEEE Trans. Info. Theory*, **IT-12**, No. 4, Oct. 1966, pp. 442–447. *(4)*

T. Kailath, The Divergence and Bhattacharyya Distance Measures in Signal Detection, *IEEE Trans. Commun. Technol.*, **COM-15**, No. 1, June 1967, pp. 52–60. *(4)*

T. Kailath, A Note on Least Squares Estimates from Likelihood Ratios, *Info. Control*, **13**, Dec. 1968, pp. 534–540. *(4, 6)*

T. Kailath, A View of Three Decades of Linear Filtering Theory, *IEEE Trans. Automatic Control*, **IT-20**, No. 2, March 1974, pp. 146–181. *(6)*

T. Kailath and R. A. Geesey, An Innovations Approach to Least Squares Estimation—Part IV: Recursive Estimation Given Lumped Covariance Functions, *IEEE Trans. Automatic Control.*, **AC-16**, No. 6, Dec. 1971, pp. 720–727. *(4, 6)*

T. Kailath and R. A. Geesey, An Innovations Approach to Least-Squares Estimation—Part V: Innovations Representations and Recursive Estimation in Colored Noise, *IEEE Trans. Automatic Control*, **AC-18**, No. 5, Oct. 1973. *(4, 6)*

L. N. Kanal and N. C. Randall, Recognition System Design by Statistical Analysis, *ACM, Proc. 19th Nat. Conf., August 1964*, pp. D2.5-1–D2.5-10. *(9)*

R. L. Kashyap and C. C. Blaydon, Recovery of Functions from Noisy Measurements Taken at Randomly Selected Points and Its Applications to Pattern Recognition, *Proc. IEEE*; **54**, No. 8, 1966, pp. 1127–1129. *(9)*

R. L. Kashyap and C. C. Blaydon, Estimation of Probability Density and Distribution Functions, *Proc. Conf. Circuits and Systems*, 1st Asilomar, Nov. 1967. *(9)*

R. L. Kaylor, A Statistical Study of Selective Fading of Super-High Frequency Radio Signals, *Bell Syst. Tech. J.*, Sept. 1953. *(7)*

W. B. Kendall and I. S. Reed, A Sequential Test for Radar Detection of Multiple Targets, *IEEE Trans. Info. Theory*, **IT-9**, No. 1, Jan. 1963, pp. 51–53. *(8)*

R. S. Kennedy and I. L. Lebow, Signal Design for Dispersive Channels, *IEEE Spectrum*, March 1964. *(4, 7)*

H. Kobayashi, Simultaneous Adaptive Estimation and Decision Algorithms for Carrier Modulated Data Transmission Systems, *IEEE Trans. Commun. Technol.*, **COM -19**. 1971, pp. 268. *(4, 6, 7)*

H. Kobayashi and D. T. Tang, A Decision Feedback Receiver for Channels with Strong Intersymbol Interference, *IBM J. Res. Develop.*, Sept. 1973, pp. 413–419. *(7)*

F. F. Krestschemer, MTI Weightings, *IEEE Trans. Aerospace Elect Syst.*, **AES-10**, No. 1, Jan. 1974, pp. 153–156. *(7)*

W. H. Kroy and A. R. Stubberud, Identification via Nonlinear Filtering, *Int. J. Control*, **6**, 1968, pp. 499–522. *(9)*

H. J. Kushner and W. N. Anakwa, An Approach to Useful but Nonoptimal Nonlinear Filters, *IEEE Trans. Automatic Control*, **AC-19**, No. 4, August 1974, pp. 398–404. *(6)*

J. S. Lee and J. H. Hughen, An Optimum Phase Synchronizer in a Partially Coherent Receiver, *IEEE Trans. Aerospcae Elect. Syst.*, **AES-7**, No. 4, July 1971, pp. 652–661. *(7)*

J. S. Lee and L. E. Miller, On the Binary DPSK Communication Systems in Correlated Gaussian Noise, *IEEE Trans. Commun.*, **COM-23**, No. 2, Feb. 1975, pp. 255–259. *(7)*

R. C. K. Lee, *Optimal Estimation, Identification and Control*, MIT Res. Monograph 28, Cambridge, Mass., 1964. *(5, 6, 9)*

S. P. Lee and J. J. Uhran, Optimum Signal Design in Underwater Acoustic Echo Ranging Systems, *IEEE Trans. Aerospace Elect. Syst.*, **AES-9**, No. 5, Sept. 1973, pp. 701–713. *(8)*

V. S. Levadi, Design of Input Signals for Parameter Estimation, *IEEE Trans. Automatic Control*, **AC-11**, 1966, pp. 205–211. *(9)*

P. B. Liebelt, *An Introduction to Optimal Estimation*, Addison-Wesley Publishing Co., Reading, Mass. 1967. *(5, 6)*

W. C. Lindsey, Error Probabilities for Rician Fading Multichannel Reception of Binary and N-ary Signals, *IEEE Trans. Info. Theory*, **IT-10**, No. 4, Oct. 1964, pp. 339–350. *(7)*

D. T. Magill, Optimal Adaptive Estimation of Sampled Stochastic Processes, *IEEE Trans. Automatic Control*, **AC-10**, No. 4, Oct. 1965, pp. 434–439. *(6)*

J. I. Marcum and P. Swerling, A Statistical Theory of Target Detection by Pulsed Radar, *IRE Trans. Info. Theory*, **IT-6**, No. 2, April 1960. *(8)*

M. B. Marcus and P. Swerling, Sequential Detection in Radar with Multiple Resolution Elements, *IRE Trans. Info. Theory,* **IT-8**, No. 2, April 1962, pp. 237–245. *(8)*

J. W. Mark, An Innovations Approach to Data Compression, *IEEE Trans. Commun.*, **COM-22**, No. 10, Oct. 1976, pp. 1618–1629. *(7)*

T. Marill and D. M. Green, Statistical Recognition Functions and the Design of Pattern Recognizers, *IRE Trans. Electron. Computers*, **C-9**, 1960, pp. 472–477. *(9)*

A. L. McBride, On Optimum Sampled Data for Demodulation, *IEEE Trans. Commun.*, **COM-21**, No. 1, Jan. 1973, pp. 40–50. *(7)*

T. P. McGarty, Azimuth Elevation Estimation Performance of Angle of Arrival Estimates, *IEEE Trans. Aerospace Elect. Syst.*, **AES-10**, No. 1, Jan. 1974, pp. 58–69. *(8)*

T. P. McGarty, The Effect of Interfering Signals on the Performance of Angle of Arrival Estimates, *IEEE Trans. Aerospace Elect. Syst.*, **AES-10**, No. 1, Jan. 1974, pp. 70–77. *(8)*

R. K. Mehra, On the Identification of Variances and Adaptive Kalman Filtering, *IEEE*, **AC-15** No. 2, April 1970, pp. 175–184. *(6, 9)*

R. K. Mehra, Approaches to Adaptive Filtering, *IEEE Trans. Automatic Control*, **AC-17**, No. 5, Oct. 1972, pp. 693–698. *(6, 9)*

J. M. Mendel, Computational Requirements for a Discrete Kalman Filter, *IEEE Trans. Automatic Control*, **AC-16**, No. 6, Dec. 1971, pp. 748–758. *(6)*

D. Middleton, *Introduction to Statistical Communication Theory*, McGraw-Hill Book Co., New York, 1960. *(3, 4, 5, 7)*

D. Middleton, On Singular and Nonsingular Optimum (Bayes) Tests for the Detection of Normal Stochastic Signals in Normal Noise, *IRE Trans. Info. Theory*, **IT-7**, No. 2, April 1961, pp. 105–113. *(4)*

D. Middleton, *Topics in Communication Theory*, McGraw-Hill, New York, 1965. *(3, 4, 5, 7)*

D. Middleton and R. Esposito, Simultaneous Optimum Detection and Estimation of Signals in Noise, *IEEE Trans. Info. Theory*, **IT-14**, No. 3, May 1968, pp. 434–444. *(4, 6)*

P. Monsen, Adaptive Equalization of the Slow Fading Channel, *IEEE Trans. Commun.*, **COM-22**, August 1974, pp. 1064–1075. *(7)*

J. B. Moore and P. Hetrakul, Optimum Demodulation of PAM Signals, *IEEE Trans. Info. Theory*, **IT-19**, No. 2, March 1973, pp. 188–196. *(7)*

M. Morf, G. S. Sidhu, and T. Kailath, Some New Algorithms for Recursive Estimation in Constant Linear Discrete-Time Systems, *IEEE Trans. Automatic Control*, **AC-19**, No. 4, pp. 315–323. *(6)*

D. R. Morgan, A Target Trajectory Noise Model for Kalman Trackers, *IEEE Trans. Aerospace Electr. Syst.*, **AES-12**, No. 3, May 1976, pp. 405–408. *(8)*

N. E. Nahi and D. E. Wallis, Jr., Optimum Inputs for Parameter Estimation in Dynamic Systems with White Observation Noise, *Joint Autom. Control Conf.*, 1969, pp. 506–513. *(9)*

D. L. Nicholson, Estimating Target Length Shorter Than the Pulse Width, *IEEE Trans. Aerospace Elect. Syst.*, **AES-11**, No. 4, July 1975, pp. 634–638. *(8)*

Z. J. Nikolic and K. S. Fu, Algorithm for Learning without External Supervision and Its Applications to Learning Control Systems, *IEEE Trans. Automatic Control*, **AC-11**, 1966, pp. 414–422. *(9)*

S. Nishikawa, Sequential *M*-ary PAM System, *IEEE Trans. Commun.*, **COM-21**, No. 1, Jan. 1973, pp. 22–33. *(7)*

A. H. Nuttall, Error Probabilities for Equicorrelated *M*-ary Signals under Phase-Coherent and Phase-Incoherent Reception, *IRE Trans. Info. Theory.*, **IT-8**, No. 4, July 1962, pp. 305–314. *(4, 7)*

J. H. Painter and L. R. Wilson, Simulation Results for the Decision-directed MAP Receiver for *M*-ary Signals in Multiplicative and Additive Gaussian Noise, *IEEE Trans. Commun.*, **COM-22**, No. 5, May 1974, pp. 649–660. *(4, 7)*

E. Parzen, Extraction and Detection Problems and Reproducing Kernel Hilbert Spaces, *J. Soc. Ind. Appl. Math., Ser. A Control*, **1**, No. 1, 1962, pp. 35–62. *(4)*

E. A. Patrick and J. C. Hancock, Nonsupervised Sequential Classification and Recognition of Patterns, *IEEE Trans. Info. Theory*, **IT-12**, 1966, pp. 362–372. *(9)*

E. A. Patrick, On a Class of Unsupervised Estimation Problems, *IEEE Trans. Info. Theory*, **IT-14**, No. 3, May 1968, pp. 407–415. *(9)*

E. A. Patrick, T. P. Costello, and F. C. Monds, Decision-directed Estimation of a Two Class Decision Boundary, *IEEE Trans. Computers*, **C-19**, No. 3, March 1970, pp. 197–205. *(9)*

J. B. Pearson and E. B. Stear, Kalman Filter Applications in Airborne Radar Tracking, *IEEE Trans. Aerospace Elect. Syst.*, **AES-10,** *No.* 3, *May* 1974 pp. 319-328. *(8)*

R. Price, Optimum Detection of Random Signals in Noise with Application to Scatter-Multipath Communications, *IRE Trans. Info. Theory*, **IT-2**, No. 4, Dec. 1956, pp. 125–135. *(7)*

V. S. Pugachev, Optimal Learning Systems, *Dokl. Akad. Nauk. SSSR*, **175**, No. 5, 1967, pp. 762–764. *(9)*

H. E. Rauch, F. Tung, and C. T. Streibel, Maximum Likelihood Estimates of Linear Dynamic Systems, *AIAA J.*, **3**, No. 8, Aug. 1965, pp. 1445–1450. *(6)*

I. B. Rhodes, A Tutorial Introduction to Estimation and Filtering, *IEEE Trans. Automatic Control*, **AC-10**, 1971, pp. 698–706. *(6)*

S. O. Rice, Mathematical Analysis of Random Noise, *Bell System Tech. J.*, **23**, 1944, pp. 282–332. *(3, 4)*

S. O. Rice, Distribution of the Duration of Fades in Radio Transmission: Gaussian Noise Model, *Bell System Tech. J.* May 1958. *(4, 7)*

A. W. Rihaczek, Radar Waveform Selection—A Simplified Approach, *IEEE Trans. Aerospace Elect. Syst.*, **AES-7**, No. 5, Nov. 1971, pp. 1078–1086. *(8)*

W. L. Root, Singular Gaussian Measures in Detection Theory, *Ann. Math. Stat.*, **34**, pp. 292–315. *(4)*

F. Russo, Bit Synchronization for Phase Noncoherent Reception of Binary Signals, *IEEE Trans. Aerospace Elect. Syst.*, **AES-12**, No. 3, May 1976, pp. 362–368. *(7)*

L. L. Scharf and D. W. Lytle, Signal Detection in Gaussian Noise of Unknown Level: An Invariance Application, *IEEE Trans. Info. Theory*, **IT-17**, No. 4, July 1971, pp. 404–411. *(4)*

L. L. Scharf and L. W. Nolte, Likelihood Ratios for Sequential Hypothesis Testing on Markov Sequences, *IEEE Trans. Info. Theory*, **IT-23**, No. 1, Jan. 1977, pp. 101–109. *(3, 7)*

F. H. Schlee, C. J. Standish, and N. F. Toad, Divergence in the Kalman Filter, *AIAA J.*, **5**, No. 6, June 1967, pp. 1114–1120. *(6)*

R. A. Scholtz and C. L. Weber, Signal Design for Phase Incoherent Communication, *IEEE Trans. Info. Theory*, **IT-12**, No. 4, Oct. 1966, pp. 456–463. *(7)*

C. C. Schooler, Optimal $\alpha-\beta$ Filters for Systems with Modeling Inaccuracies, *IEEE Trans. Aerospace Elect. Syst.*, **AES-11**, No. 6, Nov. 1975, pp. 1300–1306. *(8)*

H. Schwartz and R. W. Bass, A Computational Comparison of Several Nonlinear Filters, *IEEE Trans. Automatic Control*, **13**, No. 1, Feb. 1968, pp. 83–86. *(6)*

L. Schwartz and E. B. Stear, A Computational Comparison of Several Nonlinear Filters, *IEEE Trans. Info. Theory*, **IT-11**, No. 1, Jan. 1965, pp. 61–70. *(6)*

M. Schwartz, Effects of Signal Fluctuation on the Detection of Pulse Signals in Noise, *IRE Trans. Info. Theory*, **IT-2**, No. 2, 1956, pp. 66–71. *(7)*

S. C. Schwartz, The Estimator Correlator for Discrete Time Problems, *IEEE Trans. Info. Theory*, **IT-23**, No. 1, Jan. 1977, pp. 93–100. *(4, 7)*

A. Segall, Stochastic Processes in Estimation Theory, *IEEE Trans Info. Theory*, **IT-22**, No. 3, May 1976, pp. 275–286. *(6)*

I. Selin, The Sequential Estimation and Detection of Signals in Normal Noise, *Info. Control*, **7**, Dec. 1964, pp. 512–534; **8**, Feb. 1965, pp. 1–35. *(4, 6)*

C. Shannon, A Mathematical Theory of Communication, *Bell Sys. Tech. J.*, No. 27, July 1948, pp. 379–423, Oct. 1948, pp. 623–656. *(7)*

M. K. Simon and W. C. Lindsey, Tracking Performance of Symbol Synchronizers for Manchester Coded Data, *IEEE Trans. Commun.*, **COM-25**, No. 4, April 1977, pp. 393–408. *(7)*

D. Slepian and T. T. Kadota, Fourier Integral Equations of Detection Theory, *J. Soc. Ind. Appl. Math.*, **17**, No. 6, 1969, pp. 1102–1117. *(4)*

D. L. Snyder, Some Useful Expressions for Optimum Linear Filtering in White Noise, *Proc. IEEE*, **53**, No. 6, pp. 629–630. *(6)*

D. L. Snyder, *Random Point Processes*, Wiley-Interscience, New York, 1975. *(6)*

H. W. Sorenson, Least-Squares Estimation from Gauss to Kalman, *IEEE Spectrum*, July 1970, pp. 63–68. *(6)*

J. Spragins, Learning Without a Teacher, *IEEE Trans. Info. Theory*, **IT-12**, No. 2, April 1966, pp. 223–230. *(9)*

J. J. Stein and S. S. Blackman, Generalized Correlation of Multi-Target Track Data, *IEEE Trans. Aerospace Elect. Syst.*, **AES-11**, No. 6, Nov. 1975, pp. 1207–1217. *(8)*

M. G. Stringzis, A Solution to the Matrix Factorization Problem, *IEEE Trans. Info. Theory*, **IT-18**, No. 2, March 1972, pp. 225–232. *(6)*

P. Swerling, Detection of Fluctuating Pulsed Signals in the Presence of Noise, *IRE Trans. Info. Theory*, **IT-3**, No. 3, May 1957, pp. 175–178. *(8)*

P. Swerling, Recent Developments in Target Models for Radar Detection Analysis, *Proc. AGARD Symposium on Advanced Radar Systems*, May 1970, Istanbul, Turkey. *(8)*

J. B. Thomas and J. K. Wolf, On the Statistical Detection Problem for Multiple Signals, *IRE Trans. Info. Theory*, **IT-8**, No. 3, July 1962, pp. 274–280. *(4)*

J. B. Thomas, Non-Parametric Detection, *Proc. IEEE*, **58**, May 1970, pp. 623–631. *(3, 4)*

J. S. Thorp, Optimal Tracking of Maneuvering Targets, *IEEE Trans. Aerospace Elect. Syst.*, **AES-9**, No. 6, July 1973, pp. 512–519. *(8)*

J. T. Tou, Feature Extraction in Pattern Recognition, *Pattern Recognition*, **1**, No. 1, 1968, pp. 3–11. *(9)*

J. T. Tou, Feature Selection for Pattern Recognition Systems, in *Methodologies of Pattern Recognition*, S. Watanabe, Ed., Academic Press, New York, 1969. *(9)*

E. Tse and R. P. Wishner, Target Track Pulse Scheduling, *IEEE Trans. Aerospace Elect. Syst.*, **AES-11**, No. 6, Nov. 1971, pp. 1113–1122. *(8)*

E. T. Tsui and E. Wong, A Sequential Approach to Heart Beat Interval Classification, *IEEE Trans. Info. Theory*, **IT-21**, No. 5, Sept. 1975, pp. 596–599. *(9)*

G. L. Turin, Error Probabilities for Binary Symmetric Ideal Reception through Nonselective Slow Fading and Noise, *Proc. IRE*, **46**, Sept. 1958, pp. 1603–1619. *(4, 7)*

G. L. Turin, The Asymptotic Behavior of Ideal *M*-ary System, *Proc. IRE*, **27**, Jan. 1959, pp. 93–94. *(4, 7)*

G. L. Turin, An Introduction to Matched Filters, *IRE Trans. Info. Theory*, **IT-6**, No. 2, June 1960, pp. 311–329. *(4, 7)*

J. R. Ullman, *Pattern Recognition Techniques*, Crane-Russak, New York, 1973. *(9)*

G. Ungerboeck, Adaptive Maximum Likelihood Receiver for Carrier Modulated Data Transmission System, *IEEE Trans. Commun.*, **COM-22**, No. 5, May 1974, pp. 624–636. *(7)*

G. A. van der Spek, Detection of a Distributed Target, *IEEE Trans. Aerospace Elect. Syst.*, **AES-7**, No. 5, Sept. 1971, pp. 922–931. *(8)*

W. van Etten, Maximum Likelihood Receiver for Multiple Channel Transmission Systems, *IEEE Trans. Commun.*, **COM-24**, No. 2, Feb. 1976. pp. 276–283. *(7)*

V. Vannicola, Detection of Slow Fluctuating Targets with Frequency Diversity Channels, *IEEE Trans. Aerospace Elect. Syst.*, **AES-10**, No. 1, Jan. 1974, pp. 43–52. *(8)*

A. J. Viterbi, Phase Locked Loop Dynamics in the Presence of Noise by Fokker–Planck Techniques, *Proc. IEEE*, **51**, No. 12, Dec. 1963, pp. 1737–1753. *(7)*

T. J. Wagner, Convergence of the Nearest Neighbor Rule, **IT-17**, No. 5, Sept. 1971, pp. 566–571. *(9)*

P. Whittle, *Prediction and Regulation by Linear Least-Squares Methods*, English Universities Press, London, 1963. *(6)*

E. Wong and J. B. Thomas, On the Multidimensional Filtering and Prediction Problem and the Factorization of Spectral Matrices, *J. Franklin Inst.*, **8**, 1961, pp. 87–99. *(6)*

W. M. Wonham, Some Applications of Stochastic Differential Equations to Optimal Nonlinear Filtering, *J. SIAM Control, Ser. A*, **2**, No. 3, 1965, pp. 347–369. *(6)*

M. G. Wood, J. B. Moore, and B. D. O. Anderson, Study of an Integral Equation Arising in Detection Theory, *IEEE Trans. Info. Theory*, **IT-17**, No. 6, Nov. 1971, pp. 677–686. *(4)*

P. M. Woodward and I. L. Davies, Information Theory and Inverse Probability in Telecommunication, *Proc. IEEE.*, **99**, March 1952, pp. 37–44. *(7)*

S. J. Yakowitz, T. L. Williams, and A. D. Williams, Surveillance of Several Markov Targets, *IEEE Trans. Info. Theory*, **IT-22**, No. 6, Nov. 1976, pp. 716–724. *(8)*

T. Y. Young and T. W. Calvert, *Classification, Estimation, and Pattern Recognition*, American Elsevier Publishing Co., New York, 1974. *(9)*

T. Y. Young and Giorgio Coraluppi, Stochastic Estimation of a Mixture of Normal Density Functions Using an Information Criterion, *IEEE Trans. Info. Theory*, **IT-6**, No. 3, May 1970, pp. 258–263. *(9)*

AUTHOR INDEX

Abramson, N., 460
Agrawala, A. K., 461
Akaike, H., 462
Anderson, B. D. O., 275
Anderson, T. W., 201
Andrews, H. C., 460
Apostol, T. M., 477
Athans, M., 276, 396

Baggeroer, A., 154
Balakrishnan, A. V., 103, 201
Barrow, B., 346
Behnke, K. W., 396
Bellman, R., 477
Benedict, T. R., 397
Bennett, W. R., 345
Berlekamp, E. R., 345
Birdsall, T. G., 154
Blackman, R. B., 461
Blackwell, D., 103
Bode, H. W., 154, 275
Bordner, G. W., 397
Braverman, D., 460
Brennan, D. G., 346
Brennan, L. E., 397
Bucy, R., 275, 276

Chang, S. S. L., 275
Chen, C. T., 54
Chien, Y. T., 461
Coddington, E. A., 54
Courant, R., 55, 154
Cover, T. M., 461
Cramér, H., 201, 274

Daly, R. F., 461
Davenport, W. B., Jr., 54
DeLong, D. F., 154
Denlinger, E., 397
Deutsch, R., 201
Di Donato, A. R., 345
Di Franco, J. V., 396
Doetsch, G., 465
Dolez, M. L., 345
Doob, J. L., 54, 201
Duda, R. O., 460
Duncan, T. E., 154

Edward, J. A., 462
Eykhoff, P., 461

Fisher, R. A., 200
Fitch, S. M., 346
Fitelson, M. M., 462
Forney, G. D., 346
Fox, W. C., 154
Fralick, S. C., 461
Frankle, J. T., 346
Freeman, H., 54, 465
Frost, P. A., 275

Fu, K. S., 103, 460, 461
Fukunaga, K., 460

Gallager, R. G., 345
Gantmacher, F. R., 477
Gibson, J. D., 103
Gikhman, I. I., 55
Gonzalez, R. C., 460
Graupe, D., 462
Graybill, F. A., 200
Grenander, V., 154, 462
Grishick, M. A., 103
Gupta, S. C., 346

Hajek, J., 154
Halmos, P. R., 477
Hannan, E. J., 461
Harrington, J. V., 397
Hart, P. E., 460, 461
Heald, E. T., 345
Hecht, C., 276
Helstrom, C. W., 103, 154
Haydorn, R. P., 462
Hida, T., 55
Hilbert, D., 55, 154
Hildebrand, F. B., 469
Hoffman, K., 201, 477
Hofstetter, E. M., 154

Ito, K., 55

Jacobs, I. M., 154, 345
Jarnagin, M. I., 345
Jazwinski, A. H., 275
Johansen, D. E., 345
Joseph, P. D., 275

Kadota, T. T., 154
Kailath, T., 154, 396
Kalman, R. E., 275
Kashyap, R. L., 276, 462
Kaufman, H., 346
Keehn, D. G., 460
Kelly, E. J., 396
Klapper, J., 346
Krause, D. J., 462
Kunze, R., 201, 477
Kurz, L., 346
Kushner, K. J., 275

Larson, R. E., 396
Lawrence, R. E., 346
Leadbetter, M. R., 274
Lehmann, E. L., 103
Levinson, N., 54, 461
Lin, S., 345
Lindsey, W. C., 346
Loéve, M., 55
Lucky, R. W., 346
Luvison, A., 346

McAulay, R. J., 154, 397
McBride, A. L., 346
McGarty, T. P., 55, 155, 275
Mahalanabis, A. K., 347
Marcum, J. I., 345, 396, 397
Martin, D. L., 345
Meditch, J. S., 275
Mehra, R. K., 396, 462
Melsa, J. L., 103, 275, 461
Mendel, J. M., 460
Middleton, D., 154, 200
Miller, K. S., 154
Mood, A. M., 200
Moore, J. B., 462

Nagy, G., 460
Nasburg, R. E., 462
North, D. O., 154

Pachares, J., 397
Papoulis, A., 54
Parzen, E., 461
Patrick, E. A., 201, 461
Pearson, K., 103, 396
Peterson, W. W., 154
Pierce, J. N., 346
Pirani, G., 346
Prasad, S., 347

Ragazzini, J. R., 154
Rajasekaran, P. K., 347
Rao, C. R., 201
Reed, I. S., 396, 397
Rice, S. O., 55
Rihaczek, A. W., 396
Robinson, E. A., 461
Rogers, D. J., 460
Root, W. L., 54
Rubin, W. L., 396

Sage, A. P., 201, 275, 346, 461
Sakrison, D. J., 276, 345, 346
Salz, J., 346
Satyanarayana, N., 347
Schwartz, M., 345
Schweppe, F. C., 396, 462
Scudder, H. H., III, 461
Selin, I., 103, 154
Senne, K. D., 276
Shannon, C. E., 154, 275
Shepp, L. A., 154
Sherman, S., 201
Simon, M. K., 346
Singer, R. A., 396
Skolnik, M. I., 396
Skorokhod, A. V., 55
Smithies, F., 154
Snyder, D. L., 275, 346
Spragins, J. D., 201, 460, 461
Srinath, M. D., 347, 461
Stein, S., 345, 346
Stiffler, J. J., 346
Stratanovich, R. L., 275
Swerling, P., 201, 396, 397
Szajnowsky, W. J., 103, 201
Szego, G., 461

Tabaczynski, J. A., 276
Tanimoto, T. T., 460
Teicher, H., 460
Thomas, J. B., 154
Tou, J. T., 460, 462
Tricomi, F. G., 55
Tsypkin, Ya. Z., 461
Tukey, J. W., 461
Turin, G. L., 346

Urbano, R. H., 154

van den Bos, A., 462
Van Meter, D., 154
Van Trees, H. L., 55, 103, 154, 201, 274, 345, 346, 396
Viswanathan, M. M., 461
Viterbi, A. J., 201, 346, 347

Wald, A., 103
Watson, G. N., 396
Weldon, E. J., 346
Whalen, A. D., 55, 103, 154
Wiener, N., 275
Wishner, R. P., 276, 396
Wolfowitz, J., 103
Wong, E., 54, 275
Woodward, P. M., 154, 396
Wozencraft, J. M., 154, 345

Yaglom, A. M., 55
Yakowitz, S. J., 461
Youla, D. C., 275
Young, P. C., 461

Zadeh, L. A., 154

SUBJECT INDEX

Ambiguity function, 376
  properties of, 394, 395
Analog communication, 316-336
  discrete time receiver for, 332-336
Analytic signal, 48
Anticausal function, 9
Autoregressive moving average (ARMA) process, 422
  identification of, 440
Autoregressive (AR) process, 421
  identification of, 423

Bandlimited signal:
  complex envelope of, 45-50
  Rice's representation for, 51
Bandpass filtering of complex signals, 52
Basis functions for orthogonal expansion, 34, 35
Bayes criterion, 61
Bessel Function, first kind, 286
Beta density, 191
Bhattacharya coefficient, 455
Bias in an estimate, 168
Binary integrator for radar detection, 393
Binary system, optimum, 127

Causal function, 9
Calculus of variations, 467
Cayley-Hamilton theorem, 472
Characteristic equation, of a matrix, 470
Chi-Square density, 78
Complex white noise, 51
Convex function, 165
Convolution integral, 8, 9
Convolution summation, 21
Correlation function:
  of Poisson process, 29
  of Wiener process, 28
Correlation receiver:
  binary detection, 105
  M-ary detection, 119
  performance in colored noise, 126
  performance in white noise, 114
Cost function:
  absolute value of error, 158
  for estimation, 158
  squared error, 158
  uniform, 158
Costas loop, 301
Cramer-Rao bound, 169, 170

Demodulation using MAP approach, 316-324
Demodulation using state variable approach, 326-336
Detection:
  average number of samples in Wald's test, 92
  binary, 57
  in colored noise, 120

whitening approach 123
of known signals in white noise, 105-120
Karhunen-Loéve approach, 108-114
M-ary, 118
multiple measurements, 73-79
nonparametric, 96
of signals with unknown parameters, 141
sequential (Wald's test), 88
singular, 128
time-sampling approach, 105-107
truncated sequential, 93
Diagonalization, simultaneous of two matrices, 473
Differential phase shift keying (DPSK), 291-293
Digital Communication:
baseband, 279
bit synchronization, 301-304
carrier synchronization, 297-299
coherent, 282-285
fading channel performance, 293-297
noncoherent, 285
performance comparison, 297-298
synchronous, 279
Discriminant function:
linear, 402
quadratic, 404
Distortion:
mean square due to ISI, 306
peak due to ISI, 306
Divergence in Kalman filter, 271
Diversity FSK, 341
Doppler scattering function, 354

Eigenfunctions in Karhunen-Loéve expansion, 35
Eigenvalue:
of Karhunen-Loéve expansion, 35
of matrix, 471
Eigenvector of matrix, 471
Equalizer:
Adaptive using extended Kalman filter, 314-315
Adaptive (Lucky) for ISI channel, 310-314
for ISI channel, 307
Kalman filter-type for ISI channel, 308, 309
Wiener filter for ISI channel, 342
Error of first kind, 64
Error of second kind, 64
Error function, 67
Estimate:
asymptotically efficient, 172
Bayes', 158, 159
conditional mean, 160
condition for linearity, 183-184
consistent, 169
efficient, 170
linear mean square, 182, 187, 188, 203
maximum posteriori probability (MAP), 161
maximum likelihood, 166, 167
minimum mean squared error, 160
minimum variance of error, 169
unbiased, 168
sequential, for multiple parameters, 180-182
simultaneous, of multiple parameters, 178
Estimation, nonlinear, 254
Estimator, invariance of, 165
Exponential density, 195
Extended Kalman filter:
for analog demodulation, 326-331
continuous-time, 258
discrete-time, 259
in identification, 445
Euler-Lagrange equation, 468

False alarm, 64
Feature selection, 399, 414
Filtering, in estimation, 203
Fisher information matrix, 180
Fourier expansion of stochastic process, 32
Fredholm integral equation of first kind, 131
Fredholm integral equation of second kind, 131
Frequency shift keying (FSK), 117
coherent, 283
noncoherent, 289
Fundamental matrix, 170

Gamma density, 193
Gauss-Wishart density, 407

Hilbert transform, 48
Hypothesis testing, 57
Bayes' criterion, 61
composite, 83

maximum posteriori probability criterion, 58, 63
minimax criterion, 67
minimum probability of error criterion, 65
multiple, 79-83

Ideal observer, 66
Impulse response, 8
Impulse response matrix, 19
Incomplete Gamma function, 79
Innovations process, 225
Integral equations:
Fredholm, first kind, 136
second kind, 136
solution of, 136
Intersymbol interference (ISI), 304, 305

Kalman filter:
adaptive, 273
AR parameter identification, 436-439
ARMA parameter identification, 443-445
continuous-time, 239
algorithm, 243
with discrete observations, 247
discrete-time, 227
algorithm, 234, 235
one-stage prediction, 230-234
linearized, 255, 256
radar tracking, 384
relation to Wiener filter, 249-253
smoothing, 274
uncertain observations, 271
Kalman gain:
continuous-time filter, 241
discrete-time filter, 237
one-stage prediction, 231
Karhunen-Loéve expansion, 33, 34, 111, 120, 414, 415
Kernel:
Rational of integral equation, 136
Separable of integral equation, 140

Laplace transform, 463
bilateral, 464
region of convergence, 464
Levinson algorithm, 427
Likelihood equation, 167
Likelihood ratio, 58
conditional, 83, 143
generalized, 84, 142
Likelihood ratio test (LRT), 58
Limit in the mean (L.I.M.) convergence, 34
Lipschitz condition, 13
Log-normal density, 101

Maneuver detection of radar target, 386-389
Matrix lemma, 477
Maximum likelihood estimate, with waveform observations, 173-174
Maximum likelihood identification, 448
Maximum posteriori probability criterion, 58, 63
Maximum posteriori probability (MAP) equation, 161
MAP estimate:
for amplitude modulation, 320-322
for angle modulation, 323, 324
for bit synchronization, 303
of modulated signals, 316-325
nonlinear, 262
Matched filter, 117, 118, 132
Maximum SNR criterion, 128
Minimax criterion, 67
Minimax equation, 68
Minimum probability of error criterion, 65, 104, 116
Mixture density, 409
Moving average (MA) process, 422

Neyman-Pearson criterion, 70, 104
Neyman-Pearson test, 70
power of, 70
size of, 70
Non-return-to-zero (NRZ) waveform, 280

Objective function:
Neyman-Pearson test, 70
SNR maximization, 129
Observation space, 57
On-off-keying (OOK), 117
coherent, 282
in fading channel, 294, 295
noncoherent, 285-289
Orthogonality principle, 188, 189, 204, 205

Parameter estimation, sequential, 180-182
Partial correlation coefficient, 429
Pattern classification:
decision directed, 412, 413

nonsupervised learning, 400, 408
probabilistic learning, 412, 413
statistical, 399, 400
supervised learning, 400, 404
Phase lock loop, 297
for angle demodulation, 305
Phase-reversal-keying (PRK), 284
Phase-shift-keying (PSK), 117, 284
Point target:
steady, 350
slowly fluctuating, 351
Swerling model, 352
Poisson process:
correlation function of, 29
generalized, 29
Power of Neyman-Pearson test, 70
Prediction, 203
Pre-envelope of bandlimited signal, 48
Probabilistic transition mechanism:
in decision model, 60
in estimation model, 157
Probability of detection, 65
Probability of error:
in binary detection, 65, 116
in M-ary detection, 120
Probability of miss, 64

Q function, Marcum's, 288, 363

Radar ambiguity function, 376
Radar cross section (RCS), 350
Radar detection:
fluctuating point target, 364
multiple pulse, 368
spread target, 376
steady point target, 357, 360
Radar equation, 359, 391
Radar parameter estimation, 373
Crámer-Rao bound, 375
Fisher matrix, 380
Radar target models, 349-359
Doppler-spread, 354
doubly-spread, 355
range-spread, 355
Radar tracking, 382
maneuvering target, 386
Range scattering function, 391
Rayleigh density, 195
Rayleigh fading, 144
Receiver operating characteristic (ROC), 72
Return-to-zero (RZ) waveform, 280
Reproducing density, 190
Riccati equation, matrix differential, 243
Rician fading, 339
Robust estimator, 245

Sample mean, 197
Schwartz inequality, 170
Sequential identification:
AR parameters, 436
ARMA parameters, 443
partial correlation coefficients, 434
Sequential probability ratio test (SPRT), 89
Signal design, 127
Signal-to-noise ratio (SNR), maximization of, 128-133
Size of Neyman-Pearson test, 70
Smoothing, 203, 274
Spectral factorization, 31
of rational matrices, 224, 252
for Wiener filter, 211
Split phase (Manchester) waveform, 280
Square root of positive definite matrix, 473
State equation for linear system, 13
solution of continuous-time, 17
solution of discrete-time, 23
State transition matrix, 17, 24
State variable, 10
Stochastic process, 25
Gauss-Markov, 36, 37
independent increment, 27
Markov, 26, 36
martingalc, 27
Poisson, 29
representation of, 32
separable, 26
response of linear systems, 29-31
Wiener, 27
Sufficient statistic in detection, 79, 114
Sufficient statistic in estimation, 163
Superposition, principle of, 8
System, 6
solution, 17, 23
state space characterization, 10

Tocplitz matrix, 424
Transfer function:
continuous-time system, 10
discrete-time system, 22
Transitional probability density, 38

Uniformly most powerful (UMP) test, 84

Vector space, 470

Weibull density, 101
Whitening filter, 123, 124, 176
Whitening transformation, 474
Wide-sense conditional expectation, 190
Wiener filter:
  continuous-time, 208-211
  discrete-time, 220
  irreducible error, 213
  mean square error, 212
  relation to Kalman filter, 249-253
  vector processes, 222
Wiener-Hopf equation, 209
Wishart density, 194, 406

Yule-Walker equation, 424

Z- transform, 465

δ – function, 43
σ – field, 25